Computer Methods for Engineering

Computer Methods for Engineering

YOGESH JALURIA
Mechanical and Aerospace Engineering Department
Rutgers University

Published by
Brunner-Routledge
29 West 35th Street
New York, NY 10001

Published in Great Britain
Brunner-Routledge
11 New Fetter Lane
London EC4P 4EE

Brunner-Routledge is an imprint of the Taylor & Francis Group

Computer Methods for Engineering

2 3 4 5 6 7 8 9 0 GPGP 9 8 7

A CIP catalog record for this book is available from the British Library.
The paper in this publication meets the requirements of the ANSI Standard Z39.48-1984 (Permanence of Paper)

ISBN 1-56032-547-X (Paper)

Contents

*Advanced material.

6 NUMERICAL CURVE FITTING AND INTERPOLATION 234

7 NUMERICAL INTEGRATION 294

8 NUMERICAL SOLUTION OF ORDINARY DIFFERENTIAL EQUATIONS 352

9 NUMERICAL SOLUTION OF PARTIAL DIFFERENTIAL EQUATIONS* 440

Preface

The use of computational methods in the analysis and simulation of engineering processes and systems has grown tremendously in recent years. Increasing national and international competition has made it imperative to improve existing facilities and to develop new ones for a wide variety of applications. Because of the constraints imposed on detailed experimentation needed for design and optimization of systems, due to excessive time, manpower, and financial requirements, computer simulation is extensively employed to obtain the desired information. Analytical methods are generally very restrictive in their applicability to practical problems, and numerical methods are usually necessary. In addition to the growing need for numerical solutions to engineering problems, we have also seen substantial improvements in the computational facilities available, both in software and in hardware, over the last decade. All of these changes have made it more important than ever for engineers and engineering students to develop expertise in numerical methods and to use them for solving problems of practical interest.

In recognition of the growing importance of computer methods in engineering, many courses in engineering curricula now include the numerical solution of engineering problems on the basis of numerical analysis taught earlier at the sophomore or junior level. Generally, engineering students are first exposed to the computational procedure through a course on programming, frequently employing FORTRAN as the programming language. Numerical methods are then taught at a later stage to introduce the basic concepts of numerical analysis and to allow the students to numerically solve important mathematical problems such as integration, matrix inversion, root solving, and solution of differential equations. However, since the basic purpose of the computational approach is to provide physical insight and to obtain valuable information for the analysis and design of practical systems, such courses have been integrated into the engineering curricula at most universities. This implies that the solution methodology is coupled with the computer on one hand and with the physical or chemical nature of the problem on the other. The numerical procedure, as well as the results, are considered in terms of actual problems to permit the student to develop a physical feel for the numerical approach to engineering problems.

Traditionally, numerical analysis courses have been mathematically oriented. Although this orientation brings in some very important and fundamental aspects of numerical analysis, it lacks in the application of the methodology to actual problems. It is extremely important to integrate the basic understanding of the methods with their actual use on the computer. Unless the students learn to choose and implement a computational scheme on the computers available, they will not develop a satisfactory appreciation or understanding of the numerical technique. In addition, recent advances in computational facilities, such as structured programming, interactive computer usage, and graphics output, must be introduced so that the most efficient procedure is adopted for a given problem. The incorporation of problems derived from various engineering disciplines aids in this learning process and also makes it interesting and enjoyable. In addition, it reinforces the important point that the physical or chemical background of the given problem forms an important element in the selection of the method and in the evaluation of the accuracy of the results obtained.

This book, directed at computer methods for engineering, integrates the treatment of numerical analysis with the physical background of the problems being solved and with the implementation of the methods on available computers, employing several recent advances in this field. Although a large number of books are available on numerical analysis, not many satisfactorily discuss the implementation of the methodology on the computer, and even fewer discuss the implications of the physical nature of the problem in the numerical solution. This book recognizes the need for a satisfactory incorporation of these concepts into the mathematical treatment of numerical analysis. It couples numerical methods for a variety of mathematical problems with the use of these methods for the solution of engineering problems on the computer.

Numerical methods for important mathematical operations, such as integration, differentiation, root solving, and solution of algebraic systems, are discussed in detail. The solution of differential equations, both ordinary and partial, is presented. Curve fitting, which is an important consideration in engineering problems, is also discussed. A large number of problems from basic sciences and various engineering disciplines are chosen to illustrate the use of these methods. The problems chosen are relatively simple so that they can easily be understood by students at the sophomore/junior level. However, in several cases, the basic background of the problem is outlined so as to bring the important points into proper focus. The importance of the physical or chemical background of the problem in the selection of the method, the choice of numerical parameters, the estimation of the accuracy of the results, and the overall validity of the results is discussed. The book uses mainly FORTRAN 77 to demonstrate the implementation of the numerical methods on the computer, because of the overwhelming importance of this language in engineering applications. However, a few programs in BASIC are also given to bring out the similarities between the two languages and the ease with which one may switch from one to the other. A discussion of other languages and important aspects in computational procedure is included. A large number of examples, with the corresponding programs, are given. The programs are written specifically for these examples, so that the students must develop their own programs for the large number of problems given at the end of the chapters. Several important features that are currently employed in computational procedure are demonstrated in these programs. Recent trends in this area are outlined, and their significance for engineering applications is discussed. The students are strongly encouraged in every way to develop their own computer programs, since this is an essential ingredient for learning computer methods.

Most of the material covered in this book has been employed by the author for courses at the sophomore and junior levels. Since the background of students at the sophomore level may not be sufficient for some of the topics covered, such as partial differential equations, this particular topic and a few sections marked with an asterisk may be avoided by sophomore students. The book can also be used at the senior level, if such a course is included in the curriculum at this level. The material included is quite adequate for a one-semester course. However, the best time to teach this course is probably at the junior level, so that the students can fully understand the material and then use it in courses taught at higher levels. The book is also appropriate for professional engineers in various disciplines and as a reference for courses that employ computational methods as an important element in the presentation. The book considers problems from diverse engineering applications, and the treatment is at a level appropriate for engineering students of all disciplines.

I owe tremendous gratitude to several colleagues and students who have contributed to my understanding and enjoyment of computational methods for engineering applications. First, I would like to thank Dr. Frank Kreith, who suggested that I write this book and contributed several very valuable suggestions on the presentation. I would also like to acknowledge several stimulating and interesting discussions on the subject with Professors Dave Briggs and Abdel Zebib. Professor Samuel Temkin provided me with tremendous support and encouragement. Dr. M. V. Karwe helped with the numerical solution of some problems. Also of considerable value was the support provided by the staff of Allyn and Bacon, Inc., particularly by Ray Short. The manuscript and its several versions were typed with great patience and competence by Diane Belford and Lynn Ruggiero.

I would like to dedicate this book to my parents, who have always encouraged, supported and inspired me to strive for the best I could achieve. The greatest contributions to this effort have been the encouragement and support of my wife, Anuradha, and of our children, Pratik, Aseem, and Ankur, who had to bear long hours that kept me away, working on this book, with patience and understanding.

The author extends special thanks to the following reviewers whose contributions have enriched the text:

Professor Clayton Crowe
Washington State University

Professor Rodney W. Douglass
University of Nebraska

Professor S. V. Patankar
University of Minnesota

Dr. James F. Welty
U.S. Department of Energy

Y. J.

1

Introduction

1.1 INTRODUCTORY REMARKS

In recent years, there has been a tremendous increase in the use of computers for engineering problems. This increase has been due partly to the growing need to optimize systems and processes in order to raise productivity and reduce costs. With increasing competition worldwide, it has become necessary to modernize existing engineering facilities through analysis and design. As a consequence, the past decade has seen a considerable improvement in engineering systems, particularly those related to electronic circuitry, materials processing, and energy generation. The concern with safety and with our environment has also led to detailed investigations of existing engineering processes and to substantial improvements in many of these. Because of the complexity involved in most engineering applications, analytical methods based on mathematical techniques are usually unable to provide a solution to the governing equations, and computational methods are needed to determine numerical values of the physical quantities of interest. Even though analytical solutions may be obtained in a few cases, the form of the solution itself is often involved, since the results may be expressed as a series or in terms of integrals and transcendental functions. In such cases, the computer is needed to extract the desired information from the analytical solution obtained. Also, the problem may have to be solved several times with different sets of data, making it advantageous to use the computer rather than analytical methods.

There has also been a phenomenal increase in the availability of computers over the last few years. With the advent of microcomputers, such as the personal computers from IBM and Apple, computational facilities have become widely available. There is every indication that these trends will continue, making computers even more accessible. Although most engineering problems still require larger computers (either main-frame or minicomputers), microcomputers do allow the solution of simpler problems and are also useful in testing numerical procedures that may subsequently be employed on larger machines. The availability of microprocessors has also substantially affected the control and operation of systems through automation. Along with the revolution in computer hardware, there has inevitably been one in the available software as well, making the use of computers for scientific and engineering problems easier than ever. Thus, in a wide variety of problems, the programs available in the computer library may be used effectively. However, it is often necessary to understand the basic techniques involved in order to modify the program for satisfactory application to a given problem. In industrial systems, the use of available programs is particularly important, since the processes are often very involved and interest lies in obtaining the needed information as rapidly as possible. For simpler problems, such as those related to individual physical and chemical processes that constitute the overall system, it is often easier and more desirable to personally write the computer program, rather than use an available program. Consequently, it is important to understand the computational methods relevant to engineering applications and to use them in physical problems that are of interest to various disciplines.

Computer-aided design and computer-aided manufacturing are two important areas that have grown substantially in the very recent past. These areas have arisen from the need to optimize on one hand and the growing availability of the computers on the other. Both areas are interdisciplinary in nature, particularly computer-aided design, which is of interest in such diverse fields as electronic systems and structural design. The basic approach in this case is to numerically solve the governing equations, choose physical parameters to simulate existing processes and systems, and finally vary these parameters to optimize the design for existing and future systems. Several other similar applications of computer methods have arisen in recent years, making it imperative to link the computational approach to the physical or chemical aspects of the problem under consideration.

In view of the growth of computer usage and availability in the last decade, it is surprising that much of the mathematical background underlying numerical analysis and computer logic has been available for several centuries. Binary logic operations, which use 2 as the base, instead of 10 employed in the decimal system, and which form the basis for most present digital computing, have been known and used for quite some time. Francis Bacon used binary codes in the early seventeenth century to transmit secret messages. In 1804, Joseph Marie Jacquard used punched cards with binary code and logic to operate looms. A mathematical theory for binary logic was developed by George Boole during the nineteenth century. Similarly, adding machines and mechanical calculators were developed centuries ago, such as the one developed by Blaise Pascal in the seventeenth century. Charles Babbage designed the first automatic digital computer in 1833, with several features similar to those of modern computers. However, this machine was never constructed.

Modern digital computers were developed largely after the second world war. A high-speed electronic digital computer was developed during the period from 1945 to 1952 under the direction of John von Neumann at the Institute for Advanced Study in

Princeton. Binary digits, which can be represented by the opening or closing of a switch, were stored electrostatically in cathode-ray tubes. Several thousand vacuum tubes were used for computer memory, which had to be again stored about a thousand times per second due to the decay of electrostatic charge. Much of the logic behind this machine has persisted in modern computers. The major advancement has been in electronic hardware, particularly in the development of transistors and microelectronics. As a result, there has been a considerable reduction in size and cost of electronic digital computers and also a substantial increase in their capability, speed, and reliability. The availability of personal computers has brought computational techniques within easy access for a wide variety of problems, both for students and for the professional engineer. Therefore, the coming years may be expected to improve the available computational facilities even further through the advancement in both computer software and hardware. It is also evident that personal computers, with an interface with larger machines for more complicated problems, will continue to grow in availability and usage. Thus, it is important to learn the computational techniques relevant to engineering problems on the basis of the currently available computational facilities, while considering expected future trends as well.

Several important and useful features have been incorporated in the modern computer systems. Among the most important of these is an interactive use of the computer, rather than in the batch operation mode. Frequently, an interpretive compiler is used so that each program statement entered into the computer is screened for syntax errors and a message issued if any error has been committed. The interactive mode allows one to enter variables and make changes in the program, as the need arises after each run of the program. The execution may also be stopped to make modifications and then continued. Therefore, the interactive mode is very well suited for the initial stages of program development, when the testing and debugging of the program is being done, and for obtaining the trends for a wide range of input parameters. For instance, if the roots of a nonlinear equation $f(x) = 0$ are to be determined, the interactive mode may be used very effectively to obtain the general behavior of $f(x)$ over the range of interest in x. Various values of x may be entered and the corresponding value of $f(x)$ obtained. The information obtained may then be used to select the method for finding the roots and also to obtain suitable initial guesses for the roots. Figure 1.1 shows a few examples where the plot of $f(x)$ versus x would be particularly useful in root finding.

The batch operation mode involves feeding the complete job into the computer and then running it with no interaction with the operator until the job is executed. This mode is appropriate for obtaining the numerical results for different parametric values after the program has been developed and debugged, particularly for large programs. Other important features available with present computer systems are graphics facilities, which plot the computed results, and interfacing between various computers, which allows program development to be carried out on small computers in the interactive mode. Once the program has been completed, debugged, and tested, the numerical code may be transmitted to a larger computer, which would generally

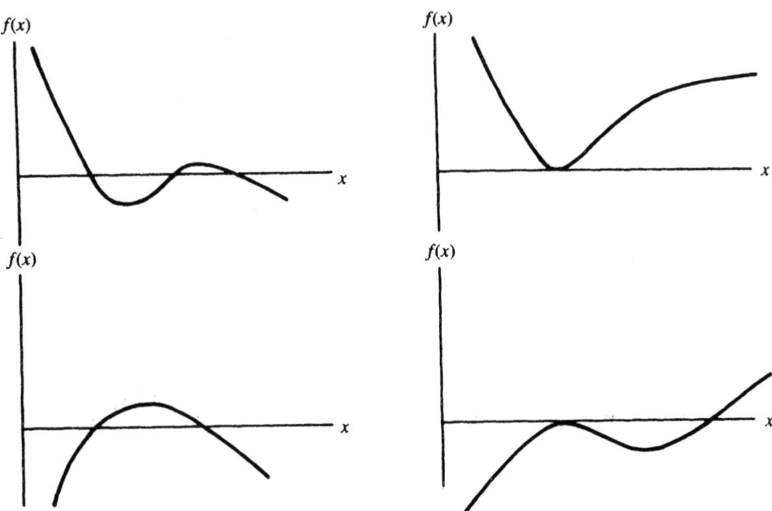

Figure 1.1 Some examples of the plots of the function $f(x)$ versus x to determine the approximate values of the roots of the equation $f(x) = 0$.

be more efficient for computing, and run in the batch mode to obtain the desired computed results. Chapter 2 discusses many of these features and considerations in greater detail. Also, many terms commonly used to describe a computer system and its sequence of operations are briefly mentioned in this chapter. These terms are presented again in Chapter 2, with examples and further discussion.

1.2 NUMERICAL SOLUTION

The development of a computational procedure, or algorithm, to solve a given problem requires a knowledge of both the available numerical methods and the methodology to interface with the computer. Since several methods are generally available for a given application, it is important to understand the applicability and advantages of each method compared to those of the other methods. For instance, a system of linear equations may be solved by a wide variety of methods, including direct methods, which give a solution in a definite number of steps, and iterative methods, which involve a repeated solution of the equations until a chosen convergence criterion is satisfied. The choice of the method for a given problem depends mainly on the nature and number of the equations. Direct methods are suitable for smaller systems and iterative methods for large sets of equations. Also, if the same system of equations must be solved several times with different constants on the right-hand side of the equality sign, methods based on matrix inversion are often preferable since the different solutions may be obtained easily once the coefficient matrix has been inverted. Similarly, in curve fitting, the method to be adopted is strongly dependent on the nature and form of the given data. If the data have been provided at uniform intervals of the independent variable, certain specialized methods may be used, taking advantage of the uniform distribution of data.

Sometimes, several methods are applicable for a given problem, and the selection of the method becomes a matter of personal choice. Previous experience with the particular method may often be an important consideration in the selection. Also, the availability of certain programs in the computer library may make it advantageous to choose a given method. It is necessary to understand the limitations of the various methods so that a proper selection may be made. Many specialized methods have been developed for specific applications. Such methods are often very limited in their applicability, although they may be the most efficient ones when applied to the problem for which they are particularly suited. For instance, Bairstow's method for finding the roots of an algebraic equation is based on the iterative factorization of polynomials. It is, therefore, applicable only to polynomial equations and is a popular choice for this application. It cannot be used for other types of algebraic equations, say, transcendental equations. Similarly, direct methods for solving systems of equations apply only for linear equations. Iterative methods are generally necessary for a system of nonlinear equations.

It is evident from the above discussion that the selection of the method for solving a given problem is a very important consideration and is generally based on the nature of the problem. Once the method has been selected, one proceeds to implement it on the computer. The program is written in a programming language that is available on the given computer system. Although FORTRAN, with its various versions, such as FORTRAN 77, is generally used in engineering applications on most minicomputers and main-frame systems, BASIC is often used on personal computers. BASIC has, therefore, become an important programming language in recent years. It has several advantages over FORTRAN, particularly in the simplicity of writing the program and in the input/output operations. It is usually implemented as an interpreter, so that program execution can be interrupted at any point, changes may be made in the program and in the input values, the output may be printed, and the execution resumed. Because of its growing importance, several improvements have recently been made on the BASIC available on most computers, making it quite suitable for many problems of engineering interest. However, for most engineering applications, which generally involve large systems of equations, FORTRAN is still the most common language. There are several other languages, which are often particularly suited for specific types of problems. For instance, LISP and PROLOG are useful languages for computer logic and are often used for programs related to artificial intelligence in engineering systems.

The program written in the chosen programming language is converted into machine language by the computer. This process, known as compilation of the program, is achieved by using the relevant software, termed the compiler, available on the computer. An operating system is used for the control of the program and the computer resources. The editing of the program, for making changes and corrections, is done with the help of the editing system available on the given computer. The compilation, editing, and execution of the program are governed by the operating system of the computer and therefore vary with the machine. Similarly, the job control language, which interfaces the programmer with the computer, depends on the computer system. For those who may not be familiar with the terms mentioned here, Chapter 2 outlines the basic features of a computer system.

The interpretation of the numerical results obtained is an extremely important consideration, since it relates to the accuracy and the correctness of the numerical solution. The computational scheme may be employed to yield results for a wide range of input variables, so that the results may be considered in terms of the physical or chemical nature of the problem being investigated. If possible, a comparison is made with the available analytical results in order to determine the accuracy of the computed results. The verification of the numerical scheme is particularly important if an available computer program is being employed to solve a given problem. It is also important to determine the range of governing parameters over which the scheme can be used to yield accurate numerical results. These considerations are discussed in the following sections. Once the accuracy and validity of the results have been verified, the desired results may be obtained in a tabulated or graphical form.

1.3 IMPORTANCE OF ANALYTICAL RESULTS

The governing equations that arise in most engineering problems are too complicated to be solved analytically, and computational techniques must be used to obtain the numerical values needed. Analytical solutions are often obtained only in very simplified circumstances. Also, as mentioned earlier, analytical results are frequently given in terms of convergent series, integrals and complicated functions (such as transcendental functions), Bessel functions, and so on. In engineering, we are interested mainly in numerical values corresponding to given input data. Therefore, the computer is frequently used to obtain the desired numerical information from a given analytical solution. However, analytical results, whenever available, are extremely important in evaluating the accuracy of the numerical scheme. Similarly, analytical results may be used to study the convergence characteristics of the numerical method and to decide if the correct solution has been obtained.

As an example, let us consider the solution of the differential equation that governs the variation, with time τ, of the charge q of a capacitor in an electrical circuit that also contains a voltage source and a resistance. If the initial charge in the capacitor is Q and the voltage input, resistance, and capacitance are denoted by E, R, and C, respectively, the governing equation is obtained as follows (Sears et al., 1981):

$$R\frac{dq}{d\tau} + \frac{q}{C} = E \tag{1.1}$$

If R, C, and E are constants, the above equation may be easily solved to obtain

$$q = Qe^{-\tau/RC} + EC(1 - e^{-\tau/RC}) = EC + (Q - EC)e^{-\tau/RC} \tag{1.2}$$

The physical problem and the analytical solution are sketched in Fig. 1.2. The charge q decreases from the initial value of Q to a steady-state value of EC, if $EC < Q$. Similarly, q increases to a steady charge of EC, if $EC > Q$.

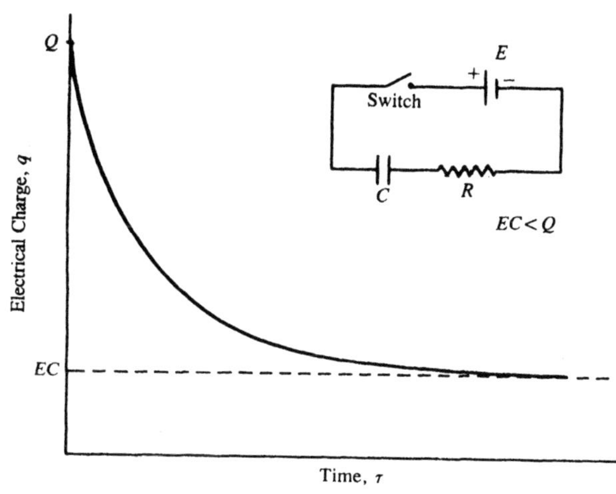

Figure 1.2 Variation with time of the charge q in a capacitor, which is originally at charge Q, due to the closing of the switch in the electrical circuit shown.

Several other physical problems are governed by equations similar to Eq. (1.1). The temperature $T(\tau)$ of a small, heated metal block being cooled by a stream of air, the moisture content of a wet body drying in air, and the pressure of gas in a container with an opening are often governed by equations of the same form as Eq. (1.1). However, in actual practice, the parameters, such as R, C, and E, may be the nonlinear functions of the charge or voltage and may, in some cases, also vary independently with time. For instance, nonlinear conductors, such as vacuum tubes, do not obey Ohm's law, and heat and mass transfer processes operating at the surface of a given object generally depend on the temperature, concentration, and pressure, making the differential equation nonlinear. The governing equation may, in general, be written as $d\phi/d\tau = -H(\phi,\tau)\phi + B$, where ϕ is the dependent variable, $H(\phi,\tau)$ is a parameter, and B is a constant. If q is replaced by ϕ in Eq. (1.1), then $H(\phi,\tau) = 1/RC$ and $B = E/R$. This equation is linear in ϕ, or q, since H and B are constants, resulting in only the first power of ϕ to appear in the equation.

If H is not a constant but a function of ϕ as $H(\phi)$, an analytical solution is often not obtained because of the nonlinear expression $-H(\phi)\phi$ that arises on the right-hand side of the differential equation. In such circumstances, a numerical solution of the differential equation may be obtained by choosing a time step $\Delta\tau$ and advancing time to compute ϕ as a function of time, starting with the given initial condition. This computation is done until a small change is observed in $\phi(\tau)$ from one time level to the next, thereby indicating that the temperature has reached steady state, given by $d\phi/d\tau = 0$. However, since an analytical solution is available for the simplified circumstance of Eq. (1.1), the numerical scheme should first be used to solve the problem with H taken as a constant and the computed results compared with the analytical solution. This comparison will allow a determination of the anticipated accuracy of the numerical results and will also check the correctness of the procedure. Such a comparison is particularly valuable in complicated problems where an error in the numerical scheme may go undetected. Fortunately, many physical and chemical problems can be formulated in terms of idealized circumstances which lead to simplified equations that can be solved analytically. Chapter 8 discusses several methods for solving ordinary differential equations and demonstrates again the importance of available analytical results.

Similarly, in numerical differentiation and integration, the computational scheme may be tested by employing simple functions whose derivatives and integrals can be obtained analytically. In radiative heat transfer, for instance, integration over the wavelength λ of the radiation is frequently needed to determine the total energy lost or gained, Q, per unit area, at a surface. The expression for Q is

$$Q = \int_0^\infty f(\lambda)\, d\lambda \tag{1.3}$$

where $f(\lambda)$ is known as the monochromatic emissive power and is often a fairly complicated function of the wavelength λ, generally obtained from a curve fit of experimental measurements. However, the radiation from a blackbody, which is an idealized circumstance, is given by Planck's law, which gives $f(\lambda)$ as

$$f(\lambda) = \frac{c_1}{\lambda^5[\exp(c_2/\lambda T) - 1]} \tag{1.4}$$

where T is the surface temperature on the Kelvin scale and c_1, c_2 are known constants. Figure 1.3 shows the variation of $f(\lambda)$ with λ for the ideal surface of a blackbody, for a real surface, and for a gray body for which $f(\lambda)$ is a constant fraction of that for a blackbody at all λ.

For a blackbody, the integral in Eq. (1.3) has been evaluated analytically and is given by

$$Q = \sigma T^4 \tag{1.5}$$

where σ is known as the Stefan-Boltzmann constant and whose numerical value is given in the literature as 5.67×10^{-8} W/m^2·K^4. Therefore, the computational scheme developed for numerically determining Q for a wide variety of surfaces, and thus different $f(\lambda)$, may first be applied to blackbody radiation and the results compared with the analytical solution given by Eq. (1.5) to determine the accuracy and validity of the numerical method.

The numerical solution of large systems of linear or nonlinear equations is often needed in engineering problems. Since small sets of equations, typically three or four equations, can be solved analytically, the numerical procedure for solving systems of

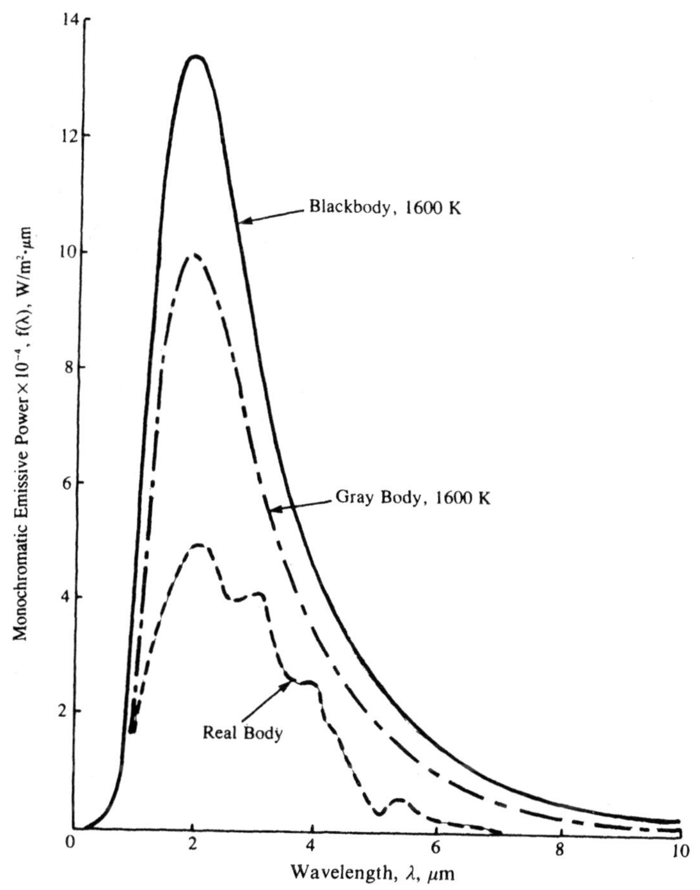

Figure 1.3 The variation of the emissive power $f(\lambda)$ with the wavelength λ for thermal radiation by a blackbody, by a gray body and by a real surface.

simultaneous algebraic equations may be employed for a small number of equations and the numerical results compared with the analytical values, to determine the accuracy and correctness of the numerical solution.

In numerical methods based on iteration, a convergence criterion ε is employed to decide when to terminate the iteration. Generally, the convergence criterion is applied to a physical variable in the problem, and computation is stopped when the change from one iteration to the next is less than the chosen value of ε. A relationship between ε and the accuracy of the numerical results may be obtained by a comparison of the computed values with the analytical solution that may be available for a simplified circumstance. This information can then be employed in the choice of the convergence criterion. If analytical results are not available, an extensive testing of the

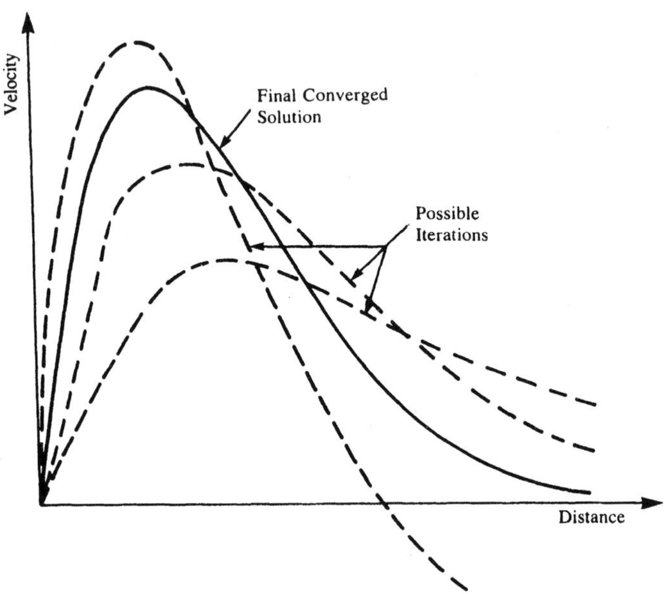

Figure 1.4 Typical iterative solutions in the numerical computation of the velocity profile in a natural convection flow.

numerical procedure, over wide ranges of the initial guess, convergence criterion, and time step $\Delta\tau$, for example, in the problem given by Eq. (1.1), must be carried out to ensure that the numerical results are essentially independent of the values chosen and that the desired accuracy level has been achieved. Figure 1.4 sketches typical computed iterative and converged solutions to the ordinary differential equations that govern a particular flow circumstance. The questions related to iterative convergence and to the choice of the numerical parameters, such as ε and $\Delta\tau$, are extremely important and are discussed in detail in the next chapter.

1.4 PHYSICAL AND CHEMICAL CONSIDERATIONS

The physical or chemical considerations that give rise to a given mathematical expression or equation can often be used very effectively in selecting the numerical method, in choosing an acceptable solution from several that may be obtained, and in testing the method for accuracy and correctness. In most engineering problems, the nature of the desired solution is known, along with the range in which it lies. Let us consider, for example, the free fall of a body of arbitrary shape in air. A terminal velocity is attained due to the balancing of the gravitational force by the frictional drag force (Halliday and Resnick, 1986). Depending on the size and shape of the body, an expression for drag may be obtained from considerations of air flow around the body. For a flat plate, a commonly employed expression for the frictional force is $(AV^{13/7} - BV)$, where V is the speed at which the plate is moving in stationary air and A and B are constants that depend on the length of the plate and the properties of air at the given temperature. Then, if m is the mass of the plate and g the magnitude of gravitational acceleration, the terminal velocity is the root of the equation

$$AV^{13/7} - BV = mg \tag{1.6}$$

From a physical consideration of the problem, we know that the terminal velocity must have a unique, positive value. The range in which the value lies may also be estimated from the available results for other bodies, for example, the sphere. A similar equation is obtained for bodies of other shapes and sizes. In many cases, the expression for drag is obtained from a curve fit of experimental results and is given as a fairly complicated function of the velocity V. A solution of the resulting force balance equation will then yield the terminal velocity for the given body. The method for solving the above equation may be selected knowing that the root is real, distinct, and positive. As discussed in Chapter 4, the secant method and the Newton-Raphson method are two efficient computation schemes that may be employed for this problem. If a method that determines all possible roots of the equation is used, the physical considerations are employed in choosing the correct solution. Since the solution is expected to be unique, the other roots must be complex numbers, negative or beyond the expected range of values.

The physical or chemical background of the mathematical problem being solved numerically is particularly important in the solution of nonlinear equations, such as the polynomial equation, Eq. (1.6), or transcendental equations. Some examples of the latter are as follows:

$$\tan x = \frac{B}{x} \tag{1.7}$$

$$\log x + 2x^2 = 4 \tag{1.8}$$

$$e^x + x^2 - 2x = 2 \tag{1.9}$$

Nonlinear equations arise very frequently in engineering problems, such as those related to fluid flow, heat transfer, chemical reactions, and dynamics of bodies. The problems encountered may involve finding the roots of a given nonlinear equation or solving a system of nonlinear equations. Since the characteristics of nonlinear equations are generally much more complicated than those of linear equations and since several solutions are feasible, the physical or chemical aspects of the problem are used in the development of the computational procedure and in deciding which solutions are acceptable. Even for solving a system of linear equations by iterative methods, physical and chemical considerations are often important in obtaining the starting values. Linear and nonlinear equations are also frequently obtained in the numerical solution of partial differential equations. The physical or chemical nature of the quantities to be computed is usually employed in the choice of the method, the initial guess, the distribution of grid points where the finite difference form of the equation is written, the desired accuracy level, and the convergence criterion for the termination of the numerical scheme. Since analytical solutions are rarely available, the numerical results obtained are generally considered in terms of the fundamental nature of the problem in order to determine the validity of the numerical scheme.

Curve fitting is another area in which the physical or chemical considerations underlying the given problem are of particular importance in developing the computational scheme. Numerical methods are generally used to obtain the best fit to a given set of data. In such cases, it is important to know the expected trends on the basis of the physical or chemical aspects of the problem, so that the best fit obtained is a true representative of the process involved. Consider, for example, the mean daily ambient air temperature at a given location. We wish to obtain a mathematical expression from the 365 data points that represent the measurements of the average daily temperature over a year. We could obtain a 364th-order polynomial from the given data. However, to do so would involve a substantial computational effort, both in obtaining the polynomial and in the subsequent usage of the polynomial in relevant problems. Moreover, the air temperatures fluctuate due to environmental disturbances. Consequently, we are interested in obtaining an expression that represents a best fit to the data and also characterizes the variation over the year. Since we know that the variation is periodic, with a time period of 365 days, we may try to fit the measurements to a sinusoidal variation. Examples of some of the distributions that may be employed are as follows:

$$T_a = A \sin[\omega(\tau - a)] \tag{1.10}$$

$$T_a = A \sin \omega t + B \cos \omega \tau \tag{1.11}$$

$$T_a = A \sin \omega \tau + B \sin 2\omega \tau \tag{1.12}$$

where T_a is the ambient temperature; ω is the frequency, given as $2\pi/365$; τ is the time in days; and A, B, and a are constants to be determined numerically from a best fit. The first equation is frequently used, with fairly satisfactory results. Figure 1.5 shows the resulting curve fit qualitatively. Similar considerations are employed in obtaining empirical correlations from experimental data and for representing material property data, such as those of interest in thermodynamics, by a best fit.

Numerical simulation of engineering systems is important in design and optimization. It involves the mathematical modeling of components and physical or chemical processes that comprise the given problem, followed by a numerical solution of the governing equations obtained. The input parameters, initially chosen on the basis of available data, are varied until a close agreement between the physical system and the numerical model is obtained. Once an existing system or process has been numerically simulated, the effects of variations in design on the performance of the system may be studied numerically, leading to optimization. At various stages in such a study, the physical or chemical aspects of the problem are employed. In fact, the comparison between the numerical model and the actual system forms the basis for

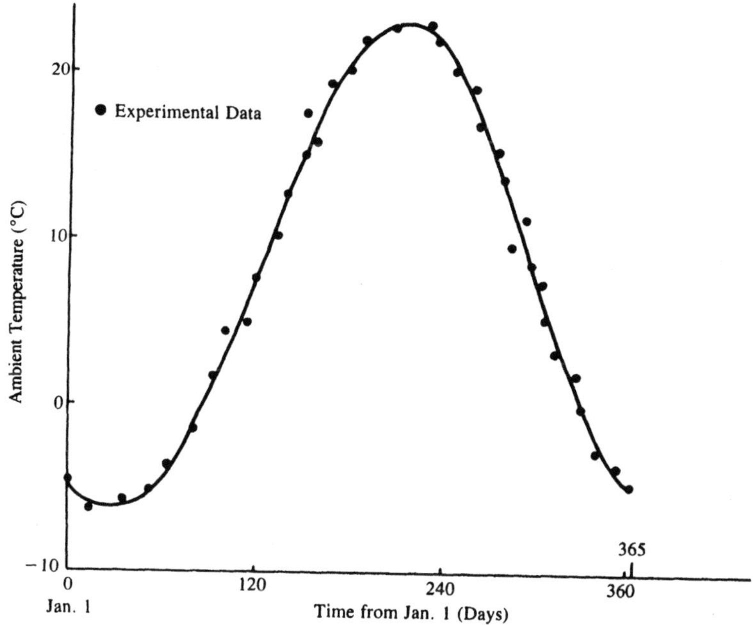

Figure 1.5 Sketch of the best curve fit to the experimental data on the ambient temperature variation over the year.

the development of the numerical scheme and for the study of the numerical results obtained.

Therefore, in the presentation of numerical methods for engineering problems, actual problems need to be considered, in order to demonstrate the importance of the physical background of the problem in the selection of the method and of the acceptable numerical results. The general features of the various methods are important and must also be studied in detail. However, some of the important aspects can be best understood in terms of the underlying physical or chemical considerations. Therefore, simple examples from several areas of engineering interest are employed in this book. Also, for conciseness, the physical or chemical aspects related to a given problem are frequently referred to simply as the physical characteristics.

1.5 APPLICATION OF COMPUTER METHODS TO ENGINEERING PROBLEMS

Computational techniques are used in engineering for a wide variety of applications. Several examples of the problems that would generally be solved on the computer have been given in the preceding discussion. The numerical methods for engineering application may best be considered in terms of the various mathematical problems that commonly arise in engineering. Computer methods for the solution of these problems may then be considered, using examples of mathematical expressions and equations from various engineering disciplines. This approach would allow a consideration of the various methods that may be employed for obtaining the numerical solution of a particular mathematical problem, say, integration, while employing examples from engineering to bring out the importance of physical or chemical considerations.

Various types of equations are encountered in engineering applications, including linear and nonlinear algebraic equations and ordinary and partial differential equations. Frequently, systems of equations, which are linked with each other through the unknown variables, are obtained. Partial differential equations arise in areas such as heat transfer, fluid mechanics, elasticity, electrostatics, and combustion. Such equations are usually solved by finite difference or finite element methods, which convert the problem into a system of algebraic equations by applying the partial differential equations at a finite number of grid points or integrating them over finite regions. Therefore, the solution of a system of algebraic equations is extremely important in engineering applications, and many methods have been developed to solve the different types of equations that are frequently encountered. Sets of algebraic equations are also directly obtained in many physical problems, such as those of interest in thermodynamics, economics, vibrations, structural analysis, and electrical networks. Although linear systems are particularly important, many engineering problems result in systems of nonlinear equations, which must be solved iteratively to obtain the solution. In most cases, nonlinear systems are formulated so that the methods for linear equations may be employed iteratively to converge to the desired solution.

In several problems, the roots of a nonlinear algebraic equation, transcendental or polynomial, are to be determined. Such problems arise, for instance, in the determination of the temperature of a body from an energy balance, of the terminal velocity of a body falling under gravity, of the density of a gas from its equation of state, and of vibration frequencies from the characteristic equation of a given system. Again, various methods are available, some of which are applicable only to polynomial equations, while others may be used for finding the real or complex roots of other types of equations. Depending on the nature of the problem, the appropriate method may be selected. If not much prior information is available on the nature and approximate magnitude of the roots, the general behavior of the function $f(x)$ that constitutes the given equation, $f(x) = 0$, where x is the unknown, may be investigated numerically. The numerical method for the solution may then be chosen on the basis of the information obtained.

Ordinary differential equations are important in several areas of engineering interest, such as heat and mass transfer, dynamics, fluid flow, chemical reactions, electrical circuit analysis, and elasticity. In a few cases, partial differential equations can also be transformed into ordinary differential equations. Frequently, several ordinary differential equations that are coupled through the unknowns are to be solved simultaneously. The solution procedure depends on the nature of the problem, particularly the order of the equation and the boundary conditions. For instance, the following second-order ordinary differential is obtained for a resonant electrical circuit:

$$A\frac{d^2V}{d\tau^2} + B\frac{dV}{d\tau} + V = 0 \tag{1.13}$$

where V is the voltage across a capacitor and A and B are constants that depend on the resistance, inductance, and capacitance in the circuit. If the initial conditions are given as

$$V = V_0 \quad \text{and} \quad \frac{dV}{d\tau} = 0 \qquad \text{at } \tau = 0 \tag{1.14}$$

we have an initial-value problem, in which the integration of the equation may be started at the given time $\tau = 0$ and incremented to larger time to obtain the solution. If one of the conditions is given at a different time or location, a boundary-value problem is obtained, in which a correction scheme is needed to satisfy the given boundary conditions. Then, iteration is generally employed to converge to the solution.

Besides algebraic and differential equations, several other mathematical problems arise in engineering. Numerical differentiation and integration are needed in many cases, often as part of a more complicated problem. Numerical integration over time is needed, for instance, in determining the total energy lost by a body. Similarly, integration of velocity across a cross section gives the total flow rate. Numerical differentiation is needed, for example, in the determination of the acceleration of a particle from the measured variation of its velocity with time. Rate processes are important in engineering, and numerical differentiation is frequently employed for obtaining the rates of change of various physical quantities. Numerical techniques are also needed in interpolation and extrapolation, employing curve fitting of given data. In some cases, an exact fit which yields the exact value at the given data points is appropriate. However, more frequently, a best fit of the data is employed so that the general features of the results may be represented by a correlating equation, without forcing the curve to pass through each data point. Computer graphics can be employed advantageously with the computer solution of engineering problems to present the numerical results and to study the computational scheme.

Therefore, a consideration of numerical methods for engineering application involves a wide variety of mathematical problems, as outlined above. It is important to understand the advantages and limitations of a particular method for solving a given problem. The numerical procedure and the results obtained must also be related to the physical or chemical background of the problem in order to test the validity of the computational scheme and to choose the acceptable solutions. Similarly, a comparison between the numerical and analytical results must be made, whenever possible, to check the accuracy of the results obtained. The development of the numerical scheme for a given problem may be discussed in several ways. A practical approach is to take the mathematical problem arising from the actual circumstance, give the computer program, and discuss the numerical results in terms of the physical aspects of the problem and available analytical results. It is this approach that is followed here. The computer languages chosen are FORTRAN 77 and BASIC, which are presently the most important languages in the application of computer methods to engineering problems. However, other languages may also be employed by suitably modifying the given programs, as discussed in Chapter 2. Of particular importance in the use of numerical techniques for solving engineering problems is the need to test the computational scheme for accuracy and to interpret the numerical results obtained. In this text, these and the other aspects mentioned above will be considered in terms of various examples taken from several engineering disciplines, including aeronautical, chemical, civil, electrical, and mechanical engineering.

1.6 OUTLINE AND SCOPE OF THE BOOK

1.6.1 Basic Features

This book presents the mathematical background as well as the application of computational techniques to problems of engineering interest. The material is developed by the derivation of the formulas for each method, followed by a discussion of the accuracy, computational effort, and range of applicability of the method. For each problem area considered, for example, root solving, several methods are discussed, emphasizing the ones that are most extensively employed. A comparison between various methods applicable for a particular type of mathematical problem is made, in order to indicate the advantages and disadvantages of a given method. Of particular interest in such a comparison are the associated errors, ease in programming, computing time needed, and flexibility in the application to a wide variety of problems. The circumstances under which a given method would be the preferred one are outlined. This consideration is a very important one, since several methods are frequently available for problems that arise in engineering applications, such as the solution of a system of linear algebraic equations and the numerical integration of a given function. The choice of the most appropriate method is highly desirable, in order to minimize the computing time and obtain the required accuracy level.

Following a detailed discussion of the mathematical background and the derivation of the relevant formulas for each numerical method, the computational procedure for applying the technique is discussed. The important considerations underlying the development of the numerical scheme are discussed, along with the difficulties that may be encountered. Finally, an example based on an actual engineering or mathematical problem is given, for most of the methods considered, and the computer program is developed. Again, the important features of the program are outlined and the numerical results obtained are presented and discussed. The emphasis is on presenting the basic algorithm of the method in terms of its application to an actual physical, chemical, or mathematical problem. Although the program is written as part of the example and is therefore geared to the solution of the specific problem considered, a few modifications in the program can easily be made to use it for the solution of other problems of a similar nature. This approach of writing a problem-oriented computer program, rather than a general program that can be used for a wide variety of problems, is adopted in order to present the program simply as a sample and to encourage the reader to write his or her own program on the basis of the information given, making the program as efficient as possible and employing the ongoing improvements in the available computational facilities.

1.6.2 Computer Programs

A wide variety of features are incorporated in the computer programs given. Both interactive and batch operation modes are utilized. Although most programs are written in FORTRAN 77, some are also given in BASIC, in order to indicate the similarities and differences between the two languages and to demonstrate the ease with which one could switch from one language to the other. Subroutines are very valuable, particularly for the complicated programs needed for the solution of differential equations, and are employed wherever appropriate. In some cases, the programs include statements that allow the results to be stored in data files for obtaining the output in graphical form. This approach is very commonly employed and, depending on the computer system available, can be used very advantageously with the computational procedure. Similarly, an interactive use of the computer, so that the input data are fed and the results obtained interactively by the operator, is preferred, particularly during program development, to the batch mode, in which the entire program is entered with the input data and the computer gives the output after the complete run. Once the program has been developed and debugged, the batch mode is more appropriate, particularly for complicated programs, as discussed in greater detail in Chapter 2.

1.6.3 Examples/Problems

The examples and problems considered in this book are derived from topics of interest in the major engineering disciplines and in the basic sciences. The physical or chemical background of the problems is outlined in order to enable the reader to follow the relevance of these considerations in the choice and testing of a particular numerical technique. Also, a selection of problems that arise in practical circumstances makes the discussion interesting and relevant to engineering applications. As discussed earlier in this chapter, numerical solutions must be considered not only in terms of the basic nature of the given problem but also in terms of any analytical solutions available, even if they are for very simple situations. These aspects are stressed in evaluating the numerical results for accuracy and validity. In solving problems of engineering interest, the available information on the given system or process must form the basis for the development of the numerical scheme and for the verification of the results obtained.

Both the problems and the examples tend to expand on the material covered, so that they contribute to an increased understanding of the discussion given in the text. Several new physical and chemical phenomena are also introduced in the problems to indicate the application of the methods presented to a much wider spectrum of engineering processes. Although the emphasis is, obviously, on the numerical solution, several problems are also directed at the mathematical background, particularly at the errors involved and the mathematical formulation for a numerical solution. In addition, many problems can be solved on a calculator in order to study a given numerical scheme.

Much of the material presented in this book has been used in courses taught at the sophomore and junior levels in engineering. A few of the topics covered may be somewhat advanced for sophomore students and are indicated by an asterisk in the book. Similarly, the physical background of the problems may not be familiar to some of the readers. Consequently, a brief discussion of the important aspects of the problem or example under consideration is included. In some cases, reference is also made to books that may be consulted for a more detailed coverage of the topic. A background in programming, such as a freshman-level, one-semester course, is assumed, although some of the important aspects are covered in Chapter 2 for completeness.

1.6.4 A Preview

The presentation of the numerical techniques for engineering application starts with Chapter 2 on the basic considerations in computer methods. This chapter outlines the important elements in computational procedure, including program development, numerical errors, accuracy, convergence, and other basic aspects. Although some of the discussion will be quite familiar to those experienced in computer programming, many of the aspects considered in this chapter are important in obtaining an accurate and valid solution to a problem of engineering interest. This chapter also outlines the current trends in computational methods and facilities, with respect to both the software development and the growing capability of computer systems.

The Taylor series, which forms an important element in the estimation of numerical truncation errors, is presented in Chapter 3, along with the numerical approximation of derivatives. Several methods for differentiation are presented, and many of the results presented here are employed in later chapters. Methods for finding the roots of nonlinear algebraic equations are discussed in Chapter 4. Several methods which are based on the sign change, at the root, of the function $f(x)$ in the given equation $f(x) = 0$ are first considered. Efficient methods such as the secant and Newton's methods, which converge very rapidly, although they may also diverge in certain cases, are discussed in detail. Specialized methods for equations in which $f(x)$ is a polynomial are also discussed. Finally, a comparison between the various available methods is made.

The solution of simultaneous linear or nonlinear algebraic equations is an important problem in engineering applications and forms the subject of Chapter 5. Direct as well as iterative numerical methods are discussed, the latter being the inevitable approach for most nonlinear equations. Eigenvalue problems are also considered and the available methods outlined. Numerical methods for curve fitting of data are presented in Chapter 6, considering both the exact fit as well as the best fit approach. Various techniques for interpolation are discussed, emphasizing popular methods such as Lagrange and Newton's interpolating polynomials. The least-squares method for a best fit is discussed in detail, and various forms of the function for curve fitting are considered.

Numerical integration forms the subject of Chapter 7, and several important methods, such as the trapezoidal and Simpson's rules, Romberg integration, and Gaussian quadrature, are discussed. The advantages of each method, its limitations, and the conditions under which it is preferred are considered in some detail. The associated errors and the resulting accuracy are also discussed. The numerical integration of improper integrals, whose limits of integration may be infinite or the integrand may become singular over the range of integration, is also presented.

The solution of differential equations is a very important subject in engineering. Because of the complexity of the problems commonly encountered, numerical methods are generally needed. Ordinary differential equations (ODEs) are considered in Chapter 8 and partial differential equations (PDEs) in Chapter 9. Both self-starting methods, such as Euler's and Runge-Kutta methods, and multistep methods, such as predictor-corrector methods, are considered for ODEs. Also, the associated errors, accuracy, stability, and convergence of these methods are considered, along with their efficiency in terms of the computational effort required. Several types of equations, including initial-value, boundary-value, and systems of equations, are considered and the relevant numerical techniques presented. Again, a critical comparison between the various methods is made in order to guide the choice of the most suitable scheme for a given problem. Finite difference methods, derived from the numerical approximation of derivatives, given in Chapter 3, are also outlined for ODEs.

Partial differential equations are included in this book largely for junior- and senior-level students and also for professional engineers. With the introductory background presented, the material could also be used for less advanced students. The material covered in Chapter 9 considers mainly linear equations of parabolic, elliptic, and hyperbolic type. The basic nature of the equations is discussed in detail, and important numerical methods for their solution are presented. The questions of accuracy, convergence, and stability are again considered. Finite difference methods are largely considered, with a brief introduction to finite element methods, since the former is easier to understand and can be developed on the basis of the material presented in Chapter 3. The methods for treating different types of boundary conditions are also outlined.

In all the topics considered here, a large numer of examples and problems are given, so as to provide a strong physical and numerical base for the computational study of engineering problems. Since the best way to learn numerical methods is by applying the techniques available to different problems and developing one's own computer code, almost all the examples and most of the problems demand the development of the relevant program and its use for obtaining the desired numerical results. Although a calculator may be used in several cases to study the computational steps in a given method, the readers are strongly encouraged to write computer programs for the problems given, using the discussion, formulas, and examples given in the text.

2

Basic Considerations in Computer Methods

2.1 INTRODUCTION

In the numerical solution of engineering problems, there are several important aspects that need to be considered in order to ensure the correctness of the chosen approach for a given problem and the accuracy of the results obtained. The computational procedure involves a consideration of the methods available for solving the given problem, of the appropriate programming language, of the computer and its operating system, and so forth, before proceeding to the development of the numerical scheme and the corresponding program. Since these considerations are fundamental to most computer methods, this chapter discusses the general approach to the development of the computational scheme. Also considered are the interfacing with available computer software and the verification of the numerical results by a comparison with available analytical and experimental results, as discussed in Chapter 1.

The consideration of numerical errors and the accuracy of the results is important in the numerical solution of a given problem. The various types of errors that arise in the computational approach are discussed, along with methods that may be employed for reducing the error. The accuracy of the solution may often be estimated by comparing the numerical results with those from the analytical solution, for simpler problems, since the analytical solution of the given problem is presumably not available. The question of accuracy is a very important one, and frequently no satisfactory analytical results may be available for comparison. In such cases, the numerical scheme itself is employed to check the accuracy of the numerical results by ensuring that numerical parameters, such as the chosen time step and grid size, do not significantly affect the results. Also, the physical or chemical nature of the problem being solved can often be employed as a check on the validity of the numerical scheme and the correctness of the results obtained. The accuracy of the numerical results can frequently be evaluated by substituting the solution obtained back into the algebraic or partial differential equation being solved to determine how closely it satisfies the equation. Several other similar procedures are generally employed to check the accuracy of the numerical solution.

Consider, for example, the dynamics of a moving body whose displacement x is governed by the ordinary differential equation $dx/d\tau = F(x,\tau)$, where τ is time and $F(x,\tau)$ is a given function. We may assume that the analytical solution is not available, since if it were, there would be no need to solve the problem numerically. However, the numerical scheme may be employed to solve a simpler equation, say, $dx/d\tau = -ax + b$, where a and b are constants. The analytical solution to this equation can easily be obtained as $x = ce^{-a\tau} + b/a$, where c is a constant to be determined by applying the initial condition, that is, by using the given value x_0 of the displacement at time $\tau = 0$ or at any given specified time; see Fig. 2.1. The accuracy of the numerical method may be determined by comparing the numerical solution for such a simple problem with the analytical solution. For a more complicated function $F(x,\tau)$, the following considerations may be used. The physical nature of the problem demands that the displacement be real and positive. Also, it would often be known whether it is periodic, or whether it must increase, or decrease, with time. This information may be employed to select the correct solution in case multiple solutions arise and also to check the validity of the numerical scheme. Once the numerical solution $x(\tau)$ is obtained, numerical differentiation may be used to determine $dx/d\tau$ at a few selected values of τ. These may then be employed to check if the numerical values of x do indeed satisfy the equation $dx/d\tau = F(x,\tau)$, to the desired accuracy level. Finally, the step size $\Delta\tau$ employed in the numerical scheme must be reduced until a

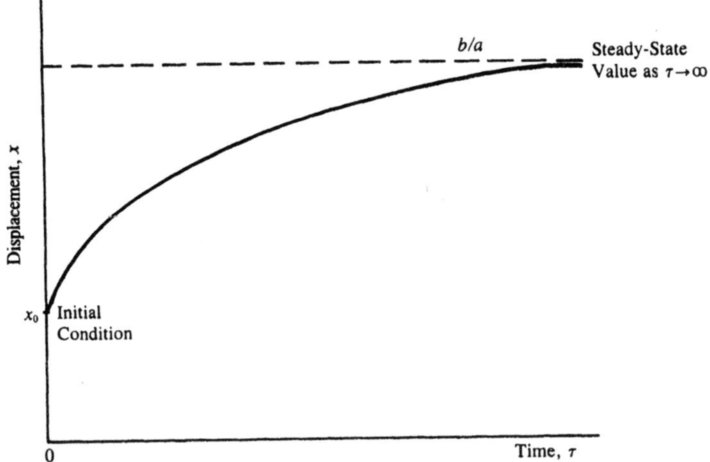

Figure 2.1 Sketch of the analytical solution of the differential equation $dx/d\tau = -ax + b$, where a and b are constants and $x = x_0$ at $\tau = 0$.

further reduction in $\Delta\tau$ does not significantly affect the numerical results.

The numerical methods for the solution of a wide variety of problems are based on an iterative approach, in which the solution is gradually improved, starting with an initial, guessed value until the change in the solution from one step to the next becomes less than a chosen small quantity, known as the *convergence criterion* or parameter. In such cases, the convergence of the iterative procedure is an important consideration, and it is necessary to determine the conditions under which the scheme may diverge. If a particular method diverges for a given problem, the problem can sometimes be reformulated so that the scheme converges. Otherwise, a different method must be employed. Numerical stability is another important consideration that guides the selection of the method and of the grid, or step, size in the numerical scheme. Again, it is necessary to determine when numerical instability might arise and to take steps to avoid it.

This chapter discusses many of these considerations which are basic to most numerical methods. The general approach to the development of a numerical scheme is outlined, indicating various important aspects that need to be taken into account. The concepts of error, accuracy, iteration, convergence, and stability are discussed in general terms, by taking examples from various topics, such as root solving, numerical differentiation and integration, curve fitting, and solution of algebraic and differential equations, considered in greater detail in later chapters. Several characteristic features of these topics are needed for presenting the material in this chapter. Thus, the relevant aspects are outlined, leaving the details for subsequent chapters. The discussion in this chapter forms the basis for the development, application, and verification of the numerical procedures for these and other topics of interest in engineering applications.

2.2 COMPUTATIONAL PROCEDURE

The general approach to the development and application of the computational procedure for solving a given problem is discussed in this section, indicating the important aspects that generally need to be considered for an efficient and accurate scheme. Although some of the basic considerations outlined here may not be applicable to a particular circumstance, it is important to recognize the important steps that lead to a successful numerical method. Most of the items included here are fairly straightforward and are quite familiar to those who have done a significant amount of numerical work. However, the systematic approach given here is helpful, particularly for those who are relatively less experienced in computer methods, in investigating the relevant aspects that determine the efficiency, accuracy, and validity of the numerical procedure. It is assumed that the mathematical formulation of the given physical or chemical problem has been completed and that an analytical solution is not easily obtainable, so that it has been decided to solve the problem numerically.

2.2.1 Method Selection

Frequently, several methods are available for the numerical solution of a given problem. The selection of the method to be employed, from among the several applicable methods, is a very important consideration and is generally based on many relevant criteria, such as the following:

1. Accuracy
2. Efficiency
3. Numerical stability
4. Programming simplicity
5. Versatility
6. Computer storage requirements
7. Interfacing with available software
8. Previous experience with a given method.

The accuracy of a given method is an important consideration in its selection for solving a particular problem. The evaluation of the accuracy of a method may be based on a comparison of the numerical results with available analytical results, as outlined in the preceding section, on an estimation of the associated numerical errors, or on various methods for checking the correctness of the numerical solution, such as substitution of the numerical results back into the equation being solved to determine the accuracy to which the numerical solution satisfies it. All these aspects, particularly the numerical errors that arise in computational methods, are discussed in detail later in this chapter.

The efficiency of a given method is generally based on the number of arithmetic operations needed per computational step or the total number needed for solving the entire problem. One could also solve a given problem with different methods and determine the computer time needed in each case. However, the number of arithmetic operations, which include addition, subtraction, multiplication, and division, can often be determined by noting down the various mathematical manipulations performed, per step, in a given numerical scheme. For example, it can be shown that the solution of n simultaneous, linear, algebraic equations involves $n^3/3$ arithmetic operations by the Gaussian elimination method and $n^3/2$ operations by the Gauss-Jordan elimination method, as discussed in Chapter 5. This implies that the former method is more efficient. Similarly, predictor-corrector methods for the solution of ordinary differential equations will be shown in Chapter 8 to generally involve a smaller number of computations per step than the Runge-Kutta methods. A higher efficiency of the method also implies shorter computer time and, thus, lower computational cost.

Numerical instability refers to the unbounded growth of numerical errors as computation proceeds. It is of particular concern in the solution of differential equations and, if present, can lead to an erroneous and unacceptable numerical solution. Therefore, it is important to determine the stability characteristics of the various methods that are applicable to a given problem. Frequently, the numerical scheme may be conditionally stable; that is, it may be stable within certain constraints that often limit the grid or step size. In the solution of parabolic partial differential equations, for instance, the explicit schemes, which are generally simpler to use, often restrict the step size to very small values, making these schemes inefficient. Then the implicit methods, which usually do not have such constraints resulting from stability considerations, are preferred. Thus, the numerical instability of the method is an important consideration in its selection.

Several other considerations also play an important role in the selection of the method. These include simplicity in programming, versatility of the method, computer storage needed, and interfacing with available software. In engineering applications, the simplicity and versatility of the method are very important, since interest often lies in solving a wide variety of problems with the least amount of effort. Frequently, some sacrifice is made with respect to accuracy and efficiency in order to select a simpler and more versatile method. An example of this is the Runge-Kutta method, for solving ordinary differential equations. This method is often chosen over predictor-corrector methods, which are more efficient than the former but are also more complicated to program.

The computer storage requirements of the method are generally important in the simulation of large systems that are of interest in engineering applications. For example, the Jacobi method for solving a system of linear algebraic equations involves the storage of the matrices of the unknowns at two iterative steps, the present and the previous one, whereas the Gauss-Seidel method requires the storage of only the latest values. Thus, the latter method requires only about half the storage needed by the first method. It is also more efficient on conventional computers and is preferred.

The interfacing of the numerical method with the computer software is particularly important when available programs are being employed. For instance, if a matrix inversion program is available, methods based on the inverse of the matrix for solving a system of linear equations may be chosen. Similarly, prior personal experience with a given method would be an important consideration in its selection.

2.2.2 Programming Language

After the numerical method for the solution of the given problem has been selected, the next step is the development of the computer program or code that allows one to interface with the computer system. However, before proceeding with the code development, one must select the programming language and the computer system to be used and become fully conversant with the selections made. The programming languages, often termed *high-level languages*, allow one to write the step-by-step instructions for the computer in a form that is quite similar to ordinary English and algebra. The computer itself interprets and executes statements only in the machine language, and a compiler is employed by the computer to achieve the translation from the programming language to the machine language. The machine language program is then stored, providing direct access for immediate or later execution.

FORTRAN. Several high-level programming languages have been developed over the years. The most widely used among these, for engineering and science, is FORTRAN, which stands for *formula translation.* It is available in many versions, such as FORTRAN IV and FORTRAN 77. The latter is extensively used today and is available on most computer systems. It has greater flexibility than the older version and, thus, results in a simpler program. However, the main advantage of FORTRAN 77 over earlier versions is in the control statements, particularly the availability of the IF-THEN-ELSE statements in this version. These statements permit the development of a structured program in which control flows from top to bottom. It is much easier to read a structured program, rather than one in which control is transferred from one point in the program to another in a seemingly random fashion. This structure also makes it simpler to validate the program. In this book, we will predominantly use FORTRAN 77 to demonstrate the numerical solution of engineering problems because of its present importance. Both FORTRAN IV and FORTRAN 77 are widely used by engineers. The latter version may not be available on some systems. However, the two versions are fairly similar in form, and it is a simple matter to go from one to the other. Several books are available on programming in FORTRAN and may be referred to for details on the language. See, for instance, the books by Merchant (1981) and Friedman and Koffman (1981), listed in the References at the end of this book.

BASIC. There are several other programming languages that are employed for solving problems in science and engineering. These include BASIC, PASCAL, C, APL, PL/I, LISP, and others. Among these, BASIC, which stands for *beginner's all-purpose symbolic instruction code*, is also a widely used language, particularly on personal computers, which are usually called *microcomputers.* BASIC is generally simpler to use than FORTRAN and is well suited for small programs. However, it is not as versatile as FORTRAN and is often inconvenient for large, complex programs. At the initial stages of the development of personal computers, BASIC was the only language available on these machines because of the simplicity in programming and smaller memory requirements, as compared to, say, FORTRAN. BASIC continues to be an important programming language, and, because of its wide use on personal computers, most students are exposed to BASIC before they move on to other, more powerful languages. Because of these considerations, several programs are also given in BASIC in this book.

Many versions of BASIC have been developed in recent years, with each version often being particularly suited to a given computer. Many of the constraints that existed in the earlier versions, such as the limitation of only two characters in variable names, lack of format for input/output of data, necessity of labeling each line in the program, and difficulty of writing a structured program, have been eliminated in more recent versions. Among these are BetterBASIC, QuickBASIC, TrueBASIC, and WATCOM BASIC. Besides several improvements in the language that allow structured programming, BASIC is now available in the traditional interpreted version, which applies the editor to each line as it is entered, and in the compiled/linked version, which puts the code in memory before execution. Hybrid versions have also been developed to obtain the advantages of both an interpreter and a compiler.

Computer programming in FORTRAN is quite similar in form to that in BASIC, and it is not difficult to convert a given program in one language to the corresponding program in the other language. In fact, the computer code may be written in the simpler BASIC language and tested on a personal computer before it is written in FORTRAN for a main-frame computer. However, there are many structural differences between the two languages, as illustrated in Example 2.1. In most common versions of BASIC, each statement is numbered, the unnumbered ones being executed as soon as they are entered, whereas in FORTRAN, only those statements that need to be identified are labeled with a number. The input/output of data is generally done without formatting in BASIC, whereas standard FORTRAN usually requires a FORMAT statement. A FORTRAN program requires adherence to specific columns; for example, the first column is used to denote a comment card, the labeling of a statement is done with a number in the second to fifth columns, and the sixth column is used to indicate continuation. In contrast, spacing is usually not important in BASIC. The control statements are also quite different in the two languages, as seen in Example 2.1, with BASIC closer to common English. For further details on the similarities and differences between BASIC and FORTRAN, the books on FORTRAN, referenced above, and those on BASIC, such as the ones by Coan (1978) and Miller (1981), may be consulted.

Other Languages. The various other languages, mentioned earlier, are also often employed in engineering problems, although much less frequently than FORTRAN, which continues to be the main programming language in science and engineering at the present time. PASCAL is a structured language developed in the last decade. Its main attractions lie in the ease with which information can be arranged for processing in its structured form, a feature that has been incorporated in FORTRAN 77, and in the ease with which the implementation of a PASCAL program may be transferred from one computer to another. However, a PASCAL program generally takes larger computer time than one in FORTRAN, used with a compiler. It is also, in general, a more complicated language to use than BASIC or FORTRAN. See, for instance, the books by Kernighan and Plauger (1981), Seiter and Weiss (1982), and Weiss and Seiter (1984).

Similarly, other programming languages have their special advantages and limitations. An important emerging language is C, which is a general-purpose programming language developed in the last few years. It is a relatively low-level language, implying that it is closer to assembly language than high-level languages such as FORTRAN. As a result, it is more difficult to move the program from one computer system to a different one. However, the language has several advantageous features in control flow and data structures. One such feature is the use of pointers which contain the addresses of objects and allow easy access to these. Because of such powerful features and lack of restrictions, it is expected that the usage of the C programming language will grow in the future. For details on the language, the books by Kernighan and Ritchie (1978, 1984), Costales (1985), Purdum (1985), and Waite et al. (1986) may be consulted.

Several other programming languages have gained considerable importance in the last few years. Among these are languages that allow symbolic manipulation, that is, languages in which words, sentences, and expressions can be employed for programming. LISP, which takes its name from *list programming*, is one such language that is important in the development of intelligence in computers. Similarly, PROLOG and SMALL TALK are languages used in generating artificial intelligence in engineering systems. For details on these languages, several references are available. See, for instance, the books by Winston and Horn (1984), Clocksin and Mellish (1984), Clark and McCabe (1984), and Brulé (1986).

Implementation. The computer program, written in a high-level language such as FORTRAN or BASIC, is implemented on the computer by means of an interpreter or a compiler. An interpreter examines each line of the program and checks it for the rules of the language before it is executed. BASIC is often executed on microcomputers this way. The interpreted approach is very valuable during program development, since error messages are given as soon as a statement is entered. However, it is very slow in the execution of the program. A compiler, on the other hand, organizes the entire program into a set of machine instructions and locations. Several compilers are available for BASIC, FORTRAN, and other languages. The compiler is often written for a given computer system and is generally a completely separate process undertaken before the program is run. Once the machine code has been produced by the compiler, the compiled program is stored and the program may be executed with a separate command. A single command that compiles and executes the program may also be used. The use of a compiler thus reduces the computer time for a given problem. Various compilers have their particular advantages and characteristics. For instance, WATFIV is a compiler that is widely used with FORTRAN during program development, since it is particularly good at providing diagnostic error messages.

From the above brief discussion of the various programming languages widely employed for engineering problems, it is obvious that the trend has been toward structured programming and interactive use of the computer, through an interpreter, which responds almost immediately, or an interactive compiler. Substantial improvements and modifications continue to be made in the available languages to simplify programming and to increase the versatility and capability of the language. Although it is difficult to keep up with all the advancements in the high-level languages and in the available interpreters and compilers, it is important to determine what is available on a given minicomputer or a main-frame computer. For programming on a microcomputer, a wide variety of software is available, and, if possible, the version most suitable to the particular needs of the problem may be chosen.

In general, an interactive use of the computer is preferable during program development, since the parameters of the problem may be entered by the operator at the terminal. The program may be compiled and executed to obtain the output as the program continues to execute. If the results are unacceptable, the execution may be stopped at any stage, and the input parameters varied and execution resumed. In the batch operation mode, the input parameters are part of the program, and the execution of the program must be completed before any changes can be made. Thus, at the initial stages of program development, interactive computer usage is particularly valuable. Once the program has been satisfactorily developed, detailed numerical results are best obtained by the batch operation mode on the computer.

Example 2.1

Write computer programs in BASIC and in FORTRAN to compute the sum S of the series

$$S = 1 + x + x^2 + x^3 + \cdots + x^n + \cdots \tag{2.1.1}$$

where x is a variable whose value is to be entered into the program interactively. In order for the series to be convergent, $|x| < 1$. This series represents the binomial expansion of $1/(1 - x)$, which therefore gives the exact value of S. Compare the exact and computed values of S to determine the numerical error. Discuss the dependence of the sum S on the number of terms n taken in the series.

Solution

Figure 2.1.1 shows the required computer programs in BASIC and FORTRAN. The value of x is entered when the computer asks for it, and the program is written to enter three different values in sequence. The basic considerations relevant to convergence are discussed in detail later in this chapter. However, it will suffice to mention here that each term in the series, given in Eq. (2.1.1), is larger than the next term, for $|x| < 1$. Thus, the contribution of each additional term to the sum decreases as n is increased. This relationship is used as a check on the convergence, since it is not possible to take an infinite number of terms and since it is desirable to have the least number of terms that give S within the acceptable error. Here, SN represents the nth term and S the sum of the series up to and including this term. Then the condition $SN/S < \varepsilon$, where ε is a chosen small quantity, taken as 10^{-6} in the program, can be employed to check the convergence and to terminate the computation if this condition is satisfied.

The program yields the number of terms needed for the above convergence criterion to be satisfied, the computed sum S of the series, and the percentage error $100(SE - S)/SE$, where $SE = 1/(1 - x)$ is the exact sum of the series. Figure 2.1.2 presents the results obtained. Clearly, the error is a function of ε, which may be chosen to keep the error within an acceptable value. This aspect is discussed in detail in Section 2.5. Also, note that the number of terms needed increases with the value of x. This result is expected, since convergence is slower at the larger value of x, as discussed in most textbooks on advanced calculus; see, for instance, Amazigo and Ruhenfeld (1980) and Kaplan (1984).

The differences in the statements involved in the two programming languages are evident in Fig. 2.1.1. The input/output statements are quite different, as are the

```
100 REM                  PROGRAM SERIES SUMMATION
110 REM
120 REM       HERE S IS THE SUM OF THE SERIES UP TO AND INCLUDING THE NTH
130 REM       TERM, SN IS THE NTH TERM, SX IS THE EXACT VALUE OF THE
140 REM       FUNCTION F(X)=1/(1-X), WHICH IS REPRESENTED BY THE SERIES,
150 REM       AND ER IS THE ERROR
160 REM
170 REM
180 REM       ENTER INPUT QUANTITIES
190 REM
200     FOR I=1 TO 5
210     PRINT "ENTER THE VALUE OF X"
220     INPUT X
230     N=0
240     S=0
250 REM
260 REM       SUM THE SERIES
270 REM
280     SN=X^N
290     S=S+SN
300 REM
310 REM       CONVERGENCE CHECK
320 REM
330     IF (SN/S) < .000001 THEN 360
340     N=N+1
350     GOTO 280
360     PRINT "X=";X
370     PRINT "THE REQUIRED NUMBER OF TERMS=";N
380     PRINT "THE SUM OF THE SERIES=";S
390 REM
400 REM       COMPUTE THE ANALYTICAL VALUE OF THE SUM AND THE ERROR
410 REM
420     SX=1/(1-X)
430     ER=((SX-S)/SX)*100
440     PRINT "THE ERROR=";ER;"  PERCENT"
450     PRINT
460     NEXT I
470     END
```

(a)

Figure 2.1.1 Computer programs in (a) BASIC and (b) FORTRAN for the summation of the series given in Example 2.1.

control statements. Note that BASIC is closer to common English; see, for instance, the commands for the DO loop. Still, the programs in the two languages are quite similar, and it is not very difficult to go from one programming language to the other. The simplicity of BASIC and the usual implementation of the language with an interpreter make it attractive during code development. For the final execution of the program, a compiler may be used. For more complex programs, various elements of the program may be tested in BASIC, with the final program being written in FORTRAN, which remains the dominant language for the numerical simulation of engineering problems.

```
C                 PROGRAM SERIES SUMMATION
C
C     HERE S IS THE SUM OF THE SERIES UP TO AND INCLUDING THE NTH
C     TERM, SN IS THE NTH TERM, SX IS THE EXACT VALUE OF THE
C     FUNCTION F(X) = 1.0/(1.0-X), WHICH IS REPRESENTED BY THE
C     SERIES, AND ER IS THE ERROR.
C
C
C     ENTER INPUT QUANTITIES
C
        IMPLICIT REAL (A-H,O-Z)
        DO 5 I=1,5
        PRINT *, 'ENTER THE VALUE OF X'
        READ *, X
        N=0
        S=0.0
C
C     SUM THE SERIES
C
  1     SN=X**N
        S=S+SN
C
C     CONVERGENCE CHECK
C
        IF ((SN/S) .GT. 1E-06)THEN
        N=N+1
        GO TO 1
        ELSE
  6     WRITE (1,2)X
  2     FORMAT(2X,'X=',F6.3)
        WRITE(1,7)N
  7     FORMAT(2X,'THE REQUIRED NUMBER OF TERMS=',I5)
        WRITE(1,3)S
  3     FORMAT(2X,'THE SUM OF THE SERIES=',F12.6)
C
C     COMPUTE THE ANALYTICAL VALUE OF THE SUM AND THE ERROR
C
        SX=1.0/(1.0-X)
        ER=((SX-S)/SX)*100.0
        WRITE(1,4)ER
  4     FORMAT(2X,'THE ERROR=',E10.5,'PERCENT'/)
        END IF
  5     CONTINUE
        STOP
        END
```

(b)

Figure 2.1.1 Continued

2.2.3 Computer System

The next consideration in the numerical solution of a given problem pertains to the computer system. Frequently, several systems, ranging from microcomputers to mainframe computers, are available to the engineer. Supercomputers may also be

```
ENTER THE VALUE OF X
? 0.1
X= .1
THE REQUIRED NUMBER OF TERMS= 6
THE SUM OF THE SERIES= 1.111111
THE ERROR= 1.072884E-05   PERCENT

ENTER THE VALUE OF X
? 0.3
X= .3
THE REQUIRED NUMBER OF TERMS= 12
THE SUM OF THE SERIES= 1.428571
THE ERROR= 2.503395E-05   PERCENT

ENTER THE VALUE OF X
? 0.5
X= .5
THE REQUIRED NUMBER OF TERMS= 19
THE SUM OF THE SERIES= 1.999998
THE ERROR= 9.536743E-05   PERCENT

ENTER THE VALUE OF X
? 0.7
X= .7
THE REQUIRED NUMBER OF TERMS= 36
THE SUM OF THE SERIES= 3.333328
THE ERROR= 1.716614E-04   PERCENT

ENTER THE VALUE OF X
? 0.9
X= .9
THE REQUIRED NUMBER OF TERMS= 110
THE SUM OF THE SERIES= 9.999912
THE ERROR= 8.583071E-04   PERCENT
```

(*a*) *From the Program in BASIC*

```
 ENTER THE VALUE OF X
0.1
  X= 0.100
  THE REQUIRED NUMBER OF TERMS=     6
  THE SUM OF THE SERIES=     1.111110
  THE ERROR=.62254E-04PERCENT

 ENTER THE VALUE OF X
0.3
  X= 0.300
  THE REQUIRED NUMBER OF TERMS=    12
  THE SUM OF THE SERIES=     1.428570
  THE ERROR=.10797E-03PERCENT

 ENTER THE VALUE OF X
0.5
  X= 0.500
  THE REQUIRED NUMBER OF TERMS=    19
  THE SUM OF THE SERIES=     1.999998
  THE ERROR=.95367E-04PERCENT

 ENTER THE VALUE OF X
0.7
  X= 0.700
  THE REQUIRED NUMBER OF TERMS=    36
  THE SUM OF THE SERIES=     3.333320
  THE ERROR=.42597E-03PERCENT

 ENTER THE VALUE OF X
0.9
  X= 0.900
  THE REQUIRED NUMBER OF TERMS=   110
  THE SUM OF THE SERIES=     9.999821
  THE ERROR=.17691E-02PERCENT
```

(*b*) *From the Program in FORTRAN*

Figure 2.1.2 Numerical results obtained for Example 2.1.

accessible for large-scale simulations of engineering systems. If several computers are available, the selection of the most appropriate one for a given problem is very important. Once this selection has been made, or if only one computer system is available, one proceeds to obtain detailed information on the various elements of the system, such as the languages available, the operating system, the input/output facilities, the memory/storage constraints, and the job control language, so as to implement the computer program being developed on the system.

As mentioned earlier, there are two main steps in the numerical solution of an engineering problem. The first involves the development of the computer code, and the second involves repeated execution of the program for a wide variety of input conditions and governing parameters to generate the numerical data needed for, say, the design and analysis of a given engineering system such as a furnace, a boiler, electronic equipment, or a chemical reactor. The computer requirements are usually quite different for these two steps. Code development involves frequent changes in the program and is thus best suited to an interactive use of the computer, preferably with an interpreter. The operating system, examples of which are UNIX and CP/M, controls the interaction with the computer, particularly the editor, and is very important. A screen editor, such as EMACS, which is available on many microcomputers and minicomputers, allows one to effect changes in the program very rapidly by moving the cursor to the desired location and making the needed modification. A line editor, on the other hand, allows changes to be made line by line, or in a collection of lines, and is much slower. The speed of the central processing unit (CPU), which gives the execution time of the program, is not a very important consideration during code development. Similarly, the output facilities are not as important as at the second stage when computational results are being obtained, in tabular or graphical form.

Thus, during the development of the computer program, a good screen editor, which allows frequent changes and corrections in the program, is desirable. Also, the interpreter or compiler should provide adequate error diagnostics. Microcomputers and several minicomputers are particularly suited to code development because of the availability of most of the desirable features mentioned above.

Once the computer program has been developed, the desired numerical results for wide ranges of the governing parameters are obtained by repeatedly running the program with minor changes to enter the appropriate parametric values. Clearly, a rapid execution, with good output facilities, particularly graphics, is desirable at this stage. The editor and error diagnostics are not important. Also, an interactive use of the computer is not necessary. Thus, a batch execution of the developed program on a main-frame computer, or on a supercomputer, is the best method. The program is loaded, compiled, and linked with computer memory before execution, which then proceeds rapidly.

It is evident from the above discussion that microcomputers and minicomputers, which allow interactive computing with an editor particularly suitable for changing and correcting the program, are ideal for program development. Once the code has been developed, debugged, and validated, it can be transmitted to a larger machine to be run in the batch mode. The numerical results may again be downloaded onto a minicomputer or a microcomputer and printed, tabulated, or plotted at convenience. This general procedure is employed for many practical problems today. However, depending on the complexity of the given problem, the entire program development and execution may be carried out on a chosen computer system. This approach is followed here, since most of the problems considered are relatively simple compared to full-scale engineering problems.

2.2.4 Program Development

Algorithm. After the selection and the consideration of the important aspects of the method of solution, the programming language, and the computer system, one proceeds to the development of the computer program. However, before the program can be written, a step-by-step procedure, known as an *algorithm*, must be developed.

The method of solution is generally expressed in terms of the mathematical formulas involved in the computation. However, the computer must be programmed to follow a definite, logical, step-by-step procedure to perform the desired computation. The algorithm may be written as a sequence of steps to be followed. More frequently, the algorithm is represented graphically by means of a flow chart, which shows the steps in the form of a block diagram. Generally, a flow chart is used to outline the computational procedure, without giving the details of the actual computational steps, which are eventually entered into the actual program. Thus, a flow chart serves to indicate the logical sequence of programming steps and is frequently drawn before the program is developed.

The flow chart follows an accepted collection of symbols to represent input/output, decision, terminal, and computation. For example, let us consider the determination of the maximum of a function $f(x)$. In the optimization of engineering systems, one is frequently concerned with maximization or minimization of functions, under specified constraints. Let us assume that it is known that the given function $f(x)$ has a maximum in the range $x_1 < x < x_2$, where x is the independent variable. We know that at the maximum, df/dx is zero and d^2f/dx^2 must be negative. Employing these characteristics of a maximum, one may write the algorithm as a sequence of steps, shown in Fig. 2.2, or represented by a flow chart, shown in Fig. 2.3.

For this problem, the computational procedure involves entering x_1 and x_2, advancing x with a chosen step size Δx, and computing the derivative df/dx. If the

---

STEP 1. Start the calculation.
2. Input the limits x_1 and x_2 on x and the definition of the function $f(x)$.
3. Select the numerical parameters: Step size Δx and the convergence parameter ε.
4. Initialize: Take $x_i = x_1$.
5. Calculate the first derivative $f'(x_i)$.
6. Check whether the magnitude of the derivative is within ε.
7. If $|f'(x_i)| > \varepsilon$, then advance x_i by Δx and check whether $x_i < x_2$. If $|f'(x_i)| < \varepsilon$, then go to Step 10.
8. Stop the calculation if $x_i > x_2$.
9. Calculate $f'(x_i)$ and again compare its magnitude with ε. Continue with Step 7 if $|f'(x_i)| > \varepsilon$.
10. If $|f'(x_i)| < \varepsilon$, then calculate the second derivative $f''(x_i)$.
11. If $f''(x_i)$ is positive or zero, advance x_i by Δx. Go to Step 8.
12. If $f''(x_i)$ is negative, a maximum is indicated.
13. Print the required results: x_i and $f(x_i)$.
14. Stop the calculation.

Figure 2.2 Representation of the algorithm for determining the value and location of the maximum of a given function $f(x)$ as a sequence of steps to be followed by the computer.

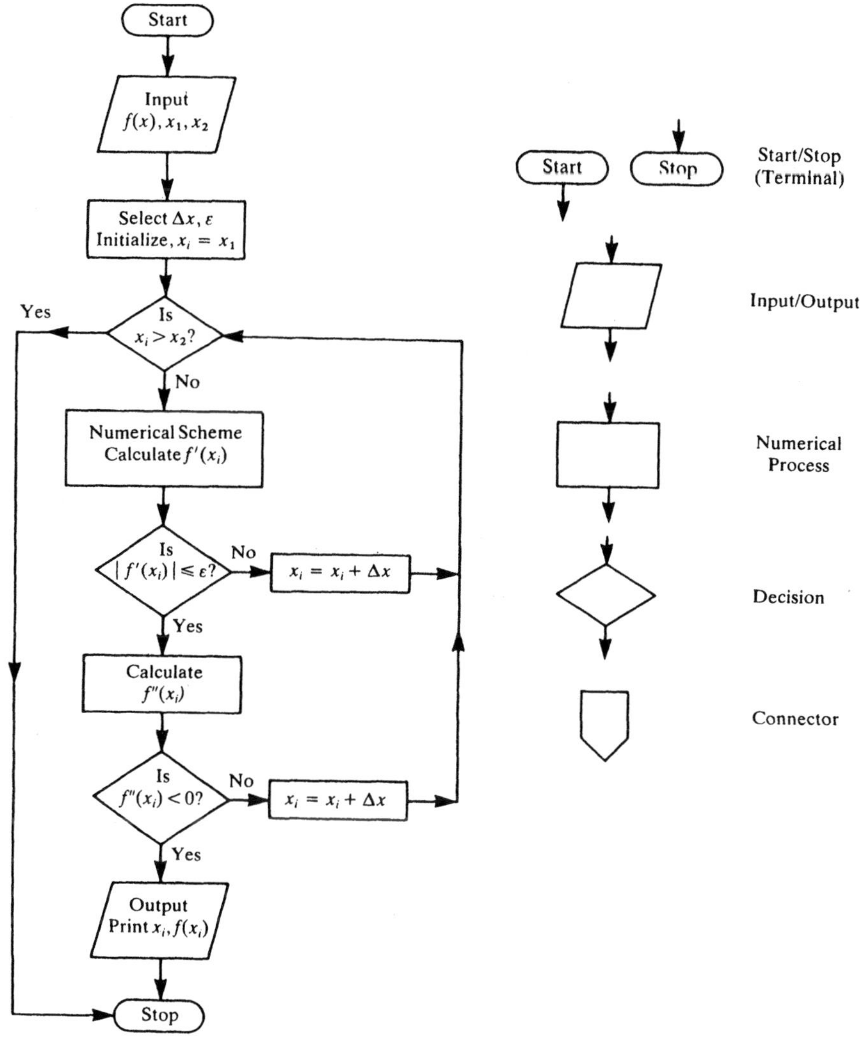

Figure 2.3 Flow chart representation of the algorithm outlined in Fig. 2.2.

derivative is close enough to a chosen small quantity ε, a maximum or a minimum is indicated. Then the second derivative d^2f/dx^2 is computed. A maximum is obtained if d^2f/dx^2 is negative. In this case, the computation is terminated and the output printed. However, if d^2f/dx^2 is positive, a minimum is indicated, and the computation of df/dx is again carried out by advancing x until a maximum is obtained or until the upper limit on x (that is, $x = x_2$) is attained. If a maximum is not obtained in the given domain and if $f(x)$ is known to have a maximum there, a larger value of ε may be selected and the procedure repeated. In fact, both ε and Δx must be varied to ensure that the location of the maximum is essentially independent of the values chosen. This aspect is considered in detail in Section 2.5.

As shown in Figs. 2.2 and 2.3, a flow chart is a more convenient representation of an algorithm. The various symbols used for indicating the type or nature of a given step are also shown in Fig. 2.3. The flow chart is a useful tool as long as it is used to give an outline of the overall process and not the detailed representations of individual steps. The numbered sequence of steps, given in Fig. 2.2, can also be used instead, depending on the personal preference of the programmer. However, with experience, one could form a mental picture of the various steps in the algorithm, particularly for relatively simple problems, and proceed directly to computer programming.

As outlined in the preceding section, the development of the computer program for the chosen algorithm is best done with an interactive operating system, particularly one with the convenience of a screen editor. The availability of an interpreter would also facilitate writing the program. Using a structured language, such as FORTRAN 77, one can write the program for easy comprehension and thus simpler debugging. Several compilers, such as WATFIV, are particularly good with error messages and may be used during program development, switching to a FORTRAN compiler for running the program after the errors have been removed. Again, program development may be carried out on a microcomputer, which allows an easy testing of various elements of the program, and a few runs may be carried out to ensure that the program is working satisfactorily. Then the program may be transmitted to a mainframe computer for repeated executions to obtain the desired numerical results.

Available Programs. Frequently, computer software for the numerical solution of specific problems is available on the computer system or is available commercially. Examples include programs for the inversion of a matrix, for the solution of an ordinary differential equation by the Runge-Kutta method, for curve fitting of data, for numerical integration, and for the solution of a system of linear algebraic equations. The computer programming would, therefore, be considerably simplified if such available programs were used. In engineering applications, the use of available software is quite prevalent, since interest usually lies in obtaining the desired numerical results as fast as possible. If a particular program has been successfully employed in the past, it is a good idea to use it for future applications. However, it is necessary to understand the algorithm adopted by the available software so that modifications, if needed, can be made and the input/output adjusted to the particular problem under consideration. It is also important to be aware of the limitations of the program and the expected accuracy of the numerical results.

Several computer programs, such as those for solving ordinary differential equations and linear algebraic systems, are available in the public domain and may simply be adapted to a given computer system. Software, which is commercially available, often comes with instructions on how to use it, but seldom with any significant details on the numerical procedure employed. Although available programs appear attractive in terms of the wide variety of problems they are often claimed to be capable of solving, one must judge each available program very carefully and choose the one suitable for a given application. A particular software may be evaluated on the basis of the information available on the accuracy, flexibility, efficiency, ease with which modifications may be made, algorithm used, and, of course, cost. Several books and review articles often list the programs available for various types of engineering problems; see, for instance, the book by Shoup (1979).

In this text we are interested mainly in learning how to develop the software for solving various mathematical problems of engineering interest, such as algebraic equations, differential equations, integration, differentiation, and curve fitting. Thus, different methods for solving a given problem will be considered, followed by the development of the relevant algorithm and computer program. Available programs will not be used, since to do so would defeat the purpose of this book. However, we should be aware of the importance of available software, as outlined above, and may find them quite useful in practical problems of engineering interest. The information presented in this book would allow one to develop the computer program itself or to evaluate the merits of an available program for use in a given practical circumstance.

Validation. The final stage in the development of the computer program for solving a given problem is verification or validation of the numerical scheme. As discussed in Section 1.3, validation is done by comparison of the numerical results with available analytical solutions. However, the analytical solution of the problem being solved numerically is obviously not available, at least in a convenient form, making a numerical solution necessary. Therefore, the numerical scheme is generally validated by comparison with the analytical solution available for simpler problems. For example, the algorithm shown in Fig. 2.3 may be used with a simple analytic function whose maximum can easily be determined mathematically. Thus, a function such as $f(x) = 5 + 4x - 3x^3$, which can easily be shown to have a maximum at $x = 2/3$, may be chosen for the testing of the numerical scheme. The numerically obtained value may be compared with the analytical one to verify that the scheme is performing satisfactorily. Other, more complicated expressions may also be employed, if the corresponding analytical results are known, for the validation of the computer program. Several of the important considerations in program development, outlined above, are demonstrated in Example 2.2.

2.2.5 Serial versus Parallel Computing

In this book, it is assumed that at a given instant only one computational step is being carried out on the computer. This assumption applies to essentially all the traditional computers that are presently employed for engineering calculations. The computational procedure in which the required calculations are performed sequentially, with each step being undertaken by the machine after the previous one is over, is known as *serial* or *sequential* computing. Thus, a single central processing unit is involved in the computation. However, in recent years, computers with multiple processors that allow concurrent calculations have been developed. Often termed *parallel computers*, these machines represent the new generation of computing and are expected to be very important in the numerical simulation of complicated processes and systems in the future.

In order to fully utilize these machines with multiple processing units, one must write the algorithm so as to employ the feature of parallel computing. Thus, statements must be given to direct various calculation steps to different units. The Jacobi iteration for solving simultaneous linear equations, for instance, is ideally suited for parallel computing, since each calculation step is independent of the others in each iteration, as seen in Chapter 5. The Gauss-Seidel method, which performs the various steps sequentially, cannot be solved as efficiently with parallel computing. Thus, parallel computing involves developing algorithms that allow concurrent calculations for greater efficiency. This developing area is beyond the scope of this book, where sequential or serial computing, which is available on traditional computers, is assumed. For details on parallel computing, see Hockney and Jesshope (1981).

Example 2.2

A firm needs to borrow \$50,000 to undertake improvements in its existing facilities. For the repayment of the loan, the firm wishes to pay only \$1000 each month, beginning at the end of the first month after taking the loan, toward the principal and the interest. Considering possible interest rates as 8%, 10%, and 12%, determine the time required to pay off the loan for these three cases. Write the computer program in both FORTRAN and BASIC for this problem. Compute the time required and the future worth, on the day the repayment is completed, of the money paid toward the loan. Also, determine the amount by which the final payment must be reduced to pay off the loan exactly.

Solution

Let x denote the percent interest rate, so that an annual compounding yields an interest of x on \$100. Then the annual interest on each dollar is $x/100$, denoted by x_1. Therefore, the future worth of an amount P after n years is $P(1 + x_1)^n$, due to this interest which is compounded annually. Similarly, the present worth of an amount R paid at the end of n years is $R/(1 + x_1)^n$. The concepts of present and future worth are very important in economic analysis; see, for instance, Stoecker (1980). First, we need to consider the present worth of a series of uniform annual amounts R, paid at the end of each year starting at the end of the first year. If n is the total number of years, the present worth (PW) of such a series of amounts is

$$\text{PW} = \frac{R}{(1 + x_1)^1} + \frac{R}{(1 + x_1)^2} + \frac{R}{(1 + x_1)^3} + \cdots + \frac{R}{(1 + x_1)^n} \tag{2.2.1}$$

$$\text{PW} = R\,\frac{(1 + x_1)^n - 1}{x_1(1 + x_1)^n} \tag{2.2.2}$$

where $x_1 = x/100$ (since x is given as a percent).

Thus Eq. (2.2.2) follows from the fact that the future worth (FW) of an amount P is given by $FW = P(1 + x_1)^n$ and from the summation of the series. Now, if we consider monthly payments, the total number of payments become m, where $m = 12n$, and the interest rate becomes x_m, where $x_m = x/(12 \times 100)$. Thus,

$$PW = R\frac{(1 + x_m)^m - 1}{x_m(1 + x_m)^m} \tag{2.2.3}$$

The future worth (FW) of this series of amounts is obtained by simply multiplying the present worth by $(1 + x_m)^m$. Therefore,

$$FW = R\frac{(1 + x_m)^m - 1}{x_m} \tag{2.2.4}$$

Now, R is given as \$1000 and x as 8%, 10%, or 12%. We wish to compute the time, in months m, needed to repay the loan, and the future worth of the total payment. The present worth is \$50,000. Thus, m is to be computed from Eq. (2.2.3), and the future worth may then be obtained from Eq. (2.2.4). The determination of m from Eq. (2.2.3) is a root-solving problem, as presented in Chapter 4. Here, we shall use a very simple approach, since root-solving methods have not been discussed yet. For a given value of x_m, the value of m may be increased in steps of 1, starting with $m = 1$, and the PW computed from Eq. (2.2.3), until the value of \$50,000 is reached. The computation stops when PW exceeds this amount, since a fixed payment of \$1000 is made each month. In practice, the monthly payment is adjusted to an appropriate value close to \$1000, so that the loan is paid off exactly.

Figure 2.2.1 shows the algorithm to be employed, in terms of a flow chart. The computational scheme is very simple for this problem and is based on a comparison between the present worth of \$50,000 and the sum of the series in Eq. (2.2.3), employing an increasing number of terms m. Once the latter exceeds the PW, the loan is paid off and the number of months needed is printed. Also, the future worth, on the date when the loan is paid off, of the total payment made is computed from Eq. (2.2.4). The present worth of the total payment exceeds \$50,000, and the last payment may be reduced to avoid this excess payment or the monthly payments may be adjusted, as mentioned above. The future worth of the loan is \$50,000 $(1 + x_m)^m$, and if this amount is subtracted from the computed FW of the payments, we obtain the amount by which the final payment must be reduced to pay off the loan exactly.

Figures 2.2.2 and 2.2.3 show the corresponding computer programs in FORTRAN and BASIC, respectively. Figure 2.2.4 presents the numerical results obtained. The structural differences between the two programming languages, as discussed earlier, are again shown here. The computed number of months m is the same in the two cases. However, the values for PW and FW are slightly different. This difference is a consequence not of the language, but of the round-off error in the two computer systems employed. These errors are considered in detail in the next section. For the solution of the present problem, a minicomputer (PRIME 950) was used for the FORTRAN program, and a microcomputer (IBM PC-XT) was used for the BASIC program. The latter was in double precision and, thus, retained a larger

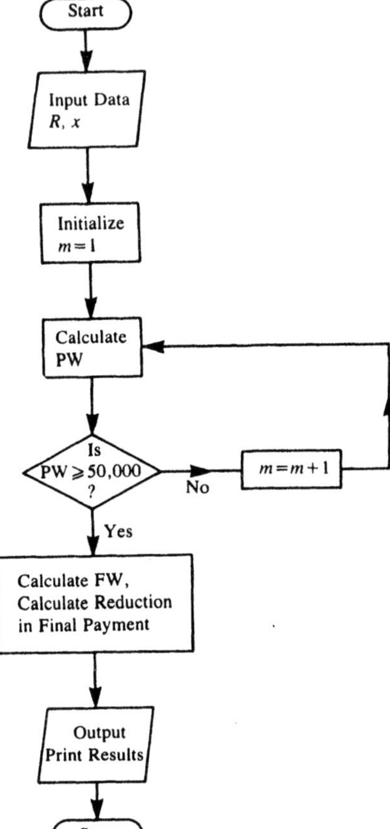

Figure 2.2.1 Flow chart for the problem in economics considered in Example 2.2.

number of digits before rounding off, resulting in greater accuracy. Two separate computer systems were purposely chosen here to demonstrate this difference. Of course, double precision could also be employed on the PRIME computer. However, the number of digits retained varies from computer to computer.

As shown in Fig. 2.2.4, the number of months needed to repay the loan increases with the interest rate, as expected. Also, the future worth increases. Note also that, since the monthly payment is kept constant, the total payment is more than the loan. To determine the amount needed to repay the loan exactly, subtract the future worth of the loan from the future worth of the total payment. This amount is the overpayment and is subtracted from the last month's payment of $1000 to obtain the reduction in the final payment if the loan is to be paid off exactly.

```
C                    PROGRAM ECONOMICS
C
C   R IS THE MONTHLY PAYMENT, X THE ANNUAL PERCENT INTEREST,
C   XM THE MONTHLY INTEREST PER DOLLAR, M THE NUMBER OF MONTHS,
C   PW THE PRESENT WORTH OF THE PAYMENTS, FW THE FUTURE WORTH
C   OF THE PAYMENTS, F THE ACTUAL FUTURE WORTH OF THE LOAN AND
C   RL THE REDUCTION IN THE FINAL PAYMENT IN ORDER TO PAY OFF
C   THE LOAN EXACTLY
C
C
C   ENTER INPUT VARIABLES
C
      IMPLICIT REAL (A-H,O-Z)
      DO 5, I=1,3
      PRINT *, 'ENTER MONTHLY DEPOSIT'
      READ (1,*)R
      PRINT *, 'ENTER INTEREST RATE'
      READ (1,*)X
      XM=X/(12.0*100.0)
      M=0
  1   M=M+1
C
C   COMPUTE PRESENT WORTH AND CHECK IF LOAN IS PAID OFF
C
      PW=R*((1.0+XM)**M-1.0)/(XM*(1.0+XM)**M)
      IF(PW.LT.50000.0)THEN
      GO TO 1
      ELSE
      WRITE(1,2)R,X
  2   FORMAT(/2X,'MONTHLY DEPOSIT=',F9.4,4X,'INTEREST RATE=',F6.3)
      WRITE(1,3)PW,M
  3   FORMAT(2X,'PRESENT WORTH=',F12.3,4X,'NUMBER OF MONTHS=',I5)
C
C   COMPUTE THE FUTURE WORTH AND REDUCTION IN FINAL PAYMENT
C
      FW=PW*(1.0+XM)**M
      F=50000*(1.0+XM)**M
      RL=FW-F
      WRITE(1,4)FW
  4   FORMAT(2X,'FUTURE WORTH=',F12.3)
      WRITE(1,9)RL
  9   FORMAT(2X,'REDUCTION IN FINAL PAYMENT=',F9.4//)
      END IF
  5   CONTINUE
      STOP
      END
```

Figure 2.2.2 Computer program in FORTRAN for the problem in Example 2.2.

```
100 REM                         PROGRAM ECONOMICS
110 REM
120 REM      R IS THE MONTHLY PAYMENT, X THE ANNUAL PERCENT INTEREST,
130 REM      XM THE MONTHLY INTEREST PER DOLLAR, M THE NUMBER OF MONTHS,
140 REM      PW THE PRESENT WORTH OF THE PAYMENTS, FW THE FUTURE WORTH
150 REM      OF THE PAYMENTS, F THE ACTUAL FUTURE WORTH OF THE LOAN AND
160 REM      RL THE REDUCTION IN THE FINAL PAYMENT IN ORDER TO PAY OFF
170 REM      THE LOAN EXACTLY
180 REM
190 REM
200 REM      ENTER INPUT VARIABLES
210 REM
220     FOR I = 1 TO 3
230     PRINT "ENTER MONTHLY DEPOSIT"
240     INPUT R
250     PRINT "ENTER INTEREST RATE"
260     INPUT X
270     XM = X/(12*100)
280     M=0
290     M=M+1
300 REM
310 REM      COMPUTE PRESENT WORTH AND CHECK IF LOAN IS PAID OFF
320 REM
330     PW = R*((1+XM)^M - 1)/(XM*(1+XM)^M)
340     IF PW < 50000! THEN 290
360     PRINT "MONTHLY DEPOSIT=";R;"   ";"INTEREST RATE=";X
370     PRINT "PRESENT WORTH=";PW;"   ";"NUMBER OF MONTHS=";M
380 REM
390 REM      COMPUTE THE FUTURE WORTH AND REDUCTION IN FINAL PAYMENT
400 REM
410     FW = PW*(1+XM)^M
420     F = 50000!*(1+XM)^M
430     RL = FW-F
440     PRINT "FUTURE WORTH=";FW
450     PRINT "REDUCTION IN FINAL PAYMENT=";RL
460     PRINT
470     PRINT
480     NEXT I
490     END
```

Figure 2.2.3 Computer program in BASIC for the problem in Example 2.2.

```
 ENTER MONTHLY DEPOSIT
1000.0
 ENTER INTEREST RATE
8.0

  MONTHLY DEPOSIT=1000.0000     INTEREST RATE= 8.000
  PRESENT WORTH=    50647.547    NUMBER OF MONTHS=    62
  FUTURE WORTH=    76466.453
  REDUCTION IN FINAL PAYMENT= 977.6354


 ENTER MONTHLY DEPOSIT
1000.0
 ENTER INTEREST RATE
10.0

  MONTHLY DEPOSIT=1000.0000     INTEREST RATE=10.000
  PRESENT WORTH=    50029.789    NUMBER OF MONTHS=    65
  FUTURE WORTH=    85801.844
  REDUCTION IN FINAL PAYMENT=  51.0732


 ENTER MONTHLY DEPOSIT
1000.0
 ENTER INTEREST RATE
12.0

  MONTHLY DEPOSIT=1000.0000     INTEREST RATE=12.000
  PRESENT WORTH=    50168.523    NUMBER OF MONTHS=    70
  FUTURE WORTH=   100676.328
  REDUCTION IN FINAL PAYMENT= 338.1790
```

(a) From the Program in FORTRAN

```
ENTER MONTHLY DEPOSIT
? 1000
ENTER INTEREST RATE
? 8
MONTHLY DEPOSIT= 1000   INTEREST RATE= 8
PRESENT WORTH= 50647.56   NUMBER OF MONTHS= 62
FUTURE WORTH= 76466.5
REDUCTION IN FINAL PAYMENT= 977.6641


ENTER MONTHLY DEPOSIT
? 1000
ENTER INTEREST RATE
? 10
MONTHLY DEPOSIT= 1000   INTEREST RATE= 10
PRESENT WORTH= 50029.64   NUMBER OF MONTHS= 65
FUTURE WORTH= 85801.4
REDUCTION IN FINAL PAYMENT= 50.82032


ENTER MONTHLY DEPOSIT
? 1000
ENTER INTEREST RATE
? 12
MONTHLY DEPOSIT= 1000   INTEREST RATE= 12
PRESENT WORTH= 50168.5   NUMBER OF MONTHS= 70
FUTURE WORTH= 100676.3
REDUCTION IN FINAL PAYMENT= 338.1328
```

(b) From the Program in BASIC

Figure 2.2.4 Numerical results obtained for Example 2.2.

2.3 NUMERICAL ERRORS AND ACCURACY

A very important consideration in the solution of a given mathematical, chemical, or physical problem by numerical methods is the accuracy of the numerical results obtained. The true measure of inaccuracy, or error, in the numerical solution is the difference between the numerical and the exact, analytical results. However, the analytical solution of the given problem is presumably not available, making it necessary to solve it numerically. Thus, alternative methods for estimating the errors involved and the accuracy of the numerical solution are needed. The dependence of the errors on the various parameters associated with the numerical procedure must also be determined, so that the accuracy of the solution may be improved, if desired, by varying these parameters.

There are several types of errors that arise in a computational solution. The two most important are the round-off and the truncation errors. The former is related to the computer system used and to the number of significant figures retained in mathematical operations. An error is introduced since a finite number of significant figures or decimal places are retained and all real numbers are rounded off by the computer. In single precision, the number of significant figures retained ranges from 7 to about 14, depending on the computer system. The truncation error results from the replacement of an exact mathematical expression or equation by a numerical approximation. It refers to the difference between an exact expression and the corresponding truncated form, employed in the numerical solution. The resulting error in the solution, assuming the round-off error to be negligible, is known as *discretization* error. All these errors are discussed in detail here, along with a few other errors that arise in certain numerical schemes.

2.3.1 Round-Off Error

The round-off error introduced in a given computation depends on the computer system used. The number of significant figures, and thus the number of decimal places retained, varies with the computer. In most cases, the last digit is rounded off to take into account the value of the digit after it. For example, the last retained digit is usually rounded up if the first discarded digit is 5 or larger. Otherwise, it is unchanged. Thus, if only four significant figures are to be retained, 4.3757 is rounded off to 4.376, and 4.3752 to 4.375. However, on some machines, the digits, beyond the ones that are to be retained, are simply chopped off. For many calculations, the round-off error is relatively unimportant, being much smaller than the truncation error, discussed below. However, it can affect the accuracy of the numerical solution and can be extremely important in certain problems.

The round-off error is fairly random in nature. If the last retained digit is rounded up, the error, obtained by subtracting the approximate value from the true value, is negative. If digits are discarded, the error is positive. Because of this random nature of the error, it does not cancel out in a given computation but rather tends to accumulate if later calculations are based on earlier ones. Thus, if a particular numerical scheme requires a large number of arithmetic operations, the cumulative effect of the round-off error can be quite significant.

It is very difficult to determine the round-off error in a given numerical method. However, the error increases with the total number of arithmetic operations. Frequently, a count of the arithmetic operations in a computational step, or procedure, may be made. For example, the multiplication of two $n \times n$ matrices can be shown to involve arithmetic operations of order n^3, usually written as $O(n^3)$. Thus, if a problem can be numerically solved by two methods, the one that requires a smaller number of arithmetic manipulations will have a smaller round-off error. It will also be more efficient, since the computational effort required is less. An example of such a consideration is the solution of a system of linear algebraic equations by Gaussian or Gauss-Jordan elimination methods, mentioned earlier and discussed in Chapter 5. The former requires $n^3/3$ arithmetic operations and the latter $n^3/2$. Thus, Gaussian elimination is more efficient and has smaller round-off error.

Frequently, the numerical scheme involves dividing a given computational region into a finite number of subdivisions. For example, the length L of a rod may be subdivided into n divisions, where $n = L/\Delta x$ and Δx is termed the *step*, or *grid*, size along the x direction, which coincides with the rod axis in this case; see Fig. 2.4(a). Thus, the total number of finite regions, or steps, is inversely proportional to Δx, implying that the number of arithmetic operations varies as $1/\Delta x$. Therefore, as Δx is reduced, the round-off error is expected to increase. This consideration is very important since it indicates that the grid size may not be reduced indefinitely. In mathematical analysis, such as differentiation and integration, the desired results are obtained by taking the limiting condition of $\Delta x \to 0$. In numerical methods, however, an extremely small Δx would lead to a very large number of arithmetic operations and to unacceptably high round-off error, as shown qualitatively in Fig. 2.4(b).

There are several circumstances under which the round-off error can be particularly important. For instance, in ill-conditioned matrices, discussed in Chapter 5, a small error in the computation due to round-off can lead to a large error in the solution. Similarly, in the solution of ordinary differential equations, considered in Chapter 8, round-off error can accumulate and lead to erroneous results. If numerical instability is present in the scheme, the solution may be completely disrupted, as outlined later in this section. Consider, for example, the ordinary differential equation $dy/dx = -1/x^2$, whose solution is $y = 1/x$, or $dy/dx = -1/2x^2y$, whose solution is $y = 1/\sqrt{x}$, if y is given as 1.0 at $x = 1.0$. In both cases, the solution decreases as x is increased and approaches zero as $x \to \infty$. Thus, at large x, y is small and the round-off error can affect the solution very substantially. Depending on the step size Δx, the value of x to which the solution is obtained, and the numerical scheme, the numerical solution may deviate significantly from the expected variation at large x, as shown in Fig. 2.5. The accumulated round-off error is large, compared to the true solution, at these values of x. Thus, extending the computation to large x must be avoided in such cases. If numerical instability exists, the error could increase at a very rapid rate, often resulting in overflow and disruption of the solution. In many of these cases, double precision may be used to avoid the problems arising due to round-off error.

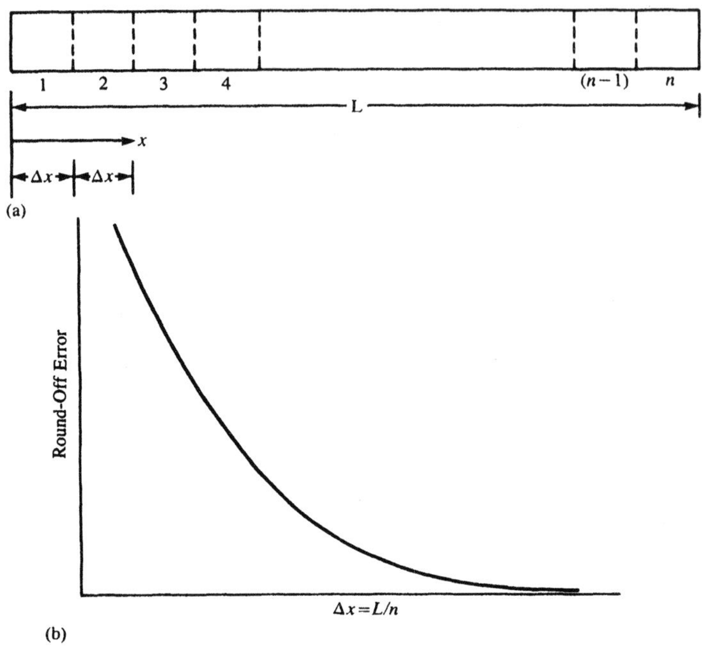

Figure 2.4 (a) Subdivision of a rod of length L into n intervals, each of length Δx, for a numerical scheme based on finite differences. (b) Qualitative representation of the variation of round-off error with the step size Δx.

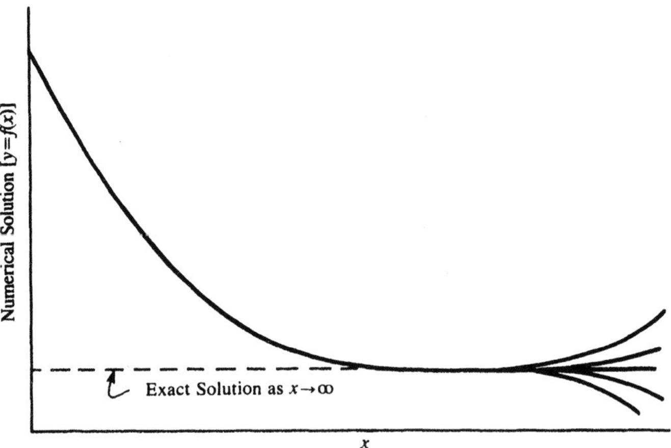

Figure 2.5 Possible effect of the round-off error, at large x, on the numerical solution of a differential equation, whose exact solution decays with increasing x to attain a constant value as $x \to \infty$.

2.3.2 Truncation Error

Truncation error is a function of the approximations used in the numerical scheme and is independent of the computer system. It arises because a function, which may be represented by an infinite series, is truncated after a finite number of terms for approximating it numerically on the computer. The nature of such an approximation and the resulting error are discussed in greater detail in Chapter 3, on the basis of the Taylor series expansion of analytic functions. However, some of the important considerations are outlined here, in order to discuss the effect on accuracy and the methods to reduce the total error.

Consider, as an example, the binomial expansion of $1/(1 - x)$, as given by Eq. (2.1.1). Then, for $|x| < 1$,

$$\frac{1}{1-x} = 1 + x + x^2 + x^3 + \cdots + x^n + \cdots \tag{2.1}$$

The variation of the function $f(x) = 1/(1 - x)$ versus x for $0 \leqslant x \leqslant 0.8$ is sketched in Fig. 2.6. Now, the function $f(x)$ is also represented by the above infinite series. However, if the series is to be entered on a computer for representing the function, only a finite number of terms can be retained. The discarded terms, thus, give rise to the truncation error, which is the difference between the exact value of the function and its approximate value, obtained after truncation. Figure 2.6 shows the approximations if one, two, three, or four terms in the series are retained. Clearly, as expected, the approximation improves as a larger number of terms is retained.

Similarly, as discussed in the next chapter, the function $f(x) = e^x$ may be represented by the following infinite series, which is known as the Taylor series expansion for the function about $x = 0$:

$$e^x = 1 + x + \frac{x^2}{2!} + \frac{x^3}{3!} + \frac{x^4}{4!} + \cdots \tag{2.2}$$

Again, a truncation error arises if a finite number of terms is used to represent the function on the computer. In numerical analysis, the computational region is often divided into a finite number of subdivisions, as shown in Fig. 2.4(a). Then the numerical scheme is based on the values of a given function $f(x)$ at the finite number of grid points, and the resulting truncation error in the formulation depends on the grid size Δx. For example, if the series in Eq. (2.2) is written for $x = \Delta x$, then

$$e^{\Delta x} = 1 + \Delta x + \frac{(\Delta x)^2}{2!} + \frac{(\Delta x)^3}{3!} + \frac{(\Delta x)^4}{4!} + \cdots \tag{2.3}$$

The truncation error resulting from the retention of a finite number of terms to represent $f(\Delta x)$, where $f(x) = e^x$, may be estimated from the above series. The error is generally written on the basis of the magnitude of the first discarded term. Thus, if only the first term is retained, the error is said to be of order Δx, that is, $O(\Delta x)$. Similarly, retaining two terms gives an error of $O[(\Delta x)^2]$, retaining three terms results in an error of $O[(\Delta x)^3]$, and so on. Thus, the error may be reduced by reducing Δx or

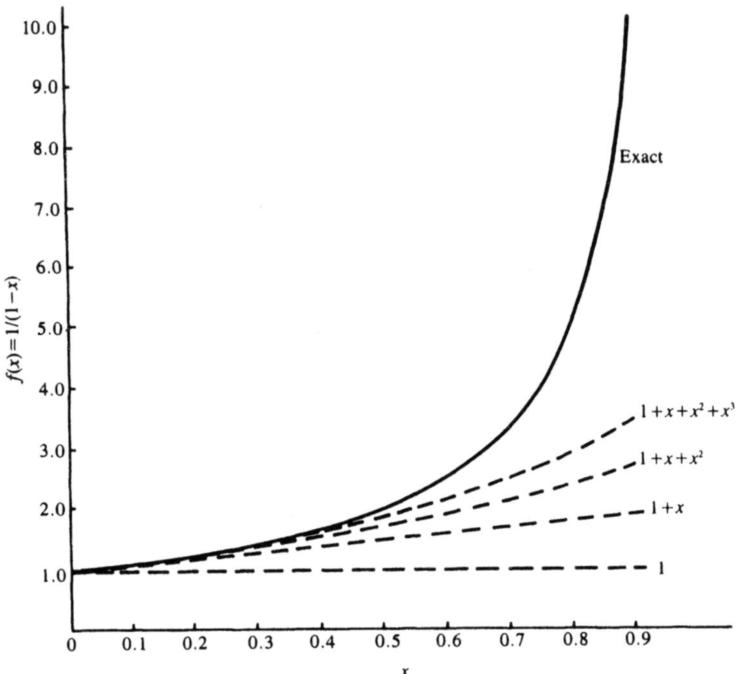

Figure 2.6 Approximations to the function represented by the series given in Example 2.1 when only one, two, three, or four terms in the series are retained. Also shown is the variation of the exact function with x.

by retaining more terms. The latter approach is generally known as higher-order approximation.

A similar approach is used to derive the truncation error that is associated with a particular numerical scheme. Such derivations are particularly important for schemes employed in numerical integration and differentiation, and in the solution of ordinary and partial differential equations. The expressions for the errors that arise in numerical differentiation are derived in the next chapter. For instance, the second derivative, d^2f/dx^2, of a function $f(x)$ will be shown to be approximated by

$$\frac{d^2f}{dx^2} = \frac{f(x+\Delta x) - 2f(x) + f(x-\Delta x)}{(\Delta x)^2} - \frac{(\Delta x)^2}{12}\frac{d^4f}{dx^4}(\xi) \tag{2.4}$$

where $x - \Delta x < \xi < x + \Delta x$. The quantities within the parentheses indicate the value of x at which the function is evaluated. From this equation, the truncation error is $O(\Delta x)^2$, and the second derivative is approximated by

$$\frac{d^2f}{dx^2} \simeq \frac{f(x+\Delta x) - 2f(x) + f(x-\Delta x)}{(\Delta x)^2} \tag{2.5}$$

Similarly, in the numerical integration of a function $h(x)$, that is, in computing $\int_a^b h(x)\,dx$, it will be shown that the total truncation errors associated with the trapezoidal rule and the Simpson method, which are two widely used schemes, are $O[(\Delta x)^2]$ and $O[(\Delta x)^4]$, respectively. Since a higher-order error term indicates the retention of a larger number of terms, $O[(\Delta x)^4]$ represents a smaller truncation error than $O[(\Delta x)^2]$. If the truncation error is $O(\Delta x)$, the scheme is said to be first-order accurate; if the error is $O[(\Delta x)^2]$, it is said to be second-order accurate; and so on.

The above brief discussion indicates the importance of truncation error in characterizing the accuracy of a given numerical scheme. The total truncation error is represented as $O[(\Delta x)^n]$, where n is the order of accuracy of the scheme, and a higher value of n signifies a more accurate scheme. Also, if Δx is reduced, the truncation error is reduced. However, truncation error indicates only the error in the formulation of the numerical scheme. The resulting error in the numerical solution, neglecting round-off error, is often termed *discretization error*. However, discretization error is much more difficult to determine than truncation error, since there is always some round-off error present and since the exact, analytical solution is generally not available. Consequently, the truncation error is generally taken as the most important measure of accuracy of a given numerical scheme.

2.3.3 Accuracy of Numerical Results

The round-off and truncation errors are the two main sources of inaccuracy in a numerical solution. However, several other errors may be present. An important one among these is the error due to incomplete convergence of an iterative solution, as discussed in the next section. The criterion used for indicating convergence must be varied to ensure that the iterative scheme has indeed converged. Inaccurate results may also be due to errors in the input data for a given problem, in the computer program itself, or in the mathematical formulation of the physical or chemical problem. Although all these errors are important, numerical methods may be studied independently, assuming that adequate care has been taken to eliminate such errors. Thus, we shall be concerned largely with the round-off and truncation errors and with the resulting total error.

As discussed in the preceding sections, a decrease in the step size Δx leads to an increase in the number of computations and, thus, to an increase in the round-off error. On the other hand, the truncation error is reduced as the step size is reduced. The total error, resulting from the summation of these two errors, will therefore initially decrease as the step size is decreased, reach an optimum, and then increase again. Figure 2.7 shows, qualitatively, the variation of these errors with step size. Clearly, a reduction in step size, or grid refinement, helps in error reduction to a point, beyond which the round-off error predominates. Thus, it is important to choose a step size that results in small truncation errors, without the associated penalty of large round-off errors.

The accuracy of a numerical solution can be best determined by a comparison between the numerical results and the analytical solution. However, the analytical

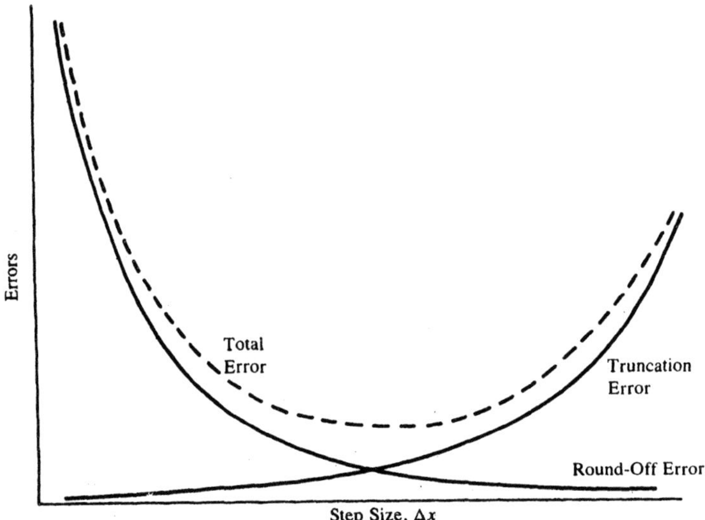

Figure 2.7 Sketch of the variation of the round-off, truncation, and total errors with step, or grid, size Δx.

solution is generally not available for the given problem. Then such a comparison may be made by employing a problem that is simpler than the one being solved numerically and for which an analytical solution is available. For example, the numerical scheme for integrating an arbitrary function $h(x)$ may be used to integrate a simpler function, such as a polynomial, which can be integrated analytically. The numerical results can then be compared with analytical ones to quantify the accuracy of the method. Similarly, the computer program developed for solving a given ordinary differential equation, for example,

$$\frac{d^3y}{dx^3} = f\left(x, y, \frac{dy}{dx}, \frac{d^2y}{dx^2}\right)$$

may be validated by solving a simpler differential equation, for example,

$$\frac{d^2y}{dx^2} = f(x)$$

whose analytical solution can be obtained easily.

Various other methods are also employed to check the accuracy of the numerical results. One method is to put the obtained solution back into the equation being solved and check if the equation is satisfied. For example, after solving the matrix equation $AX = B$ for the unknown X, multiply the solution matrix X with the coefficient matrix A to check how closely the constant matrix B is reproduced. Similarly, in curve fitting, the computed function may be plotted along with the given data to determine if, indeed, a satisfactory fit has been achieved; see Fig. 2.8. In root solving, the computed roots x are substituted into the given equation $f(x) = 0$ to ensure that the equation is satisfied. The physical or chemical nature of the problem is also used, wherever possible, to choose between multiple solutions and to determine if the numerical results show the expended trends.

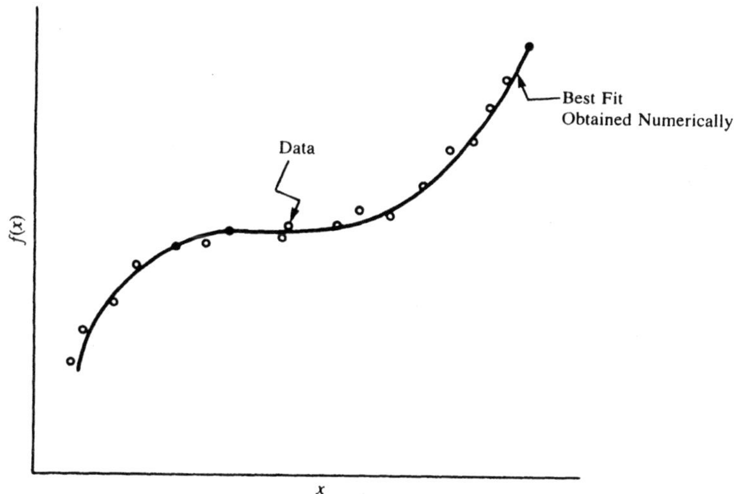

Figure 2.8 Comparison between the numerical results obtained for a best fit of given data with the data themselves, for a check on the accuracy of the results.

Thus, the accuracy of the numerical results is a very important consideration. The round-off and truncation errors, associated with the numerical method, form the basis for choosing the most appropriate method and for improving the accuracy of the numerical solution. The truncation error may be reduced by decreasing the step size or by using higher-order approximations, obtained by retaining a larger number of terms in the series representing the relevant functions. However, the round-off error can negate the effect of reducing the truncation error if the number of computations increases very substantially. Double precision may be used if round-off errors are significant and if the problem is particularly susceptible to these errors. The accuracy of the numerical solution must be studied, as outlined above, to ensure that the errors are bounded and small.

2.3.4 Numerical Stability

Another important consideration, related to the errors and accuracy of a numerical solution, is that of numerical stability. It is of particular concern in the numerical solution of ordinary and partial differential equations. Instability in a numerical scheme can lead to an unbounded growth of numerical errors that arise in the computation and thus can completely disrupt the numerical solution. If the scheme is stable, the errors are bounded and, although they accumulate as computation progresses, they do not grow to an unacceptably large level.

Let us consider, as an example, the simple ordinary differential equation $dy/dx = f(x,y) = -cy$, where x is the independent variable, y the dependent variable, and c a positive constant. The analytical solution to this equation is $y/y_0 = e^{-cx}$, where $y = y_0$ at $x = 0$. This equation may be solved by any one of the several methods discussed in Chapter 8. One of the simplest methods is Euler's method, which advances the solution from x_i to x_{i+1}, where $x_{i+1} = x_i + \Delta x$, by the recursion formula

$$y_{i+1} = y_i + \Delta x\, f(x,y) \tag{2.6}$$

Here, the subscript refers to the number of the computational step, starting with $i = 0$ at $x = 0$. Thus, $x_i = i\,\Delta x$, where Δx is the step size. The recursion formula is obtained by simply approximating the given differential equation by $\Delta y = \Delta x\, f(x,y)$, where Δy is the corresponding change in y. With $f(x,y) = -cy$,

$$y_{i+1} = y_i + \Delta x(-cy_i) = (1 - c\,\Delta x)y_i \tag{2.7}$$

The analytical solution decays exponentially with x, as sketched in Fig. 2.9. However, the numerical solution will decay with x only if $c\,\Delta x < 1$. If the step size Δx is chosen large enough to make $c\,\Delta x > 1$, the solution becomes oscillatory. Thus, the difference between the numerical and analytical results increases as Δx increases. However, the oscillations obtained for $c\,\Delta x > 1$ decay with increasing x, provided $|1 - c\,\Delta x| < 1$. But if Δx is increased still further so that $|1 - c\,\Delta x| > 1$, the numerical solution grows with increasing x and ultimately becomes very large, as x is increased to large values; see Fig. 2.9. The computer will then indicate overflow. Thus, an increasing solution is obtained instead of the decaying one given by analysis. This problem is an example of numerical instability, which must be avoided to obtain a physically realistic solution.

In this case, the scheme is conditionally stable, since if $|1 - c\,\Delta x| < 1$, an unbounded growth of the solution does not arise. Also, for the simple equation considered, a repeated application of Eq. (2.7) gives the numerical solution as

$$y_i = y_0(1 - c\,\Delta x)^i \tag{2.8}$$

Thus, the errors in the solution accumulate as x increases, and for $|1 - c\,\Delta x| > 1$, they become unbounded at large x. Such a situation arises for some of the numerical methods used for the solution of differential equations. If the method is conditionally stable, the step size must be kept small enough so that instability does not arise. A good check for instability is to solve the problem for two values of the step size that are significantly far apart. If the results obtained differ tremendously, numerical instability may be present. Frequently, numerical instability can be avoided by reducing the step size. If the scheme continues to be unstable even with small step sizes, it is best to go to some other method. For further details on numerical instability, advanced books such as those by Ferziger (1981) and Jaluria and Torrance (1986) may be consulted.

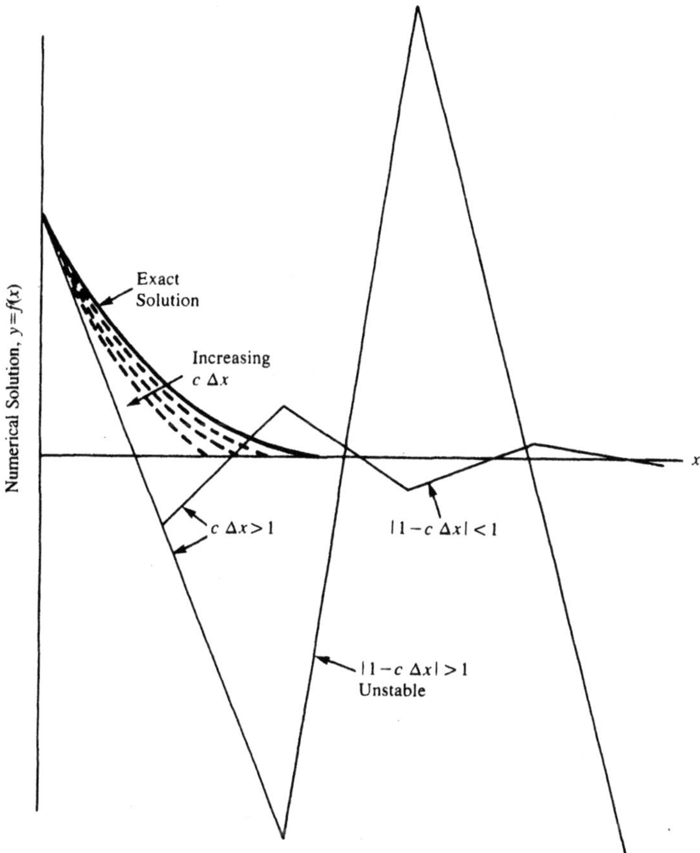

Figure 2.9 Increase in the numerical error and the onset of numerical instability as the step size Δx is increased in the solution of the differential equation $dy/dx = -cy$ by Euler's method.

2.4 ITERATIVE CONVERGENCE

Iteration is a numerical technique that is very commonly employed in the solution of a wide variety of problems. An approximation to the solution is assumed, and the approximation is gradually improved upon by iteration until no further significant variation in the solution is observed. The numerical method is then said to have *iteratively converged* to the desired numerical solution. However, iterative convergence is not always obtained, and the conditions under which the scheme converges should be determined, whenever possible, before it is used in the solution of a given problem. Some of the important considerations related to iteration and convergence are outlined in this section.

The solution of nonlinear algebraic equations is usually based on systematic iteration methods, since except for a few special cases, such as quadratic equations, the solution cannot be obtained directly by algebra. For example, in the transcendental equation $\tan x = 2/x$, or in the polynomial equation $x^4 - 11x^3 + 41x^2 - 61x + 30 = 0$, the roots, which are the values of x that satisfy the equations, are obtained by employing one of the several iterative methods given in Chapter 4. Similar considerations apply to a system of nonlinear equations. Even large systems of linear equations are often solved more effectively and more accurately by iteration than by direct algebraic methods. Such large systems frequently arise in the solution of ordinary and partial differential equations. There are several other circumstances where iterative procedures are employed to obtain the solution.

2.4.1 Conditions for Convergence

A very important consideration in the choice of an iterative method for a given problem is whether it would converge. As expected, convergence depends on the chosen, or guessed, initial approximation to the solution. A more rapid convergence usually results for an approximation that is closer to the actual solution than for one that is farther away. However, in many cases, the scheme diverges if the difference between the initial approximation and the actual solution is large. It is generally difficult to determine the region of convergence over which an arbitrary initial approximation would lead to convergence. Thus, the physical background of the problem and any available information on the solution must be used to approximate the solution as closely as possible. Still, several runs, with different starting approximations, may be needed before convergence is obtained. In some cases, the limiting values of the solution are known. Then numerical schemes that gradually reduce the region in which the solution lies and, thus, always converge may be developed. Examples of such schemes are the search, bisection, and regula falsi methods for root solving, given in Chapter 4.

In general, the conditions under which an iterative method converges must be determined. For many schemes, these conditions are known. For example, a system of linear equations, such as

$$
\begin{aligned}
a_{11}x_1 + a_{12}x_2 + \cdots + a_{1n}x_n &= b_1 \\
a_{21}x_1 + a_{22}x_2 + \cdots + a_{2n}x_n &= b_2 \\
\vdots \qquad\qquad\qquad & \;\;\vdots \\
a_{n1}x_1 + a_{n2}x_2 + \cdots + a_{nn}x_n &= b_n
\end{aligned}
\tag{2.9}
$$

where the a's and b's are constants and x's are the unknowns, may be solved by an iterative scheme, such as the Jacobi and the Gauss-Seidel methods given in Chapter 5. However, convergence is assured only if the system is diagonally dominant; that is,

$$
|a_{ii}| > \sum_{j=1,\, j \neq i}^{n} |a_{ij}| \tag{2.10}
$$

This condition requires each equation to have a dominant coefficient, which is greater in magnitude than the sum of the magnitudes of the other coefficients in the equation. The equations can then be rearranged to have the dominant coefficient along the diagonal of the corresponding matrix. Although convergence often occurs for weaker dominance than that given by Eq. (2.10), this equation gives the condition under which convergence will generally occur, unless the initial approximation to the values of x is too far from the actual solution.

Similarly, the roots of a nonlinear equation $f(x) = 0$ may often be determined by rewriting the equation as $x = g(x)$ and using iteration, starting with an initial guess for x. This method, known as the *successive substitution* method, is convergent only if $|g'(\alpha)| < 1$, where $x = \alpha$ is the desired root, and the difference between the starting approximation and α is not too large. Again, it is difficult to quantify how close to the root the approximation must be for convergence to result. However, the condition $|g'(\alpha)| < 1$ may be used in formulating the function $g(x)$ before iteration is applied. Further details are given in Chapter 4.

2.4.2 Rate of Convergence

It is also important to determine the rate of convergence, if the scheme is ascertained to be convergent. If α is the desired solution, say, the root of an equation, and x_i is the ith approximation to the solution, the magnitude of the error after the ith iteration is $(x_i - \alpha)$. Similarly, the error after the $(i + 1)$th iteration is $|x_{i+1} - \alpha|$. Then the relation between these two errors indicates how rapidly the scheme is converging. First, for the scheme to be convergent,

$$|x_{i+1} - \alpha| < |x_i - \alpha| \quad \text{as } i \to \infty \tag{2.11}$$

Also, we may write

$$|x_{i+1} - \alpha| \propto |x_i - \alpha|^n \tag{2.12}$$

where n is an exponent that depends on the numerical scheme. If $n = 1$, the scheme is said to have a *first-order convergence*, indicating that the error at a given iteration is proportional to that at the previous one. If $n = 2$, the scheme is said to have a *second-order*, or *quadratic*, *convergence*. Since the error is presumably small as i becomes large, this implies the squaring of a small quantity, resulting in a rapid reduction in error. This, in turn, results in a much more rapid convergence than that for a first-order convergence scheme. The Newton-Raphson method for root solving has quadratic convergence, whereas other methods such as bisection and regula falsi have first-order convergence. A still higher order convergence will result in an even faster convergence.

2.4.3 Termination of Iteration

The next question is when and how an iterative process should be terminated. If x_i is the approximation to the solution after the ith iteration and x_{i+1} that after the $(i + 1)$th iteration, a commonly employed criterion for deciding that convergence has been achieved and that the iteration should thus be terminated is

$$|x_{i+1} - x_i| \leqslant \varepsilon \tag{2.13}$$

where ε is a small quantity, known as the *convergence parameter* or the *convergence criterion*. Unless the solution, or the approximation x_i, is zero, ε must be small compared to the solution. Thus, the relative convergence criterion given by

$$\left|\frac{x_{i+1} - x_i}{x_i}\right| \leqslant \varepsilon \tag{2.14}$$

is also very often employed. If x_i is expected to be close to zero, the absolute convergence criterion, given by Eq. (2.13), is more appropriate, with $\varepsilon \ll 1.0$. Thus, ε is an arbitrarily chosen numerical parameter brought in to ascertain that the iteration has converged. However, if ε is too small, the computing time will be excessive; if ε is too large, the results may be in significant error. It is necessary to ensure that the numerical results are essentially independent of the chosen value of ε. These considerations are discussed in greater detail in the next section.

For an example on the use of such a convergence criterion, consider Example 2.1. We are interested in the sum S of the series. However, in a numerical scheme, we can sum only a finite number of terms. Then the error involved in neglecting the nth term, as compared to the sum S of the terms of the series up to this term, may be employed as the convergence criterion. Thus, if SN is the nth term, we have

$$\frac{SN}{S} \leqslant \varepsilon \tag{2.15}$$

as the condition for convergence. Similar considerations would apply for other iterative schemes. Unless the solution or its approximation could possibly be zero, the relative convergence condition is generally preferred, in comparison with the absolute condition, since the solution is generally not known, making it difficult to choose the value of ε in Eq. (2.13). For the relative convergence condition, Eq. (2.14), ε may be chosen to be around 10^{-4}, as the starting value, in order to obtain a reasonably small variation from one iteration to the next, in the approximation to the solution. These aspects are discussed in greater detail in the following section.

2.5 NUMERICAL PARAMETERS

The preceding sections have demonstrated that one must often introduce several arbitrarily chosen parameters into the numerical scheme in order to solve the problem. Among the most important of these chosen numerical variables are the step, or grid, size Δx, the convergence criterion ε, and the initial approximation to the solution. It is obvious that since such variables, or parameters, are chosen arbitrarily, it must be ensured that the numerical results obtained from the scheme are essentially independent of the chosen values.

2.5.1 Step Size

The effect of the step size Δx on the numerical solution has been considered earlier (see Fig. 2.7). As Δx is reduced, starting with relatively large values, the truncation error is also reduced. The round-off error generally does not become significant unless very small Δx, which involves a very large amount of computation, is employed. Thus, truncation errors dominate over much of the commonly used range of Δx, and with decreasing Δx, the numerical results tend to approach an essentially constant value or distribution. When this occurs, the effect of the step, or grid, size on the solution is negligible. Then the value of Δx may be chosen as the upper limit of the Δx range in which this effect is small; see Fig. 2.10. The largest value of Δx for which the solution is essentially independent of Δx is chosen so that both the computational effort and the round-off error are minimized. Of course, at very small Δx, the round-off error becomes significant and may substantially affect the solution, as shown in Fig. 2.10. Thus, it is important to study the dependence of the numerical results on the step, or grid, size and to choose the largest value of Δx at which the solution is essentially independent of Δx.

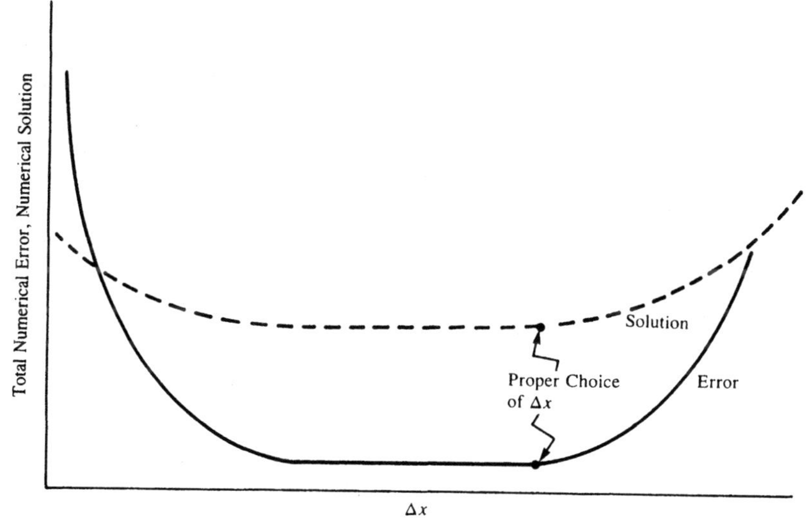

Figure 2.10 Sketch of the variations of the total numerical error and the solution with the step, or mesh, size Δx. Also indicated is the appropriate value of Δx that may be chosen for subsequent computations.

As an example, consider the numerical integration of a function $f(x)$, that is, $\int_a^b f(x)\,dx$. As discussed in detail in Chapter 6, there are several methods that may be used. One of the simplest is based on dividing the total distance of integration $(b - a)$ into n subdivisions, each of length Δx, and approximating $f(x)$ as a constant over each subdivision. Then $\Delta x = (b - a)/n$, and $f(x)$ may be taken as constant at the value on the extreme left of each subdivision; see Fig. 2.11(a). The given integral is then approximated by

$$\int_a^b f(x)\,dx \simeq \sum_{i=1}^{n} f(x_i)\,\Delta x = I \tag{2.16}$$

where $x_1 = a$ and $x_n = b - \Delta x$. The largest value of Δx is $(b - a)$, which gives the numerical value I of the integral as $(b - a)f(a)$. For the function $f(x)$ sketched in Fig. 2.11(a), this value is a gross underestimate of the integral. As Δx is reduced, or as n is increased, I increases, as shown in Fig. 2.11(b). This reduction in Δx, or grid refining, is continued until I becomes essentially independent of Δx. Then the largest value of Δx at which this condition is met is chosen. Similar considerations arise for other methods of integration and for different forms of $f(x)$.

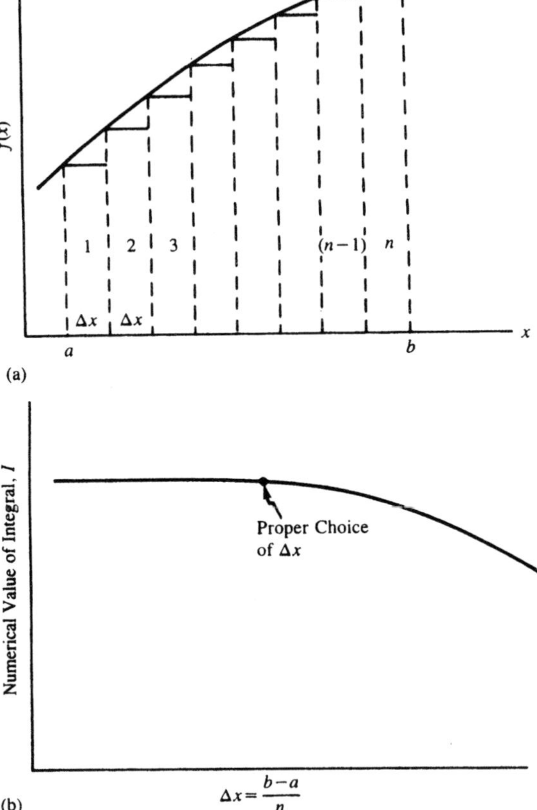

Figure 2.11 Graphical representations of (a) the integration scheme given by Eq. (2.16) and (b) the dependence of the numerical value of the integral I on the mesh size Δx.

2.5.2 Convergence Criterion

The convergence criterion, or parameter, ε must be similarly treated. A relatively large value of ε is initially employed so that a rapid convergence is achieved. Then ε is gradually reduced until the numerical results remain essentially unchanged if ε is reduced further. Since the computation involved increases with reducing ε, a continued reduction in ε will ultimately result in substantial round-off error. Thus, as before, the largest value of ε at which the dependence of the numerical solution on ε first disappears is chosen; see Fig. 2.12. Example 2.3 demonstrates the effect of ε on the solution and on the computational effort involved. Also, the convergence criterion may be applied to different variables being computed in the solution to confirm that convergence has indeed occurred.

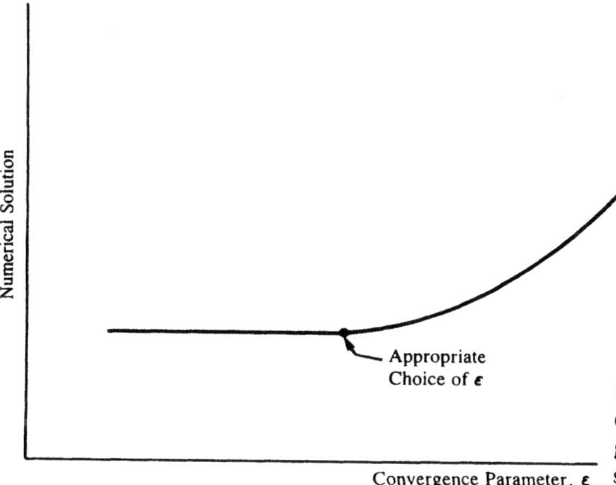

Figure 2.12 Sketch of the typical effect of a variation in the convergence parameter ε on the numerical solution.

2.5.3 Other Arbitrarily Chosen Variables

An initial approximation to the solution is needed in order to start an iteration scheme. Since convergence often depends strongly on the starting values, physical considerations and any available previous results on similar problems must be employed to choose the initial approximation. In root solving, for instance, the range of values in which the root lies is often known from the physical nature of the problem and may be used to obtain the first guess. Similarly, analytical or numerical results for similar problems are frequently used to obtain the starting values in iterative schemes for solving differential equations. However, it is important to ensure that the results are not significantly affected by the chosen initial guess. Thus, the initial approximation must be varied until the converged numerical solution is essentially independent of the starting values.

There are several other such numerical variables that must be introduced in order to obtain a numerical solution. A commonly encountered circumstance is one in which a boundary is specified as, say, x approaching infinity, that is, $x \to \infty$. This situation arises, for example, if the upper limit b in the integral of Eq. (2.16) is replaced by ∞. Similarly, boundary conditions for differential equations are often specified for $x \to \infty$ or $\tau \to \infty$, where τ is time. In all such cases, a frequently employed approach is to replace ∞ by an arbitrary large number, say, x_∞ or τ_∞, whose value is based on physical considerations or previous experience with similar problems. Again, this parameter, x_∞ or τ_∞, is varied to ensure that the numerical results are independent of the value chosen. In this case, the smallest value of the parameter at which the results become independent of a further increase is chosen. Similar considerations would apply for other numerical variables that are introduced into the numerical procedure. In all cases, it must be ensured that the numerical results are essentially independent of the values chosen.

Example 2.3

In a chemical process, the concentration C in kg/m^3 of a given species decays with time τ, in seconds, as follows:

$$C = 22.5 + 62.3\exp(-0.01\tau) \tag{2.3.1}$$

Thus, the concentration approaches a steady-state value of 22.5 kg/m^3 as time becomes large, that is, as $\tau \to \infty$. If the time τ is increased with step size $\Delta\tau$, starting with $\tau = 0$, determine the dependence of the number of steps, the time τ_{ss} required to attain steady state, and the concentration at steady state on the convergence criterion ε employed to indicate steady-state conditions.

Solution

The initial concentration, at $\tau = 0$, is $22.5 + 62.3 = 84.8$ kg/m^3. As time $\tau \to \infty$, $C \to 22.5$ kg/m^3. However, we wish to terminate the computation as soon as C is close to the steady-state value of 22.5 kg/m^3, within a chosen convergence criterion. If such a criterion is not used, the computation will proceed until C is 22.5 kg/m^3, within the round-off error of the computer, and this would generally involve a considerable wastage of computer time. Thus, we may use a condition of the form

$$|C - 22.5| \leqslant \varepsilon \tag{2.3.2}$$

where ε is the convergence parameter, in order to decide that the steady-state value has been attained and that the computation may be terminated.

The given problem is employed to demonstrate the necessity of using a convergence criterion and the effect of ε on the results. The concentration C is computed at increasing time τ, starting with $\tau = 0$, until Eq. (2.3.2) is satisfied. The step size $\Delta\tau$ determines only the values of τ at which C is computed, and thus the time τ_{ss} at which the computation is terminated is obtained within an accuracy of $\Delta\tau$. Since the exact, analytical expression for C is given, no truncation errors are involved, and round-off error arises only for each individual computation. There is no accumulation of error. Thus, the chosen value of $\Delta\tau$ has a small effect on the solution and we may focus on the effect of ε.

A simple computer program may be written to increase τ from 0, in steps of $\Delta\tau$, until Eq. (2.3.2) is satisfied. The convergence criterion ε is varied from a high value of 100, at which convergence occurs at the very first step, to very low values, of the order of 10^{-9}. The value of $\Delta\tau$ is chosen as 100 s. Thus, τ_{ss} would be obtained to an accuracy of 100 s. A smaller value of $\Delta\tau$, $\Delta\tau = 10$ s, was also considered, and the effect of this change in $\Delta\tau$ on the results at small values of ε was quite small. At steady state,

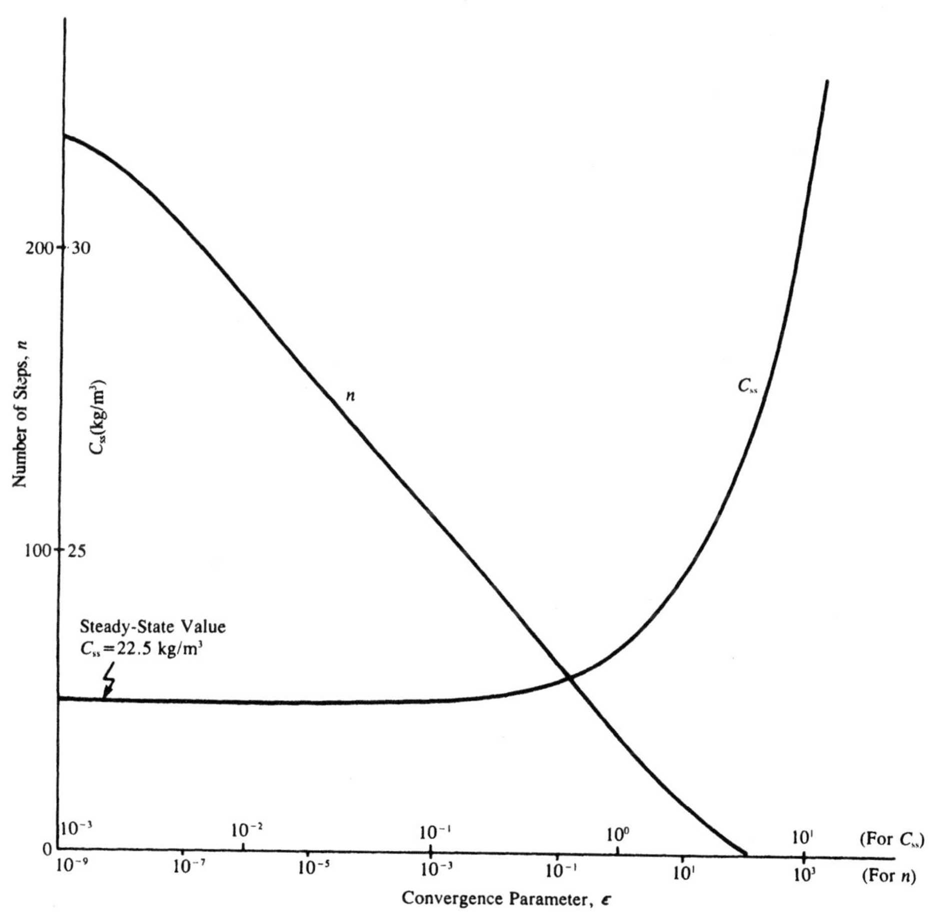

Figure 2.3.1 Dependence of the number of iterative steps to convergence n and of the steady-state solution C_{ss} on the convergence parameter ε, for the problem considered in Example 2.3.

determined by Eq. (2.3.2) being satisfied, the number of steps n, the time τ_{ss}, and the concentration C_{ss} are obtained. Here, τ_{ss} is related to n simply by $\tau_{ss} = n\,\Delta\tau$.

Figure 2.3.1 shows the dependence of the number of steps n and of the steady-state concentration C_{ss} on ε. The computational effort, as indicated by n, increases sharply as ε is reduced to very small values, whereas the solution is hardly affected as ε is reduced below about 10^{-2}. This figure indicates the importance of choosing the proper value of ε. A large value of ε results in considerable error, and a very small value leads to a very large, unnecessary computational effort. Here, a value of 10^{-2} may be chosen for ε. Such problems, in which the steady-state condition is to be determined, are frequently encountered in engineering problems. Although the first estimate of ε may be based on expected results or on previous experience with similar problems, ε must be varied to ensure that an appropriate value is chosen.

2.6 SUMMARY

This chapter discusses some of the important and fundamental considerations that form the basis for an efficient and accurate numerical scheme. The computational procedure is discussed in some detail, outlining method selection, programming language and computer system considerations, and program development. Besides indicating a systematic approach to the computational solution of a given problem, this discussion also presents some of the recent trends in the area of numerical methods for engineering applications. Although fairly straightforward for most experienced users of the computer, this discussion nevertheless focuses on several relevant aspects that need to be considered before proceeding with the development of the computer code.

Numerical errors and accuracy are of crucial importance in any computational result. The nature and characteristics of errors that arise, particularly truncation and round-off errors, are discussed, along with the methods for evaluating and improving the accuracy of the numerical solution. Numerical instability is also considered. The convergence of iterative methods, which are frequently used for various types of problems of engineering interest, is discussed in terms of a few examples. The importance of determining a criterion for deciding if convergence has occurred and of determining the conditions under which the scheme is convergent are outlined. Finally, numerical variables and parameters, which are often introduced into the numerical method in order to obtain the solution, are considered. Since such parameters are chosen arbitrarily, it is important to ensure that the numerical results are not significantly affected by a variation in the chosen values. Methods to do so and the anticipated trends are outlined. The various considerations discussed in this chapter will arise in the following chapters, and the importance of these aspects will become quite apparent as we proceed with different types of problems and solution methods.

PROBLEMS

2.1. Calculate the number of arithmetic operations involved in solving the two simultaneous linear algebraic equations $a_1x + b_1y = c_1$ and $a_2x + b_2y = c_2$, where x and y are the unknowns and a_1, a_2, b_1, b_2, c_1, and c_2 are given constants. Try different algebraic approaches to the solution, using elimination and substitution. Does your answer depend on the procedure adopted?

2.2. Write computer programs in BASIC and in FORTRAN 77 to calculate the real or complex roots of the quadratic equation $ax^2 + bx + c = 0$, where a, b, and c are given constants. Use these programs to find the roots for (a) $a = 1$, $b = -3$, $c = 2$; (b) $a = 1$, $b = -5$, $c = 6$; and (c) $a = 2$, $b = 1$, $c = -1$.

2.3. Employing the binomial series generated by $1/(1 + x)$, where $|x| < 1$, compute the sum of the series, using a finite number of terms with a convergence criterion ε, as done in Example 2.1. Write a FORTRAN program for the purpose, and study the effect of varying the convergence criterion on the numerical results.

2.4. Write computer programs in BASIC and in FORTRAN 77 to determine the maximum of the function $f(x) = 12 + 18x - 3x^2$ in the range $0 \leqslant x \leqslant 4$. Starting with the lower limit on x, advance x with a step size $\Delta x = 0.1$ until the maximum is determined. Use analytical expressions for the derivatives.

2.5. Repeat the above problem for determining the minimum of the function $f(x) = 7 - 12x^2 + 2x^3$ in the range $1 \leqslant x \leqslant 6$.

2.6. In Example 2.2, if the requirement is that the future worth (FW) of the monthly deposits of \$1000 must attain \$200,000 at an interest rate of 7.5%, compute the number of months needed to achieve this FW. Also calculate the present worth of the total money deposited. The given program may be suitably modified to solve this problem.

2.7. Employing the computer program of Example 2.2, calculate the time needed for the repayment of the loan of \$50,000 if the monthly payment is \$1500 and the interest rate is 12%. Repeat the calculation for a monthly payment of \$2000. In both cases, calculate the last payment if the loan is to be paid off exactly.

2.8. Write a computer program to study round-off errors by adding $\frac{1}{3}$ three hundred times and $\frac{1}{6}$ six hundred times. Vary the number of decimal places retained in the calculations from 1 to 8, by appropriate programming statements. Compute the round-off error and show its dependence on the number of decimal places retained, in tabular or graphical form.

2.9. Employing the expression for the second-order derivative in Eq. (2.5), compute the resulting error if the round-off errors involved in the evaluation of the function at the three values of x are equal. Repeat this calculation if the round-off errors are equal in magnitude but alternating in sign from one grid point to the next. Comment on the significance of your results.

2.10. Employing Eq. (2.6) for the numerical solution of the differential equation $dy/dx = -2y$, study the effect of varying Δx on the solution, including instability at large Δx. Confirm the trends shown in Fig. 2.9.

2.11. Consider the functions $f(x) = 2 + 3/x$ and $g(x) = 5.2 + 2.4/x^2$. Both of these approach constant values as $x \to \infty$. Employing a convergence criterion, as illustrated in Example 2.3, determine the effect of the convergence parameter ε on the value of x, x_{ss}, at which the solution has essentially attained these constant values. Does the step size Δx have any significant effect on the results?

2.12. Determine the effect of varying Δx on the computed result for the second derivative, as given by Eq. (2.5), for the function $f(x) = 5 + 10x - 4x^2 + 6x^3$. The second derivative is to be determined at $x = 1$. Write a computer program to calculate the second derivative at $x = 1$ with $\Delta x = 0.5, 0.1, 0.05$, and 0.01. Compare the results obtained with the exact value of 28.

2.13. In Example 2.3, employ a relative convergence criterion, as given by Eq. (2.14), and choose the most appropriate value by varying ε and computing its effect on the numerical results.

2.14. Using the expression for numerical integration given by Eq. (2.16), compute the integral $\int_0^2 x^2\,dx$ for $\Delta x = 2, 1, 0.5, 0.1, 0.05$, and 0.01. Compare the results obtained with the exact value of 8/3. Plot the numerical error versus the step size Δx. What value of Δx will you choose for such computations, on the basis of the results obtained?

2.15. For the problem given in Example 2.2, with a loan of \$50,000 at 12% interest, consider reducing the monthly payments; that is, instead of \$1000, the payment is, say, \$950. Compute the time needed for repaying the loan if the monthly payment is \$950. Then recompute with a monthly payment of \$900, and so on. Is a limiting value, beyond which the monthly payment cannot be decreased further for repaying the loan, indicated from your results? If so, why does such a limitation arise?

2.16. The mass transfer rate $\dot{m}$, in kg/s, at the surface in a chemical reactor at a particular time is given by the series

$$\dot{m} = 556.3 \sum_{n=1,3,5,\ldots}^{\infty} \exp(-0.04n^2)$$

where n is an odd number. Using a suitable convergence criterion, determine the number of terms needed for the numerical evaluation of $\dot{m}$ and the resulting value of the mass transfer rate.

3

The Taylor Series and Numerical Differentiation

3.1 INTRODUCTION

In several problems of engineering interest, the numerical method is based on the discrete values of a given function and its derivatives at a finite number of points in the computational domain. The need to discretize a function arises since a digital computer can generally perform only the standard arithmetic operations, employing a finite number of discrete values. In several cases, interest lies in estimating the derivatives from discrete numerical values of the function, given at specified data points. The derivatives are generally computed at these very data points or at a number of intermediate locations, employing only arithmetic operations. Similarly, the numerical integration of a function may be carried out, using the discrete values of the function. This chapter discusses the basic concepts involved in discretization as well as in the computation of the derivatives of a given function.

Numerical differentiation refers to the computational procedure for evaluating the derivatives of a function, which is given as an analytical expression or as discrete values at a finite number of points in the computational region. There are many diverse areas of engineering interest where numerical differentiation is needed. For example, in the dynamics of particles and systems, the time derivative of the displacement gives the velocity, and the second derivative gives the acceleration, which on multiplication with the mass of the body yields the force. In many engineering systems, such as robotics, the motion of the components is quite complex, and numerical differentiation is needed to determine the forces, velocities, and trajectories of the elements. Similarly, the heat transfer rate and the shear force at a surface due to fluid flow over the surface are generally proportional to the spatial derivative of the temperature and the velocity, respectively. The distributions of temperature and velocity are often too complicated to permit use of the standard analytical methods for differentiation. The numerical values of the derivatives are also needed, for example, in optimization to obtain the best solution under given constraints, in economics to obtain the effect of a change in, say, the interest rate on the financial dealings of a company, in electromagnetics to determine the wavelength interval in which the maximum energy lies. and in many other problems of practical interest.

Frequently, in engineering problems, one must solve an ordinary or partial differential equation to obtain an unknown variable or function. Again, a numerical solution involves a finite number of locations or points where the value of the variable is computed. An important class of numerical methods for the solution of differential equations is based on replacing the derivatives by their discretized forms, generally known as *finite difference approximations*, and then solving the resulting algebraic equations. The numerical analysis that forms the basis of the discretization of derivatives is often called *finite difference calculus*. Differential equations arise in many engineering areas, such as dynamics and vibrations, heat transfer and fluid flow, analysis of electrical circuitry, structural analysis, mass transfer, and neutron diffusion in nuclear reactors. The numerical methods for the solution of ordinary and partial differential equations are discussed in Chapters 8 and 9, respectively.

In this chapter, we shall obtain the finite difference formulations which allow the computation of the derivatives from the discrete values of the function given at a finite number of points. A very important consideration in finite difference calculus is the error that arises due to the use of an approximation instead of an exact mathematical expression. Some discussion on the errors associated with discretization was included in Chapter 2. These errors are considered in greater detail here. The Taylor series forms the basis for many numerical techniques and also for estimating the errors involved. The general form of the series is presented, and the error resulting from the truncation of the series after a finite number of terms is determined. There are several approaches that may be adopted for deriving the finite difference approximation of the derivatives of a function. These approaches include the direct method, based on the definition of the derivative, the Taylor series approach, and the use of a polynomial representation of the function. These three approaches are discussed, with particular emphasis on the derivation based on the Taylor series since it also yields quantitative information on the error. Finally, the corresponding approximations for partial derivatives are outlined.

3.2 THE TAYLOR SERIES

3.2.1 Basic Features

Let us consider a function $f(x)$ whose value at a given point $x = x_i$ is denoted by $f(x_i)$. The Taylor series is an infinite power series that expresses the value of the function in a region sufficiently close to $x = x_i$ as follows:

$$f(x) = f(x_i) + (x - x_i)f'(x_i) + \frac{(x - x_i)^2}{2!} f''(x_i) + \frac{(x - x_i)^3}{3!} f'''(x_i) + \cdots \tag{3.1}$$

or

$$f(x_i + \Delta x) = f(x_i) + \Delta x\, f'(x_i) + \frac{(\Delta x)^2}{2!} f''(x_i) + \frac{(\Delta x)^3}{3!} f'''(x_i) + \cdots \tag{3.2}$$

where $\Delta x\,(=x - x_i)$ is a finite increment in the independent variable x, from the given value $x = x_i$, and the primes denote differentiation with respect to x. All the derivatives are evaluated at $x = x_i$. Similarly, we may write

$$f(x_i - \Delta x) = f(x_i) - \Delta x\, f'(x_i) + \frac{(\Delta x)^2}{2!} f''(x_i) - \frac{(\Delta x)^3}{3!} f'''(x_i) + \cdots \tag{3.3}$$

It is assumed that all the derivatives of the function $f(x)$, at $x = x_i$, exist and are finite. Also, Δx must be sufficiently small so that the series is convergent. Such a power series has a radius of convergence, given in terms of the increment Δx, within which the series is convergent (Keisler, 1986). Generally, the radius of convergence is finite, and if Δx is taken as larger than this value, the series is no longer convergent and the region is not sufficiently close to $x = x_i$. However, in finite difference computations, we do have the freedom to choose the value of Δx and thus control the convergence of the series and also the accuracy of the solution, as discussed below.

If an infinite number of terms is taken in the series given by Eqs. (3.2) and (3.3), the exact value of $f(x_i + \Delta x)$, or $f(x_i - \Delta x)$, may be computed, provided the series is convergent. However, it is not possible to compute an infinite number of terms, and the practical approach to such a computation is to retain only a few terms in the series for approximating the function and to estimate the error resulting from neglecting the remaining terms. If only the first term in the series of Eq. (3.2) is retained, then $f(x_i + \Delta x) \simeq f(x_i)$, and the function $f(x)$ is taken as a constant. The retention of the first two terms gives

$$f(x_i + \Delta x) \simeq f(x_i) + \Delta x\, f'(x_i) \tag{3.4}$$

Thus, a linear approximation of the function is employed over the region from x_i to $(x_i + \Delta x)$, and the slope is taken as constant. Similarly, if the first three terms in the series are retained,

$$f(x_i + \Delta x) \simeq f(x_i) + \Delta x\, f'(x_i) + \frac{(\Delta x)^2}{2!} f''(x_i) \tag{3.5}$$

This expression allows a variation in the slope over the region and is, generally, a more accurate approximation for $f(x_i + \Delta x)$ than that given by Eq. (3.4).

Figure 3.1 shows the three circumstances of retaining one, two, or three terms in Eq. (3.2) graphically. Equation (3.4) becomes exact only if $f(x)$ is a linear function of x. Similarly, Eq. (3.5) is exact for a parabolic, or second-order, function. Therefore, for an arbitrary function $f(x)$, the accuracy of the representation by the Taylor series improves as additional terms are retained. Although an nth-order series expansion is exact for an nth-order polynomial, an infinite number of terms is, in general, needed for other differentiable and continuous functions.

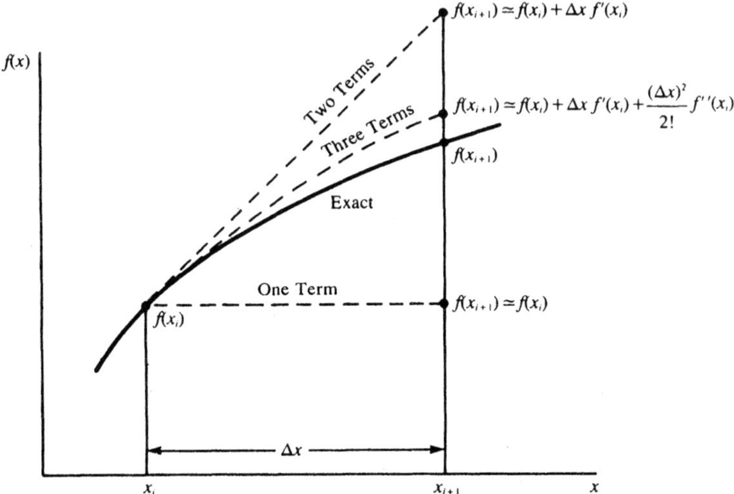

Figure 3.1 The approximation of a function $f(x)$ by a Taylor series expansion, retaining one, two, or three terms in the series.

3.2.2 Finite Difference Calculus

In finite difference calculus, the function $f(x_i + \Delta x)$ is generally written as $f(x_{i+1})$, indicating the value of the function at a neighboring point $x = x_{i+1}$, which is at an incremental distance Δx from the point $x = x_i$, about which the Taylor series expansion has been taken. Then the series in Eqs. (3.2) and (3.3) may be written as

$$f(x_{i\pm 1}) = f(x_i) \pm \Delta x\, f'(x_i) + \frac{(\Delta x)^2}{2!} f''(x_i) \pm \frac{(\Delta x)^3}{3!} f'''(x_i) + \cdots \tag{3.6}$$

This infinite series must be truncated after a few terms in order to be employed in digital computation. If the series is truncated after the $(n+1)$th term, that is, after the term containing the nth power of Δx, the neglected terms may be taken into account by means of a remainder term R_n, given by

$$R_n = \frac{d^{n+1}f}{dx^{n+1}}(\xi)\frac{(\Delta x)^{n+1}}{(n+1)!} \qquad \text{where } x_i < \xi < x_{i+1} \tag{3.7}$$

The derivative in this expression is evaluated at a point $x = \xi$ which lies within the interval from x_i to x_{i+1}. The derivation of the remainder term is given in most textbooks on calculus; see, for instance, the books by Sokolinikoff and Redheffer (1966) and Kaplan (1984).

The expression for the remainder given by Eq. (3.7) can be employed for estimating the error, known as *truncation error* and briefly considered in Chapter 2, that results from a truncation of the series. Thus, the error when the series is truncated after the term containing $(\Delta x)^n$ is less than $|d^{n+1}f/dx^{n+1}|_{max}(\Delta x)^{n+1}/(n+1)!$, where the maximum magnitude of the derivative in the interval $x_i < x < x_{i+1}$ is denoted by the subscript "max." The value of the $(n+1)$th derivative of the given function, in the entire interval, is generally not known, since this would require an analytical expression for $f(x)$, which is assumed to be unknown. If $f(x)$ is known in the interval, the Taylor series expansion will not be needed for evaluating the function at $x = x_{i+1}$. Therefore, one cannot use the remainder term to determine the error exactly. However, the term does indicate the dependence of the truncation error on Δx, and we do have control over the value of Δx.

The remainder R_n, and thus the error, is usually written as

$$R_n = O[(\Delta x)^{n+1}] \tag{3.8}$$

where, as discussed in Chapter 2, this expression implies that the truncation error is of the order of $(\Delta x)^{n+1}$. Since the quantities that multiply $(\Delta x)^{n+1}$ in Eq. (3.7) are constants over the given interval, the expression $O[(\Delta x)^{n+1}]$ may be taken to indicate that the error is proportional to the step size Δx raised to the $(n+1)$th power. Then Eq. (3.1) may be written, with the corresponding truncation error, as

$$f(x_{i+1}) = f(x_i) + \Delta x\, f'(x_i) + \frac{(\Delta x)^2}{2!} f''(x_i) + O[(\Delta x)^3] \tag{3.9a}$$

or

$$f(x_{i+1}) = f(x_i) + \Delta x\, f'(x_i) + \frac{(\Delta x)^2}{2!} f''(x_i) + \frac{(\Delta x)^3}{3!} f'''(x_i) + O[(\Delta x)^4] \tag{3.9b}$$

For a given arbitrary function $f(x)$, Eq. (3.9b), in general, yields a more accurate value of $f(x_{i+1})$, as seen in Fig. 3.1. Thus, within the radius of convergence of the series, the error term due to truncation after n terms is related to that due to truncation after $(n+1)$ terms as follows:

$$O[(\Delta x)^n] < O[(\Delta x)^{n+1}] \tag{3.10}$$

We will assume this relationship to be valid, as long as the series is convergent.

The representation of the truncation error as $O[(\Delta x)^n]$ also indicates the behavior of the error as Δx is reduced. Thus, if Δx is halved, the error becomes $1/2^n$ of the previous error. The order of the Taylor series approximation is given by the value of n. A higher value of n implies the retention of a larger number of terms in the series and, in general, a smaller truncation error. Usually, Δx is taken as sufficiently small, so that only the first few terms in the series are required to obtain a fairly accurate estimate of $f(x_{i+1})$. The characteristics of the Taylor series and of the error resulting from truncation are illustrated in the following example.

Example 3.1

(a) Derive the Taylor series expansions for e^x and $\log(1-x)$, about $x = 0$. Employing the first six terms in the series, determine the values of these functions at

$x = 0.1, 0.2, 0.3, 0.4$, and 0.5. Add the terms successively, indicating the effect of the number of retained terms on the accuracy of the numerical results.

(b) The relationship between the pressure p and temperature T of a given fluid is

$$\log p = 19.2 - \frac{5301.4}{T} \tag{3.1.1}$$

where log represents the natural logarithm, p is in kilopascals, abbreviated kPa (1 kPa $= 10^3$ newtons/m^2), and T is in kelvins. Using the Taylor series expansion for p, compute the pressure at $T = 351, 352, 355, 360$, and 370 K, given the value at 350 K, from Eq. (3.1.1), and using only five terms in the expansion.

Solution

(a) Here, the functions $f(x) = e^x$ and $\log(1 - x)$ are to be expanded in Taylor series about $x = 0$, at which location $e^x = 1$ and $\log(1 - x) = 0$. In order to obtain the series, as given by Eq. (3.1), we need to evaluate the derivatives at $x = 0$. Thus, for $f(x) = e^x$,

$$\frac{d(e^x)}{dx} = \frac{d^2(e^x)}{dx^2} = \frac{d^3(e^x)}{dx^3} = \frac{d^4(e^x)}{dx^4} = \cdots = e^x$$

At $x = 0$, all these derivatives are 1.0. Therefore, the required series for e^x, about $x = 0$, is

$$e^x = 1 + x + \frac{x^2}{2!} + \frac{x^3}{3!} + \frac{x^4}{4!} + \cdots + \frac{x^n}{n!} + \cdots \tag{3.1.2}$$

The second function, $\log(1 - x)$, yields the following derivatives:

$$\frac{d[\log(1-x)]}{dx} = -\frac{1}{1-x}$$

$$\frac{d^2[\log(1-x)]}{dx^2} = -\frac{1}{(1-x)^2}$$

$$\frac{d^3[\log(1-x)]}{dx^3} = -\frac{2}{(1-x)^3}$$

$$\vdots$$

$$\frac{d^4[\log(1-x)]}{dx^4} = -\frac{6}{(1-x)^3}$$

At $x = 0$, the derivatives of the function $f(x) = \log(1 - x)$ are

$$f'(0) = -1, \quad f''(0) = -1, \quad f'''(0) = -2, \quad f''''(0) = -6, \quad \cdots$$

```
100 REM              TAYLOR SERIES FOR EXP(X)
110 REM
120 REM      S IS THE SUM OF THE SERIES UP TO AND INCLUDING THE NTH
130 REM      TERM SN, F IS THE EXACT VALUE OF THE EXPONENTIAL OF THE
140 REM      INDEPENDENT VARIABLE X AND B IS THE FACTORIAL OF N
150 REM
160 REM
170      X=.1
180      FOR I=1 TO 5
190      S=0
200      B=1
210      PRINT "X=";X
220      FOR N=1 TO 6
230      SN = (X^(N-1))/B
240      S = S+SN
250      B = B*N
260      F = EXP(X)
270      PRINT "N=";N;":  ";"S=";S
280      NEXT N
290      PRINT "EXACT VALUE=";F
300      PRINT
310      X = X+.1
320      NEXT I
330      END
```

(*a*)

```
100 REM              TAYLOR SERIES FOR LOG(1-X)
110 REM
120 REM     S IS THE SUM OF THE TAYLOR SERIES FOR LOG(1-X) UP TO AND
130 REM     INCLUDING THE NTH TERM SN, F IS THE EXACT VALUE OF THE
140 REM     LOGARITHM OF (1-X), WHERE X IS THE INDEPENDENT VARIABLE,
150 REM     AND N IS THE NUMBER OF TERMS CONSIDERED
160 REM
170 REM
180      X=.1
190      FOR I=1 TO 5
200      S=0
210      PRINT "X=";X
220      FOR N=1 TO 6
230      SN = -(X^N)/N
240      S = S+SN
250      F = LOG(1-X)
260      PRINT "N=";N;":  ";"S=";S
270      NEXT N
280      PRINT "EXACT VALUE=";F
290      PRINT
300      X = X+.1
310      NEXT I
320      END
```

(*b*)

Figure 3.1.1 Computer programs in BASIC for the summation of the series expansions in Example 3.1(a).

Therefore, the Taylor series expansion for $\log(1 - x)$ about $x = 0$ is

$$\log(1-x) = 0 + x\cdot(-1) + \frac{x^2}{2!}\cdot(-1) + \frac{x^3}{3!}\cdot(-2) + \frac{x^4}{4!}\cdot(-6) + \frac{x^5}{5!}\cdot(-24) + \cdots$$

$$= -\left[x + \frac{x^2}{2} + \frac{x^3}{3} + \frac{x^4}{4} + \frac{x^5}{5} + \cdots + \frac{x^n}{n} + \cdots\right] \tag{3.1.3}$$

Computer programs may be easily written to sum a finite number of terms in the two series given by Eqs. (3.1.2) and (3.1.3). Starting with the first term, add additional terms successively and determine the resulting sum. Continue this process up to the sixth term, employing the various values of x given in the problem. The computer programs in BASIC are given in Fig. 3.1.1. The numerical results obtained are shown in Fig. 3.1.2. The exact values of the functions at the various x values considered are also computed and printed, along with the numerical results for comparison.

Note that the accuracy of the numerical evaluation of the functions from their respective Taylor series expansions improves as the number of terms considered increases. Six terms are found to be quite adequate at smaller x values, although more terms should be employed for x equal to or larger than 0.5 for better accuracy. The convergence is slower at larger x, as expected and as shown in most calculus textbooks. Also, the series for e^x converges at all x, whereas that for $\log(1 - x)$ converges only if $|x| < 1$. Therefore, at larger x, within $|x| < 1$ for the second case, additional terms should be included until the sum remains essentially unchanged with a further addition of terms.

(b) The given relation between p and T, Eq. (3.1.1), may be written as follows:

$$p = \exp\left[19.2 - \frac{5301.4}{T}\right] = \exp(19.2)\exp\left(-\frac{5301.4}{T}\right) \tag{3.1.4}$$

Therefore, a Taylor series expansion for $\exp(-5301.4/T)$ is needed for computing the pressure p at temperatures close to $T = 350$ K, about which the expansion must be carried out. We may write Eq. (3.1.4) as

$$p = Af(T) \qquad \text{where } f(T) = \exp\left(\frac{B}{T}\right)$$

$$= \exp\left(-\frac{5301.4}{T}\right) \quad \text{and} \quad A = \exp(19.2) \tag{3.1.5}$$

The Taylor series expansion for p is then given by

$$p = p_{350} + (T - 350)Af'(350) + \frac{(T-350)^2}{2!}Af''(350)$$

$$+ \frac{(T-350)^3}{3!}Af'''(350) + \frac{(T-350)^4}{4!}Af''''(350) + \cdots \tag{3.1.6}$$

```
X= .1
N= 1 : S= 1
N= 2 : S= 1.1
N= 3 : S= 1.105
N= 4 : S= 1.105167
N= 5 : S= 1.105171
N= 6 : S= 1.105171
EXACT VALUE= 1.105171

X= .2
N= 1 : S= 1
N= 2 : S= 1.2
N= 3 : S= 1.22
N= 4 : S= 1.221333
N= 5 : S= 1.2214
N= 6 : S= 1.221403
EXACT VALUE= 1.221403

X= .3
N= 1 : S= 1
N= 2 : S= 1.3
N= 3 : S= 1.345
N= 4 : S= 1.3495
N= 5 : S= 1.349837
N= 6 : S= 1.349858
EXACT VALUE= 1.349859

X= .4
N= 1 : S= 1
N= 2 : S= 1.4
N= 3 : S= 1.48
N= 4 : S= 1.490667
N= 5 : S= 1.491733
N= 6 : S= 1.491819
EXACT VALUE= 1.491825

X= .5
N= 1 : S= 1
N= 2 : S= 1.5
N= 3 : S= 1.625
N= 4 : S= 1.645833
N= 5 : S= 1.648438
N= 6 : S= 1.648698
EXACT VALUE= 1.648721
```

(*a*) $f(x) = e^x$

```
X= .1
N= 1 : S=-.1
N= 2 : S=-.105
N= 3 : S=-.1053333
N= 4 : S=-.1053583
N= 5 : S=-.1053603
N= 6 : S=-.1053605
EXACT VALUE=-.1053606

X= .2
N= 1 : S=-.2
N= 2 : S=-.22
N= 3 : S=-.2226667
N= 4 : S=-.2230667
N= 5 : S=-.2231307
N= 6 : S=-.2231414
EXACT VALUE=-.2231436

X= .3
N= 1 : S=-.3
N= 2 : S=-.345
N= 3 : S=-.354
N= 4 : S=-.3560251
N= 5 : S=-.356511
N= 6 : S=-.3566325
EXACT VALUE=-.356675

X= .4
N= 1 : S=-.4
N= 2 : S=-.48
N= 3 : S=-.5013334
N= 4 : S=-.5077334
N= 5 : S=-.5097814
N= 6 : S=-.510464
EXACT VALUE=-.5108257

X= .5
N= 1 : S=-.5
N= 2 : S=-.625
N= 3 : S=-.6666667
N= 4 : S=-.6822917
N= 5 : S=-.6885417
N= 6 : S=-.6911459
EXACT VALUE=-.6931473
```

(*b*) $f(x) = log\ (1 - x)$

Figure 3.1.2 Numerical results on the summation of the Taylor series expansions for e^x and log $(1 - x)$ at various values of x, as given in Example 3.1(a), along with the exact values of these functions.

```
ENTER THE TEMPERATURE STEP SIZE DT
?1.0
TEMPERATURE=351
N=1: P=57.5781331     N=2: P=60.0699267
N=3: P=60.1167256     N=4: P=60.1172157
N=5: P=60.1172187
THE EXACT VALUE OF THE PRESSURE=60.1172187

ENTER THE TEMPERATURE STEP SIZE DT
?2.0
TEMPERATURE=352
N=1: P=57.5781331     N=2: P=62.5617203
N=3: P=62.748916     N=4: P=62.7528363
N=5: P=62.7528845
THE EXACT VALUE OF THE PRESSURE=62.7528848

ENTER THE TEMPERATURE STEP SIZE DT
?5.0
TEMPERATURE=355
N=1: P=57.5781331     N=2: P=70.0371011
N=3: P=71.2070744     N=4: P=71.2683294
N=5: P=71.2702111
THE EXACT VALUE OF THE PRESSURE=71.2702421

ENTER THE TEMPERATURE STEP SIZE DT
?10.0
TEMPERATURE=360
N=1: P=57.5781331     N=2: P=82.496069
N=3: P=87.1759625     N=4: P=87.6660022
N=5: P=87.6961086
THE EXACT VALUE OF THE PRESSURE=87.6971056

ENTER THE TEMPERATURE STEP SIZE DT
?20.0
TEMPERATURE=370
N=1: P=57.5781331     N=2: P=107.414005
N=3: P=126.133579     N=4: P=130.053896
N=5: P=130.535599
THE EXACT VALUE OF THE PRESSURE=130.567704
```

Figure 3.1.3 Numerical values of the pressure p obtained from a summation of the Taylor series expansion for the function $p(T)$, as given in Example 3.1(b), for various values of the temperature T.

where p_{350} refers to the pressure at 350 K, as calculated from Eq. (3.1.4), and the quantity within the parentheses indicates the temperature T at which the evaluation is made. Now, f', f'', f''', and so on, are obtained by differentiation as

$$f'(T) = -\frac{B}{T^2}e^{B/T}$$

$$f''(T) = \left(\frac{B^2}{T^4} + \frac{2B}{T^3}\right)e^{B/T}$$

$$f'''(T) = \left(-\frac{B^3}{T^6} - \frac{6B^2}{T^5} - \frac{6B}{T^4}\right)e^{B/T}$$

$$\vdots$$

If T is replaced by 350 K in these expressions and substituted in Eq. (3.1.6), the series for p is obtained.

The value of the pressure p_{350} at $T = 350$ is calculated from Eq. (3.1.4) as 57.5781 kPa. Using this value, we calculate the pressures at $T = 351$, 352, 355, 360, and 370 K from the Taylor series given by Eq. (3.1.6). The numerical results obtained are shown in Fig. 3.1.3. The sum of the series, considering one, two, and up to five terms, is given, along with the corresponding exact value from Eq. (3.1.4). As expected, the accuracy of the numerical value for the pressure improves as the number of terms employed increases. Five terms are found to be quite satisfactory, particularly for small temperature differences $\Delta T = T - 350$. However, as $(T - 350)$ increases to large values, more terms will be needed for accurate results. Obviously, the numerical sum of the series is in considerable error if only one or two terms are retained, particularly at large ΔT.

The simple problems given in Example 3.1 illustrate practical applications of the Taylor series expansions and also bring out the important aspects that one needs to bear in mind in summing the series. It is important to ensure that the numerical results converge to a constant value as the number of terms is increased. If the results diverge as additional terms are brought in, the series is not convergent, and an alternative approach to obtain the desired numerical results must be employed. In some engineering problems, the function $f(x)$ may be too complicated to be evaluated easily in the vicinity of the x value at which it is known. Then Taylor series may be used, as outlined above. The series is also frequently employed for providing starting values in the solution of differential equations, as shown later in Chapter 8.

3.3 DIRECT APPROXIMATION OF DERIVATIVES

In many diverse engineering problems, the accurate determination of derivatives from the measured or calculated values of the function $f(x)$ at a finite number of discrete points is needed. Consider, for example, an engineer, on a test track, involved in the measurement of the location of a moving body as a function of time. The velocity and acceleration of the object are obtained from the computed values of the first and second derivatives of the displacement. Similarly, a chemical or civil engineer may measure the concentration of a pollutant in a water body as a function of the location and time and then use this information to obtain the rate of spread of chemical pollution. Heat and mass transfer processes are also concerned with the rates of transport, and the measurements of temperature and concentration are often employed for developing correlations for predicting transport rates in several practical circumstances. Thus, the approximation of derivatives is important in many practical problems and also in the solution of differential equations by the finite difference approach, as presented in Chapters 8 and 9.

A simple approach to the derivation of the finite difference approximation of the derivatives of a function $f(x)$ is based on the replacement of infinitesimal differences by discrete differences in the mathematical definition of differentiation. Finite differences are considered in the variation of the independent variable x, and the values of the function $f(x)$ at discrete points are employed in deriving the approximation. Consider the variation of $f(x)$ with x, as sketched in Fig. 3.2. Three discrete points, denoted by subscripts $i - 1$, i, and $i + 1$, are shown along the x axis. We may approximate the derivatives of $f(x)$, with respect to x, in terms of these discrete differences.

The first derivative df/dx at $x = x_i$ can be approximated by $\Delta f/\Delta x$, where the Δ's denote discrete differences. Three approximations for $(df/dx)_i$ can be written by inspection, in terms of differences between the values at the three discrete locations or nodes. These approximations are as follows:

$$\left(\frac{df}{dx}\right)_i \simeq \frac{f_{i+1} - f_i}{\Delta x} \tag{3.11}$$

$$\left(\frac{df}{dx}\right)_i \simeq \frac{f_i - f_{i-1}}{\Delta x} \tag{3.12}$$

$$\left(\frac{df}{dx}\right)_i \simeq \frac{f_{i+1} - f_{i-1}}{2\,\Delta x} \tag{3.13}$$

where the subscripts denote the nodal location, in x, where the quantity is evaluated. The first approximation is known as the *two-point forward difference approximation for* $(df/dx)_i$, since only two nodes are involved and the value of the function in the forward, or the increasing x, direction is employed. Similarly, the second approximation, Eq. (3.12), is known as the *two-point backward difference*, and the third approximation, Eq. (3.13), as the *three-point central difference approximation.*

These approximations employ the slopes of the chords to the right of, to the left

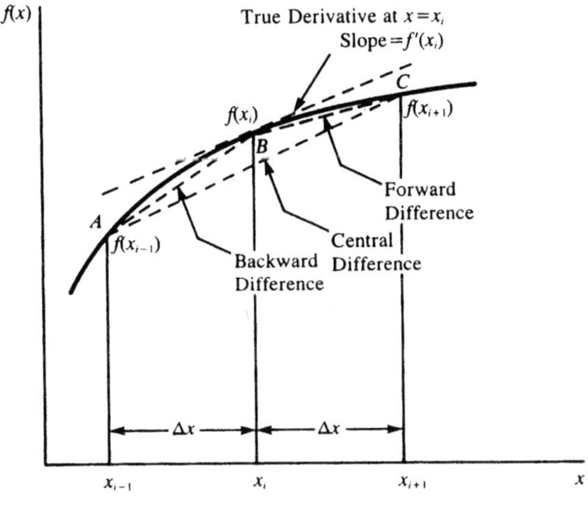

Figure 3.2 Graphical representation of the forward, backward, and central finite difference approximations of the first derivative of a function $f(x)$.

of, and centered on the node point at $x = x_i$, as shown in Fig. 3.2, to approximate the gradient of the function at $x = x_i$. Since they are only approximations to the derivative $(df/dx)_i$, an approximate equality sign ($\simeq$) is used. The central difference may be interpreted in either of the following two ways:

$$\frac{f_{i+1} - f_{i-1}}{2\,\Delta x} = \frac{1}{2}\left[\frac{f_{i+1} - f_i}{\Delta x} + \frac{f_i - f_{i-1}}{\Delta x}\right] \tag{3.14a}$$

or

$$\frac{f_{i+1} - f_{i-1}}{2\,\Delta x} = \frac{1}{\Delta x}\left[\tfrac{1}{2}(f_{i+1} + f_i) - \tfrac{1}{2}(f_i + f_{i-1})\right] \tag{3.14b}$$

The first equation represents an average of the two one-sided differences, and the second the difference based on the average values of the function at $(i + 1/2)$ and $(i - 1/2)$. Because the central difference averages out the variations on either side of the node $x = x_i$, it is expected to be a more accurate representation of the derivative. This is shown to be true on the basis of the Taylor series approach outlined in the next section.

Similarly, the finite difference approximation for the second derivative d^2f/dx^2 at $x = x_i$ may be derived. Thus,

$$\frac{d^2f}{dx^2} = \frac{d}{dx}\left(\frac{df}{dx}\right) \simeq \frac{\Delta}{\Delta x}\left(\frac{\Delta f}{\Delta x}\right)$$

$$= \frac{1}{\Delta x}\left[\frac{f_{i+1} - f_i}{\Delta x} - \frac{f_i - f_{i-1}}{\Delta x}\right]$$

or

$$\left(\frac{d^2f}{dx^2}\right)_i \simeq \frac{f_{i+1} - 2f_i + f_{i-1}}{(\Delta x)^2} \tag{3.15}$$

Here, the difference in $\Delta f/\Delta x$ is approximated by the difference between the slopes of the two chords, on either side of the node at $x = x_i$. In fact, $(f_{i+1} - f_i)/\Delta x$ represents the central difference approximation of the derivative at $x = x_i + (\Delta x/2)$, since it uses the values on either side of this location, with discrete differences of $\Delta x/2$ in x. Similarly, $(f_{i-1} - f_i)/\Delta x$ represents the central difference approximation of the derivative at $x = x_i - (\Delta x/2)$. Therefore,

$$\left(\frac{d^2f}{dx^2}\right)_i \simeq \frac{\Delta}{\Delta x}\left(\frac{\Delta f}{\Delta x}\right)$$

$$= \frac{1}{\Delta x}\left[\left(\frac{\Delta f}{\Delta x}\right)_{i+1/2} - \left(\frac{\Delta f}{\Delta x}\right)_{i-1/2}\right]$$

$$\left(\frac{d^2f}{dx^2}\right)_i = \frac{1}{\Delta x}\left[\frac{f_{i+1} - f_i}{\Delta x} - \frac{f_i - f_{i-1}}{\Delta x}\right] = \frac{f_{i+1} - 2f_i + f_{i-1}}{(\Delta x)^2} \tag{3.16}$$

The above finite difference approximation of the second derivative is known as the *three-point central second difference approximation.* Other approximations for the second derivative may also be obtained by employing other finite difference representations in the above derivation. However, the central second difference is the most frequently employed approximation. Similarly, finite difference approximations for higher-order derivatives may be derived. Again, several representations are usually possible, with central differences being more accurate than one-sided differences, if the same nodal points are used in the two cases.

The direct approximation of the derivatives thus allows one to derive the required finite difference representations. The approach is based on the mathematical interpretation of differentiation, and therefore provides a physical background for the formulation of finite differences. However, it does not give any information on the accuracy of a particular representation. For an estimation of the error involved, we must resort to the Taylor series approach as described in the next section.

3.4 TAYLOR SERIES APPROACH AND ACCURACY

The Taylor series expansions about a given nodal point $x = x_i$ may be employed to derive the finite difference approximations of the derivatives of a function $f(x)$. Since the error resulting from the truncation of the series after a finite number of terms can be estimated from the remainder term, given by Eq. (3.7), the errors associated with the various finite difference approximations of the derivatives, obtained by the Taylor series approach, may also be estimated. Using this approach, one can derive one-sided, forward and backward, and central difference approximations.

3.4.1 Finite Difference Approximation of the First Derivative

Consider the variation of the function $f(x)$ with x, as shown in Fig. 3.2. If the function $f(x)$ is sufficiently smooth, it may be expanded in a Taylor series in the neighborhood of $x = x_i$. Assuming that the points x_{i-1} and x_{i+1} lie within the region of convergence of the series, the function $f(x)$ at these points is given by Eq. (3.6) as

$$f_{i+1} = f_i + \Delta x\, f_i' + \frac{(\Delta x)^2}{2!} f_i'' + \frac{(\Delta x)^3}{3!} f_i''' + \cdots \tag{3.17}$$

$$f_{i-1} = f_i - \Delta x\, f_i' + \frac{(\Delta x)^2}{2!} f_i'' - \frac{(\Delta x)^3}{3!} f_i''' + \cdots \tag{3.18}$$

where the subscripts again denote the nodal locations, in x, where the function is evaluated, and the primes denote differentiation with respect to x.

If Eq. (3.17) is solved for the first derivative f_i', we obtain

$$f_i' = \frac{f_{i+1} - f_i}{\Delta x} - \frac{\Delta x}{2} f_i'' - \frac{(\Delta x)^2}{6} f_i''' + \cdots$$

When the infinite series is replaced by the remainder term, this equation becomes

$$f_i' = \frac{f_{i+1} - f_i}{\Delta x} - \frac{\Delta x}{2} f''(\xi) \qquad \text{where } x_i < \xi < x_{i+1}$$

$$= \frac{f_{i+1} - f_i}{\Delta x} + O(\Delta x) \tag{3.19}$$

This equation gives the forward difference approximation of the first derivative, given by Eq. (3.11), along with the truncation error in the approximation.

Similarly, the two-point backward difference for the first derivative may be obtained by solving Eq. (3.18) for f_i' as follows:

$$f_i' = \frac{f_i - f_{i-1}}{\Delta x} + \frac{\Delta x}{2} f_i'' - \frac{(\Delta x)^2}{6} f_i''' + \cdots$$

which gives

$$f_i' = \frac{f_i - f_{i-1}}{\Delta x} + \frac{\Delta x}{2} f''(\xi) \qquad \text{where } x_{i-1} < \xi < x_i$$

$$= \frac{f_i - f_{i-1}}{\Delta x} + O(\Delta x) \tag{3.20}$$

The truncation error is of the same order as that in the forward difference approximation. Since the error terms are included in Eqs. (3.19) and (3.20), the approximate equality signs of Eqs. (3.11) and (3.12) are not needed here.

A more accurate finite difference approximation of the first derivative is obtained by subtracting Eq. (3.18) from Eq. (3.17), to yield

$$f_i' = \frac{f_{i+1} - f_{i-1}}{2\,\Delta x} - \frac{(\Delta x)^2}{6} f'''(\xi) \qquad \text{where } x_{i-1} < \xi < x_{i+1}$$

$$= \frac{f_{i+1} - f_{i-1}}{2\,\Delta x} + O[(\Delta x)^2] \tag{3.21}$$

This result is the three-point central difference approximation of the first derivative, as given earlier in Eq. (3.13). The truncation error is of order $(\Delta x)^2$, and therefore this representation is more accurate than the forward and backward differences. Graphically, this expression approximates the derivative of the function $f(x)$ at x_i as the slope of the line AC in Fig. 3.2. The forward and backward differences approximate the derivative by the slopes of the chords BC and AB, respectively. Also note from the above expressions that if the step size Δx is halved, the truncation error is also approximately halved for the forward and backward differences, whereas the error becomes one-fourth for the central difference.

3.4.2 Second Derivative

A finite difference approximation for the second derivative f_i'' may also be derived by adding Eqs. (3.17) and (3.18), yielding

$$f_i'' = \frac{f_{i+1} - 2f_i + f_{i-1}}{(\Delta x)^2} - [\tfrac{1}{12}(\Delta x)^2 f_i'''' + \cdots]$$

Therefore,

$$f_i'' = \frac{f_{i+1} - 2f_i + f_{i-1}}{(\Delta x)^2} - \frac{(\Delta x)^2}{12} f''''(\xi) \qquad \text{where } x_{i-1} < \xi < x_{i+1}$$

$$= \frac{f_{i+1} - 2f_i + f_{i-1}}{(\Delta x)^2} + [O(\Delta x)^2] \tag{3.22}$$

This result is the second central difference which was derived in the preceding section by the direct approximation approach. The truncation error is $O[(\Delta x)^2]$, and therefore this finite difference approximation is of second order. Graphically, this expression approximates the second derivative by dividing the difference in the slopes of the chords that approximate the first derivatives at $x_{i+1/2}$ and $x_{i-1/2}$ by Δx; see Fig. 3.3. The slopes of these chords are approximated in the central difference formulation as follows:

$$f'\left(x_i + \frac{\Delta x}{2}\right) = f'_{i+1/2} = \frac{f_{i+1} - f_i}{\Delta x} + O[(\Delta x)^2] \tag{3.23}$$

and

$$f'\left(x_i - \frac{\Delta x}{2}\right) = f'_{i-1/2} = \frac{f_i - f_{i-1}}{\Delta x} + O[(\Delta x)^2] \tag{3.24}$$

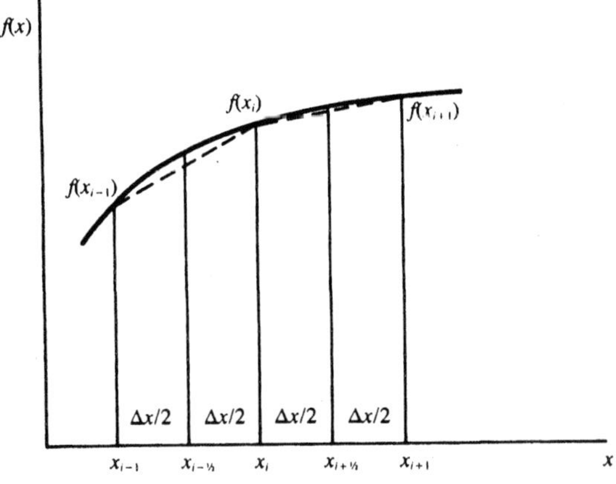

Figure 3.3 Graphical depiction of the finite difference approximation of the second derivative.

Thus,

$$f_i'' = \frac{f'_{i+1/2} - f'_{i-1/2}}{\Delta x} + O[(\Delta x)^2]$$

$$= \frac{f_{i+1} - 2f_i + f_{i-1}}{(\Delta x)^2} + O[(\Delta x)^2] \tag{3.25}$$

Similarly, one-sided forward or one-sided backward differences may be derived for f_i'' by employing points on only one side of $x = x_i$, rather than on both sides as done for the central difference. Let us consider, for example, the three points at x_i, x_{i+1}, and x_{i+2}, as shown in Fig. 3.4. The Taylor series expansion for $f(x_{i+1})$ is given by Eq. (3.17), and the expansion for $f(x_{i+2})$ is

$$f_{i+2} = f_i + (2\,\Delta x)f_i' + \frac{(2\,\Delta x)^2}{2!}f_i'' + \frac{(2\,\Delta x)^3}{3!}f_i''' + \frac{(2\,\Delta x)^4}{4!}f_i'''' + \cdots \tag{3.26}$$

Now, the first derivative f_i' may be eliminated from Eqs. (3.17) and (3.26) to yield an expression for f_i''. Thus, multiplying Eq. (3.17) by 2 and subtracting the resulting equation from Eq. (3.26) gives

$$f_i'' = \frac{f_{i+2} - 2f_{i+1} + f_i}{(\Delta x)^2} - (\Delta x\, f_i''' + \cdots)$$

$$f_i'' = \frac{f_{i+2} - 2f_{i+1} + f_i}{(\Delta x)^2} + O(\Delta x) \tag{3.27}$$

This result is the forward difference approximation of the second derivative and is accurate to within an error of order Δx.

Similarly, the backward difference approximation may be obtained by employ-

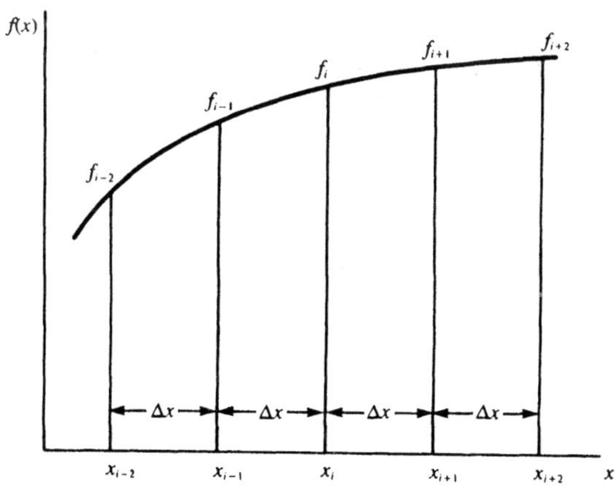

Figure 3.4 Distribution of the nodal points for deriving forward, backward, and central difference approximations for higher-order derivatives and also for higher accuracy formulas.

ing the Taylor series expansions for $f(x_{i-1})$ and $f(x_{i-2})$ as follows:

$$f_i'' = \frac{f_i - 2f_{i-1} + f_{i-2}}{(\Delta x)^2} + O(\Delta x) \tag{3.28}$$

Note again that the forward and backward difference approximations are less accurate than the central difference approximation if the same number of nodal points is used in all three cases. Higher-order approximations may be derived by employing additional points, as shown later. Even though the one-sided differences are less accurate than the central difference, they are often employed for approximating the derivatives, particularly near the boundaries of a computational domain since nodal points may be available on only one side of the boundary. Therefore, in the finite difference solution of ordinary and partial differential equations, forward and backward differences are frequently employed for obtaining the finite difference representations of the boundary conditions.

3.4.3 Higher-Order Derivatives

The finite difference approximations of higher-order derivatives may be derived by the use of Taylor series expansions, as outlined above for the first and second derivatives. However, the derivation becomes more involved as one proceeds to successively higher derivatives since an increasingly larger number of simultaneous equations must be solved. The larger number of equations is obtained by employing expansions at a larger number of nodal points. Thus, the formulas for the third and fourth derivatives may be obtained by employing the expansions for f_{i+1} and f_{i-1}, given by Eqs. (3.17) and (3.18), along with those for f_{i+2} and f_{i-2}. The expansion for f_{i+2} is given by Eq. (3.26), and that for f_{i-2} is

$$f_{i-2} = f_i - (2\,\Delta x)f_i' + \frac{(2\,\Delta x)^2}{2!} f_i'' - \frac{(2\,\Delta x)^3}{3!} f_i''' + \frac{(2\,\Delta x)^4}{4!} f_i'''' + \cdots \tag{3.29}$$

Subtracting Eq. (3.29) from Eq. (3.28), we obtain

$$f_{i+2} - f_{i-2} = 4\,\Delta x\, f_i' + \frac{8(\Delta x)^3}{3} f_i''' + O[(\Delta x)^5]$$

The substitution of the finite difference expression for f_i' from Eq. (3.21) gives

$$f_{i+2} - f_{i-2} = 4\,\Delta x \left\{\frac{f_{i+1} - f_{i-1}}{2\,\Delta x} - \frac{(\Delta x)^2}{6} f_i''' + O[(\Delta x)^4]\right\} + \frac{8(\Delta x)^3}{3} f_i''' + O[(\Delta x)^5]$$

This yields

$$f_i''' = \frac{f_{i+2} - 2f_{i+1} + 2f_{i-1} - f_{i-2}}{2(\Delta x)^3} + O[(\Delta x)^2] \tag{3.30}$$

Similarly, the finite difference approximation for the fourth derivative f_i'''' may be derived by adding Eqs. (3.28) and (3.29) and then substituting the approximation for f_i''. The resulting approximation is

$$f_i'''' = \frac{f_{i+2} - 4f_{i+1} + 6f_i - 4f_{i-1} + f_{i-2}}{(\Delta x)^4} - \frac{(\Delta x)^2}{6}\frac{d^6 f}{dx^6}(\xi) \qquad \text{where } x_{i-2} < \xi < x_{i+2}$$

$$= \frac{f_{i+2} - 4f_{i+1} + 6f_i - 4f_{i-1} + f_{i-2}}{(\Delta x)^4} + O[(\Delta x)^2] \tag{3.31}$$

Thus, the five grid points shown in Fig. 3.4 are involved in the finite difference expression for the fourth derivative f_i'''', with a truncation error of order $(\Delta x)^2$. The corresponding expression for f_i''' also involves these points, except for f_i which drops out in the derivation. By employing a still larger number of points, one may derive

---

Forward Difference Approximations of $O(\Delta x)$

$$f_i' = \frac{f_{i+1} - f_i}{\Delta x}$$

$$f_i'' = \frac{f_{i+2} - 2f_{i+1} + f_i}{(\Delta x)^2}$$

$$f_i''' = \frac{f_{i+3} - 3f_{i+2} + 3f_{i+1} - f_i}{(\Delta x)^3}$$

$$f_i'''' = \frac{f_{i+4} - 4f_{i+3} + 6f_{i+2} - 4f_{i+1} + f_i}{(\Delta x)^4}$$

Forward Difference Approximations of $O[(\Delta x)^2]$

$$f_i' = \frac{-f_{i+2} + 4f_{i+1} - 3f_i}{2\Delta x}$$

$$f_i'' = \frac{-f_{i+3} + 4f_{i+2} - 5f_{i+1} + 2f_i}{(\Delta x)^2}$$

$$f_i''' = \frac{-3f_{i+4} + 14f_{i+3} - 24f_{i+2} + 18f_{i+1} - 5f_i}{2(\Delta x)^3}$$

$$f_i'''' = \frac{-2f_{i+5} + 11f_{i+4} - 24f_{i+3} + 26f_{i+2} - 14f_{i+1} + 3f_i}{(\Delta x)^4}$$

Figure 3.5 Forward finite difference formulas, along with the truncation errors.

expressions for the fifth and sixth derivatives to the same accuracy. However, these derivations are quite involved because of the large number of equations to be solved. Another method, which is based on difference and derivative operators, may often be employed more easily for the derivation of higher-order derivatives. This approach is discussed by Salvadori and Baron (1961) and Hornbeck (1975).

Equations (3.30) and (3.31) give the central difference approximations of the third and fourth derivatives of $f(x)$, respectively. Similarly, one-sided forward or one-sided backward differences may be derived. As mentioned earlier, one-sided differences are of interest in only a few cases, such as near the boundaries of the computational region. The central differences are much more important and are

---

Backward Difference Approximations of $O(\Delta x)$

$$f_i' = \frac{f_i - f_{i-1}}{\Delta x}$$

$$f_i'' = \frac{f_i - 2f_{i-1} + f_{i-2}}{(\Delta x)^2}$$

$$f_i''' = \frac{f_i - 3f_{i-1} + 3f_{i-2} - f_{i-3}}{(\Delta x)^3}$$

$$f_i'''' = \frac{f_i - 4f_{i-1} + 6f_{i-2} - 4f_{i-3} + f_{i-4}}{(\Delta x)^4}$$

Backward Difference Approximations of $O[(\Delta x)^2]$

$$f_i' = \frac{3f_i - 4f_{i-1} + f_{i-2}}{2\,\Delta x}$$

$$f_i'' = \frac{2f_i - 5f_{i-1} + 4f_{i-2} - f_{i-3}}{(\Delta x)^2}$$

$$f_i''' = \frac{5f_i - 18f_{i-1} + 24f_{i-2} - 14f_{i-3} + 3f_{i-4}}{2(\Delta x)^3}$$

$$f_i'''' = \frac{3f_i - 14f_{i-1} + 26f_{i-2} - 24f_{i-3} + 11f_{i-4} - 2f_{i-5}}{(\Delta x)^4}$$

Figure 3.6 Backward finite difference formulas, along with the truncation errors.

Central Difference Approximations of $O[(\Delta x)^2]$

$$f_i' = \frac{f_{i+1} - f_{i-1}}{2\,\Delta x}$$

$$f_i'' = \frac{f_{i+1} - 2f_i + f_{i-1}}{(\Delta x)^2}$$

$$f_i''' = \frac{f_{i+2} - 2f_{i+1} + 2f_{i-1} - f_{i-2}}{2(\Delta x)^3}$$

$$f_i'''' = \frac{f_{i+2} - 4f_{i+1} + 6f_i - 4f_{i-1} + f_{i-2}}{(\Delta x)^4}$$

Central Difference Approximations of $O[(\Delta x)^4]$

$$f_i' = \frac{-f_{i+2} + 8f_{i+1} - 8f_{i-1} + f_{i-2}}{12\,\Delta x}$$

$$f_i'' = \frac{-f_{i+2} + 16f_{i+1} - 30f_i + 16f_{i-1} - f_{i-2}}{12(\Delta x)^2}$$

$$f_i''' = \frac{-f_{i+3} + 8f_{i+2} - 13f_{i+1} + 13f_{i-1} - 8f_{i-2} + f_{i-3}}{8(\Delta x)^3}$$

$$f_i'''' = \frac{-f_{i+3} + 12f_{i+2} - 39f_{i+1} + 56f_i - 39f_{i-1} + 12f_{i-2} - f_{i-3}}{6(\Delta x)^4}$$

Figure 3.7 Central difference approximations, with the associated truncation errors.

employed for the approximation of the derivatives in a wide variety of engineering problems. Several of the commonly used finite difference formulations are given in Figs. 3.5 through 3.7, including higher-accuracy formulas discussed below.

3.4.4 Higher-Accuracy Approximations

The Taylor series approach for the derivation of finite difference formulas may be employed for obtaining approximations of higher accuracy. All the finite difference representations derived above have a truncation error of order Δx or $(\Delta x)^2$. Although an accuracy of $O[(\Delta x)^2]$ is adequate for most problems of practical interest, higher-accuracy formulas are often employed if a given circumstance demands very accurate numerical results. Such a requirement arises, for instance, in the determination of the displacement and velocity of a projectile or of a robotic arm.

Higher-accuracy formulas can be developed by including additional terms in the Taylor series expansions. However, a larger number of grid points will also be required to generate the additional equations needed for eliminating the higher-order derivatives that arise due to the retention of additional terms. Consider, for example, the forward difference expression for the first derivative f_i'. As obtained earlier,

$$f_i' = \frac{f_{i+1} - f_i}{\Delta x} - \frac{\Delta x}{2} f_i'' - \frac{(\Delta x)^2}{6} f_i''' + \cdots$$

If instead of truncating the series after the first term, as done earlier, we retain the term of order Δx and substitute the forward finite difference expression for f_i'', we will obtain a higher-accuracy forward difference expression for f_i'. Thus, from Eq. (3.27),

$$\begin{aligned} f_i' &= \frac{f_{i+1} - f_i}{\Delta x} - \frac{\Delta x}{2}\left[\frac{f_{i+2} - 2f_{i+1} + f_i}{(\Delta x)^2} - \Delta x\, f_i''' + \cdots\right] - \frac{(\Delta x)^2}{6} f_i''' + \cdots \\ &= \frac{-f_{i+2} + 4f_{i+1} - 3f_i}{2\,\Delta x} + \frac{(\Delta x)^2}{3} f_i''' + \cdots \end{aligned}$$

$$f_i' = \frac{-f_{i+2} + 4f_{i+1} - 3f_i}{2\,\Delta x} + O[(\Delta x)^2] \tag{3.32}$$

Similarly, a backward difference expression of $O[(\Delta x)^2]$ may be obtained by retaining an additional term in the backward difference expression for f_i' and substituting the backward difference formula of $O(\Delta x)$ for f_i''. Formulations of still higher accuracy can be obtained by retaining additional terms in the series. As shown in Eq. (3.32), the value of the function at an additional grid point, x_{i+2}, is brought in to obtain the higher accuracy. Similarly, finite difference expressions for f_i' with truncation errors of order $(\Delta x)^3$ and $(\Delta x)^4$ are obtained as follows:

$$f_i' = \frac{1}{6\,\Delta x}(-2f_{i-1} - 3f_i + 6f_{i+1} - f_{i+2}) + \frac{(\Delta x)^3}{12} f_i''''(\xi) \tag{3.33}$$

$$f_i' = \frac{1}{6\,\Delta x}(f_{i-2} - 6f_{i-1} + 3f_i + 2f_{i+1}) - \frac{(\Delta x)^3}{12} f_i''''(\xi) \tag{3.34}$$

$$f_i' = \frac{1}{12\,\Delta x}(f_{i-2} - 8f_{i-1} + 8f_{i+1} - f_{i+2}) + \frac{(\Delta x)^4}{30} f_i'''''(\xi) \tag{3.35}$$

where ξ is within the range of the appropriate expansion. The first two equations are third-order correct, forward and backward, four-point differences. The third equation is a fourth-order correct, five-point central difference approximation for the first derivative at $x = x_i$. With these five points, a higher-order approximation for the second derivative is

$$f_i'' = \frac{1}{12(\Delta x)^2}[-f_{i-2} + 16f_{i-1} - 30f_i + 16f_{i+1} - f_{i+2}] + \frac{(\Delta x)^4}{90}\frac{d^6 f}{dx^6}(\xi) \qquad \text{where } x_{i-2} < \xi < x_{i+2} \tag{3.36}$$

It is evident that finite difference approximations of desired accuracy may be derived by the use of Taylor series expansions. For most practical circumstances, formulations of accuracy $O[(\Delta x)^2]$ are quite satisfactory. As shown later in Chapters 8 and 9, most finite difference solutions of ordinary and partial differential equations are based on expressions of accuracy $O[(\Delta x)^2]$. However, finite difference representations of different accuracy are also employed, depending on the special needs of a given problem. Several relevant expressions are summarized in Figs. 3.5 through 3.7.

The accuracy of the numerical results may be improved either by employing a higher-accuracy formula or by reducing the grid spacing Δx. As discussed in Chapter 2, both of these approaches are employed in practice. Grid refinement, or reducing Δx, is generally carried out until the numerical results are essentially unaffected by a further reduction. At this stage, the numerical results are as accurate as can be obtained with the chosen finite difference expression. A continued reduction in grid spacing will lead to increasing round-off error and, thus, less accurate results. Then the accuracy of the results can be increased by using a higher-accuracy formulation.

An interesting point that may be observed from all the finite difference expressions given here is that the sum of all the coefficients, which multiply the function values in the numerator, is always zero. This result arises because the derivatives must become zero if $f(x)$ is a constant. Also, if Δx approaches zero, the numerator must also approach zero so that the limiting result yields a finite value for an arbitrary continuous function $f(x)$.

Example 3.2

An engineer involved in the design of automobiles uses an experimental system for studying the motion of a wide variety of vehicular devices in a full-scale laboratory environment. One particular test involves an accurate measurement of the displacement x of the vehicle as a function of time τ. This information is then used to determine the velocity V, the acceleration A, and the rate of change of acceleration F as functions of time. In a given experiment, the displacement x was measured over a time range of 0 s to 10 s, at steps of 0.1 s. Some of the results obtained are as follows:

τ (s)	0.0	0.1	0.2	0.3	0.4	0.5
x (m)	0.0	0.8733	1.8224	2.8611	4.0032	5.2625
τ (s)		0.6	0.8	1.0	1.2	
x (m)		6.6528	9.8816	13.80	18.5184	

From these data, compute V, A, and F at $\tau = 0$ s, employing forward differences, and at $\tau = 0.3$, employing central differences, with a step size $\Delta\tau$ of 0.1 s. Repeat these calculations for $\tau = 0$ s and $\tau = 0.6$ s, with a step size $\Delta\tau$ of 0.2 s.

Solution

The velocity V, the acceleration A, and the rate of change of acceleration F are given in terms of the displacement x and time τ by

$$V = \frac{dx}{d\tau} \qquad A = \frac{d^2x}{d\tau^2} \qquad F = \frac{d^3x}{d\tau^3} \tag{3.2.1}$$

Since measurements are available only for $\tau \geqslant 0$, the values at $\tau = 0$ s can be computed only by forward differences. At $\tau = 0.3$ s, central differences can be employed with a step size of 0.1 s, and at $\tau = 0.6$ s, central differences with a step size of 0.2 s can be employed, according to the data given.

Various orders of approximation may be considered. The formulas needed for forward differences of $O(\Delta\tau)$ and $O[(\Delta\tau)^2]$ are given in Fig. 3.5. In addition, formulas of $O[(\Delta\tau)^3]$ and $O[(\Delta\tau)^4]$ may be obtained, where τ is the independent variable, instead of x, in Figs. 3.5 through 3.7. For the first derivative, Eqs. (3.34) and (3.35) give the formulas for backward differences. Similarly, for forward differences, the first derivative may be approximated, for a function $f(\tau)$, by

$$f_i' = \frac{1}{6\,\Delta x}(2f_{i+3} - 9f_{i+2} + 18f_{i+1} - 11f_i) + O[(\Delta\tau)^3] \tag{3.2.2}$$

$$f_i' = \frac{1}{12\,\Delta x}(-3f_{i+4} + 16f_{i+3} - 36f_{i+2} + 48f_{i+1} - 25f_i) + O[(\Delta\tau)^4] \tag{3.2.3}$$

These two formulas are employed, in addition to those given in Fig. 3.5, in order to demonstrate the effect of higher-order forward difference approximations on the numerical results for the velocity V. It is seen from Eq. (3.2.3) that five points, including the one at which the derivative is sought, are needed in the forward direction to obtain an accuracy of $O[(\Delta\tau)^4]$. For computing A and F by forward differences, only formulas of $O(\Delta\tau)$ and $O[(\Delta\tau)^2]$ are used. The central differencing formulas of $O[(\Delta\tau)^2]$ and $O[(\Delta\tau)^4]$ are given in Fig. 3.7. These may be employed for the computation of the first, second, and third derivatives needed in the present case.

Figure 3.2.1 shows the computer program in FORTRAN for solving this problem. The time τ at which the derivatives are to be computed is entered in terms of the integer variable I, where $I = 1$ at $\tau = 0$ s. Also, I is taken as 4 at $\tau = 0.3$ s, with $\Delta\tau = 0.1$ s, and at $\tau = 0.6$ s, with $\Delta\tau = 0.2$ s. The step size $\Delta\tau$ is also entered interactively, as are the data values needed for the computation. For forward differences, the values of x at five points, I, $I + 1, \ldots, I + 4$, are to be entered. Similarly, for central differences, the values of x at six points, $I - 3$, $I - 2, \ldots, I, \ldots,$ $I + 3$, are needed. The program first employs forward differencing to compute the

```
C     COMPUTATION OF DERIVATIVES BY FORWARD AND CENTRAL DIFFERENCES
C
C     V REPRESENTS THE VELOCITY, A THE ACCELERATION AND F THE RATE OF
C     CHANGE OF ACCELERATION. THE NUMBERS AFTER V,A AND F INDICATE THE
C     ORDER OF THE APPROXIMATION AND THE NUMBERS WITHIN PARENTHESES THE
C     INDEX I THAT LABELS THE TIME AT WHICH THE QUANTITY IS COMPUTED.
C     X IS THE DISPLACEMENT, T THE TIME AND DT THE TIME STEP.
C     N AND M INDICATE NUMBER OF SETS OF DATA FOR WHICH THE DERIVATIVES
C     ARE CALCULATED.
C
C
      IMPLICIT REAL (A-H,O-Z)
      DIMENSION X(7),V1(7),V2(7),V3(7),V4(7),A1(7),A2(7),A4(7)
      DIMENSION F1(7),F2(7),F4(7)
      PRINT *,'ENTER THE VALUE OF N'
      READ *,N
      DO 2 J=1,N
C
C     INPUT OF DATA
C
      PRINT *,'INPUT DATA:'
      PRINT *,'ENTER THE VALUES OF I, T AND DT'
      READ *,I,T,DT
      PRINT *,'ENTER THE MEASURED VALUES OF X(I) TO X(I+4)'
      READ *,X(I),X(I+1),X(I+2),X(I+3),X(I+4)
      PRINT *,' '
      PRINT *,'CALCULATED RESULTS:'
      WRITE(1,7)T,DT
 7    FORMAT(/4X,'TIME=',F9.4,6X,'TIME STEP=',F9.4)
C
C     CALCULATE DERIVATIVES FROM FORWARD DIFFERENCE APPROXIMATIONS
C
      V1(I)=(X(I+1)-X(I))/DT
      V2(I)=(-X(I+2)+4.0*X(I+1)-3.0*X(I))/(2.0*DT)
      V3(I)=(2.0*X(I+3)-9.0*X(I+2)+18.0*X(I+1)-11.0*X(I))/(6.0*DT)
      V4(I)=(-3.0*X(I+4)+16.0*X(I+3)-36.0*X(I+2)+48.0*X(I+1)
     $ -25*X(I))/(12.0*DT)
      A1(I)=(X(I+2)-2.0*X(I+1)+X(I))/(DT**2)
      A2(I)=(-X(I+3)+4.0*X(I+2)-5.0*X(I+1)+2.0*X(I))/(DT**2)
      F1(I)=(X(I+3)-3.0*X(I+2)+3.0*X(I+1)-X(I))/(DT**3)
      F2(I)=(-3.0*X(I+4)+14.0*X(I+3)-24,0*X(I+2)+18.0*X(I+1)
     $ -5.0*X(I))/(2.0*DT**3)
      WRITE (1,1) V1(I),V2(I),V3(I),V4(I)
 1    FORMAT('V1=',F9.4,4X,'V2=',F9.4,4X,'V3=',F9.4,4X,'V4=',F9.4)
      WRITE(1,3)A1(I),A2(I),F1(I),F2(I)
 3    FORMAT('A1=',F9.4,4X,'A2=',F9.4,4X,'F1=',F9.4,4X,'F2=',F9.4//)
 2    CONTINUE
C
C     INPUT DATA FOR CENTRAL DIFFERENCING
C
      PRINT *,'ENTER THE VALUE OF M'
      READ *,M
      DO 4 J=1,M
      PRINT *,'INPUT DATA:'
      PRINT *,'ENTER THE VALUES OF I, T AND DT'
      READ *,I,T,DT
```

Figure 3.2.1 Computer program for calculating the first, second, and third derivatives of the displacement x, using the data given in Example 3.2 and finite difference approximations of various accuracy.

```
      PRINT *,'ENTER THE VALUES OF X(I-3) TO X(I+3)'
      READ *,X(I-3),X(I-2),X(I-1),X(I),X(I+1),X(I+2),X(I+3)
      PRINT *,' '
      PRINT *,'CALCULATED RESULTS:'
      WRITE(1,8)T,DT
 8    FORMAT(/4X,'TIME=',F9.4,6X,'TIME STEP=',F9.4)
C
C     CALCULATE DERIVATIVES FROM CENTRAL DIFFERENCE APPROXIMATIONS
C
      V2(I)=(X(I+1)-X(I-1))/(2.0*DT)
      V4(I)=(-X(I+2)+8.0*X(I+1)-8.0*X(I-1)+X(I-2))/(12.0*DT)
      A2(I)=(X(I+1)-2.0*X(I)+X(I-1))/(DT**2)
      A4(I)=(-X(I+2)+16.0*X(I+1)-30.0*X(I)+16.0*X(I-1)-X(I-2))/
     $ (12.0*DT**2)
      F2(I)=(X(I+2)-2.0*X(I+1)+2.0*X(I-1)-X(I-2))/(2.0*DT**3)
      F4(I)=(-X(I+3)+8.0*X(I+2)-13.0*X(I+1)+13.0*X(I-1)-8.0*X(I-2)
     $ +X(I-3))/(8.0*DT**3)
      WRITE(1,5)V2(I),A2(I),F2(I)
      WRITE(1,6)V4(I),A4(I),F4(I)
 5    FORMAT('V2=',F9.4,4X,'A2=',F9.4,4X,'F2=',F9.4)
 6    FORMAT('V4=',F9.4,4X,'A4=',F9.4,4X,'F4=',F9.4//)
 4    CONTINUE
      STOP
      END
```

Figure 3.2.1 Continued

derivatives and then central differencing. The integer variables N and M are employed to indicate the number of sets of such computations to be performed; for example, in the present problem two sets each of forward and central difference computations are to be carried out.

The numerical results obtained are shown in Fig. 3.2.2. First considering forward difference results, note that the velocity V converges to 8.4 m/s as the order of the approximation is increased. A considerable error is observed at the first-order approximation, particularly for the larger $\Delta\tau$ (0.2 s), as expected. However, the third-order approximation is adequate for this problem, since essentially no change is observed by going to the fourth-order approximation. The acceleration A is given as 6.2 m/s^2 by the second-order approximation at both the mesh sizes considered. Again, the first-order approximation is in considerable error, particularly at $\Delta\tau = 0.2$ s. The computed value of F is found to be 13.8 m/s^3 at $\Delta\tau = 0.2$ s, although a variation is observed from the first-order to the second-order approximation at $\Delta\tau = 0.1$ s. In this problem, F is a constant at 13.8 m/s^3, as illustrated by the remaining results, discussed below. Thus, a higher-order approximation will not improve the accuracy if the function being considered is a polynomial of lower order. Of course, for an arbitrary function, accuracy is generally improved by employing a higher-order approximation.

The results from central differencing indicate only small changes from the second-order to the fourth-order approximations. Thus, the second-order formulas are adequate for this problem, as is often the case in most engineering problems. At $\tau = 0.3$ s, V, A, and F are obtained as 10.881 m/s, 10.34 m/s^2, and 13.8 m/s^3,

FORWARD DIFFERENCES

```
 ENTER THE VALUE OF N
2
 INPUT DATA:
 ENTER THE VALUES OF I, T AND DT
0  0.0  0.1
 ENTER THE MEASURED VALUES OF X(I) TO X(I+4)
0.0  0.8733  1.8224  2.8611  4.0032
 CALCULATED RESULTS:

    TIME=   0.0000        TIME STEP=   0.1000
V1=   8.7330    V2=   8.3540    V3=   8.4000    V4=   8.4000
A1=   7.5800    A2=   6.2000    F1=  13.7998    F2=  13.7997

 INPUT DATA:
 ENTER THE VALUES OF I, T AND DT
0  0.0  0.2
 ENTER THE MEASURED VALUES OF X(I) TO X(I+4)
0.0  1.8224  4.0032  6.6528  9.8816
 CALCULATED RESULTS:

    TIME=   0.0000        TIME STEP=   0.2000
V1=   9.1120    V2=   8.2160    V3=   8.4000    V4=   8.4000
A1=   8.9600    A2=   6.2000    F1=  13.8001    F2=  13.8001
```

CENTRAL DIFFERENCES

```
 ENTER THE VALUE OF M
2
 INPUT DATA:
 ENTER THE VALUES OF I, T AND DT
4  0.3  0.1
 ENTER THE VALUES OF X(I-3) TO X(I+3)
0.0  0.8733  1.8224  2.8611  4.0032  5.2625  6.6528
 CALCULATED RESULTS:

    TIME=   0.3000        TIME STEP=   0.1000
V2=  10.9040    A2=  10.3399    F2=  13.8004
V4=  10.8810    A4=  10.3399    F4=  13.8004

 INPUT DATA:
 ENTER THE VALUES OF I, T AND DT
4  0.6  0.2
 ENTER THE VALUES OF X(I-3) TO X(I+3)
0.0  1.8224  4.0032  6.6528  9.8816  13.8  18.5184

 CALCULATED RESULTS:

    TIME=   0.6000        TIME STEP=   0.2000
V2=  14.6960    A2=  14.4800    F2=  13.8000
V4=  14.6040    A4=  14.4799    F4=  13.7999
```

Figure 3.2.2 Numerical results obtained for the problem given in Example 3.2.

respectively. Similarly, at $\tau = 0.6$ s, V, A, and F are obtained as 14.604 m/s, 14.48 m/s^2, and 13.8 m/s^3, respectively. Again, the second-order approximations are found to be adequate.

Example 3.2 has illustrated the use of numerical differentiation in a practical circumstance. The displacement x can generally be measured very accurately as a function of time τ, and finite difference formulas can then be employed to yield velocity, acceleration, and so on. Forward and backward differences are generally used only at the start and the termination of the measurements, central differences being appropriate for other times. Although higher-order approximations may be used, second-order formulas often yield satisfactory accuracy in most problems of engineering interest.

3.5 POLYNOMIAL REPRESENTATION

Another frequently employed approach for the derivation of the finite difference approximations of the derivatives of a given function $f(x)$ is based on a polynomial fit to the values at the grid points. Depending on the order of the derivative whose approximation is to be obtained and the desired accuracy, the order of the polynomial may be chosen. For an nth-order polynomial, $(n + 1)$ grid points are needed to evaluate all the coefficients that appear in the polynomial. Curve fitting is discussed in detail in Chapter 6, and only a few simple aspects are employed here.

By way of illustration, let us consider fitting a second-order polynomial to the three grid points shown in Fig. 3.8. The function $f(x)$ is taken as

$$f(x) = A_0 + A_1 x + A_2 x^2 \tag{3.37}$$

Fitting this parabola to the three points yields

$$f_i = A_0 + A_1 x_i + A_2 x_i^2 \tag{3.38a}$$

$$f_{i+1} = A_0 + A_1(x_i + \Delta x) + A_2(x_i + \Delta x)^2 \tag{3.38b}$$

$$f_{i+2} = A_0 + A_1(x_i + 2\,\Delta x) + A_2(x_i + 2\,\Delta x)^2 \tag{3.38c}$$

From these equations, the coefficients A_0, A_1, and A_2 may be determined. The first and second derivatives of the function are given by

$$f_i' = A_1 + 2A_2 x_i \tag{3.39}$$

$$f_i'' = 2A_2 \tag{3.40}$$

Since the finite difference expression should be independent of the absolute location of the points and should depend only on the relative positions of the grid points, any arbitrary value of x_i may be taken. The algebra is simplified if x_i is taken as 0 so that $x_{i+1} = \Delta x$ and $x_{i+2} = 2\,\Delta x$. However, the resulting expressions for f_i' and f_i''

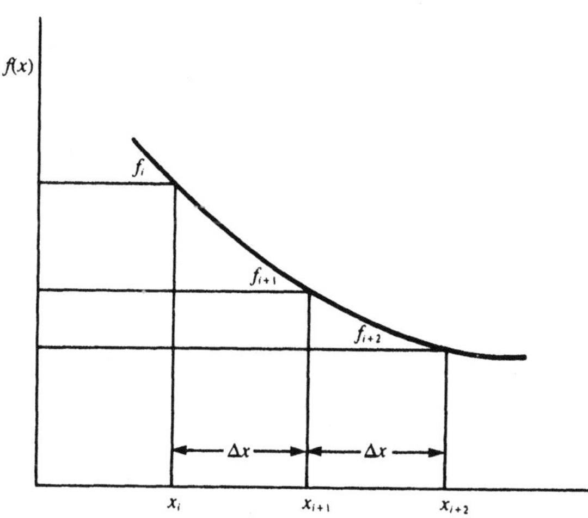

Figure 3.8 Uniform one-sided distribution of grid points used for illustrating the derivation of the finite difference approximations of the first and second derivatives by means of a second-order polynomial representation of the function $f(x)$.

are the same whatever the value of x_i. These expressions are obtained from Eqs. (3.38) through (3.40) as follows:

$$f_i' = \frac{-f_{i+2} + 4f_{i+1} - 3f_i}{2\,\Delta x} \tag{3.41}$$

and

$$f_i'' = \frac{f_{i+2} - 2f_{i+1} + f_i}{(\Delta x)^2} \tag{3.42}$$

The approximations are identical to those derived earlier from the Taylor series; see Eqs. (3.27) and (3.32). The first expression was shown to have a truncation error of $O[(\Delta x)^2]$ and the second one of $O(\Delta x)$. Both the equations give forward difference approximations because of the chosen grid points, which are on one side of $x = x_i$ in the direction of increasing x.

The error term is not explicitly given by this approach. However, since the third derivative $f'''(x)$ is zero for the second-order polynomial chosen, the order of magnitude of the error may be estimated from the discussion on the Taylor series approach. The third derivative being zero implies that the expression for f_i'' would be of $O(\Delta x)$ and that for f_i' of $O[(\Delta x)^2]$. For an accurate evaluation of the error, one must resort to the Taylor series approach.

The polynomial representation is particularly useful in the derivation of finite difference expressions for grid points that are located at nonuniform distances from each other. For instance, consider the distribution shown in Fig. 3.9. Taking $x_i = 0$, $x_{i+1} = \Delta x$, and $x_{i+2} = 3\,\Delta x$, the parabola of Eq. (3.37) gives

$$f_i = A_0 \tag{3.43a}$$

$$f_{i+1} = A_0 + A_1\,\Delta x + A_2(\Delta x)^2 \tag{3.43b}$$

$$f_{i+2} = A_0 + A_1(3\,\Delta x) + A_2(3\,\Delta x)^2 \tag{3.43c}$$

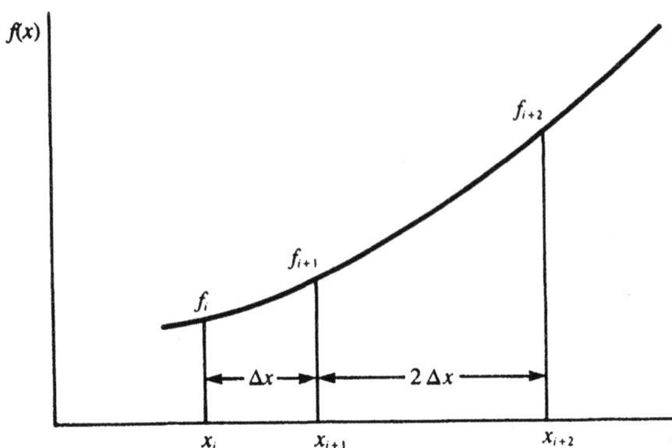

Figure 3.9 Nonuniform distribution of grid points employed for illustrating the use of a polynomial representation of $f(x)$ to derive the finite difference approximations of the derivatives.

with

$$f'(0) = A_1 \quad \text{and} \quad f''(0) = 2A_2 \tag{3.44}$$

From these equations, we obtain the finite difference expressions for f_i' and f_i' as follows:

$$f_i' = \frac{-f_{i+2} + 9f_{i+2} - 8f_i}{6\,\Delta x} \tag{3.45}$$

and

$$f_i'' = \frac{f_{i+2} - 3f_{i+1} + 2f_i}{3(\Delta x)^2} \tag{3.46}$$

Similarly, expressions for other arbitrary distributions of grid points may be derived. This approach is frequently employed for determining the derivatives from experimental data. Examples of such data are those pertaining to the variation of material properties and of physical quantities like pressure and density with an independent variable, such as temperature, while other independent variables are held constant. Such data are generally available at nonuniformly distributed values of the independent variable, and the polynomial approach provides a simple method for computing the derivatives.

3.6 PARTIAL DERIVATIVES

In the preceding sections, we have considered the numerical differentiation of an arbitrary function $f(x)$ that depends on a single independent variable x. The finite difference representations of the ordinary, or full, derivatives of the function were derived by considering the variation with x. However, in engineering problems, we frequently encounter circumstances where the dependent variable is a function of two or more independent variables. In such cases, partial derivatives arise, and the finite difference approximations of these are of interest. Since a partial derivative is defined in terms of the variation of the function with a given independent variable while the others are held constant, the finite difference approximations are completely analogous to those for the ordinary derivatives.

Consider, for instance, a function $f(x,y)$. Then the partial first derivatives of the function are $(\partial f/\partial x)$ and $(\partial f/\partial y)$, where y is kept constant in the first case and x in the second. The variables that are held constant for a particular differentiation are sometimes indicated by means of subscripts as, for instance, $(\partial f/\partial x)_y$ and $(\partial f/\partial y)_x$. However, it is understood that for the partial differentiation $\partial f/\partial x$, only the variation of $f(x,y)$ with x is under consideration, y being kept unchanged. Since two independent variables, x and y, are involved, a location in the computational domain is represented by two subscripts, instead of only one needed for ordinary derivatives. Thus, the value of the function at a grid point represented by indices (i,j) may be denoted as $f_{i,j}$, where $x = i\,\Delta x$ and $y = j\,\Delta y$, as shown in Fig. 3.10. Such a grid is employed in the solution of partial differential equations by finite difference methods, as discussed in Chapter 9.

Considering the variation of the function $f(x,y)$ with x alone, we may write the finite difference approximations of the first and second derivatives, in a manner analogous to that outlined in Section 3.3:

$$\left(\frac{\partial f}{\partial x}\right)_{i,j} \simeq \frac{f_{i+1,j} - f_{i,j}}{\Delta x} \quad \text{forward difference} \tag{3.47}$$

$$\left(\frac{\partial f}{\partial x}\right)_{i,j} \simeq \frac{f_{i,j} - f_{i-1,j}}{\Delta x} \quad \text{backward difference} \tag{3.48}$$

$$\left(\frac{\partial f}{\partial x}\right)_{i,j} \simeq \frac{f_{i+1,j} - f_{i-1,j}}{2\,\Delta x} \quad \text{central difference} \tag{3.49}$$

$$\left(\frac{\partial^2 f}{\partial x^2}\right)_{i,j} \simeq \frac{f_{i+1,j} - 2f_{i,j} + f_{i-1,j}}{(\Delta x)^2} \quad \text{second central difference} \tag{3.50}$$

Therefore, the subscript j is not varied in these expressions. Its presence indicates that the function also depends on another independent variable.

Similarly, the partial derivatives with respect to y may be obtained. Thus, the central differences yield

$$\left(\frac{\partial f}{\partial y}\right)_{i,j} \simeq \frac{f_{i,j+1} - f_{i,j-1}}{2\,\Delta y} \tag{3.51}$$

$$\left(\frac{\partial^2 f}{\partial y^2}\right)_{i,j} \simeq \frac{f_{i,j+1} - 2f_{i,j} + f_{i,j-1}}{(\Delta y)^2} \tag{3.52}$$

In this case, the subscript i is not varied. Thus, all the expressions derived in the preceding sections may easily be extended to partial derivatives.

The derivation of the finite difference approximations for partial derivatives may, again, be based on direct approximation, the Taylor series, or polynomial

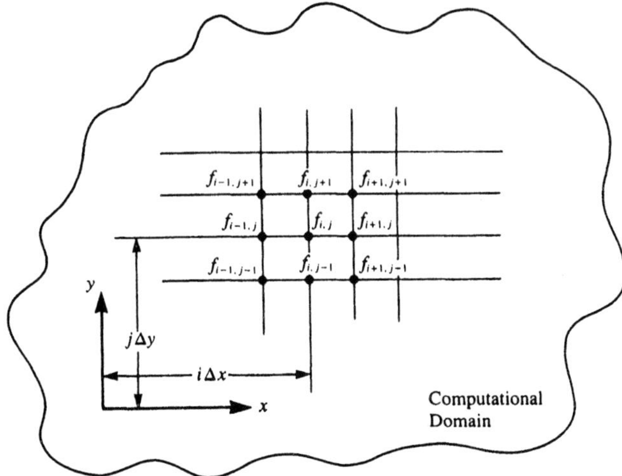

Figure 3.10 A two-dimensional grid indicating the finite number of locations at which the function $f(x,y)$ is evaluated in the computational domain.

representation. The Taylor series expansions about the point (i,j) may be written as

$$f_{i\pm 1,j} = f_{i,j} \pm \Delta x\left(\frac{\partial f}{\partial x}\right)_{i,j} + \frac{(\Delta x)^2}{2!}\left(\frac{\partial^2 f}{\partial x^2}\right)_{i,j} \pm \frac{(\Delta x)^3}{3!}\left(\frac{\partial^3 f}{\partial x^3}\right)_{i,j} + \cdots \tag{3.53}$$

$$f_{i,j\pm 1} = f_{i,j} \pm \Delta y\left(\frac{\partial f}{\partial y}\right)_{i,j} + \frac{(\Delta y)^2}{2!}\left(\frac{\partial^2 f}{\partial y^2}\right)_{i,j} \pm \frac{(\Delta y)^3}{3!}\left(\frac{\partial^3 f}{\partial y^3}\right)_{i,j} + \cdots \tag{3.54}$$

The remainder terms may also be obtained for truncation after a finite number of terms, as outlined earlier for ordinary derivatives. The remainder term for truncation after the term containing the mth power of Δx, that is, after $(m + 1)$ terms in Eq. (3.53) is

$$R_{m,x} = (-1)^{m+1}\frac{(\Delta x)^{m+1}}{(m+1)!}\left(\frac{\partial^{m+1} f}{\partial x^{m+1}}\right)_{\xi,j} \qquad \text{where } i < \xi < i + 1 \tag{3.55}$$

Similarly, the remainder term $R_{m,y}$ is obtained as $O[(\Delta y)^{m+1}]$ for truncation after $(m + 1)$ terms in Eq. (3.54). The total remainder term is the sum of $R_{m,x}$ and $R_{m,y}$, as discussed in Chapter 9.

Thus, the finite difference approximations may be derived from the Taylor series expansions, as given earlier. Sometimes cross derivatives such as $\partial^2 f/\partial x\,\partial y$ may have to be evaluated. The corresponding finite difference representations may be derived by applying the approximation twice for the two differentiations with different independent variables. A two-variable Taylor series expansion for $f(x,y)$ may also be employed for the purpose, as outlined by Jaluria and Torrance (1986).

Partial derivatives are of interest in many important engineering applications, such as those that involve fluid flow, heat and mass transfer, thermodynamics, chemical reactions, structural vibrations, and electrical fields. Obviously, these topics encompass a wide range of engineering problems, extending from aerospace and environmental problems to nuclear and chemical reactors and power plants. Partial differential equations, which are discussed in detail in Chapter 9, govern such physical phenomena. The finite difference approximations of the partial derivatives are then employed for developing the numerical procedure for solving these problems.

Example 3.3

Planck's law for blackbody radiation is given as

$$E_{b\lambda}(\lambda,T) = \frac{c_1}{\lambda^5[e^{c_2/\lambda T} - 1]} \tag{3.3.1}$$

where $c_1 = 3.7413 \times 10^8$ W·μm^4/m^2, $c_2 = 1.4388 \times 10^4$ μm·K, λ is wavelength in μm, and T is temperature in K. $E_{b\lambda}$ is a function of λ and T, and is known as the *monochromatic emissive power* of a blackbody (see Fig. 1.3). Numerically determine $\partial E_{b\lambda}/\partial\lambda$ and $\partial E_{b\lambda}/\partial T$ at $\lambda = 4\ \mu$m and $T = 1600$ K. Repeat the calculation for $\lambda = 2\ \mu$m and $T = 1000$ K. Use different values of $\Delta\lambda$ and ΔT to ensure accuracy of your results, employing the second-order formula for the derivative. Compare your results with analytical ones.

Solution

Energy transfer by radiation is of importance in several areas of engineering and physical sciences. Planck's law is of considerable value in the calculations for energy transfer since it gives the characteristics and magnitude of energy lost by an idealized surface, termed *blackbody*, as functions of the temperature T and wavelength λ. Our interest here lies in numerically evaluating the rate of change of the emissive power $E_{b\lambda}$ with these two independent variables. The results are to be obtained at the two sets of values given for λ and T. Also, the step sizes $\Delta\lambda$ and ΔT are to be varied so that we may study their effect on the results and choose the most appropriate values, as outlined earlier in Section 2.5.

The problem is fairly straightforward, and Fig. 3.3.1 gives the computer program in FORTRAN for obtaining the required numerical results. The function $E_{b\lambda}(\lambda,T)$ is defined and the constants c_1 and c_2 are specified. The input values for λ and T, at which the gradients $\partial E_{b\lambda}/\partial\lambda$ and $\partial E_{b\lambda}/\partial T$ are to be determined, are entered interactively into the computer. Although the program is written for two sets of data, more sets can easily be accommodated by changing the condition on N for termination of the computation. The starting values of the step sizes $\Delta\lambda$ and ΔT are taken as 1 μm and 500 K. These are successively halved in the program, and the corresponding derivatives are computed by the following central differencing

```
C     COMPUTATION OF PARTIAL DERIVATIVES BY CENTRAL DIFFERENCE APPROXIMATION
C
C     EBL REPRESENTS THE RADIATION INTENSITY OF A BLACKBODY, X IS THE
C     WAVELENGTH, T IS THE ABSOLUTE TEMPERATURE, C1 AND C2 ARE
C     CONSTANTS, DX AND DT ARE THE STEP SIZES IN WAVELENGTH AND
C     TEMPERATURE, AND DXEBL AND DTEBL ARE THE CORRESPONDING
C     DERIVATIVES
C
      IMPLICIT REAL (A-H,O-Z)
C
C     PLANCK'S EQUATION FOR BLACKBODY RADIATION
C
      EBL(X,T)=C1/((X**5)*(EXP(C2/(X*T))-1.0))
C
C     INITIALIZING VARIABLES
C
      C1=3.7413*10.0**8.0
      C2=1.4388*10.0**4.0
      PRINT *,'ENTER NUMBER OF POINTS WHERE DERIVATIVES ARE NEEDED'
      PRINT *,'M='
      READ *,M
      DO 6 N=1,M
      PRINT *,' '
 5    PRINT *,'INPUT VALUES'
      PRINT *,'X=','        ','T='
      READ *,X,T
      WRITE (1,1)X,T
 1    FORMAT(6X,'WAVELENGTH=',F5.2,6X,'TEMPERATURE=',F8.2)
      WRITE(1,4)
 4    FORMAT (7X,'DX',11X,'DE/DX',11X,'DT',14X,'DE/DT')
C
C     A DO LOOP SUCCESSIVELY HALVES THE STEP SIZES DX AND DT
C
      DO 2 I=0,10,1
      DX=1.0/(2.0**I)
      DT=500.0/(2.0**I)
C
C     DETERMINATION OF THE DERIVATIVES BY CENTRAL DIFFERENCING
C
      DXEBL=(EBL(X+DX,T)-EBL(X-DX,T))/(2.0*DX)
      DTEBL=(EBL(X,T+DT)-EBL(X,T-DT))/(2.0*DT)
      WRITE (1,3)DX,DXEBL,DT,DTEBL
 3    FORMAT(3X,F8.5,6X,F10.2,4X,F10.3,6X,F10.2)
 2    CONTINUE
 6    CONTINUE
      IF (N .LE. 2) GO TO 5
      STOP
      END
```

Figure 3.3.1 Computer program for evaluating the partial derivatives of the function $E_{b\lambda}(\lambda,T)$, given in Example 3.3, by using central finite difference approximations.

formulas:

$$\frac{\partial f}{\partial x} = \frac{f(x + \Delta x, y) - f(x - \Delta x, y)}{2\,\Delta x} \tag{3.3.2a}$$

$$\frac{\partial f}{\partial y} = \frac{f(x, y + \Delta y) - f(x, y - \Delta y)}{2\,\Delta y} \tag{3.3.2b}$$

where $f(x,y)$ is a given function of x and y. The various symbols employed in the computer program are defined in Fig. 3.3.1.

The numerical results obtained are presented in Fig. 3.3.2. The wavelength step size $\Delta\lambda$ is varied from 1.0 μm to 0.00098 μm, and the temperature step size ΔT from

```
 ENTER NUMBER OF POINTS WHERE DERIVATIVES ARE NEEDED
 M=
2

 INPUT VALUES
 X=        T=
4.0       1600.0
       WAVELENGTH= 4.00          TEMPERATURE= 1600.00
       DX              DE/DX                DT                DE/DT
    1.00000         -28565.48            500.000             65.96
    0.50000         -27287.71            250.000             67.32
    0.25000         -26934.26            125.000             67.65
    0.12500         -26844.16             62.500             67.74
    0.06250         -26821.57             31.250             67.76
    0.03125         -26815.87             15.625             67.76
    0.01562         -26814.12              7.812             67.77
    0.00781         -26814.11              3.906             67.76
    0.00391         -26815.02              1.953             67.76
    0.00195         -26815.08              0.977             67.76
    0.00098         -26812.19              0.488             67.76

 INPUT VALUES
 X=        T=
2.0       1000.0
       WAVELENGTH= 2.00          TEMPERATURE= 1000.00
       DX              DE/DX                DT                DE/DT
    1.00000           6308.24            500.000             97.40
    0.50000           8804.91            250.000             72.68
    0.25000           9456.30            125.000             65.68
    0.12500           9612.72             62.500             63.87
    0.06250           9651.27             31.250             63.42
    0.03125           9660.86             15.625             63.31
    0.01562           9663.27              7.812             63.28
    0.00781           9663.95              3.906             63.27
    0.00391           9663.89              1.953             63.27
    0.00195           9663.66              0.977             63.27
    0.00098           9663.03              0.488             63.27
```

Figure 3.3.2 Numerical results obtained for Example 3.3, indicating the dependence of the computed derivatives on the step sizes $\Delta\lambda$ and ΔT.

500 K to 0.488 K. As mentioned above, the step sizes are halved in each successive computation. Note that $\partial E_{b\lambda}/\partial T$ approaches a constant value, as ΔT is reduced, much more rapidly than $\partial E_{b\lambda}/\partial \lambda$, with reduction in $\Delta\lambda$. This indicates that the results are more sensitive to variations in λ, as is also evident from the λ^5 dependence in the denominator of the function $E_{b\lambda}$; see Eq. (3.3.1). From the numerical results presented in Fig. 3.3.2, $\partial E_{b\lambda}/\partial T$ is evaluated as 67.76 W/m$^2\cdot\mu$m$\cdot$K at $\lambda = 4\ \mu$m and $T = 1600$ K and as 63.27 W/m$^2\cdot\mu$m$\cdot$K at $\lambda = 2\ \mu$m and $T = 1000$ K. For $\partial E_{b\lambda}/\partial \lambda$, the computed value increases monotonically, from the starting value of $\Delta\lambda$, as $\Delta\lambda$ is halved until $\Delta\lambda = 0.00781\ \mu$m. Then it starts to decrease or oscillate. This effect is probably due to the appearance of significant round-off error at these small values of $\Delta\lambda$. Thus, choosing the computed values at $\Delta\lambda = 0.00781$ m, we evaluate $\partial E_{b\lambda}/\partial \lambda$ as $-26{,}814.11$ W/m$^2\cdot\mu$m^2 at $\lambda = 4\ \mu$m and $T = 1600$ K and as 9663.95 W/m$^2\cdot\mu$m^2 at $\lambda = 2\ \mu$m and $T = 1000$ K.

The corresponding analytical values may also be determined by differentiating $E_{b\lambda}$ successively with respect to the two independent variables λ and T. The given sets of input values may then be substituted into the mathematical expressions obtained. The analytical results thus obtained are as follows:

	$\partial E_{b\lambda}/\partial \lambda$	$\partial E_{b\lambda}/\partial T$
$\lambda = 4$, $T = 1600$:	$-26{,}813.970$	67.765
$\lambda = 2$, $T = 1000$:	9,664.069	63.268

Therefore, the numerical results obtained are very close to the analytical results at the chosen values of the step sizes. The error is within 0.007%. Obviously, at large values of the step sizes, the numerical results are in considerable error; see Fig. 3.3.2. An appropriate reduction in step sizes is, therefore, needed until the change in the numerical results is small.

3.7 SUMMARY

In this chapter, we consider the basic concepts underlying numerical differentiation and finite difference calculus. Three difference approaches, namely, direct approximation, the Taylor series, and polynomial representation, for the derivation of the finite difference approximations of the various derivatives of an arbitrary function $f(x)$ are discussed. The truncation error resulting from the retention of a finite number of terms in the Taylor series is considered in detail and is related to the accuracy of the various finite difference approximations. The general procedures for deriving finite difference expressions of higher accuracy and those for higher-order derivatives are outlined. The Taylor series approach is the preferred one since it also yields the error, which the other two methods, although relatively simpler to employ, do not. In addition, the Taylor series expansions may be conveniently employed to improve the accuracy of the finite difference approximation, if a higher level of accuracy is desired in a given application. The polynomial representation approach is particularly useful if a nonuniform distribution of grid points is employed. Finally, the chapter discusses partial derivatives and shows how the finite difference approximations may easily be obtained from those for ordinary, or full, derivatives.

PROBLEMS

3.1. Derive the Taylor series expansions for $\sin x$, $(1-x)^{-2}$, and e^{-x}, about $x = 0$. Are there any constraints on $|x|$ for the series to be convergent? Why does the series for $\log(1-x)$, derived in Example 3.1, converge only if $|x| < 1$?

3.2. Using the Taylor series for e^x and $\sin x$, obtain the series for $e^x \sin x$, about $x = 0$. Compute the value of $e^{0.2} \sin 0.2$ by a summation of the series, retaining terms so that the first neglected term is $O[(\Delta x)^4]$. Compare your result with the true value of the function $e^x \sin x$ at $x = 0.2$.

3.3. Show that the Taylor series expansion for x^5 about $x = 0$ is x^5 itself.

3.4. Calculate $e^{0.3}$ by employing the Taylor series for e^x. How many terms are needed if the error from the true value of the quantity $e^{0.3}$ is to be less than 0.01%?

3.5. The pressure p of a gas is given by the expression $\log p = 21.6 - 2420/T$, where T is the temperature in kelvins. Using the exact value of p from this expression, at $T = 400$ K, and the Taylor series expansion for $p(T)$, compute the pressures at 410, 420, and 450 K. Compare these values with the exact ones obtained from the given expression. Refer to Example 3.1(b).

3.6. Compute the value of $e^{\sin x}$ at $x = 0.25$, employing the corresponding Taylor series expansion, and compare the result with the exact value.

3.7. Consider the function $f(x) = 2x^{1/2} + 3x$. The derivatives of the function at $x = 0$ are all infinite, and therefore the Taylor series expansion about $x = 0$ cannot be obtained. Instead, obtain the series about $x = 0.1$ and also about $x = 0.2$. Compare the two and comment on the difference.

3.8. Compute the first and second derivatives of $\sin x$ and e^x at $x = 0$, employing forward and central differencing formulas of $O[(\Delta x)^2]$. Consider three values, 0.2, 0.1, and 0.01, of the step size Δx. Compare the numerical results obtained with the exact, mathematical values of the derivatives. Discuss the effect of Δx on the accuracy of the numerical results.

3.9. Calculate the numerical value of $d[(\sin x)^{e^x}]/dx$ at $x = 1$, using central difference approximations of $O[(\Delta x)^2]$. Start with $\Delta x = 0.2$ and reduce it until the numerical result remains essentially unaffected by a further reduction in Δx.

3.10. Consider the expressions for computing the present worth (PW) and future worth (FW) of a series of equal monthly payments, given by Eqs. (2.2.3) and (2.2.4), respectively. It is important for the economic planning of an engineering system to determine the effect of a change in the interest rate x on PW and FW. Compute the rate of change of these quantities with x, at $x = 10\%$, for $R = \$1000$ and $m = 240$ months. Also, refer to Example 2.2 for details on this problem. Employ an appropriate value of Δx, as discussed in Section 2.5.

3.11. In Problem 3.10, compute $d(\mathrm{FW})/dm$ and $d(\mathrm{PW})/dm$, where m is the number of months, at $x = 10\%$, $m = 240$, and $R = \$1000$.

3.12. The measured temperature distribution in a solar energy heating system may be represented by the equation

$$T(x) = 15.5 + 68.2\left[1 - \frac{\exp(-x/2.7)}{(1 + x^2)}\right]$$

where x is the distance away from the surface being heated and T is the temperature. Compute the temperature gradient dT/dx and the second derivative d^2T/dx^2 at $x = 0$. The heat transfer rate is proportional to the temperature gradient at $x = 0$. The second derivative is related to energy lost or gained by radiative transport. Employ forward differences to $O[(\Delta x)^2]$ and reduce Δx to obtain numerical results that are largely independent of the value of Δx chosen.

3.13. For the problem considered in Example 3.2, if, in addition to the data given, the displacements at $\tau = 0.7$ s and 0.9 s are given as 8.1879 m and 11.7477 m, respectively, compute the velocity and acceleration at $\tau = 0.5$ s, using forward, backward, and central differences of $O[(\Delta x)^2]$. Employ a step size $\Delta\tau$ of 0.1 s.

3.14. In a periodic mass transfer process in a chemical plant, the concentration C of the moisture is obtained from analysis as

$$C = 9.5[\exp(-0.75x)]\cos(\tau - 0.75x - 2)$$

where x is distance in meters, τ is the time in seconds, and C is in kg/m^3. The mass transfer rate is proportional to the gradient $\partial C/\partial x$ at $x = 0$. The second derivative is related to the rate of moisture addition per unit volume. Compute both $\partial C/\partial x$ and $\partial^2 C/\partial x^2$ at $\tau = 1$ s and $x = 0$, using forward differences. Start with $\Delta x = 0.1$ m and reduce it to 0.01 m to see whether there is any significant effect on the numerical results.

3.15. In fluid mechanics, a stream function ψ, which is related to the flow rate, is frequently employed for analysis. The distribution of ψ for a particular problem is obtained as

$$\psi = 1.1[\exp(-y^{1.3})]\cos\left(\frac{y^2}{3\pi}\right) + 0.9[\exp(-y^{0.7})]\sin\left(\frac{y^2}{3\pi}\right)$$

The velocity V is given by $d\psi/dy$, and the shear force generated by the flow is proportional to $d^2\psi/dy^2$. Compute both these derivatives at $y = 0.5$, with $\Delta y = 0.2$, 0.1, and 0.05. Employ second-order central differences.

3.16. It is known from analysis that the distribution of $E_{b\lambda}$, given in Example 3.3, has a maximum at $\lambda T = 2897.6$ μm·K. Confirm this by using numerical differentiation to obtain the first and second derivatives, with respect to λ, at $T = 1000$ K and $\lambda T = 2897.6$ μm·K. Remember that, at a maximum, the first derivative is zero and the second derivative is negative.

3.17. The hot-wire anemometer is an instrument used for measuring velocities or temperatures. If, during its calibration, the output signal E is measured as 0, 1.7, 3.3, and 5.6 volts at velocities V of 0, 1, 1.5, and 2 m/s, obtain the gradient dE/dV at $V = 0$ m/s, using the polynomial representation of the function $E(V)$.

3.18. Obtain the finite difference approximation for df/dx, where $f(x)$ is a given function of x, using four uniformly spaced grid points and the polynomial representation of $f(x)$.

4

Roots of Equations

4.1 INTRODUCTION

In a wide variety of engineering problems, we are frequently faced with the task of determining the values of the variable x that would satisfy a given algebraic equation, such as $x^3 - 4x^2 + 5x = 2$, or $x \tan x = 1$. Depending on the problem, defined by the equation and the range of x under consideration, these values of x, which are termed as *roots* of the equation, may be real or complex and may be finite or infinite in number. Root solving is needed, for instance, in determining the terminal velocity of a falling body, the concentration of a diffusing species at a surface subjected to mass transfer, the time needed to repay a loan at a given interest rate and with a fixed monthly payment, and the natural frequencies of vibration of a beam.

The algebraic equation to be solved is represented by the general form

$$f(x) = 0 \tag{4.1}$$

where the function $f(x)$ may designate a polynomial or a transcendental expression, such as $x \tan x - 1$, given above. The problem of finding the roots of Eq. (4.1), therefore, involves obtaining the values of x for which the function $f(x)$ is zero. Consequently, the roots are also often referred to as *zeros* of the function. Although interest usually lies in determining the real roots of equations with real coefficients, we do encounter problems, such as periodic processes, in which complex roots are of interest or in which the equation has complex coefficients. Therefore, the discussion in this chapter is initially directed at obtaining the real roots of equations with real coefficients, the other circumstances being considered at a later state in the chapter.

There are several methods available for finding the roots of algebraic equations. Some of these are applicable only to polynomial equations, which are obtained when $f(x)$ represents a polynomial to yield an equation of the form

$$f(x) = x^n + a_1 x^{n-1} + a_2 x^{n-2} + \cdots + a_{n-1} x + a_n = 0 \tag{4.2}$$

where n is the degree of the polynomial equation and $a_1, a_2, \ldots, a_n$ are real coefficients. This equation has n roots, which may be real or complex. Some of the real roots may be equal. Also, the complex roots occur in conjugate pairs; that is, a complex root $a + ib$, where a and b are real and $i = \sqrt{-1}$, will occur in conjunction with another root $a - ib$. For a linear equation, $n = 1$, the root may be found directly as $-a_1/a_0$ from the equation $a_1 + a_0x = 0$. For a quadratic equation, $n = 2$, the roots may again be determined by using the following well-known expression for the two roots, α_1 and α_2:

$$\alpha_1, \alpha_2 = \frac{-b \pm \sqrt{b^2 - 4ac}}{2a} \tag{4.3}$$

where $f(x) = ax^2 + bx + c = 0$ is the quadratic equation. Depending on whether the determinant $(b^2 - 4ac)$ is positive, zero, or negative, the roots are, respectively, real, equal, or complex. In a few cases, such formulas are available for higher-order equations too. Although these formulas provide a direct analytical method for finding the exact solution, they are usually very restrictive in their applicability and are also often quite complicated. Therefore, it is generally easier or necessary to use indirect, or iterative, methods to find the roots of a nonlinear equation numerically. Transcendental equations, which involve trigonometric and other special functions such as exponentials and logarithms, also arise in engineering problems. These equations are also generally nonlinear, and the number of roots is often unknown. Several of the methods considered in this chapter are applicable to both polynomial and transcendental equations.

In many cases of practical interest, the approximate variation of the function $f(x)$ with x, the nature of the roots, and the interval over which these are to be determined are known. For instance, if the surface temperature of a pond, resulting from the various heat transfer processes operating at the surface, is to be determined, the energy balance equation must be solved to yield a single, real, positive root in a fairly narrow range of temperatures. Similarly, if the terminal velocity of a particle moving under the action of various forces or the lowest natural frequency of vibration of a dynamic system were to be determined, one would search for real, positive roots of the corresponding equations over specified intervals.

Real, negative roots may also be obtained, for instance, when considering temperature and concentration differences or velocities that may be positive or negative, depending on the direction of motion. If no prior information is available on the function and on the roots, a rough plot of the variation of $f(x)$ with x may be obtained numerically, for real roots, to determine the behavior of the function and the approximate location of the roots. This information may then be used in the choice of the method and the interval over which the roots are sought. In some engineering problems, such as those concerned with the stability of systems and with periodic processes, complex roots are of interest. Some of the techniques discussed here may also be employed for finding such complex roots. Again, a prior knowledge of the approximate value of the roots would be useful. Therefore, the physical nature of the problem, which gives rise to the equation whose roots are to be determined, is often important in the solution of the equation.

This chapter discusses several numerical methods for finding the real roots of polynomial and transcendental equations. These include the search method, the bisection method, the regula falsi method (also known as the method of linear interpolation), the secant method, Newton's method, Newton's second-order method, and the method of successive substitution. Some of these are also applicable to complex roots, and the corresponding procedure is outlined. Other methods, including Graeffe's root-squaring method and Bairstow's iterative factorization of polynomials, are also discussed for finding the real and complex roots of polynomials. Since, in many engineering applications, the information on the physical background of the problem may be used effectively in the choice of the method and of the interval of interest, several examples of physical problems are taken in order to indicate the importance of the physical characteristics of the problem in the solution of the equations. Several different types of equations are considered in order to illustrate the various methods discussed.

4.2 SEARCH METHOD FOR REAL ROOTS

The search method is a very simple method based on the change in the sign of the function $f(x)$ as x is incremented, starting with an initial value x_0, to determine the zero crossings of the function; that is, the locations where the plot of the function $f(x)$ crosses the x-axis (see Fig. 4.1). An increment Δx is chosen, and x is successively increased by this value. If the function $f(x)$ changes sign between two successive values x_i and x_{i+1}, then $[f(x_i) \cdot f(x_{i+1})] < 0$, and the presence of a real root in the

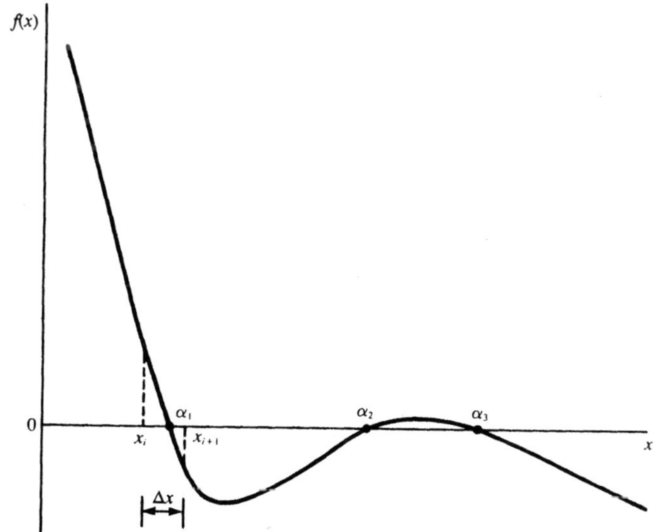

Figure 4.1 Search method for the real roots of the equation $f(x) = 0$.

interval between these values of x is indicated, as shown in Fig. 4.1. This process is repeated by starting with $x = x_i$ and taking a smaller increment to narrow in on the root. Therefore, one can make the interval in which the root lies as small as desired by successively reducing the increment Δx to search for the zero crossing of the function. Generally, one starts with a large step size Δx and successively reduces it to a small fraction, say, one-tenth of the previous value, for locating the root more accurately in the reduced interval obtained from a sign change of the function. If the initial increment is chosen as Δx and the subsequent increments for search in the successively reduced intervals are $\Delta x/n$, $\Delta x/n^2$, $\Delta x/n^3$, and so on, where n is a constant, it is evident that the root may be obtained to the desired accuracy in only a few incremental searches.

When one root, $x = \alpha_1$, has been determined and if other roots are sought, the incremental search proceeds to larger values of x, taking $x > \alpha_1$, until another sign change in $f(x)$ occurs and the above process is repeated. Generally, one takes the starting value x_0 as the smallest value in the range of interest and proceeds to larger x until all the real roots in the given range are found. The method can be used to find positive or negative real roots of a given polynomial or transcendental equation, if the function $f(x)$ crosses the x axis. It will fail to find a zero resulting from the function being tangent to the x axis, since no sign change occurs in this case.

The incremental search method is particularly suitable for obtaining the various intervals in which the real roots of the given equation $f(x) = 0$ are located. Once an interval containing a root has been found, various other, more efficient methods, discussed in later sections, may be employed for a faster convergence to the root. Consequently, the incremental search method often precedes other methods and is thus employed in conjunction with them. Frequently, the starting point, $x = x_0$, and the incremental step size Δx are chosen and $f(x)$ is determined for successively incremented values of x over the entire range of interest, thus yielding the various intervals in which real roots are located. Such a search for roots over the entire range is often known as *exhaustive search*. This procedure also allows one to obtain a rough plot of $f(x)$ versus x and thus determine the approximate behavior of the function. This information is useful in dealing with problem spots such as multiple or equal roots, obtained, for instance, if $f(x)$ does not cross the x axis but is tangent to it, and roots that are very close to each other in value. If no prior information is available on the nature of the roots and the behavior of the function, a small value of the increment Δx may be taken at the onset to ensure that no roots are missed in the search.

The search method is frequently employed with an interactive computer program so that one might choose the increment and the starting point as the interval containing the root is successively reduced, thus coupling one's previous experience and knowledge with the program. The function $f(x)$ may also be plotted, using available software for graphics, for a visual study of the behavior of the function and the approximate location of the roots. Search methods, such as the one outlined here, are frequently employed in optimization of systems, where one seeks to maximize or minimize a given function. The incremental search method can easily be extended to find the maxima or minima of a function by seeking the zeros of the derivative of the function instead of those of the function $f(x)$

The search method provides an iterative procedure for determining the roots of the given algebraic equation $f(x) = 0$. Iteration is terminated when the root has been determined to the desired accuracy level. A commonly employed criterion for convergence is specified as

$$\left|\frac{x^{(n)} - x^{(n-1)}}{x^{(n)}}\right| \leqslant \varepsilon \tag{4.4a}$$

where $x^{(n)}$ and $x^{(n-1)}$ represent the approximations to the root, i.e., the values of x at which the function changes sign, for the nth and the $(n-1)$th iterations, respectively, and ε is a specified small quantity, often taken to be in the range 10^{-5} to 10^{-3}. The above convergence criterion is thus based on the magnitude of the relative change in the approximation to the root from one iteration to the next. Another criterion, based on the increment $\Delta x^{(n)}$ for the nth iteration, is also frequently used. This criterion may be written as follows:

$$\left|\frac{\Delta x^{(n)}}{x^{(n)}}\right| \leqslant \varepsilon \tag{4.4b}$$

This condition for convergence, therefore, implies that the fractional error in the computed root is less than or equal to ε. The first criterion, Eq. (4.4), does not explicitly indicate the accuracy of the root, and therefore the second one is preferable in most cases. The above criteria are also often replaced by

$$|x^{(n)} - x^{(n-1)}| \leqslant \varepsilon \tag{4.5a}$$

or

$$|\Delta x^{(n)}| \leqslant \varepsilon \tag{4.5b}$$

where the actual values of the quantities are considered, instead of the relative magnitudes. This form is particularly useful if the expected value of the root is equal to or close to zero, since then the forms in Eq. (4.4) cannot be used because of the denominator becoming zero. The accuracy of the calculated root may be improved by taking it as the average of the two x values, x_i and x_{i+1}, between which a sign change occurs; that is, $\alpha = (x_i + x_{i+1})/2$. The following example illustrates the use of the search method in finding the real roots of an algebraic equation.

Example 4.1

The surface of a furnace wall is exposed to radiative, convective, and conductive heat transfer, as shown in Fig. 4.1.1. Under steady-state conditions, the surface temperature T is obtained from an energy balance, which yields the equation

$$\varepsilon\sigma(T_h^4 - T^4) = h(T - T_a) + \frac{k}{d}(T - T_2) \tag{4.1.1}$$

where the three terms, from the left, represent heat transfer by radiation, convection, and conduction, respectively. Here, T_h is the temperature of the hot environment radiating to the surface, T_a is the air temperature, and T_2 is the temperature at the

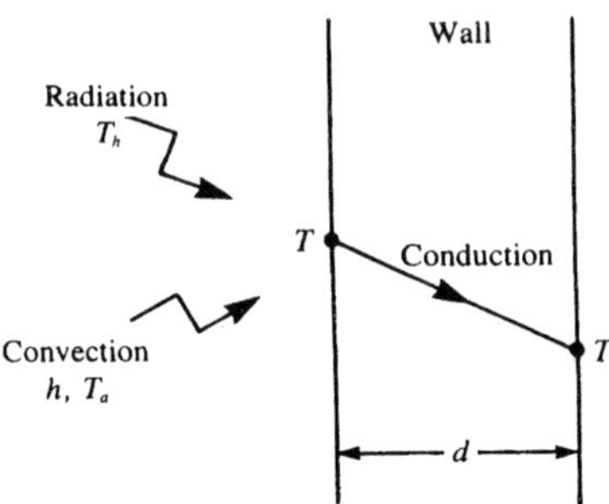

Figure 4.1.1 Heat transfer at the surface of the furnace wall considered in Example 4.1.

outer surface of the wall. Also, h is known as the convective heat transfer coefficient, k the thermal conductivity of the wall material, d the wall thickness, ε the emissivity of the surface, and σ the Stefan-Boltzmann constant, given as 5.67×10^{-8} W/m^2·K^4. Find the wall temperature by the search method, if $T_h = 1000$ K, $T_a = 500$ K, $T_2 = 300$ K, $h = 50$ W/m^2·K, $k = 25$ W/m·K, $\varepsilon = 0.8$, and $d = 0.15$ m.

Solution

The given problem reduces to finding the roots of the equation

$$f(x) = 0.8 \times 5.67 \times 10^{-8}[(1000)^4 - x^4] - 50(x - 500) - \frac{25}{0.15}(x - 300) = 0 \tag{4.1.2}$$

where x is the unknown temperature. Since this is a fourth-order polynomial, there are four roots that will satisfy the equation. However, a consideration of the physical problem, described above, indicates that a unique, positive value of the temperature must be obtained in the range 300 K to 1000 K, these being the two extreme temperatures in the problem. Therefore, we expect only one root to lie in this range. The others will be physically unacceptable, being negative, complex, or beyond the indicated range.

The general behavior of the function $f(x)$ can be first studied by obtaining a rough plot of $f(x)$ versus x. This plot may be obtained simply by incrementing x, starting with the lower limit of $x = 300$ K, and computing $f(x)$ until the upper limit of $x - 1000$ K is reached. Figure 4.1.2 shows this plot, and a single, real, positive root is observed to lie between 500 K and 550 K.

We may now proceed with the incremental search method, starting with $x = 300$ K or with a value close to 500 K, and narrow in on the root. A maximum value of $x = 1000$ K must also be specified to avoid going beyond the range, although, in the present case, the function is very well behaved, and we expect to encounter no problems. Figures 4.1.3 and 4.1.4 show the computer program in FORTRAN 77 and the output at various values of the convergence criterion EPS, which is applied to Δx, as given by Eq. (4.5b). The program increments x, with a chosen increment of 50 K and an initial value of 300 K, until the function $f(x)$ changes sign. The increment is reduced to one-tenth of the earlier value, and the initial value of x is now taken as the

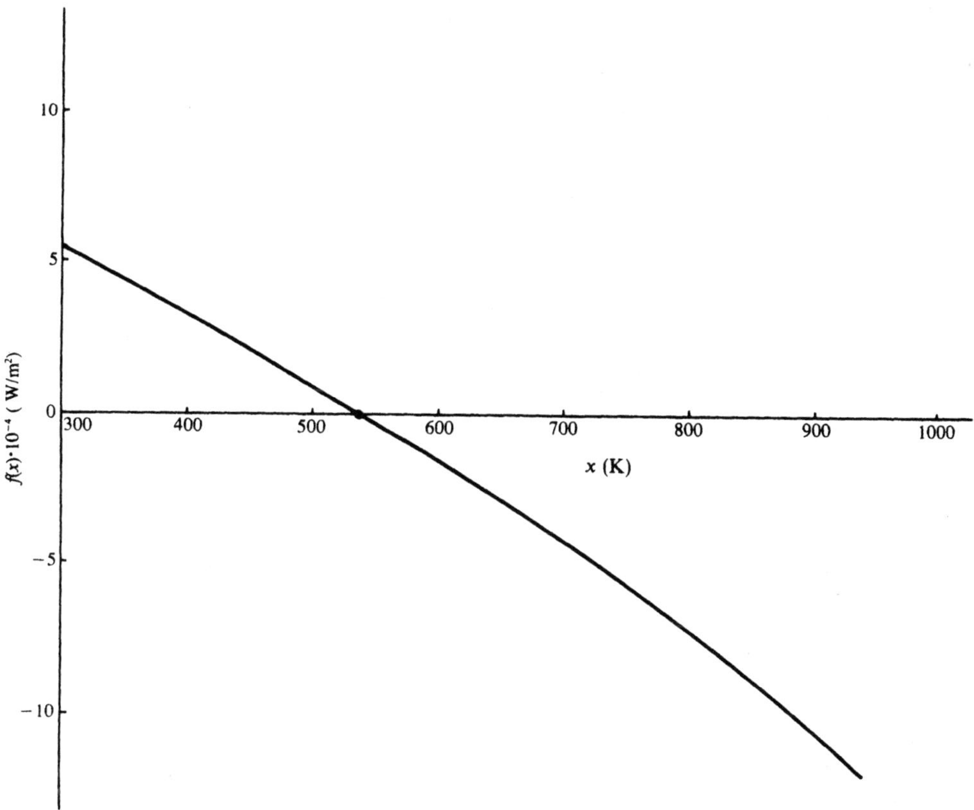

Figure 4.1.2 A rough plot of $f(x)$ versus x to determine the approximate value of the root in Example 4.1.

beginning of the step in which the sign change occurs. This process is repeated until the unknown temperature x is obtained to the desired accuracy level, given by EPS.

The results demonstrate that the number of iterations increases as EPS is decreased. Also, the value of the function $f(x)$ at the estimated root decreases toward zero. Note that a change only in the fourth decimal place occurs when EPS is reduced from 0.01 to 0.001, indicating that the first value will probably give adequate accuracy. The desired accuracy will generally be governed by the engineering application being considered. However, it is important to vary the convergence criterion EPS to ensure that the numerical results are not significantly affected by the value chosen, as discussed in Sec. 2.5. The convergence criterion may also be applied to the function $f(x)$, which represents the net energy gain at the surface.

A similar program may be written in BASIC, as shown in Fig. 4.1.5 for an IBM personal computer. The advantages of BASIC over FORTRAN in the input/output

```
C         THIS PROGRAM FINDS THE REAL ROOTS OF THE EQUATION F(X)=0
C         BY THE SEARCH METHOD
C
C      HERE X IS THE UNKNOWN, XMAX IS THE MAXIMUM VALUE OF X,
C      F1,F2 ARE THE VALUES OF THE FUNCTION F(X) AT TWO CONSECUTIVE
C      X VALUES, DX IS THE INCREMENT IN X, AND EPS IS THE CONVERGENCE
C      CRITERION ON X
C
C
C      DEFINE GIVEN FUNCTION F(X)
C
       F(X)=0.8*5.67E-8*(1000.0**4.0-X**4.0)-50.0*(X-500.0)
     $  -(25.0/0.15)*(X-300.0)
C
C      SPECIFY INITIAL PARAMETERS
C
       EPS=1.0
       DO 14 I=1,4
       XMAX=1000.0
       X=300.0
       DX=50.0
       WRITE(6,16)EPS
  16   FORMAT(2X,'EPS=',F10.4/)
       F1=F(X)
  4    X=X+DX
       F2=F(X)
       A=F1*F2
C
C      CHECK FOR CHANGE IN SIGN OF F(X)
C
       IF(A .LT. 0.0)THEN
       WRITE(6,15)X,F1,F2
C
C      CHECK FOR CONVERGENCE TO THE ROOT
C
       IF(DX .LT. EPS) GO TO 7
       X=X-DX
       DX=DX/10.0
       GO TO 4
       ELSE IF(A .EQ. 0.0) THEN
  7    WRITE(6,10)X,F1
  15   FORMAT(2X,'X=',F10.4,4X,'F1=',F10.4,4X,'F2=',F10.4)
  10   FORMAT(/2X,'TEMPERATURE =',F10.4,4X,'F(X)=',F10.4//)
       ELSE
       IF(X .GT. XMAX) STOP
       F1=F2
       GO TO 4
       END IF
C
C      VARY CONVERGENGE CRITERION
C
  14   EPS=EPS/10.0
  12   STOP
       END
```

Figure 4.1.3 Computer program in FORTRAN 77 for the search method, as applied to the problem given in Example 4.1.

```
EPS=      1.0000

X=   550.0000     F1= 9191.6560      F2=-2957.3940
X=   540.0000     F1=  727.2227      F2= -496.9961
X=   538.0000     F1=  115.6055      F2=   -6.8359

TEMPERATURE =  538.0000     F(X)=  115.6055


EPS=      0.1000

X=   550.0000     F1= 9191.6560      F2=-2957.3940
X=   540.0000     F1=  727.2227      F2= -496.9961
X=   538.0000     F1=  115.6055      F2=   -6.8359
X=   537.9980     F1=    5.8398      F2=   -6.3555

TEMPERATURE =  537.9980     F(X)=    5.8398


EPS=      0.0100

X=   550.0000     F1= 9191.6560      F2=-2957.3940
X=   540.0000     F1=  727.2227      F2= -496.9961
X=   538.0000     F1=  115.6055      F2=   -6.8359
X=   537.9980     F1=    5.8398      F2=   -6.3555
X=   537.9724     F1=    1.1133      F2=   -0.0781

TEMPERATURE =  537.9724     F(X)=    1.1133


EPS=      0.0010

X=   550.0000     F1= 9191.6560      F2=-2957.3940
X=   540.0000     F1=  727.2227      F2= -496.9961
X=   538.0000     F1=  115.6055      F2=   -6.8359
X=   537.9980     F1=    5.8398      F2=   -6.3555
X=   537.9724     F1=    1.1133      F2=   -0.0781
X=   537.9722     F1=    0.0977      F2=   -0.0195

TEMPERATURE =  537.9722     F(X)=    0.0977
```

Figure 4.1.4 Numerical results obtain for Example 4.1.

and in the simplicity of the program are clearly seen. However, for more involved problems, FORTRAN would be more efficient, as discussed in detail in Chapter 2. The output is also shown in this figure for two values of the convergence criterion on Δx. The slight difference in the values of the temperature obtained by FORTRAN and by BASIC is mainly due to the difference in the round-off error in the two machines employed. For further details on the physical problem considered here, books on heat transfer or on college physics, such as Sears et al. (1981), may be consulted.

```
100 REM  FINDING THE ROOTS OF AN EQUATION BY THE SEARCH METHOD
110 REM
120 REM  HERE X IS THE INDEPENDENT VARIABLE IN THE EQUATION F(X)=0, F1 AND
130 REM  F2 ARE THE VALUES OF THE FUNCTION AT TWO SUCCESSIVE X VALUES, DX
140 REM  IS THE INCREMENT IN X, XMAX IS THE MAXIMUM VALUE OF X AND EPS IS
150 REM  THE CONVERGENCE CRITERION APPLIED TO DX
160 REM
170 REM
180   INPUT "X=?       ",X
190   INPUT "XMAX=?    ",XMAX
200   INPUT "DX=?      ",DX
210   INPUT "EPS=?     ",EPS
220   DEF FNA(Y)=.8*5.67*1E-08*(1000^4-Y^4)-50*(Y-500)-(25/.15)*(Y-300)
230   F1=FNA(X)
240   X=X+DX
250   F2=FNA(X)
260   IF (F1*F2) = 0 THEN 360
270   IF (F1*F2) < 0 THEN 310
280   IF X > XMAX THEN 370
290   F1=F2
300   GOTO 240
310   PRINT "X=";X,"F1=";F1
320   IF (DX-EPS) < 0 THEN 360
330   X=X-DX
340   DX=DX/10
350   GOTO 240
360   PRINT "THE DESIRED ROOT OF THE EQUATION IS X=";X
370   END
```

(a)

RESULTS

```
RUN
X=?       300
XMAX=?    1000
DX=?      50
EPS=?     0.001
X= 550          F1= 9191.668
X= 540          F1= 727.2305
X= 538          F1= 115.6133
X= 537.9999     F1= 5.449219
X= 537.975      F1= .546875
X= 537.9724     F1= 6.640625E-02
THE DESIRED ROOT OF THE EQUATION IS X= 537.9724

RUN
X=?       300
XMAX=?    1000
DX=?      50
EPS=?     0.0001
X= 550          F1= 9191.668
X= 540          F1= 727.2305
X= 538          F1= 115.6133
X= 537.9999     F1= 5.449219
X= 537.975      F1= .546875
X= 537.9724     F1= 6.640625E-02
X= 537.9722     F1= .0078125
THE DESIRED ROOT OF THE EQUATION IS X= 537.9722
```

(b)

Figure 4.1.5 Computer program in BASIC for the search method and the numerical results for Example 4.1.

4.3 BISECTION METHOD

The half-interval, or bisection, method may be used for a rapid convergence to the root once the interval containing a real root has been determined by the incremental search method. Consider the function $f(x)$, shown in Fig. 4.2, which is known to have only one real root in the interval $x_1 \leq x \leq x_2$. The interval is bisected, and the function is computed at the midpoint x_0, where

$$x_0 = \frac{x_1 + x_2}{2} \tag{4.6}$$

Now the product $f(x_0) \cdot f(x_1)$ is computed. If $f(x_0) \cdot f(x_1) < 0$, then the root lies in the interval $x_1 < x < x_0$, since the function has changed sign in this half-interval. If the product is positive, then the root must be in the other half-interval $x_0 < x < x_2$. The interval containing the root is, therefore, reduced by half, and the above procedure is now applied to this reduced interval. The process is repeated until the location of the root is obtained to the desired accuracy. Since the interval containing the root is halved in each bisection, the original interval is reduced by a factor of 2^n after n bisections. If the computed root is assumed to be at the midpoint of the interval obtained after n bisections, the maximum error equals half the size of this interval. Therefore, the error ε in the root varies as $I_0/2^{n+1}$, where I_0 is the starting interval containing the root. The number of bisections needed to reduce the maximum error to ε is obtained by taking the logarithm of the equation $\varepsilon = I_0/2^{n+1}$ as follows:

$$n = \frac{\log(I_0/\varepsilon)}{\log 2} - 1 \tag{4.7}$$

where log represents the natural logorithm. Therefore, only nine bisections are needed to reduce the error to less than 0.1% of the original interval I_0.

After each interval-halving operation, the new interval containing the root is determined, and the designation of the end points of the reduced interval is changed to

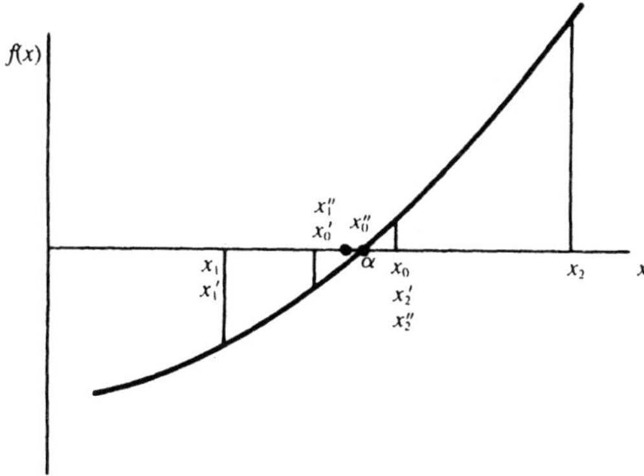

Figure 4.2 Sketch of the computational procedure for the bisection method.

x_1 and x_2, with x_0 taking the appropriate value. Figure 4.2 shows a few steps in the computational procedure, denoting the successive values of x_1, x_2, and x_0 by means of primes. If at any stage in the computation the function $f(x_0)$ is found to be zero or close to it, within a specified error, the root of the equation is taken as the corresponding value of x_0, and further computation is stopped. Otherwise, the computation may be carried out for a specified number of bisections n or until the root is obtained to the desired accuracy, the location of the root being taken as the midpoint of the successively reducing interval. Frequently, the computation is terminated when the change in the root from one bisection to the next is less than a specified error tolerance ε. In most practical problems, the value of ε may be obtained from a consideration of the accuracy needed in the determination of the physical quantity represented by the root. For example, if the pressure of a given volume of gas is obtained by solving its equation of state, ε may be chosen as a small fraction of the acceptable error in the pressure.

The bisection method will always yield a root of the given equation $f(x) = 0$ if $f(x)$ changes sign in the interval. However, if the interval contains more than one real root, one must determine the subintervals containing the roots before proceeding to bisection. An odd number of roots in the interval $x_1 < x < x_2$ will give $f(x_1) \cdot f(x_2) < 0$. If there are no roots in the interval or if there are an even number of roots, then $f(x_1) \cdot f(x_2) > 0$. Several bisections may be needed to obtain the subintervals in which $f(x)$ changes sign, and the method may converge to the same root more than once. In such cases, the search method, exhaustive or incremental, may be employed effectively to determine the approximate location of the roots and the subintervals where bisection may then be used. Similarly, bisection will not locate multiple roots that arise due to the plot of the function $f(x)$ being tangent to the x axis, without crossing it, since a sign change does not occur at the root. Again, the approximate behavior of the function may be determined from search methods. If such multiple roots arise, other methods, discussed later in this chapter, will be needed. If the multiple root arises at a location where the function $f(x)$ crosses the axis, bisection can be used to find the root. The numerical process will always converge if the interval containing a zero crossing of the function $f(x)$ is known. The convergence is faster than that obtained by the search method. However, one may use the search method before using the bisection method to obtain the intervals containing the roots, if the approximate range in which each root is located is not known. Thus the bisection method can be employed for determining a real root of an algebraic equation if a sign change in the function $f(x)$ occurs at the root.

4.4 REGULA FALSI AND SECANT METHODS

4.4.1 Regula Falsi Method

The regula falsi, or false-position, method is similar to the bisection method in that it will always yield a real root in the interval in which the function $f(x)$ changes sign. However, the convergence to the root is generally more rapid. Let us again consider an interval $x_1 < x < x_2$ found from the search method to contain a real root of the equation $f(x) = 0$. Therefore, $f(x_1)$ and $f(x_2)$ are opposite in sign. A chord is drawn joining the two end points $[x_1, f(x_1)]$ and $[x_2, f(x_2)]$. The intersection of the chord, which represents a linear approximation to the function $f(x)$, with the x axis, $x = x_3$, is taken as the first estimation of the root of the given equation. From Fig. 4.3, x_3 may be obtained by using the geometrical relationship between the two triangles formed as follows:

$$x_3 = \frac{x_1 f(x_2) - x_2 f(x_1)}{f(x_2) - f(x_1)} \tag{4.8}$$

If $f(x_3)$ is zero or close to zero, within a specified convergence criterion, the process is terminated, and the root is located at $x = x_3$. If $f(x_3)$ has the same sign as $f(x_1)$, then $f(x_3) \cdot f(x_1) > 0$, and the root lies between x_2 and x_3. Then x_2 remains unchanged and x_3 becomes the new value of x_1, thus giving the reduced interval $x_3 < x < x_2$. Similarly, if $f(x_3) \cdot f(x_1) < 0$, then the root is located in the interval $x_1 < x < x_3$. In this case, x_1 remains unaltered, and x_3 becomes the new value of x_2. Figure 4.3 shows a few steps in the computational process, denoting new values by primes.

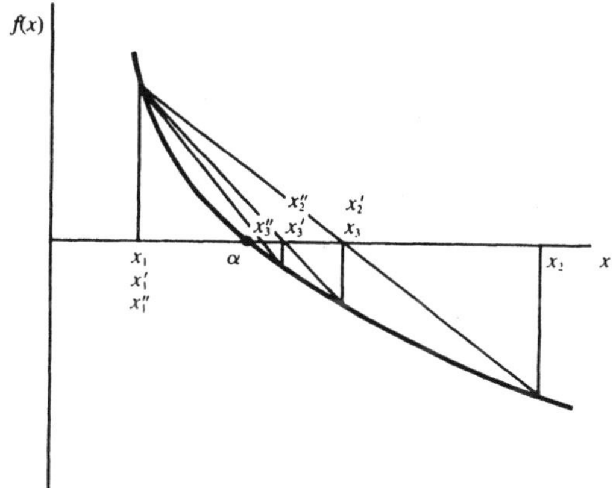

Figure 4.3 The regula falsi method for root solving.

The new values of x_1 and x_2 are employed in Eq. (4.8) to yield an improved approximation to the root. The process is continued, successively reducing the interval containing the root until $|f(x_3)|$ or the change in the root, which is approximated by x_3, from one computational step to the next is less than a specified small quantity ε, where ε is often in the range 10^{-5} to 10^{-3}. The value of the error tolerance ε is chosen on the basis of the accuracy desired in the evaluation of the root. Since, at each step, a subinterval containing the root is considered, the method will always converge if a sign change in $f(x)$ occurs in the initial interval. Multiple roots due to the plot of $f(x)$ being tangent to the x axis cannot be located by this method, which requires a sign change in $f(x)$. The rate of convergence to the root depends on the nature of the function $f(x)$ and the initial interval. Although convergence is often faster than that obtained by the bisection method, examples can be found where such is not the case. The procedure outlined above is also sometimes known as the *linear interpolation method.*

4.4.2 Secant Method

The secant method is similar to the regula falsi, or false-position, method, but it does not always consider subintervals containing the root. The preceding methods always considered the subinterval that enclosed the root and will, therefore, always yield the solution if a sign change in $f(x)$ occurs in the interval. The secant method is, therefore, not guaranteed to converge, unlike the enclosure methods discussed so far. However, when it does converge, it does so more rapidly than the previous methods. Instead of using the two values of x that bound a subinterval containing a real root, the method uses the two most recent values of x in the iterative procedure. Therefore, it employs both interpolation and extrapolation to approximate the root by the intersection of the line joining the two points $[x_1, f(x_1)]$ and $[x_2, f(x_2)]$ with the x axis. Figure 4.4 shows a few iterative steps for the secant method, using primes to denote the new values. As in the regula falsi method, a chord is drawn between the two end points of the initial interval, and the intersection with the x axis is obtained from Eq. (4.8). However, in the next step, the two most recent values of x, x_2 and x_3, are taken. The former thus becomes x_1, and the latter x_2. These are again substituted in Eq. (4.8) to yield the new intersection point, which is the next approximation to the root. Therefore, the general expression for iteration by the secant method may be written from Eq. (4.8) as

$$x_{i+1} = \frac{x_{i-1} f(x_i) - x_i f(x_{i-1})}{f(x_i) - f(x_{i-1})} \tag{4.9}$$

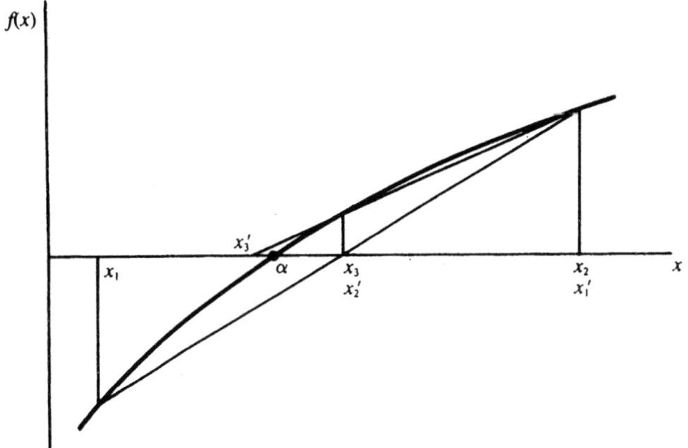

Figure 4.4 The secant method for finding the real roots of an algebraic equation.

where the subscript represents the order of the iteration, starting with $x_{i-1} = x_1$ and $x_i = x_2$ as the first two approximations to the root.

The above process is continued until $|f(x_i)| \leqslant \varepsilon$, where ε is the specified convergence criterion. Since the method employs linear extrapolation for subintervals that do not contain the root, the process may not converge. Convergence depends on the nature of the function and on the limits of the initial interval. If the initial guesses x_1 and x_2 are chosen sufficiently close to the root, $x = \alpha$, convergence of the iterative process to the root can be expected. The method may diverge if the initial guesses are not well chosen. Therefore, this method is suitable if the interval containing the root is known to a fairly good approximation from a rough plot of the function or from the search method. Both the regula falsi and the secant methods may be employed for the real roots of polynomial as well as transcendental equations. The following examples illustrate the use of these methods for finding the roots of algebraic equations.

Example 4.2

A flat plate falling freely in air is subjected to a downward gravitational force and an upward frictional drag due to air. This drag force D_f is given by the expression

$$D_f = \frac{0.2275V^2}{465.9 + (\log V)^{2.58}} - 0.017V \quad \text{for } V \geqslant 1 \text{ m/s} \tag{4.2.1}$$

where V is the vertical velocity of the plate and $\log V$ is the natural logarithm of V. A terminal velocity is attained when this drag force equals the gravitational force. The net force F acting on the plate is given by

$$F = D_f - mg \tag{4.2.2}$$

where m is the mass of the plate and g the magnitude of the gravitational acceleration. Find the terminal velocity by the regula falsi method if $m = 1$ kg and $g = 9.8$ m/s^2. Vary the convergence criterion, on F, from 1.0 to 10^{-4}, and study the effect on the numerical value obtained for the terminal velocity.

Solution

The terminal velocity is given by the root of the equation

$$F(V) = \frac{0.2275V^2}{465.9 + (\log V)^{2.58}} - 0.017V - 9.8 = 0 \tag{4.2.3}$$

Since the expression for the drag force D_f is valid for $V \geqslant 1$ m/s, we may take a value of V around 1 m/s as the lower limit, x_1, of the range in which the root is located. The upper limit, x_2, may be taken as a large value, say, 500 m/s. If the root is not found within this range, a still larger value of x_2 may be employed. A rough plot of $F(V)$ versus V may also be obtained to guide the choice of the initial range of values. The root is expected to be real, positive, and unique.

Figure 4.2.1(a) shows the FORTRAN program for solving this problem by the regula falsi method. The initial range for the unknown x is taken as 2 m/s to 500 m/s.

```
C          ROOT SOLVING WITH THE REGULA FALSI METHOD
C
C      X IS THE INDEPENDENT VARIABLE, FUN(X) IS THE GIVEN
C      FUNCTION, X1 AND X2 ARE THE TWO EXTREME VALUES OF X BOUNDING
C      THE REGION WHICH CONTAINS THE ROOT AT A GIVEN ITERATION,
C      X3 IS THE APPROXIMATION TO THE ROOT, F1, F2 AND F3
C      ARE THE CORRESPONDING VALUES OF THE FUNCTION, AND EPS
C      IS THE CONVERGENCE CRITERION
C
C
C      DEFINE THE GIVEN FUNCTION
C
       FUN(X)=0.2275*X*X/(465.9+ALOG(X)**2.58)-0.017*X-9.8
       EPS=1.0
       DO 4 I=1,5
       X1=2.0
       X2=500.0
  1    F1=FUN(X1)
       F2=FUN(X2)
C
C      COMPUTE APPROXIMATION TO THE ROOT
C
       X3=(X1*F2-X2*F1)/(F2-F1)
       F3=FUN(X3)
C
C      CHECK FOR CONVERGENCE
C
       IF (ABS(F3) .LE. EPS) GO TO 2
       IF ((F1*F3) .GE. 0.0) THEN
       X1=X3
       GO TO 1
       ELSE
       X2=X3
       GO TO 1
       END IF
  2    WRITE(6,3)EPS,X3,F3
  3    FORMAT(2X,'EPS=',F8.5,4X,'TERMINAL VELOCITY=',F10.4,4X,
     $ 'FUN(X)=',F8.4)
C
C      VARY CONVERGENCE CRITERION
C
  4    EPS=EPS/10
       STOP
       END
```

(a)

RESULTS

```
EPS= 1.00000    TERMINAL VELOCITY=  167.1907    FUN(X)= -0.7216
EPS= 0.10000    TERMINAL VELOCITY=  172.2868    FUN(X)= -0.0948
EPS= 0.01000    TERMINAL VELOCITY=  172.9755    FUN(X)= -0.0086
EPS= 0.00100    TERMINAL VELOCITY=  173.0377    FUN(X)= -0.0008
EPS= 0.00010    TERMINAL VELOCITY=  173.0431    FUN(X)= -0.0001
```

(b)

Figure 4.2.1 Computer program for root solving in Example 4.2 by the regula falsi method, with the results for various values of the convergence parameter EPS.

The intersection of the chord joining the two extreme points with the x axis is given by x_3, which is obtained from Eq. (4.8). Depending on the sign of $f(x_1)\cdot f(x_3)$, the new values of x_1 and x_2 are chosen. If $f(x_1)\cdot f(x_3) < 0$, then the root lies between x_1 and x_3. Thus, x_3 becomes the new x_2, and the procedure is repeated. Similarly, if $f(x_1)\cdot f(x_3) > 0$, x_3 becomes the new x_1. Iteration is terminated when $|f(x_3)| \leqslant \text{EPS}$, where EPS is the chosen convergence criterion. The numerical results obtained for various values of EPS are shown in Fig. 4.2.1(b). A velocity V of 173.0432 m/s is obtained for EPS $= 10^{-4}$, and a change of less than 0.003% is observed when EPS is varied from 10^{-3} to this value. The net force F on the plate is found to be negative, that is, downward, and of the order of -10^{-4}. Depending on the desired accuracy of the terminal velocity, the corresponding value of EPS may be chosen.

Example 4.3

Solve the problem of Example 4.2 by the secant method, and compare the results with those obtained by the regula falsi, or false-position, method.

Solution

In this case, the two most recent values of the unknown x are employed, with interpolation and extrapolation, to find the root. Equation (4.9) is used instead of Eq. (4.8). Therefore, x_2 is taken as the new value of x_1, and x_3 as the new value of x_2. The initial values of x_1 and x_2 are taken as 150 m/s and 200 m/s, respectively, in the program shown in Fig. 4.3.1(a). The function $F(V)$ is well behaved, with no discontinuities or sharp changes, and convergence is obtained even with a larger initial range. If x_1 is taken as 2 m/s and x_2 as 500 m/s, as employed in Example 4.2, convergence is again achieved and a smaller computer (CPU) time than that needed for the regula falsi method is required. However, a narrower initial interval for the unknown root would generally be needed for the secant method, as compared to the regula falsi method, in order to obtain convergence. The secant method is not guaranteed to converge since the interval being considered at any given stage does not necessarily contain the root.

The numerical results for the terminal velocity at various values of the convergence criterion EPS are shown in Fig. 4.3.1(b). It is interesting to note that the value obtained, for $x_1 = 150$ m/s and $x_2 = 200$ m/s, does not change when EPS is varied from 10^{-2} to 10^{-4}, for the four significant decimal places printed. The value itself is within 0.001% of that obtained in Example 4.2, and a relatively large value of EPS, 0.01 from the results shown, is found to be satisfactory. Therefore, if the appropriate range or interval containing the root is known, a rapid convergence to the root may be obtained by the secant method. The Newton-Raphson method gives an even faster convergence in most cases, as discussed later. The results for the starting values of x_1 and x_2 taken as 2 m/s and 500 m/s, respectively, are also shown in Fig. 4.3.1(b). The results are very slightly different from those obtained when these are taken as 150 m/s and 200 m/s, respectively, and are also very close to those obtained in

```
C                 ROOT SOLVING WITH THE SECANT METHOD
C
C     X IS THE INDEPENDENT VARIABLE, FUN(X) IS THE GIVEN FUNCTION,
C     X1 AND X2 ARE THE X VALUES FROM THE TWO PREVIOUS ITERATIONS,
C     STARTING WITH THE TWO POINTS BOUNDING THE REGION, X3 IS THE
C     APPROXIMATION TO THE ROOT, F1, F2 AND F3 ARE THE CORRESPONDING
C     VALUES OF THE FUNCTION, AND EPS IS THE CONVERGENCE CRITERION
C
C
C
C     DEFINE FUNCTION
C
      FUN(X)=0.2275*X*X/(465.9+ALOG(X)**2.58)-0.017*X-9.8
      X1=150.0
      X2=200.0
      WRITE(6,12)X1,X2
 12   FORMAT(/10X,'INITIAL X1=',F7.2,10X,'INITIAL X2=',F7.2//)
      EPS=1.0
      DO 2 I=1,5
  1   F1=FUN(X1)
      F2=FUN(X2)
C
C     COMPUTE THE APPROXIMATION TO THE ROOT
C
      X3=(X1*F2-X2*F1)/(F2-F1)
      F3=FUN(X3)
C
C     CHECK FOR CONVERGENCE
C
      IF (ABS(F3) .GT. EPS) THEN
      X1=X2
      X2=X3
      GO TO 1
      ELSE
 11   WRITE(6,13)EPS,X3,F3
 13   FORMAT(2X,'EPS=',F8.5,4X,'TERMINAL VELOCITY=',F10.4,4X,
     $ 'FUN(X)=',F8.4)
      END IF
C
C     VARY CONVERGENCE CRITERION
C
  2   EPS=EPS/10
      STOP
      END
```

(a)

RESULTS

```
          INITIAL X1= 150.00            INITIAL X2= 200.00


EPS= 1.00000    TERMINAL VELOCITY=  171.1760     FUN(X)= -0.2331
EPS= 0.10000    TERMINAL VELOCITY=  172.9019     FUN(X)= -0.0178
EPS= 0.01000    TERMINAL VELOCITY=  173.0445     FUN(X)=  0.0001
EPS= 0.00100    TERMINAL VELOCITY=  173.0445     FUN(X)=  0.0001
EPS= 0.00010    TERMINAL VELOCITY=  173.0445     FUN(X)=  0.0001
```

```
          INITIAL X1=    2.00             INITIAL X2= 500.00

EPS= 1.00000     TERMINAL VELOCITY=  170.7366      FUN(X)= -0.2876
EPS= 0.10000     TERMINAL VELOCITY=  173.1468      FUN(X)=  0.0129
EPS= 0.01000     TERMINAL VELOCITY=  173.0430      FUN(X)= -0.0001
EPS= 0.00100     TERMINAL VELOCITY=  173.0430      FUN(X)= -0.0001
EPS= 0.00010     TERMINAL VELOCITY=  173.0430      FUN(X)= -0.0001
```

(b)

Figure 4.3.1 Computer program for the solution of the problem in Example 4.2 by the secant method, with the corresponding results.

Example 4.2. The slight difference obviously arises because the convergence criterion is satisfied over a corresponding small range of the terminal velocity.

4.5 NEWTON-RAPHSON METHOD AND MODIFIED NEWTON'S METHOD

The Newton-Raphson method for finding the roots of polynomial and transcendental equations is very widely used because of its versatility and generally rapid convergence. The method employs an initial approximation to the root of a given equation and iteratively improves the root until convergence to the desired accuracy is achieved. However, as in the secant method, convergence is not assured. The method is not based on the plot of the function $f(x)$ crossing the x axis, and therefore it may be used for complex and multiple roots too. The modified Newton's method is an extension of the conventional method and has certain important advantages, as discussed later in this section.

4.5.1 Newton-Raphson Method

If $x = x_1$ is the first approximation to the root of the equation $f(x) = 0$, the function $f(x)$ may be expanded in a Taylor series about x_1, for x close to x_1, as given in Chapter 3 to yield

$$f(x) = f(x_1) + (x - x_1)f'(x_1) + \frac{(x - x_1)^2}{2!} f''(x_1) + \cdots \tag{4.10}$$

where the primes denote the order of the differentiation with respect to x. In order to determine the root, $x = \alpha$, we set $f(x)$ equal to zero and may then solve the resulting equation for the root. However, this gives a polynomial equation of order infinity. If only the first two terms are retained, the next approximation to the root, x_2, may be obtained. Therefore, setting $f(x) = 0$, we get

$$0 = f(x_1) + (x_2 - x_1)f'(x_1)$$

or

$$x_2 = x_1 - \frac{f(x_1)}{f'(x_1)}$$

where x_2 represents an improved estimate of the root. In the next iteration, x_1 is replaced by this new approximation x_2, and a still better approximation is obtained. The general expression for iteration by Newton's method is, therefore, written as follows:

$$x_{i+1} = x_i - \frac{f(x_i)}{f'(x_i)} \tag{4.11}$$

where x_i and x_{i+1} are the values obtained after the ith and the $(i + 1)$th iterations, respectively.

This iterative process is continued until the approximation to the root from one iteration to the next changes by less than a specified small quantity ε. As discussed for the previous methods, the convergence criterion ε is often chosen on the basis of the nature of the physical problem under consideration and may be applied to the approximate root or to the function $f(x)$, as $|f(x_i)| \leqslant \varepsilon$. Note that Newton's method is similar to the secant method, discussed in the preceding section. In the secant method, the slope $f'(x_i)$ is approximated by $[f(x_i) - f(x_{i-1})]/(x_i - x_{i-1})$. If this approximation is substituted in Eq. (4.11), the formula for iteration by the secant method, Eq. (4.9), is obtained. Consequently, the convergence characteristics of the secant method and the Newton-Raphson method are quite similar.

The derivative $f'(x_i)$ of the function is needed for using this method. In many cases, particularly for polynomial equations, the derivative may be obtained easily. However, there are problems, such as those involving transcendental functions, in which the differentiation of the function $f(x)$ may be quite complicated. The derivative may then be computed numerically by a finite difference approximation, as outlined in Chapter 3. The iterative procedure converges very rapidly, as shown in Fig. 4.5.

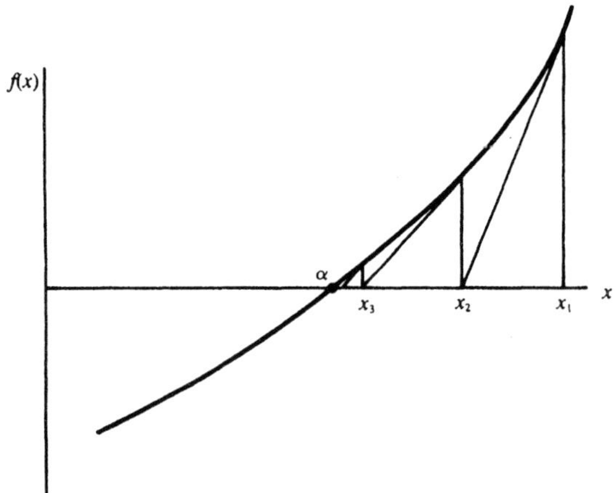

Figure 4.5 The Newton-Raphson iterative procedure for solving an algebraic equation.

However, the method may not converge if the initial guess is too far from the root and also if the derivative is close to zero or varies substantially near the root. A few cases in which the iteration does not converge are shown in Fig. 4.6. The computer program should include this possibility so that, if the method diverges or if the root is not obtained in a specified number of iterations, a new initial guess is chosen and the numerical procedure for finding the root is carried out again. A rough plot of the function, if available from the background information on the problem or from an incremental search, will be useful in the choice of the initial approximation, so that a rapid convergence to the root may be obtained.

Newton's method may also be used for determining the complex roots of the equation $f(z) = 0$, where the complex variable $z = x + iy$, i being $\sqrt{-1}$ and x and y real quantities. The function $f(z)$ may be written as $f(z) = u(x,y) + iv(x,y)$, where u and v are the real and imaginary parts of the function. Then the iterative process for Newton's method is given by the complex expression

$$z_{i+1} = z_i - \frac{f(z_i)}{f'(z_i)} \tag{4.12}$$

where z_i represents the approximation to the root after the ith iteration. Therefore, if complex arithmetic is available on the computer, complex roots may be determined by the procedure outlined above for real roots. If complex variables cannot be used, the following procedure may be used. The real and imaginary parts on the two sides of

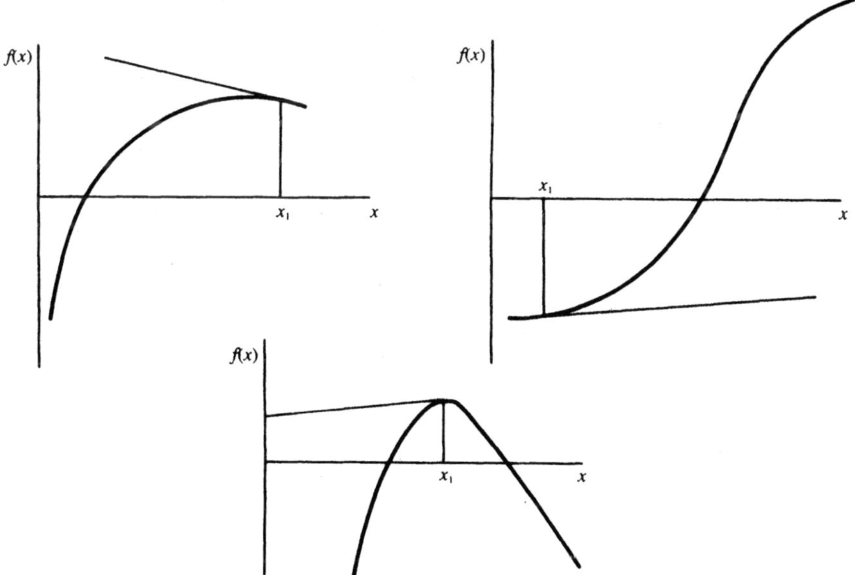

Figure 4.6 A few cases in which the Newton-Raphson method diverges.

the above equation may be equated to obtain the following (Carnahan et al., 1969):

$$x_{i+1} = x_i + \left(\frac{vu_y - uu_x}{u_x^2 + u_y^2}\right)_{x_i, y_i} \tag{4.13}$$

$$y_{i+1} = y_i + \left(\frac{-vu_x - uu_y}{u_x^2 + u_y^2}\right)_{x_i, y_i} \tag{4.14}$$

where u_x and u_y denote the partial derivatives $\partial u/\partial x$ and $\partial u/\partial y$, respectively. The subscripts x_i, y_i outside the parentheses indicate that the values are to be determined at $x = x_i$ and $y = y_i$. For obtaining the above equations, the Cauchy-Riemann equations, $u_x = v_y$ and $u_y = -v_x$, have been employed. Therefore, this method may be used for finding the zeros of complex functions whose real and imaginary parts can be separated easily. Complex functions arise in several engineering problems, such as those concerned with vibrations, stability of systems, electrical circuits with alternating current sources, periodic processes, wave phenomena, and flow fields that may be represented by a complex potential. In most computer systems, complex arithmetic is available, and Eq. (4.12) may be employed directly to determine the complex roots, as outlined in Example 4.5. A similar treatment is applicable if the coefficients of the equation are not real but complex. For further details on complex algebra, see any textbook on calculus, such as Thomas and Finney (1984). It may be mentioned that though z is used to denote an independent complex variable here for clarity, the independent variable x will, in general, be used in the following to denote a real or complex variable.

4.5.2 Modified Newton's Method

The Newton-Raphson method can also be used for multiple roots corresponding to points where the function $f(x)$ becomes tangent to the x axis. However, since $f'(x)$ also goes to zero, as $f(x)$ approaches zero at the root, the convergence is slow, and computational difficulties may arise. For such cases and for achieving a faster rate of convergence, the above procedure may be modified to obtain Newton's second-order method, which employs the second derivative $f''(x)$ of the function in the computation of the root. If three terms are retained from the Taylor series given in Eq. (4.10), instead of the two used for the Newton-Raphson method, we obtain

$$0 = f(x_1) + (x_2 - x_1)\left[f'(x_1) + \frac{(x_2 - x_1)f''(x_1)}{2}\right]$$

To avoid solving this quadratic equation for x_2, one may substitute the approximation for $(x_2 - x_1)$ from the Newton-Raphson method in the brackets above. The resulting linear equation may be solved to obtain the next approximation to the root for this method, often known as the *modified Newton's method*. Therefore,

$$0 = f(x_1) + (x_2 - x_1)\left[f'(x_1) - \frac{f(x_1)}{f'(x_1)} \cdot \frac{f''(x_1)}{2}\right]$$

or

$$x_2 = x_1 - \frac{f(x_1)}{f'(x_1) - \dfrac{f(x_1)f''(x_1)}{2f'(x_1)}} \tag{4.15}$$

This equation gives the general expression for iteration by this method as

$$x_{i+1} = x_i - \frac{f(x_i)}{f'(x_i) - \dfrac{f(x_i)f''(x_i)}{2f'(x_i)}} \tag{4.16}$$

Newton's second-order method, therefore, requires the value of the second derivative of the function. Similarly, higher-order modifications of the conventional Newton's method may be derived for better convergence characteristics. However, the applicability of the method is limited by the computational difficulty in obtaining the derivatives. If they are obtained easily from the given function $f(x)$, the method is advantageous to use. However, if the derivatives are not easy to obtain, one may need to compute the finite difference approximations of the derivatives, leading to a considerable increase in the computational effort. In most problems of engineering interest, the Newton-Raphson method is employed, instead of its higher-order modifications, because of the programming and computational simplicity of the method.

4.5.3 Convergence

As mentioned above, the Newton-Raphson method may not converge, but, if it does converge, it does so very rapidly. It can be shown that for nonzero $f'(\alpha)$, where α is a real root, convergence is guaranteed if the starting value x_1 is close enough to α (Carnahan et al., 1969). Also, once the approximation x_i to the root is near the exact value α, the error after the next iteration, $x_{i+1} - \alpha$, can be shown to be proportional to the square of the error in the present step, $x_i - \alpha$. The relationship between the two is obtained as follows (Carnahan et al., 1969, and Atkinson, 1978):

$$x_{i+1} - \alpha = (x_i - \alpha)^2 \frac{f''(\alpha)}{2f'(\alpha)} \tag{4.17}$$

where in the limit $i \to \infty$, $x_i \to \alpha$. The resulting convergence is termed *quadratic*, or *second order*, and is more rapid than that for several other methods discussed in this chapter. Most methods have a linear convergence; that is, the error after a given iteration is proportional to that obtained after the preceding iteration. Because of its high rate of convergence, applicability to a variety of equations, and simplicity in programming, the Newton-Raphson method is used extensively in engineering applications. It is also used as a correction scheme in the solution of ordinary differential equations, for satisfying the boundary conditions, and in the iterative solution of a system of nonlinear equations. These applications are considered in later chapters. The modified Newton's method is generally employed if the derivative $f'(\alpha)$ goes to zero or becomes very small in the vicinity of the root. The following examples illustrate the use of these methods.

Example 4.4

The water mass flow rate w, in kg/s, in a heating equipment that transfers energy from condensing steam to water, is to be obtained from energy balance considerations. If 250 kW of thermal energy are to be exchanged between the two fluids, the equation for the conservation of energy is given as

$$250 = 294w\left[1 - \exp\left(\frac{-1000}{21(5 + 20w)}\right)\right] \tag{4.4.1}$$

Find the root of this equation by the Newton-Raphson method. The flow rate is known to be less than 5 kg/s.

Solution

It is evident from the above outline of the physical problem under consideration that the root to be obtained is real and positive, being in the range 0 to 5 kg/s. The given equation may be written as

$$f(x) = 294x\left[1 - \exp\left(\frac{-1000}{21(5 + 20x)}\right)\right] - 250 = 0 \tag{4.4.2}$$

where x is the unknown flow rate in kg/s. To apply the Newton-Raphson method, we need a starting guess for the unknown and the value of the derivative df/dx at each approximation to the root. Although the derivative may be obtained analytically in this case, there are several problems where the differentiation may be quite involved. In such cases, numerical differentiation may be employed. The function $f(x)$ is determined numerically at two values of x, which are close to each other and are represented by x and x_N, with $x_N > x$. Then the derivative of the function at x is approximated by

$$\frac{df}{dx} \simeq \frac{f(x_N) - f(x)}{x_N - x} \tag{4.4.3}$$

Once the derivative $f'(x)$ has been evaluated, we use Eq. (4.11) to determine the next approximation to the root.

Figure 4.4.1(a) shows the FORTRAN 77 program for the problem. The difference between x_N and x is taken arbitrarily as 0.001. The initial guess for x is chosen as 0.1 kg/s. Various values of the convergence criterion EPS, as applied to the function $f(x)$, were considered, and the results for EPS $= 10^{-3}$ are shown in Fig. 4.4.1(b). A rapid convergence to the root, which is obtained as 0.9987 kg/s, is observed. Convergence was found to occur over a wide range of the starting value. A smaller value of EPS gives essentially the same flow rate. A statement for terminating the computation is included in the program if the derivative becomes very large, this being

```
C         THIS PROGRAM FINDS THE REAL ROOTS OF AN EQUATION F(X)=0
C         BY THE NEWTON-RAPHSON METHOD
C
C
C
C         HERE X IS THE INDEPENDENT VARIABLE, Y1 THE VALUE OF THE
C         FUNCTION AT X, Y2 THE FUNCTION AT X+0.001, YD THE DERIVATIVE,
C         DX THE INCREMENT IN X FOR THE NEXT ITERATION, EPS THE
C         CONVERGENCE CRITERION ON THE FUNCTION AND XMAX THE MAXIMUM
C         VALUE OF X
C
C
C         DEFINE FUNCTION AND SPECIFY INPUT PARAMETERS
C
          Y(X)=294.0*X*(1.0-EXP(-1000.0/(21.0*(5.0+20.0*X))))-250.0
          EPS=0.001
          WRITE(6,15)EPS
   15     FORMAT(2X,'EPS=',F8.4/)
          X=0.1
          XMAX=5.0
   1      Y1=Y(X)
          WRITE(6,10) X,Y1
C
C         CHECK FOR CONVERGENCE
C
          IF(ABS(Y1) .GT. EPS) THEN
          XN=X+0.001
          Y2=Y(XN)
          YD=(Y2-Y1)/0.001
C
C         CHECK IF RESULTS DIVERGE
C
          IF(YD .GE. (1.0/EPS))GO TO 20
C
C         COMPUTE NEW APPROXIMATION TO THE ROOT
C
          DX=-Y1/YD
          X=X+DX
          IF(X .GE. XMAX)GO TO 20
          GO TO 1
          ELSE
  5       WRITE(6,12) X,Y1
  12      FORMAT(/2X,'FLOW RATE X=',F8.4,4X,'FUNCTION F(X)=',F12.6)
  10      FORMAT(2X,'X=',F8.4,4X,'FUNCTION F(X)=',F12.6)
          END IF
  20      STOP
          END
```

(a)

Results

```
EPS=  0.0010

X=   0.1000     FUNCTION F(X)= -220.632600
X=   0.8529     FUNCTION F(X)=  -28.200450
X=   0.9916     FUNCTION F(X)=   -1.305893
X=   0.9987     FUNCTION F(X)=   -0.002151
X=   0.9987     FUNCTION F(X)=   -0.000031

FLOW RATE X=  0.9987     FUNCTION F(X)=    -0.000031
```

(b)

Figure 4.4.1 FORTRAN 77 program for the Newton-Raphson method, as applied to the problem in Example 4.4, and the results at EPS = 10^{-3}.

```
100 REM  FINDING THE ROOTS OF AN EQUATION BY THE NEWTON-RAPHSON METHOD
110 REM
120 REM  HERE X IS THE INDEPENDENT VARIABLE, F(X)=0 THE GIVEN EQUATION,
130 REM  XMAX THE MAXIMUM VALUE OF X, Y1 AND Y2 THE VALUES OF THE FUNCTION
140 REM  F(X) AT X AND X+0.001, RESPECTIVELY, YD THE DERIVATIVE OF THE
150 REM  FUNCTION AND DX THE INCREMENT IN X FOR THE NEXT ITERATION
160 REM
170 REM
180   INPUT "X=?      ",X
190   INPUT "XMAX=?   ",XMAX
200   INPUT "EPS=?    ",EPS
210   DEF FNA(V)=294*V*(1-EXP(-1000/(21*(5+20*V))))-250
220   Y1=FNA(X)
230   IF ABS(Y1) < EPS THEN 320
240   PRINT "X=";X,"F(X)=";Y1
250   XN=X+.001
260   Y2=FNA(XN)
270   YD=(Y2-Y1)/.001
280   DX=-Y1/YD
290   X=X+DX
300   IF X > XMAX THEN 340
310   GOTO 220
320   PRINT "THE ROOT OF THE EQUATION IS X=";X
330   PRINT "FUNCTION F(X)=";Y1
340   END
```

(a)

RESULTS

```
RUN
X=?      0.1
XMAX=?   5
EPS=?    0.001
X= .1          F(X)=-220.6327
X= .852936     F(X)=-28.1926
X= .9915845    F(X)=-1.313461
X= .998702     F(X)=-2.639771E-03
THE ROOT OF THE EQUATION IS X= .9987164
FUNCTION F(X)= 1.525879E-05

RUN
X=?      0
XMAX=?   5
EPS=?    0.0001
X= 0           F(X)=-250
X= .8504101    F(X)=-28.70673
X= .9913754    F(X)=-1.352066
X= .9987009    F(X)=-2.822876E-03
THE ROOT OF THE EQUATION IS X= .9987163
FUNCTION F(X)= 0

RUN
X=?      2
XMAX=?   5
EPS=?    0.001
X= 2           F(X)= 133.9172
X= .6038186    F(X)=-83.39636
X= .9510526    F(X)=-8.903916
X= .9979448    F(X)=-.1417847
THE ROOT OF THE EQUATION IS X= .9987166
FUNCTION F(X)= 3.051758E-05
```

(b)

Figure 4.4.2 Program in BASIC for the Newton-Rapson method for Example 4.4 and the results obtained.

specified by its numerical value becoming greater than 1/EPS. Alternatively, a new starting value may be chosen for the computation if the derivative becomes too large.

If the derivative can easily be obtained by differentiation, the scheme may be modified to evaluate the derivative directly, instead of by the numerical differentiation procedure given here. It can easily be shown that the convergence is faster by the Newton-Raphson method, as compared to that obtained by the other methods discussed previously. Because of the ease in programming and often faster convergence, this method is very frequently employed in engineering applications. The corresponding program in BASIC for this example and the numerical results obtained are shown in Fig. 4.4.2. For such a small computational problem, a microcomputer may be employed quite effectively. Also, from the two programs presented, note that only a few modifications are needed to go from one programming language to the other.

Example 4.5

Several fluid flow circumstances of interest in mechanical and civil engineering problems can be represented in terms of a complex variable x, known as the *complex potential*. The complex potential x for a flow is governed by the polynomial equation

$$f(x) = x^4 - 4x^3 + 7x^2 - 6x + 2 = 0 \tag{4.5.1}$$

Using the Newton-Raphson method, find the complex roots of this equation. The zeros of the polynomial represent certain locations of symmetry in the flow.

Solution

The given polynomial equation has four roots, which may be real or complex. The complex roots arise in conjugate pairs. A rough plot of the function $f(x)$, shown in Fig. 4.5.1, indicates the possibility of a multiple real root around $x = 1$, as confirmed

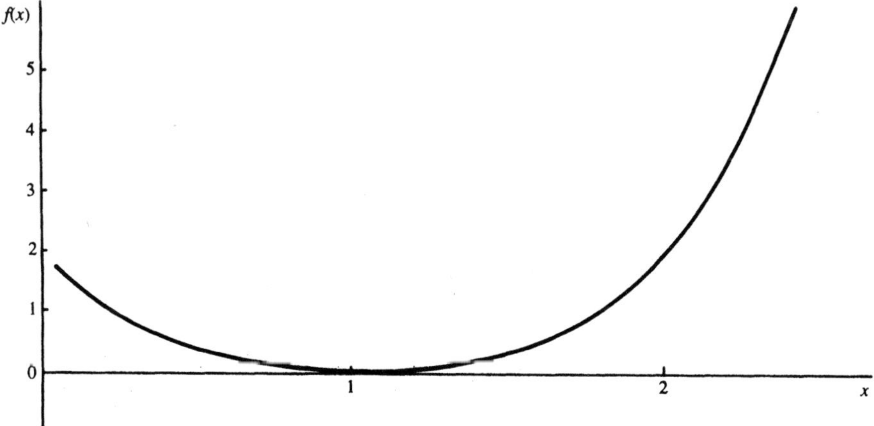

Figure 4.5.1 Rough plot of the function $f(x)$ of Example 4.5.

```
C       THIS PROGRAM FINDS THE  COMPLEX ROOTS OF AN EQUATION F(X)=0
C       BY THE NEWTON-RAPHSON METHOD
C
C
C
C       HERE X IS THE INDEPENDENT COMPLEX VARIABLE, Y1 THE VALUE OF
C       THE FUNCTION AT X, YD THE DERIVATIVE, DX THE INCREMENT IN X
C       FOR THE NEXT ITERATION, AND EPS THE CONVERGENCE CRITERION
C       ON THE FUNCTION F(X)
C
C
        COMPLEX X,Y1,YD,DX
C
C       SPECIFY INPUT PARAMETERS
C
        X=CMPLX(1.5,2.0)
        EPS=0.001
        WRITE(6,1)EPS
  1     FORMAT(2X,'EPS=',F8.4/)
  2     Y1=X**4-4.0*X**3+7.0*X**2-6.0*X+2.0
        WRITE(6,5) X,Y1
        A1=REAL(Y1)
        B1=AIMAG(Y1)
        C1=SQRT(A1**2+B1**2)
C
C       CHECK FOR CONVERGENCE
C
        IF(C1 .GT. EPS) THEN
        YD=4.0*X**3-12.0*X**2+14.0*X-6.0
C
C       COMPUTE NEW APPROXIMATION TO THE ROOT
C
        DX=-Y1/YD
        X=X+DX
        GO TO 2
        ELSE
        WRITE(6,4) X,Y1
  4     FORMAT(/2X,'THE SOLUTION IS X=',F8.4,2X,F8.4,4X,
     $  'FUNCTION=',F8.4,2X,F8.4)
  5     FORMAT(2X,'X=',F10.4,2X,F10.4,4X,'FUNCTION=',F10.4,2X,F10.4)
        END IF
        STOP
        END
```

(a)

Results

```
EPS=  0.0010

X=     1.5000      2.0000     FUNCTION=     6.3125    -13.0000
X=     1.3546      1.5644     FUNCTION=     1.8370     -4.0419
X=     1.2360      1.2624     FUNCTION=     0.4722     -1.2370
X=     1.1327      1.0734     FUNCTION=     0.0715     -0.3617
X=     1.0423      0.9908     FUNCTION=    -0.0268     -0.0804
X=     0.9987      0.9954     FUNCTION=    -0.0091      0.0026
X=     1.0000      1.0000     FUNCTION=     0.0001     -0.0001

THE SOLUTION IS X=  1.0000     1.0000     FUNCTION=  0.0001    -0.0001
```

(b)

Figure 4.5.2 Program for finding the complex roots of the polynomial equation of Example 4.5 by the Newton-Raphson method and the results obtained at EPS $= 10^{-3}$.

from the results obtained in Example 4.6. Therefore, a conjugate pair of complex roots is sought. The mathematics for complex variables available on the computer is employed, with the function $f(x)$, the unknown x, the derivative, and the increment Δx, in x, for the next iteration being defined as complex. The derivative is obtained simply as

$$\frac{df}{dx} = 4x^3 - 12x^2 + 14x - 6 \tag{4.5.2}$$

The convergence criterion EPS is applied to the magnitude of the function, given by $\sqrt{u^2 + v^2}$, where u and v are the real and imaginary parts of the function $f(x)$. The real part represents the velocity potential, and the imaginary part the stream function which is related to the flow rate. The convergence criterion used ensures that, at convergence, the magnitudes of both of these are less than a given small quantity.

The computer program for finding the complex roots of the polynomial equation, Eq. (4.5.1), is shown in Fig. 4.5.2. The starting value of x is taken as $1.5 + 2.0i$, where $i = \sqrt{-1}$. The values of the function and its derivative are computed at this value of x. We then use Eq. (4.12) to determine the next approximation to the root and repeat the process until convergence is achieved. The results for EPS $= 10^{-3}$ are shown in Fig. 4.5.2, indicating a complex root at $(1 + i)$. Therefore, the conjugate root is $(1 - i)$. Again, the convergence to the root is quite rapid, although each iteration involves a larger computational effort as compared to that for real variables. The roots may also be found analytically in this case. The values are found to be identical to those obtained numerically. Complex roots also arise in stability analyses, periodic processes, and alternating-current electrical circuits. The Newton-Raphson method is well suited for most of these problems.

Example 4.6

Find the real roots of the polynomial equation given in Eq. (4.5.1) by the modified Newton's method, and compare the convergence to the root with that obtained by the Newton-Raphson method.

Solution

As seen from the rough plot of the given function, shown in Fig. 4.5.1, a multiple root is expected in the neighborhood of $x = 1$. The second derivative of the function is obtained as

$$\frac{d^2f}{dx^2} = 12x^2 - 24x + 14 \tag{4.6.1}$$

An initial guess of $x = 0.1$ is taken. The function and the first and second derivatives are computed at this value of x. Equation (4.16) is then employed to obtain the next approximation to the root. The convergence criterion, EPS, on the function is

```
C       THIS PROGRAM FINDS THE REAL ROOTS OF AN EQUATION F(X)=0
C       BY MODIFIED NEWTON'S METHOD
C
C
C
C      HERE X IS THE UNKNOWN, Y1 THE VALUE OF THE FUNCTION AT X,
C      YD1 THE FIRST DERIVATIVE, YD2 THE SECOND DERIVATIVE, DX THE
C      INCREMENT IN X FOR THE NEXT ITERATION, EPS THE CONVERGENCE
C      CRITERION ON THE FUNCTION AND XMAX THE MAXIMUM VALUE OF X
C
C
C
C      ENTER INPUT PARAMETERS
C
       EPS=0.00001
       WRITE(6,1)EPS
  1    FORMAT(2X,'EPS=',F10.6/)
       X=0.1
       XMAX=5.0
C
C      COMPUTE VALUE OF THE FUNCTION F(X) AT X
C
  2    Y1=X**4-4.0*X**3+7.0*X**2-6.0*X+2
       WRITE(6,4) X,Y1
C
C      CHECK FOR CONVERGENCE TO THE ROOT
C
       IF(ABS(Y1) .GT. EPS) THEN
C
C      COMPUTE FIRST AND SECOND DERIVATIVES
C
       YD1=4.0*X**3-12.0*X**2+14.0*X-6.0
       YD2=12.0*X**2-24.0*X+14.0
C
C      APPLY MODIFIED NEWTON'S METHOD
C
       DX=-(Y1/(YD1-((YD2*Y1)/(2.0*YD1))))
       X=X+DX
       IF(X .GE. XMAX) GO TO 5
       GO TO 2
       ELSE
       WRITE(6,3) X,Y1
  3    FORMAT(/2X,'THE SOLUTION IS X=',F10.6,4X,'FUNCTION=',F10.6/)
  4    FORMAT(2X,'X=',F10.6,4X,'FUNCTION=',F10.6)
       END IF
  5    STOP
       END
```

Figure 4.6.1 Computer program for the modified Newton's method as applied to Example 4.6.

taken as 10^{-5}, although only a small difference in the results was observed when EPS was varied between 10^{-5} and about 10^{-3}. Figure 4.6.1 shows the computer program. The numerical results are shown in Fig. 4.6.2. The analytical solution yields a multiple root at $x = 1$, and the computed value is 0.998419. Because the plot of the function is tangent to the x axis at this value of x, both $f(x)$ and $f'(x)$ go to zero at the root,

```
MODIFIED NEWTON'S METHOD
EPS=   0.000010

X=  0.100000     FUNCTION=  1.466100
X=  0.606555     FUNCTION=  0.178761
X=  0.863423     FUNCTION=  0.019002
X=  0.954440     FUNCTION=  0.002081
X=  0.984829     FUNCTION=  0.000231
X=  0.994977     FUNCTION=  0.000026
X=  0.998419     FUNCTION=  0.000003

THE SOLUTION IS X=   0.998419     FUNCTION=  0.000003

NEWTON-RAPHSON METHOD
EPS=     0.000010

X=  0.100000     FUNCTION=  1.466100
X=  0.410878     FUNCTION=  0.467519
X=  0.645094     FUNCTION=  0.141824
X=  0.804693     FUNCTION=  0.039600
X=  0.898886     FUNCTION=  0.010329
X=  0.948940     FUNCTION=  0.002614
X=  0.974405     FUNCTION=  0.000656
X=  0.987205     FUNCTION=  0.000165
X=  0.993650     FUNCTION=  0.000040
X=  0.996804     FUNCTION=  0.000010
X=  0.998446     FUNCTION=  0.000003

THE SOLUTION IS X=   0.998446     FUNCTION=  0.000003
```

Figure 4.6.2 The results for Example 4.6 from the modified Newton's method and the Newton-Raphson method.

resulting in a slower convergence as compared to the case where the function crosses the x axis.

The numerical results obtained by an application of the Newton-Raphson method to this problem are also shown in Fig. 4.6.2. The values of the root obtained in the two cases are quite close, but the convergence for the Newton-Raphson method is much slower. The retention of the additional term, involving a nonzero $f''(x)$, in the modified Newton's method accelerates the convergence. Therefore, for multiple roots, arising due to the plot of the function $f(x)$ being tangential to the x axis, the modified Newton's method gives a faster convergence to the root.

4.6 SUCCESSIVE SUBSTITUTION METHOD

In this method, the equation $f(x) = 0$ is rewritten to obtain an expression for the independent variable x as

$$x = g(x) \tag{4.18}$$

so that $\alpha = g(\alpha)$ if $f(\alpha) = 0$, where α is a root of the original equation. Thus, if x_1 is an initial approximation to a root, the successive approximations to the root may be obtained from the recursion relation

$$x_{i+1} = g(x_i) \tag{4.19}$$

Therefore, a successive substitution of the approximation x_i to a root into the function $g(x)$ yields a sequence of iterations that may converge to the root. The equation $x = g(x)$ can be obtained from the original equation $f(x) = 0$ in an unlimited number of ways. In many cases, the equation may contain a linear expression in x, for instance, $f(x) = x^4 - 6.5x^3 + 7x^2 - 11.5x + 3 = 0$. Then the equation for the successive substitution method may be obtained simply by isolating the linear expression to give $x = g(x) = (x^4 - 6.5x^3 + 7x^2 + 3)/11.5$. Similarly, we may rewrite the equation as $x = g(x) = (6.5x^3 - 7x^2 + 11.5x - 3)^{1/4}$. Convergence is often sensitive to the choice of $g(x)$ and may not occur for a chosen form of the function, as seen in Example 4.7. In the above example, for instance, the first formulation is appropriate if the root is less than 1.0, and the second formulation is better if the root is larger than 1.0, as shown below.

We can also employ the following recursion equation:

$$x_{i+1} = (1 - \beta)x_i + \beta g(x_i) \tag{4.20}$$

where β is a constant and may be chosen to improve the convergence characteristics of the successive substitution method. This equation is obtained from the consideration that if $\alpha = g(\alpha)$, then α also satisfies the equation $\alpha = (1 - \beta)\alpha + \beta g(\alpha)$. The choices for $g(x)$ and β are dependent on the behavior of the function $f(x)$. Because of the arbitrariness in the choice of $g(x)$ and the usual strong dependence of convergence on the function $g(x)$, the method is not used very frequently for root solving. However, in certain practical circumstances, the method is employed for the solution of simultaneous, nonlinear algebraic equations governing the performance of engineering systems. The successive substitution method then provides a relatively simple computational technique for obtaining the values of the physical variables that satisfy the given system of equations. This application of the method is discussed in Chapter 5.

It can be shown that the successive substitution method will converge if for $|x - \alpha| < |x_1 - \alpha|$, the function of $g(x)$ possesses a derivative $g'(x)$ such that $|g'(x)| < 1$. Here, x_1 is the initial approximation to the root. This condition implies that the derivative is less than 1.0 in the computational region. When x_i is close to the root α, the next approximation x_{i+1} can be shown to be given by the approximate relation

$$x_{i+1} - \alpha \simeq g'(\alpha)(x_i - \alpha) \tag{4.21}$$

Therefore, if $|g'(\alpha)| < 1$, the method converges to the root in a region near the root. The derivative $g'(\alpha)$ is often termed the *asymptotic convergence factor*. The Newton-Raphson method, discussed in the previous section, may also be considered in terms of the successive substitution method to obtain the convergence characteristics, as outlined by Carnahan et al. (1969).

The major problem with the successive substitution method is the frequent divergence of the iteration for a given choice of the function $g(x)$. The above condition for convergence, $|g'(x)| < 1$, can sometimes be employed in the formulation of the function $g(x)$, as mentioned for the polynomial considered above. Frequently, convergence occurs over a very narrow range of the starting value, and one may need to try several values before the iteration converges. The method is very easy to program, and, as shown by Eq. (4.21), a linear convergence is obtained when x_i is close to the root. The following example illustrates the use of the successive substitution method in finding the roots of an algebraic equation and also demonstrates the convergence characteristics of the method.

Example 4.7

The gas flow rate R, in m^3/s, through a duct in a chemical reactor due to a fan is given in terms of the pressure P, in N/m^2, by the equation

$$R = 15 - 75 \times 10^{-6} \times P^2 \tag{4.7.1}$$

where

$$P = 80 + 10.5R^{5/3} \tag{4.7.2}$$

Employing the successive substitution method, find the gas flow rate at which the system operates.

Solution

The problem involves finding the roots of the equation

$$R = 15 - 75 \times 10^{-6} \times (80 + 10.5R^{5/3})^2 \tag{4.7.3}$$

Both the pressure P and the flow rate R are real and positive quantities. It is also obvious from Eq. (4.7.1) that R must be less than 15 m^3/s, since P is zero if $R = 15$ m^3/s and imaginary if R is larger than this value. Thus, R lies between 0 and 15 m^3/s, the two extreme values being excluded, since nonzero values of R and P are expected.

Equation (4.7.3) is already in the form of Eq. (4.18), and the successive substitution method may be applied to this equation. However, it is found that the method does not converge, mainly because of the large exponent of R on the right-hand side which makes even a small error in the numerical solution grow from one iteration to the next. In fact, $|g'(\alpha)|$, defined in Eq. (4.21), is found to be larger than 1.0 for R larger than about 3.5. Consequently, the problem may be reformulated in terms of a smaller exponent of R as

$$R = \left(\frac{P - 80}{10.5}\right)^{3/5} \quad \text{where } P = \left(\frac{15 - R}{75 \times 10^{-6}}\right)^{1/2} \tag{4.7.4}$$

```
100 REM  SUCCESSIVE SUBSTITUTION METHOD FOR ROOT SOLVING
110 REM
120 REM  HERE X AND Z ARE THE APPROXIMATIONS TO THE ROOT AFTER TWO
130 REM  SUCCESSIVE ITERATIONS, XMAX IS THE MAXIMUM VALUE OF THE
140 REM  UNKNOWN AND EPS IS THE CONVERGENCE CRITERION APPLIED TO
150 REM  THE APPROXIMATION TO THE ROOT
160 REM
170 REM
180   INPUT "X=?        ",X
190   INPUT "XMAX=?   ",XMAX
200   INPUT "EPS=?    ",EPS
210   Z=((((15-X)/7.500001E-05)^.5-80)/10.5)^.6
220   IF Z > 10^6 THEN 290
230   IF ABS(Z-X) < EPS THEN 280
240   PRINT "X=";X,"Z=";Z
250   X=Z
260   IF X > XMAX THEN 290
270   GOTO 210
280   PRINT "THE REQUIRED ROOT IS X=";X
290   END
```

(a)

RESULTS

```
RUN
X=?       0.5
XMAX=?   15
EPS=?    0.001
X= .5           Z= 8.3339
X= 8.3339       Z= 6.173267
X= 6.173267     Z= 6.907465
X= 6.907465     Z= 6.675186
X= 6.675186     Z= 6.750351
X= 6.750351     Z= 6.726206
X= 6.726206     Z= 6.73398
X= 6.73398      Z= 6.73148
THE REQUIRED ROOT IS X= 6.73148
```

```
RUN
X=?       1
XMAX=?   15
EPS=?    0.001
X= 1            Z= 8.227138
X= 8.227138     Z= 6.213559
X= 6.213559     Z= 6.895107
X= 6.895107     Z= 6.679227
X= 6.679227     Z= 6.749059
X= 6.749059     Z= 6.726623
X= 6.726623     Z= 6.733844
X= 6.733844     Z= 6.731522
THE REQUIRED ROOT IS X= 6.731522
```

```
RUN
X=?       0.5
XMAX=?   15
EPS=?    0.0001
X= .5           Z= 8.3339
X= 8.3339       Z= 6.173267
X= 6.173267     Z= 6.907465
X= 6.907465     Z= 6.675186
X= 6.675186     Z= 6.750351
X= 6.750351     Z= 6.726206
X= 6.726206     Z= 6.73398
X= 6.73398      Z= 6.73148
X= 6.73148      Z= 6.732283
X= 6.732283     Z= 6.732023
THE REQUIRED ROOT IS X= 6.732023
```

```
RUN
X=?       1
XMAX=?   15
EPS=?    0.0001
X= 1            Z= 8.227138
X= 8.227138     Z= 6.213559
X= 6.213559     Z= 6.895107
X= 6.895107     Z= 6.679227
X= 6.679227     Z= 6.749059
X= 6.749059     Z= 6.726623
X= 6.726623     Z= 6.733844
X= 6.733844     Z= 6.731522
X= 6.731522     Z= 6.732268
X= 6.732268     Z= 6.732029
THE REQUIRED ROOT IS X= 6.732029
```

(b)

Figure 4.7.1 Computer program in BASIC for the solution of the problem in Example 4.7 by the successive substitution method, along with the results for various values of the convergence parameter and the initial estimate of the unknown root.

This gives the equation for R as

$$R = \left[\frac{\left(\frac{15 - R}{75 \times 10^{-6}} \right)^{1/2} - 80}{10.5} \right]^{3/5} \tag{4.7.5}$$

The successive substitution method is now applied to this formulation of the problem. Since $0 < R < 15$, the minimum and maximum values of the unknown may be suitably specified. Figure 4.7.1 shows the program in BASIC for this problem. Also shown are the corresponding outputs for two starting values, 0.5 and 1.0, of the flow rate, denoted by X in the program. The results are obtained for convergence criterion EPS of 10^{-3} and 10^{-4}, the criterion being applied to (Z − X), where X and Z are the values of the unknown R after two successive iterations.

The flow rate is computed as 6.732 m^3/s, this value being only slightly changed by a variation in the convergence criterion EPS. A large number of iterations are needed at the smaller value of EPS, as expected, and the starting value has a negligible effect on the converged solution. As shown by this example, convergence may often be achieved in successive substitution by rewriting the algebraic equation in a different way, if the method diverges when applied to the given equation. It can be verified that $|g'(\alpha)|$ is indeed less than 1.0 near the root for the formulation given in Eq. (4.7.5).

4.7 GRAEFFE'S METHOD*

All the methods discussed in the preceding sections can be used for finding the real roots of a polynomial. However, there are problems when multiple roots arise, due to the graph of the function $f(x)$ becoming tangent to the x axis, or when complex roots are sought. The Newton-Raphson method is widely used for these cases as well, but there are a few methods which are particularly suitable for polynomials and which determine both the real and the complex roots. Among these are Graeffe's root-squaring method and Bairstow's method of iterative factorization, discussed in Section 4.8. These methods are discussed in order to indicate the basic characteristics of the roots of polynomial equations and present specialized methods that may be employed for root solving.

Let us consider the nth degree polynomial equation

$$f(x) = x^n + a_1 x^{n-1} + a_2 x^{n-2} + \cdots + a_{n-1} x + a_n = 0 \tag{4.2}$$

This equation has n roots, which may be real, multiple, or complex, with the complex roots appearing in conjugate pairs. Graeffe's method is based on obtaining a new polynomial, which is of the same degree as the original polynomial and whose roots are some large, even power of the roots of the original equation. The roots of the derived equation are first obtained, and these then yield the required roots of the given equation. Let us first consider real and distinct roots.

The given polynomial equation may be written as

$$f(x) = (x - \alpha_1)(x - \alpha_2)\dots(x - \alpha_n) \tag{4.22}$$

where $\alpha_1, \alpha_2, \alpha_3, \dots, \alpha_n$ are the roots. A new function $F(x)$ may be defined as

$$F(x) = (-1)^n f(x) f(-x) \tag{4.23}$$

which gives

$$F(x) = (x^2 - \alpha_1^2)(x^2 - \alpha_2^2)\dots(x^2 - \alpha_n^2) \tag{4.24}$$

Therefore, $F(x)$ contains only even powers of x, and a function $f_2(x)$ may be defined as

$$f_2(x) = F(\sqrt{x}) = (x - \alpha_1^2)(x - \alpha_2^2)\dots(x - \alpha_n^2) \tag{4.25}$$

Therefore, the roots of the derived equation $f_2(x) = 0$ are squares of the roots of the original equation. The process may be repeated to obtain a sequence of polynomials $f_4, f_6, f_8, \dots$, so that a derived polynomial $f_m(x)$ is obtained, where

$$f_m(x) = (x - \alpha_1^m)(x - \alpha_2^m)\dots(x - \alpha_n^m) = 0 \tag{4.26}$$

The roots of the above equation are a large, even power m of the roots of the original equation. If $|\alpha_1| > |\alpha_2| > \dots |\alpha_n|$, then the ratios of the roots of the derived equation, $|\alpha_2^m/\alpha_1^m|, |\alpha_3^m/\alpha_2^m|, \dots, |\alpha_n^m/\alpha_{n-1}^m|$, may be made as small as desired by making m large. The derived polynomial $f_m(x)$ may be written as follows:

$$f_m(x) = x^n - (\alpha_1^m + \alpha_2^m + \cdots)x^{n-1} + (\alpha_1^m\alpha_2^m + \alpha_1^m\alpha_3^m + \alpha_2^m\alpha_3^m \cdots)x^{n-2}$$
$$- (\alpha_1^m\alpha_2^m\alpha_3^m + \alpha_1^m\alpha_2^m\alpha_4^m + \cdots)x^{n-3} + \cdots + (-1)^n\alpha_1^m\alpha_2^m\dots\alpha_n^m \tag{4.27a}$$

or

$$f_m(x) = x^n - A_1x^{n-1} + A_2x^{n-2} + \cdots + (-1)^nA_n \tag{4.27b}$$

Then the magnitude of the roots may be approximated by

$$\alpha_1^m \simeq A_1, \alpha_2^m \simeq \frac{A_2}{A_1}, \dots, \alpha_n \simeq \frac{A_n}{A_{n-1}}$$

if only the leading, or dominant, terms within the parentheses in Eq. (4.27a) are retained. The values of the roots $\alpha_1, \alpha_2, \dots, \alpha_n$ of the original equation may be obtained by taking the mth root of the above equations; that is, $\alpha_1 \simeq \pm A_1^{1/m}$, $\alpha_2 \simeq \pm(A_2/A_1)^{1/m}$, and so on. The signs of the roots are not determined and must be obtained by substitution in the original equation or from any previous information on the roots, based on the physical nature of the problem.

The coefficients of the derived equation may be obtained by successively performing the polynomial multiplication $(-1)^n f(x)f(-x)$ and, at each step, substituting x^2 by y to obtain the new polynomial. The derived polynomial $F(y)$, where $y = x^2$, is obtained from the original polynomial $f(x)$ as

$$F(y) = y^n - (a_1^2 - 2a_2)y^{n-1} + (a_2^2 - 2a_1a_3 + 2a_4)y^{n-2}$$
$$- (a_3^2 - 2a_2a_4 + 2a_1a_5 - 2a_6)y^{n-3} + \cdots + (-1)^na_n^2 = 0 \tag{4.28}$$

In the next step, y is replaced by x and the computation repeated to obtain the polynomial whose roots are the fourth power of the roots of the original equation. The process is continued until consecutive applications of the root-squaring procedure yield coefficients $A_1, A_2, \ldots, A_n$, of the derived polynomial, that are essentially squares of the corresponding values of the preceding step. This indicates that the required polynomial $f_m(x)$ has been obtained. The coefficients A_i of the derived polynomial at a given iteration (l) may be obtained from those of the polynomial at the previous iteration, using Eq. (4.28), as follows:

$$A_i^{(l)} = (-1)^i \left[(A_i^{(l-1)})^2 + 2 \sum_{j=1}^{i} (-1)^j A_{i+j}^{(l-1)} A_{i-j}^{(l-1)} \right] \tag{4.29}$$

where the superscript denotes the iteration number. Note that $A_0 = 1$ from Eq. (4.27b). If $i + j > n$, zero is to be used for that coefficient. The coefficients for $l = 0$ are those of the given polynomial, Eq. (4.2), and the subsequent polynomials are obtained from Eq. (4.29). If the roots of the original equation are real and distinct, all the coefficients of $f_m(x)$ become essentially squares of the corresponding coefficients of $f_{m-1}(x)$ as m becomes large. Then the roots $\alpha_1, \alpha_2, \ldots, \alpha_n$, are

$$\alpha_1 \simeq (A_1)^{1/m},\ \alpha_2 \simeq \left(\frac{A_2}{A_1}\right)^{1/m}, \ldots, \alpha_n \simeq \left(\frac{A_n}{A_{n-1}}\right)^{1/m} \tag{4.30}$$

and the signs of the roots are obtained by substituting in the original equation or by using the available information on the nature of the roots.

If the original polynomial equation has real and equal roots, the regular relationship between the coefficients of successive polynomials, as mentioned above, is not obtained. Since the method does not determine the sign of the root, equal roots, in Graeffe's method, are those that have the same absolute value. If the roots α_i and α_{i+1} are taken as equal, it can be shown from the above analysis that the coefficient A_i of the polynomial $f_m(x)$ is essentially equal to half the square of the corresponding coefficient in the polynomial $f_{m-1}(x)$ for large m. The other coefficients are squares of the corresponding preceding values if the remaining roots are real and distinct. If three equal roots are present, say, α_i, α_{i+1}, and α_{i+2}, the coefficients A_i and A_{i+1} become one-third of the corresponding preceding values. The corresponding relationship between the roots and the coefficients of the polynomial may be obtained from Eq. (4.27). See Example 4.8 for further details.

Graeffe's method may also be used for complex roots, which appear in conjugate pairs. The conjugate pair may be taken as $(u + iv)$ and $(u - iv)$, and the above analysis may be applied to such roots. It was shown that, at large m, the real and distinct roots give rise to coefficients of the polynomial $f_m(x)$ that are essentially squares of the corresponding coefficients of $f_{m-1}(x)$. The presence of complex roots is indicated by a fluctuation in the sign of a coefficient, since a trigonometric function $\cos m\theta$, where m is a constant and the complex roots are written as $Re^{i\theta}$ and $Re^{-i\theta}$, appears in the relationships. If the sign of the coefficient A_i fluctuates, the roots α_i and α_{i+1} are a conjugate pair of complex roots. The magnitude R of the roots is determined from the coefficients A_{i-1} and A_{i+1} as

$$R = \left(\frac{A_{i+1}}{A_{i-1}}\right)^{1/m} \tag{4.31}$$

By expanding Eq. (4.23), we find that the sum of the roots is given by

$$\alpha_1 + \alpha_2 + \alpha_3 + \cdots + \alpha_n = -a_1 \tag{4.32}$$

Also,

$$R = u^2 + v^2 \tag{4.33}$$

From these equations, the complex roots may be determined after obtaining the real roots. If more than one conjugate pair of complex roots is present, correspondingly more coefficients in the polynomial $f_m(x)$ fluctuate in sign.

Therefore, Graeffe's method provides a means of determining all the roots of a polynomial equation, whether they are real, equal, or complex. However, despite this attractive feature of the method, it has not become very popular mainly because of the need to make decisions that considerably complicate the programming. The round-off error introduced at any stage of the process accumulates in the computation and affects the accuracy of the roots obtained. Also, the coefficients frequently exceed the floating-point range of the computer, particularly if there are two roots which are close to each other and which, therefore, require a large value of m for the separation of the roots. However, this last problem may be avoided by the scaling of the polynomial, which involves dividing the roots by a scale factor. The roots of the modified equation yield the scaled roots, from which the desired roots are obtained. In most practical problems, the nature of the roots is generally known. For finding the applicable interest rates in financial transactions, for instance, we know that the roots are real and distinct and Graeffe's method may be suitably programmed. Similarly, in the analysis of the stability of flows, we seek complex roots which indicate whether or not the flow is stable.

Probably the best procedure for employing Graeffe's method is to work interactively with the computer. Such an interactive program would allow one to make decisions as the computation proceeds and make the necessary changes in the process. Although Graeffe's method is not widely used, it does have the attractive aspect of evaluating all the roots of a polynomial equation. The method is discussed here since it indicates a different approach, as compared to the methods outlined earlier in this chapter, to root solving and may form the basis for solving certain complicated equations of engineering interest that cannot be solved by other methods. The use of the method is illustrated by the following examples.

Example 4.8

(a) The characteristic equation for the vibration of a mechanical system is given as

$$\lambda^4 - 10\lambda^3 + 35\lambda^2 - 50\lambda + 24 = 0 \tag{4.8.1}$$

Using Graeffe's method, find the real and positive roots, which represent the frequencies of vibration of the system and are known to be real and distinct. Such equations also arise in several analogous electrical systems where the oscillations in the current correspond to the vibration of a mechanical system.

(b) Using the scheme developed for this problem, also find the roots of the following equations which arise in other systems:

$$x^4 - 3x^3 + x^2 + 3x - 2 = 0 \tag{4.8.2}$$

$$x^4 - 5x^2 + 4 = 0 \tag{4.8.3}$$

$$x^4 - 7x^3 + 17x^2 - 17x + 6 = 0 \tag{4.8.4}$$

Solution

All the given equations are fourth-order polynomial equations. Therefore, each equation has four roots, which may be real or complex. Roots that are equal in magnitude may also arise. The nature and magnitude of the roots may be obtained from the coefficients of the polynomial derived from Graeffe's root-squaring method. These coefficients are obtained from Eq. (4.29). Since the roots of Eq. (4.8.1) are given to be real and distinct, we may use Eq. (4.30) directly to obtain the four roots for this circumstance.

Figure 4.8.1(a) shows the computer program for determining the coefficients of the derived polynomials, starting with those of the given polynomial equation. The process is continued until the coefficients are approximately square of those obtained after the previous iteration. At this stage, Eq. (4.30) is applied to obtain the magnitude of the roots α_1, α_2, α_3, and α_4. The computer output in Fig. 4.8.1(b) shows that only four root-squaring steps are adequate in obtaining a high degree of accuracy. The difference in the results after the fourth and fifth steps is quite small. The numerical values of the roots are 4, 3, 2, and 1, which are identical to those obtained analytically. In order to determine the sign of these roots, we substitute both positive and negative values in Eq. (4.8.1). It is found that all the roots are positive and are, therefore, acceptable, since the frequency of vibration must be positive.

If the nature of the roots is not known at the onset, the coefficients of the derived polynomials must be tabulated and the values after two successive steps compared, for large m, to determine if multiple or complex roots are involved. For Eq. (4.8.1), the coefficients at increasing values of the order m of the derived polynomial are tabulated in Fig. 4.8.2. It can be easily seen that the coefficients for $m = 32$ are approximately square of the corresponding values for $m = 16$. The results for the other three polynomial equations, Eqs. (4.8.2) through (4.8.4), are also shown in Fig. 4.8.2. In these cases, the regular relationship between the coefficients after two successive root-squaring steps, i.e., the squaring of the coefficients, is not obtained.

From the numerical results for Eq. (4.8.2), we obtain

$$B_1 = (A_1)^2 \quad B_2 = \frac{(A_2)^2}{3} \quad B_3 = \frac{(A_3)^2}{3} \quad B_4 = (A_4)^2 \tag{4.8.5}$$

```
C         ROOT SOLVING BY GRAEFFE'S ROOT SQUARING METHOD
C
C
C      HERE THE A'S REPRESENT THE COEFFICIENTS OF THE POLYNOMIAL
C      OF THE PREVIOUS ITERATION AND THE B'S ARE THOSE OF THE
C      PRESENT ITERATION. THE R'S REPRESENT THE ROOTS.
C
C
C      ENTER COEFFICIENTS OF GIVEN POLYNOMIAL
C
       A1=-10.0
       A2=35.0
       A3=-50.0
       A4=24.0
       S=1.0
       DO 1 I=1,5
       S=S*2.0
C
C      COMPUTE COEFFICIENTS OF NEW POLYNOMIAL
C
       B1=A1**2-2.0*A2
       B2=A2**2-2.0*A1*A3+2.0*A4
       B3=A3**2-2.0*A2*A4
       B4=A4**2
       M=2**I
       A1=B1
       A2=B2
       A3=B3
       A4=B4
C
C      COMPUTE THE ROOTS
C
       R1=B1**(1.0/S)
       R2=(B2/B1)**(1.0/S)
       R3=(B3/B2)**(1.0/S)
       R4=(B4/B3)**(1.0/S)
       WRITE(6,11)M,R1,R2,R3,R4
 11    FORMAT(2X,'M=',I2,2X,'R1=',F8.3,2X,'R2=',F8.3,2X,'R3=',
     $ F8.3,2X,'R4=',F8.3/)
 1     CONTINUE
       STOP
       END
```

(a)

RESULTS FOR EQ. (4.8.1)

```
M= 2  R1=  5.477  R2=  3.017  R3=  1.733  R4=  0.838

M= 4  R1=  4.338  R2=  2.941  R3=  1.917  R4=  0.981

M= 8  R1=  4.050  R2=  2.979  R3=  1.990  R4=  0.999

M=16  R1=  4.002  R2=  2.998  R3=  2.000  R4=  1.000

M=32  R1=  4.000  R2=  3.000  R3=  2.000  R4=  1.000
```

(b)

Figure 4.8.1 Program for root solving in Example 4.8 by Graeffe's method and the corresponding results.

EQ. (4.8.1)

M	B1	B2	B3	B4
M= 2	B1=0.3000E 02	B2=0.2730E 03	B3=0.8200E 03	B4=0.5760E 03
M= 4	B1=0.3540E 03	B2=0.2648E 05	B3=0.3579E 06	B4=0.3318E 06
M= 8	B1=0.7235E 05	B2=0.4485E 09	B3=0.1105E 12	B4=0.1101E 12
M=16	B1=0.4338E 10	B2=0.1852E 18	B3=0.1212E 23	B4=0.1212E 23
M=32	B1=0.1845E 20	B2=0.3418E 35	B3=0.1468E 45	B4=0.1468E 45

EQ. (4.8.2)

M	B1	B2	B3	B4
M= 2	B1=0.7000E 01	B2=0.1500E 02	B3=0.1300E 02	B4=0.4000E 01
M= 4	B1=0.1900E 02	B2=0.5100E 02	B3=0.4900E 02	B4=0.1600E 02
M= 8	B1=0.2590E 03	B2=0.7710E 03	B3=0.7690E 03	B4=0.2560E 03
M= 16	B1=0.6554E 05	B2=0.1966E 06	B3=0.1966E 06	B4=0.6554E 05
M= 32	B1=0.4295E 10	B2=0.1288E 11	B3=0.1288E 11	B4=0.4295E 10
M= 64	B1=0.1845E 20	B2=0.5534E 20	B3=0.5534E 20	B4=0.1845E 20
M=128	B1=0.3403E 39	B2=0.1021E 40	B3=0.1021E 40	B4=0.3403E 39

EQ. (4.8.3)

M	B1	B2	B3	B4
M= 2	B1=0.1000E 02	B2=0.3300E 02	B3=0.4000E 02	B4=0.1600E 02
M= 4	B1=0.3400E 02	B2=0.3210E 03	B3=0.5440E 03	B4=0.2560E 03
M= 8	B1=0.5140E 03	B2=0.6656E 05	B3=0.1316E 06	B4=0.6554E 05
M=16	B1=0.1311E 06	B2=0.4295E 10	B3=0.8590E 10	B4=0.4295E 10
M=32	B1=0.8590E 10	B2=0.1845E 20	B3=0.3689E 20	B4=0.1845E 20
M=64	B1=0.3689E 20	B2=0.3403E 39	B3=0.6806E 39	B4=0.3403E 39

EQ. (4.8.4)

M	B1	B2	B3	B4
M= 2	B1=0.1500E 02	B2=0.6300E 02	B3=0.8500E 02	B4=0.3600E 02
M= 4	B1=0.9900E 02	B2=0.1491E 04	B3=0.2689E 04	B4=0.1296E 04
M= 8	B1=0.6819E 04	B2=0.1693E 07	B3=0.3366E 07	B4=0.1680E 07
M=16	B1=0.4311E 08	B2=0.2821E 13	B3=0.5642E 13	B4=0.2821E 13
M=32	B1=0.1853E 16	B2=0.7959E 25	B3=0.1592E 26	B4=0.7959E 25
M=64	B1=0.3434E 31	B2=0.6334E 50	B3=0.1267E 51	B4=0.6334E 50

Figure 4.8.2 Coefficients obtained by Graeffe's method for the equations of Example 4.8.

where the B's represent the coefficients of the highest derived polynomial, and the A's those of the preceding one. Therefore, three roots that are equal in magnitude are indicated. Since B_2 and B_3 are one-third of the squares of the corresponding preceding coefficients, the first root α_1 is distinct from the other three, which are equal and are denoted by α_2. Following the treatment given in Section 4.7, we obtain the relationships between the roots and the coefficients of the derived polynomial from

Eq. (4.27a) for $m = 128$ as follows:

$$\begin{aligned} B_1 &= \alpha_1^m \\ B_2 &= 3\alpha_1^m\alpha_2^m \\ B_3 &= 3\alpha_1^m\alpha_2^{2m} \\ B_4 &= \alpha_1^m\alpha_2^{3m} \end{aligned} \tag{4.8.6}$$

Using these equations, we obtain α_1 and α_2 as

$$\alpha_1 = (B_1)^{1/m} \tag{4.8.7}$$

$$\alpha_2 = \left(\frac{B_2}{3B_1}\right)^{1/m} = \left(\frac{B_3}{3B_1}\right)^{1/2m} = \left(\frac{B_4}{B_1}\right)^{1/3m} \tag{4.8.8}$$

The numerical values are computed as 2 and 1. Therefore,

$$\alpha_1 = \pm 2 \quad \text{and} \quad \alpha_2 = \pm 1$$

When these values are substituted in Eq. (4.8.2), we obtain the roots as 2, 1, and -1, which are the same as the exact values of these roots. A multiple root exists at $x = +1$. Note that Graeffe's method indicates that there are three equal roots at $x = \pm 1$, since the method gives only the magnitude of the roots.

Similarly, the results for Eq. (4.8.3) give

$$B_1 = \frac{(A_1)^2}{2} \quad B_2 = (A_2)^2 \quad B_3 = \frac{(A_3)^2}{2} \quad B_4 = (A_4)^2 \tag{4.8.9}$$

Therefore, there are two multiple roots, α_1 and α_2. We have

$$\begin{aligned} B_1 &= 2\alpha_1^m \\ B_2 &= \alpha_1^{2m} \\ B_3 &= 2\alpha_1^{2m}\alpha_2^m \\ B_4 &= \alpha_1^{2m}\alpha_2^{2m} \end{aligned} \tag{4.8.10}$$

Using these equations, we obtain $\alpha_1 = \pm 2$ and $\alpha_2 = \pm 1$. By substitution in Eq. (4.8.3), we find that all the four values, 2, 1, -1, and -2, satisfy the equation, giving these as the desired roots. The roots for Eq. (4.8.4) may be similarly analyzed to yield a multiple root at $x = +1$ and two distinct roots at $x = +2$ and $x = +3$. Therefore, the method may be used for finding real, distinct, or multiple roots. Complex roots may also be obtained, as outlined in Section 4.7.

4.8 ITERATIVE FACTORIZATION OF POLYNOMIALS*

Analytically, the roots of a polynomial equation can often be obtained by factorization and equating each factor to zero. A similar approach may be employed for root solving by numerical methods. Several methods are based on the iterative factorization of the given polynomial and can be used to obtain factors of arbitrary degree. Generally, linear or quadratic factors are determined so that the roots may be obtained directly from these factors. Let us first consider Bairstow's method, which iteratively determines quadratic factors of the form $x^2 + bx + c$, so the roots are given by Eq. (4.3) as

$$\alpha_1, \alpha_2 = \frac{-b \pm \sqrt{b^2 - 4c}}{2}$$

The polynomial given in Eq. (4.2) may be written as

$$f(x) = (x^2 + bx + c)(d_0 x^{n-2} + d_1 x^{n-3} + d_2 x^{n-4} + \cdots + d_{n-3}x + d_{n-2}) + \text{remainder} \tag{4.34}$$

where the d's are functions of b and c and are obtained from a comparison with the original polynomial of Eq. (4.2) as

$$\begin{aligned} d_0 &= 1 \\ d_1 &= a_1 - b \\ d_2 &= a_2 - d_1 b - c \\ d_3 &= a_3 - d_2 b - d_1 c \\ &\vdots \\ d_i &= a_i - d_{i-1} b - d_{i-2} c \end{aligned} \tag{4.35}$$

and the remainder is $(x + b)d_{n-1} + d_n$.

To extract the quadratic factor from the polynomial, we must reduce the remainder to zero, within a specified error tolerance. We do so by iteratively reducing d_{n-1} and d_n to zero. Since both of these are functions of b and c, we may use Taylor's expansion for a function of two variables. If only the linear terms are retained, we obtain

$$\begin{aligned} d_n(b + \Delta b, c + \Delta c) &\approx d_n(b,c) + \frac{\partial d_n}{\partial b}\Delta b + \frac{\partial d_n}{\partial c}\Delta c = 0 \\ d_{n-1}(b + \Delta b, c + \Delta c) &\simeq d_{n-1}(b,c) + \frac{\partial d_{n-1}}{\partial b}\Delta b + \frac{\partial d_{n-1}}{\partial c}\Delta c = 0 \end{aligned} \tag{4.36}$$

where Δb and Δc are increments in b and c. We set the equations equal to zero in order to obtain the next approximation to b and c so that d_n and d_{n-1} become zero.

The set of equations for the d's may be differentiated to obtain a similar sequence of equations for their partial derivatives. The corresponding expressions are as follows:

$$\frac{\partial d_1}{\partial b} = -1 = e_0 \qquad \frac{\partial d_1}{\partial c} = 0$$

$$\frac{\partial d_2}{\partial b} = b - d_1 = e_1 \qquad \frac{\partial d_2}{\partial c} = -1 = e_0$$

$$\frac{\partial d_3}{\partial b} = -d_2 - e_1 b - c e_0 = e_2 \qquad \frac{\partial d_3}{\partial c} = b - d_1 = e_1 \tag{4.37}$$

$$\vdots \qquad \vdots$$

$$\frac{\partial d_{n-1}}{\partial b} = -d_{n-2} - e_{n-3} b - e_{n-4} c = e_{n-2} \qquad \frac{\partial d_{n-1}}{\partial c} = e_{n-3}$$

$$\frac{\partial d_n}{\partial b} = -d_{n-1} - e_{n-2} b - e_{n-3} c = e_{n-1} \qquad \frac{\partial d_n}{\partial c} = e_{n-2}$$

which give

$$e_i = -d_i - e_{i-1} b - e_{i-2} c \quad \text{for } i = 2, 3, \ldots, (n-1)$$

Also,

$$e_0 = -1 \quad \text{and} \quad e_1 = b - d_1$$

Therefore, the partial derivatives for the remainder terms may be obtained. From Eqs. (4.36),

$$\begin{aligned} d_n &= -e_{n-1}\,\Delta b - e_{n-2}\,\Delta c \\ d_{n-1} &= -e_{n-2}\,\Delta b - e_{n-3}\,\Delta c \end{aligned} \tag{4.38}$$

Solving these simultaneous linear equations, we find that Δb and Δc are given by

$$\begin{aligned} \Delta b &= \frac{d_{n-1} e_{n-2} - d_n e_{n-3}}{e_{n-1} e_{n-3} - (e_{n-2})^2} \\ \Delta c &= \frac{d_n e_{n-2} - d_{n-1} e_{n-1}}{e_{n-1} e_{n-3} - (e_{n-2})^2} \end{aligned} \tag{4.39}$$

To apply Bairstow's method, initial guessed values of b and c are taken and the corresponding d's and e's are determined. The increments Δb and Δc are obtained for the next approximation of b and c. The recursion formula is

$$b_{i+1} = b_i + \Delta b \quad \text{and} \quad c_{i+1} = c_i + \Delta c \tag{4.40}$$

where i is the iteration number. The iterative process for the determination of b and c is continued until $|\Delta b|$ and $|\Delta c|$ are less than a specified convergence criterion. The quadratic factor thus obtained yields two roots of the equation. The reduced

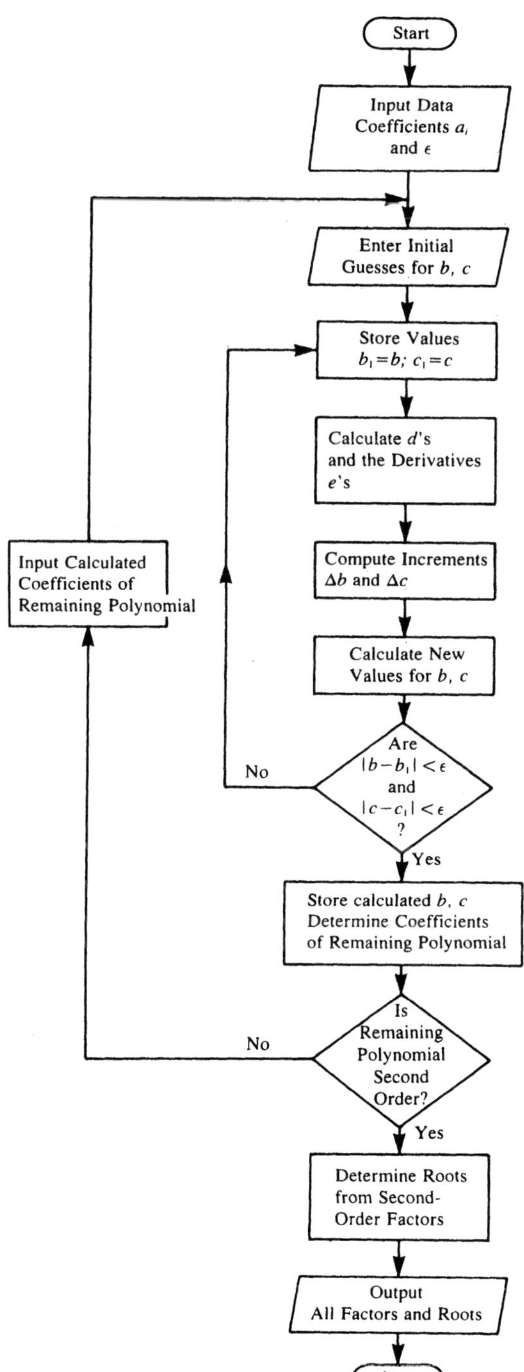

Figure 4.7 Flow chart for the solution of a polynomial equation by Bairstow's method.

polynomial of degree $(n - 2)$ is next considered to obtain the remaining roots. The algorithm is shown in terms of a flow chart in Fig. 4.7.

Bairstow's method can be used for finding real, equal, or complex roots of a polynomial. Although the analysis appears to be complicated, the method may be programmed for the computer without too much difficulty. However, convergence cannot be guaranteed for an arbitrary choice of initial values. If there is some prior information available on the roots or on the coefficients of a factor, the method may be used very effectively to improve the accuracy of the roots. Since divergence may occur with an arbitrary choice of the initial values of b and c, one may restrict the total number of iterations in the program and choose the starting values again if divergence occurs.

Several other methods have been developed based on the extraction of factors from polynomials. Synthetic division by a linear or quadratic factor allows one to obtain equations for the coefficients of the reduced polynomial and for the remainder, as discussed above. Bairstow's method uses the Newton-Raphson method for the solution of simultaneous nonlinear equations to iteratively reduce the remainder to zero [see Eq. (4.36)]. If the successive substitution method for simultaneous nonlinear equations is employed, instead of the Newton-Raphson method, to reduce the remainder to zero, the procedure is known as *Lin's method*. This method provides a simpler, although less efficient, iterative procedure for obtaining the quadratic factors of a polynomial of degree greater than two. Both the Newton-Raphson and the successive substitution methods for simultaneous algebraic equations are similar to those presented in this chapter for solving single equations, and they retain their basic characteristics for multiple equations, as discussed in Chapter 5. The extraction of linear factors from the polynomial may also be carried out by using synthetic division. However, quadratic factors are the most desirable ones since they allow the direct determination of real and complex roots. The methods based on the iterative factorization of polynomials also have the attractive feature of obtaining all the real, multiple, and complex roots. Therefore, despite their complexity, they are frequently used, particularly for problems of engineering interest in which the nature and magnitude of the roots are not known. The following example illustrates the use of Bairstow's method for solving the problem given in Example 4.8.

Example 4.9

Use Bairstow's method for the iterative factorization of polynomials to find a quadratic factor and the roots of the characteristic equation given in Example 4.8(a) as

$$\lambda^4 - 10\lambda^3 + 35\lambda^2 - 50\lambda + 24 = 0$$

Solution

The quadratic factor to be determined is taken as $x^2 + bx + c$, where b and c are to be obtained from Bairstow's method. Equations (4.35) and (4.37) give the coefficients of the remaining polynomial and the derivatives of these coefficients with respect to b

```
C            BAIRSTOW'S METHOD FOR FINDING THE ROOTS OF A
C            POLYNOMIAL EQUATION
C
C        A'S ARE THE COEFFICIENTS OF THE GIVEN POLYNOMIAL,
C        B AND C ARE THE CONSTANTS OF THE DESIRED QUADRATIC FACTOR,
C        D'S ARE THE COEFFICIENTS OF THE REMAINING POLYNOMIAL
C        E'S ARE THE DERIVATIVES OF THE CORRESPONDING D'S, DB IS THE
C        INCREMENT IN B FOR THE NEXT ITERATION, DC THE INCREMENT IN C
C        AND EPS IS THE CONVERGENCE CRITERION
C
C
C        ENTER COEFFICIENTS OF GIVEN POLYNOMIAL
C
          A1=-10.0
          A2=35.0
          A3=-50.0
          A4=24.0
C
C        ENTER INITIAL GUESSES
C
          B=-2.0
          C=1.0
          WRITE(6,2)B,C
 2        FORMAT(2X,'THE INITIAL GUESS: B=',F8.4,4X,'C=',F8.4/)
          EPS=0.001
 1        B1=B
          C1=C
C
C        COMPUTE THE COEFFICIENTS OF THE REMAINING POLYNOMIAL
C
          D0=1.0
          D1=A1-B
          D2=A2-D1*B-C
          D3=A3-D2*B-D1*C
          D4=A4-D3*B-D2*C
C
C        COMPUTE THE CORRESPONDING DERIVATIVES
C
          E0=-1.0
          E1=B-D1
          E2=-D2-E1*B-C*E0
          E3=-D3-E2*B-E1*C
C
C        CALCULATE THE NEW VALUES OF THE CONSTANTS B AND C
C
          DB=(D3*E2-D4*E1)/(E3*E1-E2*E2)
          DC=(D4*E2-D3*E3)/(E3*E1-E2*E2)
          B=B+DB
          C=C+DC
          WRITE(6,10)B,C
 10       FORMAT(2X,'B=',F10.6,4X,'C=',F10.6)
C
C        CONVERGENCE CHECK
C
          IF(SQRT((B-B1)**2+(C-C1)**2) .GT. EPS) THEN
          GO TO 1
          ELSE
          WRITE(6,11)B,C
 11       FORMAT(/2X,'THE QUADRATIC FACTOR :B=',F10.6,3X,'C=',F10.6)
          WRITE(6,12)D1,D2
 12       FORMAT(/2X,'REMAINING POLYNOMIAL:D1=',F10.6,3X,'D2=',F10.6)
          END IF
          STOP
          END
```

Figure 4.9.1 Program for Bairstow's method for finding the quadratic factors and the roots of the polynomial equation of Example 4.9.

and c. The iterative procedure given by Eqs. (4.39) and (4.40) is employed to converge to the desired values of the constants b and c, using a convergence criterion of 10^{-3}. Figure 4.9.1 shows the computer program written for this problem. The d's and e's are obtained and employed to determine the increments in b and c for the next iteration. The starting values are taken as $b = -2$ and $c = 1$. The iterative process converges to $b = -3$ and $c = +2$, which gives the quadratic factor as $x^2 - 3x + 2$. The remaining polynomial is $x^2 - 7x + 12$; see Fig. 4.9.2.

```
THE INITIAL GUESS: B= -2.0000     C=  1.0000

B= -2.776470     C=  1.352941
B= -3.162417     C=  1.922668
B= -2.825542     C=  1.902719
B= -2.969001     C=  1.973022
B= -2.999051     C=  1.998892
B= -3.000000     C=  1.999998
B= -3.000003     C=  2.000003

THE QUADRATIC FACTOR :B= -3.000003    C=  2.000003

REMAINING POLYNOMIAL:D1= -7.000000    D2= 12.000000
```

(a)

```
THE INITIAL GUESS: B=  0.0000     C=  5.0000

B= -1.120000     C=  2.200001
B= -2.010739     C=  1.453996
B= -2.633198     C=  1.577560
B= -2.957543     C=  1.865094
B= -3.005317     C=  1.999321
B= -2.999916     C=  1.999945
B= -2.999983     C=  1.999984

THE QUADRATIC FACTOR :B= -2.999983    C=  1.999984

REMAINING POLYNOMIAL:D1= -7.000084    D2= 12.000400
```

(b)

```
THE INITIAL GUESS: B=  2.0000     C= -2.0000

B= -1.967654     C=-13.789750
B= -4.887068     C=-14.755200
B= -4.943721     C= -4.912420
B= -4.972162     C= -0.009018
B= -4.986652     C=  2.395084
B= -4.994331     C=  3.505467
B= -4.998444     C=  3.918154
B= -4.999864     C=  3.996880
B= -4.999969     C=  3.999967
B= -4.999969     C=  3.999973
```

(c)

Figure 4.9.2 The numerical results for Example 4.9, employing three different starting values.

Therefore, the roots of the given equation, Eq. (4.8.1), may be obtained by solving the two quadratic equations

$$x^2 - 3x + 2 = 0 \quad \text{and} \quad x^2 - 7x + 12 = 0 \tag{4.9.1}$$

The four roots of the given equation are found to be 4, 3, 2, and 1. It is evident that the quadratic factor derived from the given polynomial equation will depend on the starting values of b and c. Six different quadratic factors are possible, since the first root may be combined with any one of the three remaining ones, the second root with two remaining ones, and the third with the fourth one, to yield a quadratic factor each. The convergence of the method is quite rapid, as shown in Fig. 4.9.2. The convergence criterion EPS must also be varied to ensure a negligible dependence of the results on the value chosen. Results are also shown for two different sets of starting values and convergence to a different quadratic factor is observed in the last case.

Note that the method is not very difficult to program, and, unless the initial guesses are very bad, the convergence is quite rapid. Thus, the method is attractive for problems, such as the one considered here, for which the approximate range in which the roots lie is not known. If multiple or complex roots exist, this method can again be used conveniently. The Newton-Raphson method can also be used for all of these cases, and it is therefore extensively used in engineering applications because of its relative simplicity. However, it diverges in many cases, unless an initial guess that is fairly close to the root is employed. Then methods such as Graeffe's and Bairstow's methods may be employed advantageously.

4.9 SUMMARY

In this chapter, several methods for finding the roots of polynomial and transcendental equations have been presented. The discussion considers some of the most important and widely used methods, outlining their limitations and advantages. There are obviously many more methods available in the literature. Many of these are based on considerations and techniques quite similar to those discussed here. Among those that may be mentioned are Brent's method, Muller's method for polynomials, and Bernoulli's method (Carnahan et al., 1969, and Atkinson, 1978).

An important class of methods have been developed using minimization principles. Ward's method uses these principles to find the roots of a complex polynomial equation $f(z) = 0$, where z is a complex variable and

$$f(z) = u(x, y) + iv(x, y) \quad \text{and} \quad z = x + iy \tag{4.41}$$

Ward's method seeks to minimize the function $p(x, y)$, where

$$p(x, y) = |u(x, y)| + |v(x, y)| \tag{4.42}$$

The method is iterative, and at each step the value of $p(x, y)$ is compared with the values at $(x + \Delta x, y)$, $(x - \Delta x, y)$, $(x, y + \Delta y)$, and $(x, y - \Delta y)$, where Δx and Δy are increments in x and y. If a smaller value of $P(x,y)$ is found at any of these four points, that point becomes the new (x,y) location. If the four points do not give a smaller value of p, Δx and Δy may be reduced until they do. Otherwise, the minimum has been reached. Therefore, the method simply moves in the direction of decreasing $p(x,y)$. The root is obtained when u and v become zero, or smaller than a chosen convergence parameter. The search for a minimum value of p is, therefore, expected to lead to the root. However, convergence does not necessarily occur. Similarly, other functions, such as $(u^2 + v^2)$, may be minimized. Other minimization techniques are also available from mathematical procedures that have been developed for the optimization of systems. Search methods are frequently used in optimization, the above procedure being generally termed as *lattice search* (Stoecker, 1980). In engineering problems, one may encounter equations that are so complicated that the various methods discussed in this chapter may not be convenient to use. In such cases, one may resort to minimization methods to obtain the desired roots.

The selection of the method for solving a given problem depends on the nature of the equation and of the roots. The physical characteristics of the problem, if known, are useful in choosing the method and in determining the interval over which the roots are sought. The incremental search method may be used to yield the approximate nature of the function $f(x)$ and the approximate location of the roots. Once this information has been obtained, one may switch to a method, such as the Newton-Raphson method, that converges more rapidly. In most problems of practical interest, prior information on the nature and location of roots is available. This information should be built into the computer program to obtain rapid convergence to the roots and to reject unacceptable values. The convergence parameter ε must be varied, say, to half its value, to ensure that the results obtained are not significantly affected by the value of ε chosen. Several of these considerations, related to the computational procedure, were also discussed earlier in Chapter 2 and may be employed in developing the computer program. Some of the methods presented here can also be extended to the solution of simultaneous equations, as discussed in the next chapter. Further details on the various methods considered in this chapter may be obtained from the discussions given by Traub (1964), Ostrowski (1966), Householder (1970), Brent (1973), and Atkinson (1978). Ralston and Rabinowitz (1978), Rice (1983), and Gerald and Wheatley (1984) may also be consulted for the mathematical background of some of the methods outlined here.

PROBLEMS

4.1. We wish to find the cube root of 17, that is, $17^{1/3}$, by root solving. Set up the equation to be solved, and outline a method to compute the desired value.

4.2. For the following equation, find the first two positive roots, which represent the lowest frequencies of natural vibration of a mechanical system:

$$\tan x = \tanh x$$

Use the search method to obtain an accuracy of order 10^{-3} on the roots.

4.3. The root of an algebraic equation is known to be between 0 and 800 m/s. This root is to be determined to an accuracy of ± 0.1 m/s by the bisection method. Derive Eq. (4.7) and use it to determine the number of bisections needed to achieve this accuracy.

4.4. The voltage x at a given junction in an electrical circuit is given by the first positive root of the equation

$$f(x) = \log_{10} x + x^2 - 6 = 0$$

Employ the bisection method, following the determination of the interval containing the root by the search method, to obtain the voltage.

4.5. The equation that governs the frequency of vibration of a cantilever beam is of the form

$$\cos x \cosh x = -1$$

Use the search method to obtain the approximate locations of the first two positive roots of this equation. Then use the regula falsi method to converge to the roots.

4.6. If the derivative $f'(x_i)$ in Eq. (4.11) is replaced by its backward finite difference approximation, Eq. (3.12), obtain the resulting recursive formula, and compare it with that for the secant method.

4.7. The roots of the equation

$$\tan x = \frac{B}{x}$$

where B is a constant, are needed in a series representation of the temperature field in a conduction heat transfer problem. Use the search method to obtain the approximate values of the first three positive roots of this equation, and then use the Newton-Raphson method to obtain these more accurately. Consider two values, 1.0 and 2.0, of B. Also use the successive substitution method, and compare the results with those obtained by Newton's method.

4.8. For the physical problem discussed in Example 4.1, take the values of T_1, h, and k as 1500 K, 10 W/m²K, and 50 W/mK. Using the Newton-Raphson method, find the resulting surface temperature. The temperature is to be determined to an accuracy of 0.01 K.

4.9. Explain what is meant by the statement that the Newton-Raphson method has second-order convergence. Obtain the general form of the corresponding convergence formulas for search, bisection, and successive substitution methods. Compare these with that for the Newton-Raphson method, and discuss the resulting difference in convergent rate.

4.10. Use the Newton-Raphson method for root solving to find the nth root of a number N, that is, $x^n = N$, in order to derive the formula

$$x_{i+1} = \frac{x_i}{n}\left[n - 1 + \frac{N}{x_i^n}\right]$$

4.11. The temperature of an electrically heated wire is to be determined from its energy balance. If the energy input into the wire due to the electric current is 1000 W/m², the resulting equation is obtained as

$$1000 = 0.5 \times 5.67 \times 10^{-8} \times [T^4 - (300)^4] + 10 \times (T - 300)$$

Determine the temperature T of the wire by employing the search method. Since the equation is a fourth-order polynomial equation, there are four roots. How would you choose the correct solution?

4.12. In a manufacturing process, a spherical piece of metal is subjected to radiative and convective heat transfer, resulting in the energy balance equation

$$f(T) = 0.6 \times 5.67 \times 10^{-8} \times [(850)^4 - T^4] - 40 \times (T - 350) = 0$$

Obtain a rough plot of the function $f(T)$ versus T, and use the secant method to find the real root in the range $350 < T < 850$.

4.13. The Planck distribution for the emission of radiation from a blackbody is

$$E = \frac{c_1}{\lambda^5 \left[\exp\left(\frac{c_2}{\lambda T} \right) - 1 \right]}$$

where E is termed the monochromatic emissive power, λ is the wavelength of radiation, T is the temperature, and c_1, c_2 are constants. We wish to find the value of λ at which E is a maximum. Therefore, the first positive real root of the equation

$$\frac{dE}{d\lambda} = 0$$

is to be determined. Find this value and compare it with the result given in the literature as $\lambda_{max} = 2897.6/T$, where λ is in μm and T is in K. Take the constants c_1 and c_2 as 3.741×10^8 and 1.439×10^4 in the appropriate units.

4.14. In the turbulent flow of a fluid through a smooth pipe, the frictional force on the fluid is represented in terms of a friction factor f, which is positive and less than 0.1. The equation for f is

$$\frac{1}{f^{1/2}} = 2 \log_{10}(\text{Re}\, f^{1/2}) - 0.8$$

where Re is a constant, termed the Reynolds number, which varies with the fluid properties, flow rate, and tube diameter. Obtain the approximate value of the friction factor by the search method, and then use the Newton-Raphson method to converge to the root for $\text{Re} = 10^4$ and 10^6.

4.15. If the fluid, in the physical circumstance of the above problem, flows through a rough pipe, whose roughness is given by a parameter ε/D, where ε is the physical size of the surface protrusions and D is the pipe diameter, the friction factor is given by

$$\frac{1}{f^{1/2}} = -2 \log_{10}\left(\frac{\varepsilon/D}{3.7} + \frac{2.51}{\text{Re}\, f^{1/2}} \right)$$

Obtain the friction factor f for $\varepsilon/D = 10^{-4}$ at $\text{Re} = 10^6$ and also for $\varepsilon/D = 4 \times 10^{-4}$ at $\text{Re} = 10^7$, using the search method.

4.16. The equation of state for a gas is given by the Van der Waals equation

$$\left(P + \frac{a}{v^2} \right)(v - b) = RT$$

where P is the pressure, v is the specific volume, T is the temperature, R is the gas constant, and a, b are constants that depend on the gas. For $P = 70$ atm, $T = 200$ K, $R = 0.08205$ liter atm/mole K, $a = 3.59$, and $b = 0.0427$, the specific volume is given in liters/mole. Find this value using the Newton-Raphson method, after obtaining the approximate value by the search method.

4.17. Use Graeffe's method to obtain all the roots of the following polynomial equations:

$$x^4 - 10x^3 + 35x^2 - 50x + 24 = 0$$

$$x^4 - 5x^3 + 5x^2 + 5x - 6 = 0$$

Also use the search method for the real roots. Compare the values obtained by the two methods.

4.18. Solve the problem discussed in Example 4.4 by the regula falsi and secant methods. Compare the results obtained and the iterations needed for convergence with those for the Newton-Raphson method.

4.19. Solve the problem considered in Example 4.7 by the Newton-Raphson method, and compare the value of the root and the convergence characteristics with those discussed for the successive substitution method. Comment on the observed differences.

4.20. A cylindrical probe of diameter D is placed in a stream of air, and the energy transfer from it is measured as 100 W. If the energy balance equation is obtained as

$$\left[\frac{60}{D} D^{0.466} + 50\right] \pi D = 100$$

find the diameter D of the probe using the bisection method. Also write the equation as $f(D) = 0$, draw an approximate plot of $f(D)$ versus D, and discuss the behavior of the function as D increases from zero to 0.01 m.

4.21. A loan of \$5000 is taken from a bank that charges a nominal annual interest rate i, compounded monthly. A payment of \$200 is made each month, starting at the end of the first month, toward the loan. If it takes 36 months to pay off the loan, the rate of interest i may be determined from the following equation, which is obtained by summing the present worth of the monthly payments (see Example 2.2):

$$\frac{\left(1 + \frac{i}{12}\right)^{36} - 1}{\frac{i}{12}\left(1 + \frac{i}{12}\right)^{36}} \times 200 = 5000$$

Find this interest rate, by any method of your choice. An accuracy of 10^{-3} on i is adequate.

4.22. A bond of \$1000 yields 8% interest annually and has 7 years to maturity. It is sold for \$500 due to the prevailing higher interest rate i. If the buyer achieves the current interest rate on his investment, the equation governing the transaction is obtained by equating the monetary value of the bond before and after the sale (Stoecker, 1980) as follows:

$$1000 + 1000 \times 0.08 \times \frac{(1 + i)^7 - 1}{i} = 500 \times (1 + i)^7$$

Find the prevailing interest rate i from this equation, using any suitable method.

4.23. A function $y(x)$ is given as

$$y(x) = \frac{\log x \cdot \sin \frac{x^2}{25}}{x}$$

where log represents the natural logarithm. Determine the maximum value of y for $x > 1.5$, using the search method.

4.24. Use the Newton-Raphson and the second-order Newton's methods for finding the nonzero real roots of the equations

$$f(x) = e^x - 5 - x^2 = 0$$

$$f(x) = e^{-x} + 1 - x^2 = 0$$

Compare the results and the rate of convergence obtained by the two methods. Obtain the roots for two values of ε, 10^{-3} and 10^{-5}, where ε is the convergence criterion applied to the two functions represented by $f(x)$.

4.25. Find all the roots of the polynomial obtained in Example 4.1 by Bairstow's method, and show that only one is acceptable because of the physical considerations of the problem.

4.26. Using Graeffe's method, obtain the four roots corresponding to the polynomial in Problem 4.11. Again, show that three of them are not acceptable.

4.27. For the following polynomial equations, use both Graeffe's and Bairstow's methods to determine all the roots:

$$x^4 - 8x^3 + 22x^2 - 24x + 9 = 0$$

$$x^5 - 5x^4 - 16x + 80 = 0$$

Compare the results obtained by these methods and the computational effort needed in the two cases.

4.28. Use the search method to determine the approximate location of the real roots and the behavior of the polynomial functions in the above problem. Once this information has been obtained, how would you choose the method for finding the real roots more accurately?

4.29. Use the Newton-Raphson method for finding the real roots, in the range $0 < x < 1.5$, of the following polynomial equation, which represents the variation of the force on a vertical structure with distance x:

$$f(x) = x^4 - 3x^3 + x^2 + 3x - 2 = 0$$

Also use the modified Newton's method for the problem and compare the convergence in the two cases. Obtain a rough plot of $f(x)$ versus x to guide the choice of the starting value.

4.30. Using Graeffe's method, find all the roots of the following polynomial equations:

$$x^7 - 3x^6 + 2x^5 - 32x^2 + 96x - 64 = 0$$

$$x^5 - 15x^4 + 85x^3 - 225x^2 + 274x - 120 = 0$$

$$x^6 - 21x^5 + 175x^4 - 735x^3 + 1624x^2 - 1764x + 720 = 0$$

Also use the Newton-Raphson method to find the real roots in the range $0 < x < 1.5$ in these three cases. Compare the convergence and the accuracy obtained by the two methods. The real roots give the frequencies of vibration of systems represented by these equations.

4.31. Use all the applicable methods, discussed in the text, for finding the first positive real root of the following transcendental equations:

$$\cosh x = 4x$$

$$\sin x = \cos^2 x$$

$$2e^x + 5x - 4 = 0$$

Compare the convergence of these methods to the root and the computational effort involved.

4.32. The calibration curve for a temperature-measuring device is given by

$$T = 15 + 3.5V + 0.6V^2 + 0.5V^3 + 0.1V^4$$

where T is the temperature in °C and V is the voltage signal. Determine the voltage output at $T = 30$°C and 60°C. Use any suitable method.

4.33. Obtain a rough plot of the function $f(x)$ where

$$f(x) = e^{-x/3}(4 - x) - 2 - x$$

and determine the real roots by the bisection method. How many bisections are needed to locate the roots with a convergence criterion of $\varepsilon = 10^{-5}$, where ε is the change in the root from one iteration to the next?

4.34. Determine the effect of a variation in the convergence criterion ε on the value of the root obtained in Problem 4.8. Take ε varying from 1 to 10^{-5}.

4.35. The critical load for the buckling of a vertical column is governed by the transcendental equation

$$\tan x = x$$

where $\sqrt{x}$ represents the critical load. Solve this equation by the modified Newton's method to obtain a real positive root, starting with an initial guess of $x = 4.0$. Also try the successive substitution method. Discuss your results.

4.36. The real and complex roots of the following polynomial equation are related to the stability of a body subjected to a system of forces:

$$x^4 + 3x^3 + 6x^2 + 7x + 3 = 0$$

where x represents the complex amplification factor for the disturbance. Find these roots, using the Newton-Raphson method. Also use Newton's second-order method for the real root, which is a multiple root at $x = -1$. Compare the convergence by the two methods.

4.37. The decomposition of carbon dioxide into oxygen and carbon monoxide is governed by the equation

$$\left(\frac{P}{E^2} - 1\right)x^3 + 3x - 2 = 0$$

where P is the pressure in atmospheres, E is the temperature-dependent equilibrium constant, and x is the fractional decomposition of CO_2. Using any suitable method, find x for $P = 1$ atm and $E = 1.65$.

4.38. The vapor pressure P of a material is given in terms of the temperature T as

$$\log P = a + \frac{b}{T} + c \log T$$

where log is the natural logarithm and a, b, and c are constants that depend on the material. If their values are given as 17.5, -2.2×10^4, and -0.9, respectively, find the temperatures at pressures of 0.01 and 0.1 atmospheres, using the search method to determine the approximate value of the root, followed by the Newton-Raphson method to converge to the root.

4.39. When water vapor is heated to very high temperatures, it dissociates to give oxygen and hydrogen. Then the mole fraction, x, of water that dissociates is given by the equation

$$S = \frac{x}{1 - x}\sqrt{\frac{2P}{2 + x}}$$

where S is the equilibrium constant of the reaction and P is the pressure of the mixture. If the pressure P is given as 3.2 atmospheres and S as 0.055, compute the value of x that satisfies the above equation, using any suitable method.

4.40. The current I in an electrical circuit containing resistances R_1 and R_2, inductance L, and voltage source E is given by the equation

$$I = \frac{E}{R_1}\left[1 - \frac{R_2}{R_1 + R_2} e^{(-R_1/L)\tau}\right]$$

where τ is the time. If $R_2 = 10$ ohms, $L = 10$ henries, and $E = 20$ volts, find the resistance R_1 which gives a current of 1.4 amperes at $\tau = 0.5$ second.

5

Numerical Solution of Simultaneous Algebraic Equations

5.1 INTRODUCTION

Systems of simultaneous algebraic equations are frequently encountered in engineering applications such as those concerned with electrical networks, structural analysis, heat transfer, fluid flow, optimization, vibrations, chemical reactions, and data analysis. The numerical solution of ordinary and partial differential equations also often reduces to a solution of a set of algebraic equations, as discussed in Chapters 8 and 9. A system of n simultaneous equations, with $x_1, x_2, \ldots, x_n$ as the n unknowns, may be written as

$$\begin{aligned} f_1(x_1, x_2, \ldots, x_n) &= 0 \\ f_2(x_1, x_2, \ldots, x_n) &= 0 \\ \vdots \qquad\quad & \vdots \\ f_n(x_1, x_2, \ldots, x_n) &= 0 \end{aligned} \tag{5.1}$$

where $f_1, f_2, \ldots, f_n$ denote n different functions of the n independent variables. Various methods have been developed to solve this system of equations to obtain the values of the variables $x_1, x_2, \ldots, x_n$. The choice of a particular method for a given problem generally depends on the nature of the equations and the number of unknowns n.

In many circumstances, the equations are linear in the unknown variables. Such a system of linear equations has the general form

$$\begin{aligned} a_{11}x_1 + a_{12}x_2 + \cdots + a_{1n}x_n &= b_1 \\ a_{21}x_1 + a_{22}x_2 + \cdots + a_{2n}x_n &= b_2 \\ \vdots \qquad\qquad\qquad & \vdots \\ a_{n1}x_1 + a_{n2}x_2 + \cdots + a_{nn}x_n &= b_n \end{aligned} \tag{5.2}$$

where the a's represent n^2 coefficients and the b's similarly represent n constants. In matrix notation, this system may be written more concisely as

$$AX = B \tag{5.3}$$

where A is a square matrix of the coefficients, X is a column matrix of the unknowns, and B is a column matrix of the constants that appear on the right-hand side of the equations. From Eq. (5.2), a_{ij} represents an element of the matrix A and b_i an element of the vector B. If the column matrix is zero, that is, if the b's are all zero, the set of equations is homogeneous, and nontrivial solutions can be obtained only if all the equations are not independent, as discussed later in this chapter. If the b's are not all zero, the set of equations is nonhomogeneous. In this case, all the equations must be independent in order to yield unique values of the unknowns.

A system of linear algebraic equations may be solved by employing Cramer's rule which gives the unknown x_i, as

$$x_i = \frac{\text{Det } A_i}{\text{Det } A} \tag{5.4}$$

where A_i is the matrix A with its ith column replaced by the column vector B and Det represents the determinant of the corresponding matrices. It can be shown that the number of basic arithmetic operations needed to solve for all the unknowns in a set of equations by Cramer's rule, employing expansion by minors for obtaining the determinants, is $(n + 1)!$ Therefore, this method is satisfactory only for a small number of equations, generally less than 5. For the large sets of equations generally encountered in engineering problems, the time required to solve the equations using Cramer's rule is very large and the method is quite impractical, as compared to other methods discussed in this chapter. However, Eq. (5.4) indicates some important points regarding the solution of linear simultaneous algebraic equations.

If Det $A = 0$, the matrix A is termed *singular*, and no unique solution can be obtained if the numerator is nonzero. However, if the column matrix B is also zero, then Det $A_i = 0$, since one entire column of the matrix A_i is zero. In this case, nontrivial solutions can be obtained. This is the circumstance of homogeneous equations, which give rise to eigenvalue problems, as discussed later in this chapter. If Det $A \neq 0$, the equations are said to be *all independent*, and unique solutions may be obtained for nonhomogeneous equations. A very brief outline of matrix algebra needed for the following discussion is given here. For further details, textbooks on matrices such as Lancaster (1969) and Bronson (1970) may be consulted.

The solution to the set of linear equations given by Eq. (5.3) may also be written as

$$X = A^{-1}B \tag{5.5}$$

where A^{-1} is the inverse of the matrix A, which must be nonsingular for the inverse to exist. Then $A^{-1}A = I$, where I is the identity, or unit, matrix, which is a square matrix consisting of zeros everywhere except at the diagonal, where all the elements are unity.

Several methods for the solution of simultaneous linear equations are based on obtaining A^{-1} as an intermediate step. This is particularly advantageous if solutions are to be obtained for many systems of equations in which A is unchanged and only the column vector B varies. Also, many computer systems have available programs to invert a matrix. These programs may often be employed for solving a set of equations. However, most methods, which are used for solving simultaneous algebraic equations, do not solve for A^{-1} as an intermediate step and solve only for X, unless there is a particular advantage in obtaining A^{-1}, such as those mentioned above, or unless the information is needed to study the nature of the equations.

There are two different types of methods, direct and iterative, that may be adopted for solving Eq. (5.3) for the unknown X. Direct methods solve the equations exactly, except for the computational round-off error, in a finite number of operations. The methods based on finding the inverse matrix A^{-1} to obtain the solution from Eq. (5.5) fall under direct methods, as do several other methods discussed in this chapter. These direct techniques are particularly useful when the number of equations to be solved is typically less than 40. However, in recent years, a few special methods have been developed for particular types of equations. These methods may be used advantageously even for a much larger number of equations. Among these are the fast Fourier transform method and the cyclic reduction method, both of which are particularly suited for the large number of algebraic equations obtained from finite difference approximations of partial differential equations, as discussed in Chapter 9. The second class of methods is based on iteration. Iterative methods are appropriate for large systems of algebraic equations, typically of the order of 100 or more equations, in which the sparseness of the unknowns in the equations often makes iterative computation more efficient. Again, these methods are of particular interest in the finite difference and finite element solutions of partial differential equations.

In this chapter, both direct and iterative methods for the numerical solution of systems of linear algebraic equations are discussed. Most of the direct methods are based on matrix inversion or on elimination and reduction so that the given set of equations is obtained in a form that is amenable to a direct solution. Among those discussed here are the Gaussian elimination, Gauss-Jordan elimination, Crout's decomposition, and matrix inversion methods. The iterative methods discussed here include the Jacobi, Gauss-Seidel, and relaxation methods. The solution of homogeneous linear equations, which often result in eigenvalue problems, is also discussed. The methods outlined for obtaining the eigenvalues and the corresponding eigenvectors include the Gauss-Jordan method, the power method, the Jacobi method, and Householder's method, used in conjunction with the LR, QR, and QL algorithms. Nonlinear algebraic equations are also of interest in many engineering applications, and the solution of these equations is outlined. In most cases, the equations are linearized in order to employ the methods applicable to linear equations for solving systems of nonlinear equations. For small systems of nonlinear equations, methods based on those discussed in Chapter 4 for single nonlinear algebraic equations, such as the Newton-Raphson and the successive substitution methods, may be employed.

5.2 GAUSSIAN ELIMINATION

Gaussian elimination is a direct method for solving a system of linear algebraic equations and is very frequently employed in a wide variety of engineering problems. By a process of elimination of the unknowns, the method reduces the given set of n equations to a triangular set, so that one of the equations has only one unknown. This unknown is determined and the remaining unknowns are obtained by the process of back-substitution. This method is of particular interest since several other direct methods are based on it.

5.2.1 Basic Approach

Let us consider a general system of three linear equations, given as

$$\begin{aligned} a_{11}x_1 + a_{12}x_2 + a_{13}x_3 &= b_1 \\ a_{21}x_1 + a_{22}x_2 + a_{23}x_3 &= b_2 \\ a_{31}x_1 + a_{32}x_2 + a_{33}x_3 &= b_3 \end{aligned} \tag{5.6}$$

As a first step, eliminate x_1 from the second equation by adding it to the equation obtained by multiplying the first equation by $-a_{21}/a_{11}$. Similarly, multiply the first equation by $-a_{31}/a_{11}$ and add the third equation to it to eliminate x_1 from the third equation as well. The resulting system of equations is

$$\begin{aligned} a_{11}x_1 + a_{12}x_2 + a_{13}x_3 &= b_1 \\ a_{22}^{(1)}x_2 + a_{23}^{(1)}x_3 &= b_2^{(1)} \\ a_{32}^{(1)}x_2 + a_{33}^{(1)}x_3 &= b_3^{(1)} \end{aligned} \tag{5.7}$$

where the superscripts indicate new values of the coefficients after the first step.

The first equation, which has been used to eliminate the unknown x_1 from the equations that follow, is known as the *pivot equation*, and the coefficient a_{11} of the eliminated unknown is the *pivot coefficient* or element. In the next step, multiply the second equation by $-a_{32}^{(1)}/a_{22}^{(1)}$ and add it to the third equation, in order to eliminate x_2 from the latter. The result is the triangular set

$$\begin{aligned} a_{11}x_1 + a_{12}x_2 + a_{13}x_3 &= b_1 \\ a_{22}^{(1)}x_2 + a_{23}^{(1)}x_3 &= b_2^{(1)} \\ a_{33}^{(2)}x_3 &= b_3^{(2)} \end{aligned} \tag{5.8}$$

where $a_{33}^{(2)}$ and $b_3^{(2)}$ arise from the second step in the elimination process. Now, x_3 is directly obtained as $b_3^{(2)}/a_{33}^{(2)}$ from the last equation. This value may be substituted in the second equation to obtain x_2, which, along with x_3, is substituted in the first equation to obtain x_1. This process, known as *back-substitution*, can be employed easily with a triangular set of equations to obtain the unknowns.

The above process can easily be extended to n equations, with n unknowns. Employing successive pivot equations, the elimination procedure, outlined above, is carried out until the original system of equations is reduced to a triangular set. Back-substitution then yields the unknowns. To illustrate the Gaussian elimination method, let us consider the following set of equations:

$$\begin{aligned} 3x + 5y + z &= 16 \\ x + 4y + 2z &= 15 \\ 2x + 2y + 3z &= 15 \end{aligned} \tag{5.9}$$

where x, y, and z are the unknowns. With the first equation as the pivot equation, x is eliminated from the other two equations to obtain

$$\begin{aligned} 3x + 5y + z &= 16 \\ \frac{7y}{3} + \frac{5z}{3} &= \frac{29}{3} \\ -\frac{4y}{3} + \frac{7z}{3} &= \frac{13}{3} \end{aligned} \tag{5.10}$$

Now, we use the second equation as the pivot equation to eliminate y from the third equation. This results in the following triangular set of equations:

$$\begin{aligned} 3y + 5y + z &= 16 \\ \frac{7y}{3} + \frac{5z}{3} &= \frac{29}{3} \\ \frac{69z}{21} &= \frac{207}{21} \end{aligned} \tag{5.11}$$

The fractional coefficients in the resulting equations may be avoided by multiplying on both side by the largest denominator. The value of z is obtained as 3 from the third equation. Back-substitution then yields the values of the remaining unknowns. The result is

$$x = 1 \qquad y = 2 \qquad z = 3 \tag{5.12}$$

5.2.2 Computational Procedure

The Gaussian elimination method is very well suited for digital computation. The system of linear equations is written in matrix form as

$$AX = B \tag{5.3}$$

We may consider an augmented matrix C of this set, defined as follows:

$$C = \begin{bmatrix} a_{11} & a_{12} & \cdots & a_{1n} & a_{1,n+1} \\ a_{21} & a_{22} & \cdots & a_{2n} & a_{2,n+1} \\ \vdots & & & & \vdots \\ a_{n1} & a_{n2} & \cdots & a_{nn} & a_{n,n+1} \end{bmatrix} \tag{5.13}$$

where the $(n+1)$th column consists of the b's, with $a_{1,n+1} = b_1$, $a_{1,n+2} = b_2$, and so on. The computational procedure is then concerned with reducing this matrix to the following upper triangular augmented matrix after $(n-1)$ elimination steps:

$$C^{(n-1)} = \begin{bmatrix} a_{11} & a_{12} & a_{13} & \cdots & a_{1n} & a_{1,n+1} \\ 0 & a_{22}^{(1)} & a_{23}^{(1)} & \cdots & a_{2n}^{(1)} & a_{2,n+1}^{(1)} \\ \vdots & & & & & \vdots \\ 0 & 0 & 0 & \cdots & a_{nn}^{(n-1)} & a_{n,n+1}^{(n-1)} \end{bmatrix} \tag{5.14}$$

See Fig. 5.1 for a few special types of matrices that are of particular interest in the numerical solution of linear systems.

The first step in the reduction of the matrix C to the augmented triangular

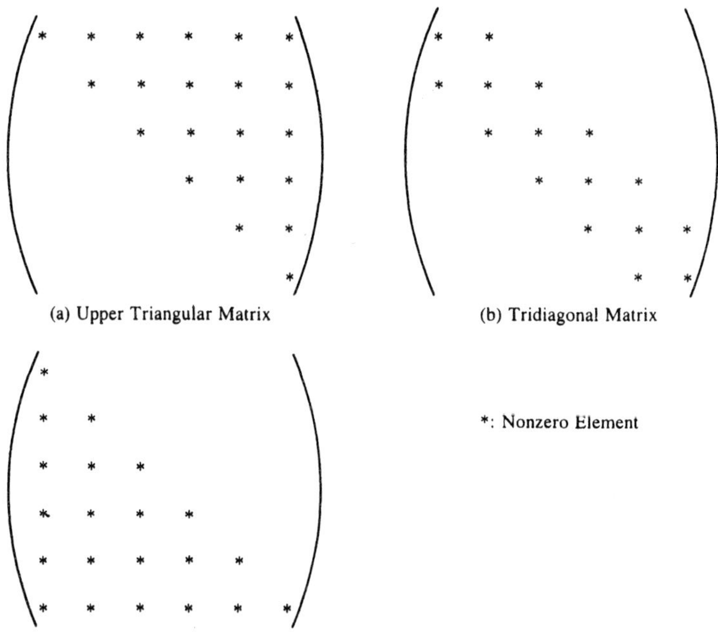

Figure 5.1 Sketch of a few special types of matrices that are of interest in the solution of simultaneous algebraic equations.

matrix $C^{(n-1)}$ is obtained from a generalization of the procedure outlined above. Therefore, the elements $a_{ij}^{(1)}$ of the matrix obtained after the first elimination step are given by

$$a_{ij}^{(1)} = a_{ij} - \frac{a_{i1}}{a_{11}}(a_{1j}) \qquad \text{where } 2 \leq i \leq n \quad \text{and} \quad 1 \leq j \leq n+1 \tag{5.15}$$

Here, a_{ij} are the elements of the original matrix C. The first row is the pivot row and a_{11} is the pivot element. Note that the first element of each row, except the first one, becomes zero after this computational step.

Similarly, the complete elimination procedure may be generalized by recognizing that the pivot row varies from the first to the $(n-1)$th row as elimination proceeds and that the pivot element for each step is a_{rr}, where r denotes the number of the pivot row. Therefore, the general procedure is written as

$$a_{ij}^{(r)} = a_{ij}^{(r-1)} - \frac{a_{ir}^{(r-1)}}{a_{rr}^{(r-1)}}[a_{rj}^{(r-1)}] \qquad \text{where } r+1 \leq i \leq n \quad \text{and} \quad r \leq j \leq n+1 \tag{5.16}$$

The superscripts within parentheses indicate the elimination step, with r varying from 1 to $(n-1)$. For $r = 1$, Eq. (5.15) is obtained, with the superscript (0) simply denoting the coefficients of the original matrix. Again, note that the element in the rth column of all the rows that follow the pivot row becomes zero. Therefore, an upper triangular matrix, augmented with the modified values of the b's which were the given constants on the right-hand side of the given system of equations, is obtained in the form represented by the matrix $C^{(n-1)}$ after successively applying the above procedure to all the pivot rows, going from the first to the $(n-1)$th row. Once this reduced matrix has been obtained, as shown in Fig. 5.1(a), the unknown x_n is obtained from the elements of the matrix as

$$x_n = \frac{a_{n,n+1}}{a_{nn}} \tag{5.17}$$

The other unknowns are obtained from back-substitution as

$$x_i = \frac{a_{i,n+1} - \sum_{j=i+1}^{n} a_{ij}x_j}{a_{ii}} \qquad \text{where } i = n-1, n-2, \ldots, 2, 1 \tag{5.18}$$

where the superscripts have been dropped for convenience. In a computer program, the old elements may be replaced by new ones as elimination proceeds. No superscripts are needed if j is varied from $(r+1)$ to $(n+1)$ in Eq. (5.16) and the elements in the rth column are simply replaced by zero. Back-substitution yields the unknowns in the reverse order, starting with x_n and ending with x_1, as given by Eq. (5.18).

Let us apply the generalized procedure outlined above to the example given earlier, whose augmented matrix C is

$$C = \begin{bmatrix} 3 & 5 & 1 & 16 \\ 1 & 4 & 2 & 15 \\ 2 & 2 & 3 & 15 \end{bmatrix} \tag{5.19}$$

If Eq. (5.16) is successively applied, with the first row as the pivot row in the first step and then the second row, the matrices obtained in the two steps are

$$\begin{bmatrix} 3 & 5 & 1 & 16 \\ 0 & 7/3 & 5/3 & 29/3 \\ 0 & -4/3 & 7/3 & 13/3 \end{bmatrix} \quad \text{and} \quad \begin{bmatrix} 3 & 5 & 1 & 16 \\ 0 & 7/3 & 5/3 & 29/3 \\ 0 & 0 & 69/21 & 207/21 \end{bmatrix}$$

Therefore, Eq. (5.17) yields $x_3 = 3$, and back-substitution, given by Eq. (5.18), gives $x_2 = 2$ and $x_1 = 1$. The generalized procedure can easily be applied on the computer, as discussed in Example 5.1.

It may be mentioned here that the solution of n linear equations by Gaussian elimination can be shown to require about $n^3/3$ arithmetic operations, as compared to n^3 for the multiplication of two $n \times n$ matrices (Atkinson, 1978). The method can also be used for matrix inversion and for the evaluation of a determinant, as outlined later in this section. We now proceed to a consideration of the accuracy of the results obtained by Gaussian elimination.

5.2.3 Solution Accuracy

In the above example, the Gaussian elimination method yielded the exact solution of the given system of linear equations, because the number of equations was small and only whole numbers and exact fractions were involved. On the computer, however, fractions are replaced by decimals, retaining a limited number of significant places. As a consequence, a round-off error is introduced in dealing with quantities that have a larger number of significant decimal digits than those retained in the computation. The round-off error and its effect on accuracy and convergence were considered in detail in Chapter 2. Note from the earlier discussion that the round-off error may substantially affect the accuracy of the solution if a large number of equations are involved. Consequently, Gaussian elimination is generally used if the number of equations is typically less than 20 if most of the unknowns arise in each equation. Such a system gives rise to a dense coefficient matrix and, consequently, each element has to be considered at each elimination step. If, however, only a few unknowns are present in each equation, a sparse coefficient matrix is obtained. In certain special cases, several elimination steps are avoided and a larger number of equations may be solved by Gaussian elimination, while obtaining an acceptable accuracy level. An important example of such a sparse system is the *tridiagonal matrix*, which has nonzero elements only at the diagonal and on either side of the diagonal; see Fig. 5.1(b). This system arises in the numerical solution of ordinary and partial differential equations and is discussed in greater detail in Example 5.2.

Ill-Conditioned Set. The accuracy of the solution is also substantially influenced by the conditioning of the given system of linear equations. The system is said to be *ill-conditioned* if a relatively small change in one of the coefficients results in a relatively large change in the solution. Similarly, if there are elements in the inverse A^{-1} of the matrix that are several orders of magnitude larger than those in the original matrix A, then the matrix is probably ill-conditioned. The main problem with ill-conditioning is that the round-off error may cause slight changes in the coefficients which may, in turn, result in a large variation in the solution. To test whether the round-off error is significant for a given problem, one may use the solution vector computed to determine the coefficient vector B from the equation $AX = B$ and compare it with the original column matrix. Also, A may be inverted twice and compared with A, or the product $A^{-1}A$ may be compared with the identity matrix I to determine whether the round-off error is large for a given problem. An example of an ill-conditioned set of equations is

$$\begin{aligned} x - 1.9999y &= 0 \\ x - 1.9998y &= 1.0 \end{aligned} \tag{5.20}$$

for which the values of x and y are 19999 and 10^4, respectively. A small change in the coefficients of y can result in a large effect on the solution. Since the round-off error can cause such a small change, the solution obtained may be quite inaccurate. To keep the error small, one may use double precision on the computer. Double precision reduces the speed of computation but is necessary for an ill-conditioned system.

Error Correction. We may also improve the accuracy of the solution obtained by applying an error correction. If X' is the solution vector obtained from numerical computation, the constant vector B' may be found by the substitution of X' into the given system of equations, as $B' = AX'$. Since the solution obtained is not exact due to the round-off error, B' will differ from the original constant matrix B. If X is the exact solution, the error vector E is defined as $X - X'$. Therefore, from the original system of equations, the following equations, known as *error equations*, are obtained:

$$A(X - X') = B - B' \quad \text{or} \quad AE = B - B' \tag{5.21}$$

Thus, the error vector may be computed from this set, which differs from the original system of equations only in the constant vector. For applying this error correction, Gaussian elimination is employed to solve the given system of equations in order to obtain X', from which B' is computed. The multipliers in the elimination process are $a_{ir}^{(r-1)}/a_{rr}^{(r-1)}$. We may retain these to solve other systems of equations with the same coefficient matrix A and a different constant vector B, by applying Eq. (5.16) to the $(n + 1)$th column, which represents the new constant vector. Therefore, the error correction vector E may be obtained. Then a more accurate solution $\bar{X}$ to the set of equations is obtained from

$$\bar{X} = X' + E \tag{5.22}$$

The process may be repeated with the improved solution to obtain a still greater accuracy. However, the exact solution is not obtained at any iterative stage because of the presence of round-off errors.

Pivoting. In the solution of a system of linear equations by Gaussian elimination, we may encounter steps for which the pivot element is zero. In some cases, the pivot element may theoretically be zero but may acquire a small, nonzero value in the computational process due to the round-off error. The use of such a pivot element would lead to very inaccurate results. In fact, the accuracy of the solution is considerably affected by the magnitude of the pivot element, which is employed in all the arithmetic operations for elimination in a given step. Greater accuracy is obtained if reduction is carried out with the row that contains the largest pivot element. A process, known as *partial pivoting*, in which the rows are interchanged at each step, to employ the row with the largest pivot element as the pivot row, is very commonly employed for more accurate results, particularly for large systems of equations. This procedure also avoids the problem with a zero pivot element during the elimination process. In some cases, complete pivoting, with both rows and columns being interchanged to obtain the largest pivot element, is employed, as outlined later for homogeneous equations. Partial pivoting can be easily incorporated in the computer program, as demonstrated in Example 5.1, given at the end of this section.

5.2.4 Matrix Inversion and Determinant Evaluation

Gaussian elimination may also be employed to obtain the inverse of a matrix. The inverse A^{-1} is generally not needed in the solution of a set of linear equations. However, as discussed later, the matrix may be needed for studying the nature of the equations and for solving several systems of equations that have the same coefficient matrix A but a different constant vector B. Finding A^{-1} is equivalent to solving the equation $AX = I$, where X is now an $n \times n$ unknown matrix. An augmented matrix C is formed with the elements of A and I, placing the elements of I on the right-hand side of matrix A, as those of vector B were placed earlier. By applying Gaussian elimination to A, the identity matrix is replaced by the solution matrix, which is the required inverse A^{-1}. The computation of the inverse by this method requires about $(4/3)n^3$ arithmetic operations, which is approximately four times the number of operations required for solving a set of linear equations (Atkinson, 1978). Other methods for calculating the inverse of a matrix are discussed in Section 5.5.

In some engineering problems, such as those that arise in vibrations and in stability analysis, it may be necessary to evaluate a determinant. Although not needed for solving linear equations, since Cramer's rule is rarely applied, the value of the determinant may be required for studying the nature of the equations and for determining whether the inverse of a matrix exists. Gaussian elimination may again be used for the evaluation of a determinant. The value of the determinant of a matrix of an upper triangular form is simply the product of the diagonal elements. Therefore, Gaussian elimination may be used, as discussed earlier for a matrix, to obtain a given determinant in this form. This method is applicable since the value of a determinant is not altered if a constant multiple of the elements of a row or column are added to or subtracted from the elements of another row or column. However, if any two rows or columns are interchanged, the sign of the determinant is changed. Therefore, if partial pivoting is used, to achieve greater accuracy or to avoid a zero pivot element, this change in sign must be taken into account.

Let us consider the determinant of the coefficient matrix of the system of equations given by Eq. (5.9). Then the determinant is

$$\begin{vmatrix} 3 & 5 & 1 \\ 1 & 4 & 2 \\ 2 & 2 & 3 \end{vmatrix}$$

We apply Gaussian elimination to this determinant by using Eq. (5.16) with $r = 1$ and then with $r = 2$ to obtain the upper triangular form

$$\begin{vmatrix} 3 & 5 & 1 \\ 0 & 7/3 & 5/3 \\ 0 & 0 & 69/21 \end{vmatrix}$$

The value of this determinant is given by a product of the diagonal elements and is, therefore, obtained as

$$3 \times \frac{7}{3} \times \frac{69}{21} = 23$$

5.2.5 Tridiagonal Systems

The direct solution of systems of linear equations that have tridiagonal, or banded, coefficient matrices, as shown in Fig. 5.1(b), is important in many practical problems, particularly in the numerical solution of partial differential equations, as discussed in Chapter 9. The set of equations, in this case, may be written as

$$\begin{bmatrix} b_1 & c_1 & 0 & 0 & & \cdots & & & & 0 \\ a_2 & b_2 & c_2 & 0 & & \cdots & & & & 0 \\ 0 & a_3 & b_3 & c_3 & 0 & 0 & \cdots & & & 0 \\ \vdots & & & & & & & & & \vdots \\ 0 & 0 & \cdots & & 0 & 0 & a_{n-1} & b_{n-1} & & c_{n-1} \\ 0 & 0 & & \cdots & & & & 0 & a_n & b_n \end{bmatrix} \begin{bmatrix} x_1 \\ x_2 \\ \vdots \\ x_{n-1} \\ x_n \end{bmatrix} = \begin{bmatrix} d_1 \\ d_2 \\ \vdots \\ d_{n-1} \\ d_n \end{bmatrix} \quad (5.23)$$

Therefore, the matrix A is tridiagonal if only a_{ii}, $a_{i,i-1}$, and $a_{i,i+1}$ are nonzero. This implies that $a_{ij} = 0$ for $|i - j| > 1$. If Gaussian elimination is applied to this system, only one of the a's is eliminated from the column containing the pivot element in each step, since the remaining elements below the diagonal are zero. Therefore, only one elimination process is employed at each step. The original zero elements are kept unchanged, and the resulting system, after completion of the elimination procedure, which is often known as the *Thomas algorithm*, is of the form

$$\begin{bmatrix} b_1 & c_1 & 0 & 0 & \cdots & & & 0 \\ 0 & b_2' & c_2' & 0 & \cdots & & & 0 \\ 0 & 0 & b_3' & c_3' & \cdots & & & \vdots \\ \vdots & & & & & & & \\ 0 & & \cdots & & 0 & 0 & b_{n-1}' & c_{n-1}' \\ 0 & \cdots & & & 0 & 0 & 0 & b_n' \end{bmatrix} \begin{bmatrix} x_1 \\ x_2 \\ \vdots \\ x_{n-1} \\ x_n \end{bmatrix} = \begin{bmatrix} d_1 \\ d_2' \\ \vdots \\ d_{n-1}' \\ d_n' \end{bmatrix} \tag{5.24}$$

where the primes indicate new values. From this system of equations, the unknowns may easily be obtained by back-substitution.

The recursion formulas for the above system may be written in terms of the elements a_i, b_i, and c_i, where i denotes the row in the coefficient matrix. Therefore, the new elements are given by

$$\begin{aligned} a_i' &= 0 \qquad & b_i' &= b_i - c_{i-1}\frac{a_i}{b_{i-1}'} \\ c_i' &= c_i \qquad & d_i' &= d_i - d_{i-1}'\frac{a_i}{b_{i-1}'} \end{aligned} \tag{5.25}$$

with the calculations being carried out for increasing values of i, from $i = 2$ to n. Once the reduced system, given by Eq. (5.25), has been obtained, back-substitution may be applied as follows:

$$x_n = \frac{d_n'}{b_n'} \qquad x_i = \frac{d_i' - c_i' x_{i+1}}{b_i'} \tag{5.26}$$

On the computer, the new elements simply replace the old ones, and primes are not needed. The number of operations needed for solving a tridiagonal system is of order n, $O(n)$, as compared to $O(n^3)$ for a system with a dense coefficient matrix. Therefore, much smaller computing times and, consequently, much smaller round-off errors arise in the solution of such systems. Thus, large tridiagonal systems are generally solved by this method. Example 5.2 illustrates the use of Gaussian elimination for solving systems of equations of the tridiagonal form.

Example 5.1

The specific volume v of saturated steam in m^3/kg is given at six dimensionless temperature T values of 1, 2, 3, 4, 5, and 6, where 1 represents 10°C in physical terms, as, respectively, 106.4, 57.79, 32.9, 19.52, 12.03, and 7.67 by Reynolds and Perkins (1977). Using the Gaussian elimination method, obtain a fifth-order polynomial that passes through these data points.

Solution

The required polynomial is of the form

$$v = x_1 + x_2 T + x_3 T^2 + x_4 T^3 + x_5 T^4 + x_6 T^5 \tag{5.1.1}$$

where T is the dimensionless temperature and $x_1, x_2, \ldots, x_6$ are the unknowns to be determined. The following system of six equations is obtained if the given data are substituted in this equation:

$$x_1 + x_2 + x_3 + x_4 + x_5 + x_6 = 106.4 \quad (5.1.2)$$

$$x_1 + 2x_2 + 2^2x_3 + 2^3x_4 + 2^4x_5 + 2^5x_6 = 57.79 \quad (5.1.3)$$

$$x_1 + 3x_2 + 3^2x_3 + 3^3x_4 + 3^4x_5 + 3^5x_6 = 32.9 \quad (5.1.4)$$

$$x_1 + 4x_2 + 4^2x_3 + 4^3x_4 + 4^4x_5 + 4^5x_6 = 19.52 \quad (5.1.5)$$

$$x_1 + 5x_2 + 5^2x_3 + 5^3x_4 + 5^4x_5 + 5^5x_6 = 12.03 \quad (5.1.6)$$

$$x_1 + 6x_2 + 6^2x_3 + 6^3x_4 + 6^4x_5 + 6^5x_6 = 7.67 \quad (5.1.7)$$

This set of linear equations is to be solved to obtain the unknowns x_i, where $i = 1, 2, \ldots, 6$. The augmented matrix consists of 6 rows and 7 columns, where the seventh column contains the constants on the right-hand side of the equations. It may be mentioned here that the dimensionless temperature T is employed simply for convenience. The actual temperatures may also be used if so desired.

Figure 5.1.1(a) shows the computer program in FORTRAN 77 for this problem. The problem is written for a system of up to 10 equations. The number of equations and the augmented matrix are given as input data. At each elimination step, the row with the largest pivot element is found, considering the rows below and including the pivot row. If another row has a pivot element larger than that in the pivot row, it is interchanged with the pivot row, making it the pivot row for the next elimination step. This partial pivoting improves the accuracy of the solution and also avoids problems if a zero pivot element arises. Gaussian elimination is applied to reduce the given matrix to an upper triangular one. Then $X(6)$, which represents the unknown x_6, is computed directly from Eq. (5.17). The other unknowns are determined by back-substitution, using Eq. (5.18).

Figure 5.1.1(b) shows the computed results. The program also computes the constants B_i, where $i = 1, 2, \ldots, 6$, using the equation $B = AX$, where X is the computed vector of the unknowns. These computed constants, denoted by $B(I)$ in the program, may be compared with the given constants of Eqs. (5.1.2) through (5.1.7) to determine the accuracy of the numerical results. Note from the results presented that the computed values of the constants are close to the given values. In fact, they are identical if we retain the same number of significant figures as those in the given data. Therefore, a high level of accuracy in the computed results is indicated.

Problems similar to the one discussed here arise in several practical applications that involve curve fitting of experimental data, as discussed in Chapter 6. Some similar problems are also included in the exercises at the end of this chapter. Curve fitting of property data is important in the computational study of several chemical and mechanical engineering processes. The example considered here is of interest in the numerical simulation of power plants. Similarly, property data for gases are needed in the design and analysis of chemical reactors. Once the polynomial given in Eq. (5.1.1)

```
C     GAUSSIAN ELIMINATION METHOD FOR A SYSTEM OF LINEAR EQUATIONS
C
C
C     A(I,J) REPRESENTS THE ELEMENTS OF THE AUGMENTED MATRIX BEING
C     REDUCED BY THE GAUSSIAN ELIMINATION METHOD, A1(I,J) ARE THE
C     ELEMENTS OF THE ORIGINAL AUGMENTED MATRIX, X(I) ARE THE
C     UNKNOWN VARIABLES, N IS THE NUMBER OF EQUATIONS, M IS THE
C     NUMBER OF COLUMNS IN THE AUGMENTED MATRIX, K REPRESENTS THE
C     NUMBER OF THE PIVOT ROW AND B(I) REPRESENTS THE CONSTANTS ON
C     THE RIGHT-HAND SIDE OF THE GIVEN SYSTEM OF EQUATIONS
C
C        DIMENSION A(10,11),A1(10,11),X(10)
C
C     ENTER THE COEFFICIENT MATRIX
C
         READ(5,*)N
         M=N+1
         READ(5,*)((A(I,J),J=1,M),I=1,N)
         DO 1 I=1,N
         DO 1 J=1,M
  1      A1(I,J)=A(I,J)
C
C     CALL SUBROUTINE TO SOLVE THE SYSTEM OF EQUATIONS
C
         CALL GAUSS(N,A,X)
         WRITE(6,9)
  9      FORMAT(2X,'THE SOLUTION TO THE EQUATIONS IS:'//)
         DO 10 I=1,N
  10     WRITE(6,11)I,X(I)
  11     FORMAT(2X,'X(',I1,')=',F12.5)
         WRITE(6,12)
  12     FORMAT(//2X,'THE CONSTANT VECTOR OF THE EQUATIONS IS:'//)
C
C     CALCULATE THE CONSTANT VECTOR B USING THE SOLUTION
C     OBTAINED TO CHECK THE ACCURACY OF THE RESULTS
C
         DO 13 I=1,N
         Y=0.0
         DO 14 J=1,N
  14     Y=Y+X(J)*A1(I,J)
         WRITE(6,15)I,Y
  15     FORMAT(2X,'B(',I1,')=',F12.5)
  13     CONTINUE
         STOP
         END
C
C
         SUBROUTINE GAUSS(N,A,X)
         DIMENSION A(10,11),X(10)
         N1=N-1
         M=N+1
C
C     FIND THE ROW WITH THE LARGEST PIVOT ELEMENT
C
         DO 2 K=1,N1
         K1=K+1
         K2=K
```

Figure 5.1.1 FORTRAN computer program for solving the system of linear equations of Example 5.1 by the Gaussian elimination method, along with the numerical results obtained.

```
      B0=ABS(A(K,K))
      DO 3 I=K1,N
      B1=ABS(A(I,K))
      IF((B0-B1) .LT. 0.0) THEN
      B0=B1
      K2=I
      END IF
3     CONTINUE
      IF((K2-K) .NE. 0) THEN
C
C   INTERCHANGE ROWS TO OBTAIN THE LARGEST PIVOT ELEMENT
C
      DO 5 J=K,M
      C=A(K2,J)
      A(K2,J)=A(K,J)
5     A(K,J)=C
      END IF
      DO 2 I=K1,N
C
C    APPLY THE GAUSSIAN ELIMINATION ALGORITHM
C
      DO 6 J=K1,M
6     A(I,J)=A(I,J)-A(I,K)*A(K,J)/A(K,K)
2     A(I,K)=0.0
C
C   APPLY BACK SUBSTITUTION
C
      X(N)=A(N,M)/A(N,N)
      DO 7 I1=1,N1
      I=N-I1
      S=0.0
      J1=I+1
      DO 8 J=J1,N
8     S=S+A(I,J)*X(J)
7     X(I)=(A(I,M)-S)/A(I,I)
      RETURN
      END
```

(a)

```
RESULTS

THE SOLUTION TO THE EQUATIONS IS:

X(1)=   201.26010
X(2)=  -128.82130
X(3)=    40.67448
X(4)=    -7.42302
X(5)=     0.74085
X(6)=    -0.03108

THE CONSTANT VECTOR OF THE EQUATIONS IS:

B(1)=   106.39990
B(2)=    57.79021
B(3)=    32.90009
B(4)=    19.52042
B(5)=    12.03053
B(6)=     7.67023
```

(b)

Figure 5.1.1 Continued

has been determined, the specific volume at intermediate points in the range $1 \leqslant T \leqslant 6$ may be determined by interpolation, as outlined in Chapter 6.

Example 5.2

The temperature $T(x)$ that arises due to steady-state heat conduction in a bar 30 cm long is governed by the following ordinary differential equation, if uniform temperature is assumed across any cross section:

$$\frac{d^2T}{dx^2} - GT = 0 \tag{5.2.1}$$

where T is the temperature difference from the ambient medium, which is at 20°C, x is the axial coordinate distance, and G is a constant that depends on the surface heat transfer rate. As discussed in greater detail in Chapter 8, this equation may be replaced by a finite difference approximation, using the second central difference given in Chapter 3, as

$$\frac{T_{i+1} - 2T_i + T_{i-1}}{(\Delta x)^2} - GT_i = 0 \tag{5.2.2}$$

where $x = i\,\Delta x$. Considering 30 subdivisions of the length of the rod, as shown in Fig. 5.2.1, find the temperature differences T_i, where $i = 1, 2, \ldots, 29$. The temperatures differences T_0 and T_{30} are given as 100°C, and the constant G as $(0.071)^2\ \text{cm}^{-2}$. Use Gaussian elimination.

Solution

The system of equations to be solved by Gaussian elimination is

$$T_{i+1} - [2 + G(\Delta x)^2]T_i + T_{i-1} = 0 \tag{5.2.3}$$

or

$$-T_{i+1} + (2 + S)T_i - T_{i-1} = 0 \quad \text{for } i = 1, 2, \ldots, 29 \tag{5.2.4}$$

where

$$S = G(\Delta x)^2 = (0.071)^2(1.0)^2 \tag{5.2.5}$$

The constants F_i on the right-hand side of Eq. (5.2.4) are all zero except in the two

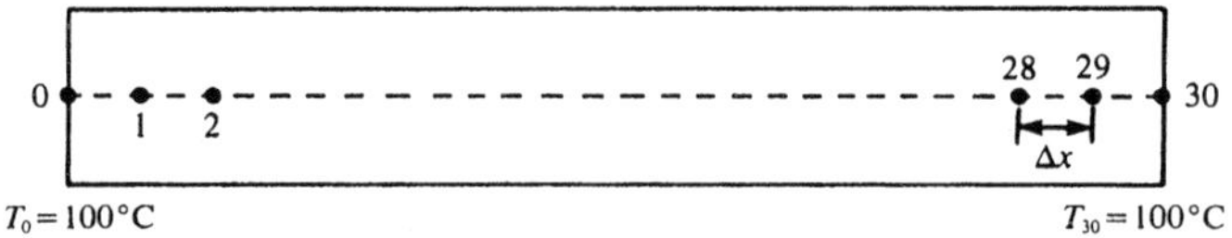

Figure 5.2.1 Physical problem considered in Example 5.2.

```
C           GAUSSIAN ELIMINATION FOR A TRIDIAGONAL SYSTEM
C
C
C        A(I), B(I) AND C(I) ARE THE THREE ELEMENTS IN EACH ROW
C        OF THE GIVEN SYSTEM OF EQUATIONS, F(I) REPRESENTS THE
C        CONSTANTS ON THE RIGHT-HAND SIDE OF THE EQUATIONS, T(I)
C        ARE THE TEMPERATURE DIFFERENCES TO BE COMPUTED, S IS A
C        PARAMETER DEFINED IN THE GIVEN PROBLEM, N IS THE NUMBER
C        OF EQUATIONS AND TT REPRESENTS THE PHYSICAL TEMPERATURES
C
C
C        SPECIFY INITIAL PARAMETERS
C
         DIMENSION A(30),B(30),C(30),T(30),F(30)
         N=29
         NN=N-1
         S=0.071**2
C
C        ENTER THE CONSTANTS ON THE RIGHT-HAND SIDE OF THE EQUATIONS
C
         F(1)=100.0
         F(N)=100.0
         DO 1 I=2,NN
  1      F(I)=0.0
C
C        ENTER THE MATRIX COEFFICIENTS
C
         DO 2 I=1,N
  2      B(I)=2.0+S
         DO 3 I=1,NN
  3      C(I)=-1.0
         DO 4 I=2,N
  4      A(I)=-1.0
C
C        COMPUTE THE NEW MATRIX COEFFICIENTS
C
         DO 5 I=2,N
         D=A(I)/B(I-1)
         B(I)=B(I)-C(I-1)*D
  5      F(I)=F(I)-F(I-1)*D
C
C        APPLY BACK SUBSTITUTION
C
         T(N)=F(N)/B(N)
         DO 6 I=1,NN
         J=N-I
  6      T(J)=(F(J)-C(J)*T(J+1))/B(J)
C
C        COMPUTE THE ACTUAL TEMPERATURES FROM THE TEMPERATURE
C        DIFFERENCES T(I)
C
         WRITE(6,7)
  7      FORMAT(2X,'THE REQUIRED TEMPERATURES ARE'/)
         DO 8 I=1,N
         TT=T(I)+20.0
  8      WRITE(6,9)I,TT
  9      FORMAT(2X,'T(',I2,')=',F10.4)
         STOP
         END
```

(a)

Figure 5.2.2 Computer program for solving the tridiagonal system of Example 5.2, along with the computed temperatures.

```
THE REQUIRED TEMPERATURES ARE

T( 1)=  114.6583
T( 2)=  109.7937
T( 3)=  105.3816
T( 4)=  101.3998
T( 5)=   97.8283
T( 6)=   94.6491
T( 7)=   91.8461
T( 8)=   89.4051
T( 9)=   87.3139
T(10)=   85.5620
T(11)=   84.1405
T(12)=   83.0422
T(13)=   82.2616
T(14)=   81.7948
T(15)=   81.6395
T(16)=   81.7948
T(17)=   82.2616
T(18)=   83.0421
T(19)=   84.1404
T(20)=   85.5620
T(21)=   87.3140
T(22)=   89.4052
T(23)=   91.8462
T(24)=   94.6493
T(25)=   97.8286
T(26)=  101.4001
T(27)=  105.3819
T(28)=  109.7939
T(29)=  114.6584
```

(b)

Figure 5.2.2 Continued

equations corresponding to $i = 1$ and 29, where T_0 and T_{30} appear and yield $F_1 = 100$ and $F_{29} = 100$. A tridiagonal system is obtained, which may be written as follows:

$$\begin{bmatrix} 2+S & -1 & 0 & 0 & \cdots & & 0 \\ -1 & 2+S & -1 & 0 & \cdots & & 0 \\ 0 & -1 & 2+S & -1 & \cdots & & 0 \\ \vdots & & & & & & \vdots \\ 0 & 0 & \cdots & & 0 & -1 & 2+S \end{bmatrix} \begin{bmatrix} T_1 \\ T_2 \\ \vdots \\ T_{29} \end{bmatrix} = \begin{bmatrix} 100 \\ 0 \\ \vdots \\ 0 \\ 100 \end{bmatrix} \tag{5.2.6}$$

This tridiagonal system of equations can easily be solved by the Gaussian elimination method, as outlined in the text. The three nonzero elements in each row are denoted by $A(I)$, $B(I)$, and (I), where $I = 1, 2, \ldots, 29$. The constants on the right-hand side are denoted by $F(I)$. The coefficients and constants are given as input data. Gaussian elimination is used to eliminate the left-most element in each row in one traverse from the top row to the bottom row. The temperature difference T_{29} is then computed as $F(29)/B(29)$, where both F and B are the new values after reduction. Back-substitution then yields the remaining temperature differences.

Figure 5.2.2(a) shows the FORTRAN program for solving a tridiagonal system containing 30 equations. It can easily be extended to a larger number of equations.

```
C       TRIDIAGONAL MATRIX ALGORITHM
C
C       A, B AND C ARE THE THREE ELEMENTS IN EACH ROW, WITH B AT THE
C       DIAGONAL, F IS THE CONSTANT ON THE RIGHT-HAND SIDE OF EACH
C       EQUATION, N IS THE NUMBER OF EQUATIONS AND T IS THE VARIABLE
C       TO BE COMPUTED
C
        SUBROUTINE TRIDIA(A,B,C,F,N,T)
        DIMENSION A(400),B(400),C(400),F(400),T(400)
C
C       ELIMINATE THE A'S BY GAUSSIAN ELIMINATION AND DETERMINE
C       THE NEW COEFFICIENTS
C
        DO 5 I=2,N
        D=A(I)/B(I-1)
        B(I)=B(I)-C(I-1)*D
  5     F(I)=F(I)-F(I-1)*D
C
C       BACK SUBSTITUTION
C
        T(N)=F(N)/B(N)
        DO 6 I=1,N-1
        J=N-I
  6     T(J)=(F(J)-C(J)*T(J+1))/B(J)
        RETURN
        END
```

Figure 5.2.3 Tridiagonal matrix algorithm, presented as a subroutine.

The matrix coefficients and the constants F_i are given, and the unknown temperature differences T_i are computed by Gaussian elimination. Finally, the ambient temperature of 20°C is added to T_i to yield the physical temperatures, denoted by $TT(I)$. It is evident that the program is much simpler than the corresponding program for a system that is not tridiagonal. The total arithmetic operations are $O(n)$, where n is the number of equations, for a tridiagonal system, as compared to $O(n^3)$ for a different matrix. Since tridiagonal systems arise very frequently in the numerical solution of differential equations, the above computational procedure is of considerable importance. Because of the associated small number of arithmetic operations, the Gaussian elimination method of solving a tridiagonal system results in smaller computer time and smaller round-off error than do most of the other methods discussed in this chapter. The computed temperature distribution is shown in Fig. 5.2.2(b). The temperatures $TT(0)$ and $TT(30)$ at the two ends are 120°C. The problem is symmetric about $T(15)$, and the computational procedure may be simplified by employing only 16 points and taking $T(16) = T(14)$ from symmetry.

The tridiagonal matrix algorithm (TDMA), also known as *Thomas algorithm*, is generally employed as a subroutine to solve the system of linear equations generated in the main program. Figure 5.2.3 shows the program as a subroutine which may easily be appended to the main program, as demonstrated for the finite difference solution of partial differential equations in Chapter 9.

5.3 GAUSS-JORDAN ELIMINATION

In the Gauss-Jordan elimination method, which is a variation of Gaussian elimination, the original matrix A of the coefficients is reduced to an identity matrix I so that the unknowns $x_1, x_2, \ldots, x_n$ are found directly, without back-substitution. At each step involved in the elimination of an unknown using a pivot equation, the unknown is eliminated from the equations above the pivot equation as well as from those below it. The pivot equation is normalized by dividing it throughout by the pivot element, so that the diagonal elements are finally obtained as unity, resulting in a reduced system of equations of the form $IX = B'$, where B' now gives the solution vector. Since the procedure is similar to Gaussian elimination, partial or complete pivoting may be employed to improve the accuracy of the solution and to avoid the use of a pivot element that is zero, or close to it.

5.3.1 Mathematical Procedure

Let us consider the following set of linear equations for solving by the Gauss-Jordan elimination method:

$$\begin{aligned} 2x + y + 3z &= 9 \\ 3x - 4y + 4z &= 7 \\ x + 4y - z &= 3 \end{aligned} \tag{5.27}$$

We normalize the first equation by dividing it by 2, which is the pivot element. Then we eliminate x from the other two equations by multiplying the normalized pivot equation by the coefficients of x in the other equations and subtracting the equations thus obtained from the original equations, to yield

$$\begin{aligned} x + \frac{1}{2}y + \frac{3}{2}z &= \frac{9}{2} \\ -\frac{11}{2}y - \frac{1}{2}z &= -\frac{13}{2} \\ \frac{7}{2}y - \frac{5}{2}z &= -\frac{3}{2} \end{aligned} \tag{5.28}$$

We repeat the above step with the second equation as the pivot equation, which we first normalize by dividing it by the pivot element $-11/2$. Thus, y is eliminated from the first and third equations to give

$$\begin{aligned} x + \frac{16}{11}z &= \frac{43}{11} \\ y + \frac{1}{11}z &= \frac{13}{11} \\ -\frac{31}{11}z &= -\frac{62}{11} \end{aligned} \tag{5.29}$$

In the final step, z is eliminated from the first two equations, to give

$$\begin{aligned} x &= 1 \\ y &= 1 \\ z &= 2 \end{aligned} \tag{5.30}$$

5.3.2 Computational Scheme

To implement this method on the computer, we follow the same procedure as that outlined earlier for Gaussian elimination. The augmented matrix C of a system of linear equations is given by Eq. (5.13). The first row is the pivot equation for the first elimination step, and the elements of the matrix obtained after this step are

$$a_{1j}^{(1)} = \frac{a_{1j}}{a_{11}} \qquad \text{where } 1 \leqslant j \leqslant n+1 \tag{5.31a}$$

$$a_{ij}^{(1)} = a_{ij} - a_{i1}a_{1j}^{(1)} \qquad \text{where } 1 \leqslant j \leqslant n+1 \quad \text{and} \quad 2 \leqslant i \leqslant n \tag{5.31b}$$

We can easily generalize this procedure by noting that Eq. (5.31b) is not applied to the pivot row, in a given step, and that the columns to the left of the one containing the pivot element are not affected. Therefore, if r denotes the pivot row and, hence, the elimination step, the Gauss-Jordan elimination method is given by the following recursion formulas:

$$a_{rj}^{(r)} = \frac{a_{rj}^{(r-1)}}{a_{rr}^{(r-1)}} \qquad \text{where } r \leqslant j \leqslant n+1 \tag{5.32a}$$

$$a_{ij}^{(r)} = a_{ij}^{(r-1)} - a_{ir}^{(r-1)}a_{rj}^{(r)} \qquad \text{where } r \leqslant j \leqslant n+1 \quad \text{and} \quad 1 \leqslant i \leqslant n, \text{ except } i = r \tag{5.32b}$$

In most computational schemes, the new elements simply replace the old ones as they are computed, to avoid additional storage. However, the $j = r$ column must be stored separately at each step to provide the $a_{ir}^{(r-1)}$ needed in Eq. (5.32b). This situation may be avoided by varying j from right to left or from $(r + 1)$ to $(n + 1)$. In the latter case, the $j = r$ column is not computed but may simply be inserted after the completion of the step, with 1 at the pivot element and zeros above and below it in the column. Then the storage required is the same as that for the original system. This approach is demonstrated in Example 5.3, using partial pivoting.

Let us now apply the above generalized procedure to the system of equations considered earlier, with the augmented matrix C given by

$$C = \begin{bmatrix} 2 & 1 & 3 & 9 \\ 3 & -4 & 4 & 7 \\ 1 & 4 & -1 & 3 \end{bmatrix} \tag{5.33}$$

Applying the recursion formulas of Eq. (5.32), successively, for $r = 1$, 2, and 3, we find that the resulting matrices are, respectively,

$$\begin{bmatrix} 1 & \frac{1}{2} & \frac{3}{2} & \frac{9}{2} \\ 0 & -\frac{11}{2} & -\frac{1}{2} & -\frac{13}{2} \\ 0 & \frac{7}{2} & -\frac{5}{2} & -\frac{3}{2} \end{bmatrix} \quad \begin{bmatrix} 1 & 0 & \frac{16}{11} & \frac{43}{11} \\ 0 & 1 & \frac{1}{11} & \frac{13}{11} \\ 0 & 0 & -\frac{31}{11} & -\frac{62}{11} \end{bmatrix} \quad \begin{bmatrix} 1 & 0 & 0 & 1 \\ 0 & 1 & 0 & 1 \\ 0 & 0 & 1 & 2 \end{bmatrix}$$

Thus the required solution is $x = 1$, $y = 1$, and $z = 2$. As shown above, the columns to the left of the pivot element, at each step, are unaffected, and the elements above and below the pivot element, in the same column, become zero. The solution is given by the $(n+1)$th column. No back-substitution is needed.

If several systems of equations with the same coefficient matrix A and different constant vector B are given, they can all be solved by application of the above procedure once. The various constant vectors are simply added as columns to the augmented matrix, and the reduction process is carried out. At the completion of the elimination, the solutions for the different systems is given by the corresponding columns in the reduced matrix. It can be shown that the solution of a system of n linear equations by the Gauss-Jordan elimination method requires about $n^3/2$ arithmetic operations. Therefore, this method takes a larger computing time and has a larger round-off error than Gaussian elimination and is not preferred for solving linear systems. However, it can be used to develop a method for matrix inversion employing minimum storage. For inverting a matrix A, the equation to be solved is $AX = I$, where X becomes A^{-1} when Gauss-Jordan elimination transforms A into I. This procedure is discussed in greater detail in Section 5.5. Also, if several systems of equations with the same coefficient matrix A and different constant vector B are to be solved, as is the case, for instance, in engineering problems where the boundary conditions are varied while the governing equations remain unchanged, Gauss-Jordan elimination is more advantageous to use than Gaussian elimination.

Example 5.3

Consider the electrical network shown in Fig. 5.3.1(a) and compute the electric currents $I_1, I_2, \ldots, I_6$ through the six resistances. Also, solve the problem, employing three loop currents I_1, I_2, and I_3 in the three closed circuits shown in Fig. 5.3.1(b). Use Gauss-Jordan elimination for this problem.

Solution

A system of six linear equations may be written for the six unknowns x_i, where x_i represents the current through a given resistance and $i = 1, 2, \ldots, 6$. By Kirchhoff's laws, the sum of the currents entering a node is equated to the sum of the currents

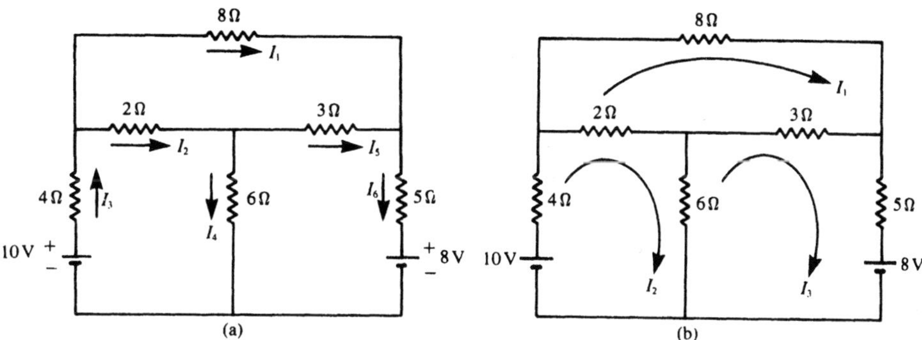

Figure 5.3.1 Electrical network considered in Example 5.3.

leaving it. Thus,

$$x_1 + x_2 - x_3 = 0 \tag{5.3.1}$$

$$x_2 - x_4 - x_5 = 0 \tag{5.3.2}$$

$$x_1 + x_5 - x_6 = 0 \tag{5.3.3}$$

Also, the voltage change as one goes around each loop is equated to zero to yield

$$2x_2 + 4x_3 + 6x_4 - 10 = 0 \tag{5.3.4}$$

$$-6x_4 + 3x_5 + 5x_6 + 8 = 0 \tag{5.3.5}$$

$$8x_1 - 2x_2 - 3x_5 = 0 \tag{5.3.6}$$

Therefore, six independent linear equations are obtained for determining the six currents x_i. If other nodes are considered, only linear combinations of the first three equations will be obtained; see, for instance, Sears et al. (1981).

The Gauss-Jordan elimination method, with partial pivoting, is used for solving this system of equations. The FORTRAN 77 program, which is shown in Fig. 5.3.2(a), is quite similar to that for Gaussian elimination. However, in this case, the elements in the column containing the pivot element are reduced to zero in rows both above and below the pivot row. Also, at each step, the pivot row is taken as the row with the largest pivot element, considering the rows below the pivot row for the preceding step. The elements of this row are normalized by dividing throughout by the pivot element. Therefore, the reduced matrix is an identity matrix, instead of the upper triangular matrix obtained in Gaussian elimination. The augmented matrix, with the constant vector B taken as the last column, is supplied to the program as input data. This column vector becomes the solution after the Gauss-Jordan elimination process has been completed. Therefore, no back-substitution is necessary, and the last column yields the solution.

```
C       THE GAUSS JORDAN ELIMINATION METHOD
C
C
C    A(I,J) REPRESENTS THE ELEMENTS OF THE AUGMENTED MATRIX,
C    X(I) DENOTES THE UNKNOWN VARIABLES, K IS THE NUMBER OF
C    THE PIVOT ROW, N IS THE NUMBER OF EQUATIONS, AND M IS N+1.
C
        DIMENSION A(10,11),X(10)
        READ(5,*)N
        M=N+1
C
C    READ COEFFICIENTS OF THE AUGMENTED MATRIX
C
        READ(5,*)((A(I,J),J=1,M),I=1,N)
        N1=N-1
        DO 6 K=1,N
        K1=K+1
        K2=K
C
C    SEARCH FOR ROW WITH LARGEST PIVOT ELEMENT
C
        B0=ABS(A(K,K))
        DO 1 I=K,N
        B1=ABS(A(I,K))
        IF((B0-B1) .LT. 0.0)THEN
        B0=B1
        K2=I
        END IF
   1    CONTINUE
C
C    DECIDE IF ROW INTERCHANGE IS NEEDED FOR MAXIMUM PIVOT ELEMENT
C
        IF((K2-K) .NE. 0)THEN
C
C    INTERCHANGE ROW FOR OBTAINING LARGEST PIVOT ELEMENT
C
        DO 2 J=K,M
        C=A(K2,J)
        A(K2,J)=A(K,J)
   2    A(K,J)=C
        END IF
C
C    APPLY GAUSS JORDAN ELIMINATION
C
   3    DO 4 J=K1,M
   4    A(K,J)=A(K,J)/A(K,K)
        A(K,K)=1.0
        DO 6 I=1,N
        IF (I .NE. K) THEN
        DO 5 J=K1,M
   5    A(I,J)=A(I,J)-A(I,K)*A(K,J)
        A(I,K)=0.0
        END IF
   6    CONTINUE
C
C    DETERMINE THE UNKNOWNS
C
        DO 7 I=1,N
```

Figure 5.3.2 Program for the solution of the linear system of Example 5.3 by the Gauss-Jordan elimination method, along with the computed results.

```
7       X(I)=A(I,M)
        WRITE(6,8)
8       FORMAT(2X,'THE SOLUTION TO THE EQUATIONS IS:'//)
        DO 9 I=1,N
9       WRITE(6,10)I,X(I)
10      FORMAT(2X,'X(',I1,')=',F12.5)
        WRITE(6,11)
11      FORMAT(//2X,'THE REDUCED MATRIX IS'//)
        DO 13 I=1,N
        WRITE(6,12)(A(I,J),J=1,N)
12      FORMAT(10F10.3)
13      CONTINUE
        STOP
        END
```

(a)

RESULTS

```
THE SOLUTION TO THE EQUATIONS IS:

X(1)=     0.05135
X(2)=     0.66351
X(3)=     0.71487
X(4)=     0.96892
X(5)=    -0.30541
X(6)=    -0.25405

THE REDUCED MATRIX IS

  1.000     0.000     0.000     0.000     0.000     0.000
  0.000     1.000     0.000     0.000     0.000     0.000
  0.000     0.000     1.000     0.000     0.000     0.000
  0.000     0.000     0.000     1.000     0.000     0.000
  0.000     0.000     0.000     0.000     1.000     0.000
  0.000     0.000     0.000     0.000     0.000     1.000
```

(b)

Figure 5.3.2 Continued

Figure 5.3.2(b) gives the computed results in terms of the unknowns, denoted here by $X(I)$, where $I = 1, 2, \ldots, 6$. The reduced matrix is also shown to indicate the successful completion of the procedure. As done for Example 5.1, the constant vector B may be computed from the obtained values of the currents and compared with the values given in Eqs. (5.3.1) through (5.3.6), in order to evaluate the accuracy of the solution.

If the loop currents of Fig. 5.3.1(b) are considered instead, the voltage change around each loop may be equated to zero:

$$-2I_1 + 12I_2 - 6I_3 - 10 = 0 \tag{5.3.7}$$

$$-3I_1 - 6I_2 + 14I_3 + 8 = 0 \tag{5.3.8}$$

$$13I_1 - 2I_2 - 3I_3 = 0 \tag{5.3.9}$$

This system was also solved, using the program given in Fig. 5.3.2(a). The resulting values of I_1, I_2, and I_3 were obtained as 0.05135, 0.71486, and −0.25405 amperes, which are almost identical to the values of $X(1)$, $X(3)$, and $X(6)$, as expected from the nomenclature of Fig. 5.3.1. Once these currents have been obtained, the other currents and desired voltages may be computed from Kirchhoff's laws.

5.4 COMPACT METHODS*

5.4.1 Matrix Decomposition

There are several numerical methods for the solution of simultaneous linear equations that are based on the decomposition of the coefficient matrix A into an upper triangular matrix U and a lower triangular matrix L, as shown in Fig. 5.1, such that

$$A = LU \tag{5.34}$$

In Gaussian elimination, we obtain an upper triangular matrix U of the form

$$U = \begin{bmatrix} a_{11} & a_{12} & \cdots & a_{1n} \\ 0 & a_{22}^{(1)} & \cdots & a_{2n}^{(1)} \\ \vdots & & & \vdots \\ 0 & \cdots & 0 & a_{nn}^{(n-1)} \end{bmatrix} \tag{5.35}$$

In order to obtain the above form, we use Eq. (5.16), with the multipliers m_{ir} at each elimination step being given by

$$m_{ir} = \frac{a_{ir}^{(r-1)}}{a_{rr}^{(r-1)}} \qquad \text{for } i = r+1, \ldots, n \tag{5.36}$$

These multipliers, if stored, can be used for solving different systems of equations, $AX = B$, which have the same coefficient matrix A but a different constant vector B. Usually, these multipliers are stored in place of the zero elements below the diagonal of the matrix U. Thus, m_{ij} may be stored in the space originally employed for a_{ij} below the diagonal, i.e., for $i > j$.

Let us consider the lower triangular matrix L, defined as follows:

$$L = \begin{bmatrix} 1 & 0 & 0 & \cdots & 0 \\ m_{21} & 1 & 0 & \cdots & 0 \\ \vdots & & & & \vdots \\ m_{n1} & m_{n2} & \cdots & \cdots & 1 \end{bmatrix} \tag{5.37}$$

We can show that $A = LU$, by using the above definitions of L and U and carrying out the matrix multiplication of Eq. (5.34). Therefore, the coefficient matrix A may be decomposed into the two triangular matrices L and U. If the system of equations given by Eq. (5.9) is considered, the multipliers m_{ir} may be retained to yield the matrices

$$L = \begin{bmatrix} 1 & 0 & 0 \\ 1/3 & 1 & 0 \\ 2/3 & -4/7 & 1 \end{bmatrix} \quad \text{and} \quad U = \begin{bmatrix} 3 & 5 & 1 \\ 0 & 7/3 & 5/3 \\ 0 & 0 & 69/21 \end{bmatrix} \tag{5.38}$$

It can be easily verified that $A = LU$. Recursive formulas for the elements of L and U may be obtained directly from Eq. (5.34) and employed in the solution of a system of linear equations. If the above form for L, with all the diagonal elements equal to 1, is considered, the decomposition is the one obtained from Gaussian elimination. For this circumstance, Doolittle's method gives the corresponding explicit recursive formulas for the elements l_{ij} and u_{ij} and employs them in the solution of linear systems (Atkinson, 1978). Another method, known as *Crout's method*, employs a U matrix whose diagonal elements are all equal to 1. This method is discussed in some detail here, since it is widely used in many engineering problems. It is also generally more efficient than the other elimination methods discussed earlier. Consequently, it requires less computer time and generates smaller round-off error.

5.4.2 Crout's Method

In Crout's method, Gaussian elimination is written in a more compact form, using the following decomposition, and recursive relations for the matrix elements are obtained:

$$A = LU \qquad \text{where } L = \begin{bmatrix} l_{11} & 0 & \cdots & \cdots & 0 \\ l_{21} & l_{22} & 0 & \cdots & 0 \\ \vdots & \vdots & \vdots & & \vdots \\ l_{n1} & l_{n2} & \cdots & \cdots & l_{nn} \end{bmatrix}$$

$$\text{and} \quad U = \begin{bmatrix} 1 & u_{12} & u_{13} & \cdots & u_{1n} \\ 0 & 1 & u_{23} & \cdots & u_{2n} \\ \vdots & & & & \vdots \\ 0 & \cdots & \cdots & 0 & 1 \end{bmatrix} \tag{5.39}$$

From the above decomposition, the product rule of the determinants may be employed to obtain

$$\text{Det}(A) = \text{Det}(L)\text{Det}(U) \tag{5.40}$$

For independent equations, the determinant of matrix A is nonzero, as discussed earlier. Therefore, the determinant of matrix L is also nonzero. Since L is a lower triangular matrix, its determinant is the product of its diagonal elements. This implies that if Det(L) is nonzero, all the diagonal elements l_{ii} of this matrix are nonzero. It is shown below that, if l_{ii} and a_{11} are nonzero, L and U matrices exist and their elements can be determined uniquely.

To solve a system of linear equations by Crout's method, we write A and U as augmented matrices of the form

$$\begin{bmatrix} l_{11} & 0 & \cdots & 0 \\ l_{21} & l_{22} & 0 & 0 \\ \vdots & & & \vdots \\ l_{n1} & l_{n2} & \cdots & l_{nn} \end{bmatrix} \begin{bmatrix} 1 & u_{12} & u_{13} & \cdots & u_{1n} & u_{1,n+1} \\ 0 & 1 & u_{23} & \cdots & u_{2n} & u_{2,n+1} \\ \vdots & & & & & \vdots \\ 0 & 0 & \cdots & \cdots & 1 & u_{n,n+1} \end{bmatrix}$$

$$= \begin{bmatrix} a_{11} & a_{12} & a_{13} & \cdots & a_{1n} & a_{1,n+1} \\ a_{21} & a_{22} & a_{23} & \cdots & a_{2n} & a_{2,n+1} \\ \vdots & & & & & \vdots \\ a_{n1} & a_{n2} & a_{n3} & \cdots & a_{nn} & a_{n,n+1} \end{bmatrix} \tag{5.41}$$

Recursive formulas for l_{ij} and u_{ij} may be developed by application of matrix multiplication to the above equation. Therefore,

$$l_{11} = a_{11},\, l_{11}u_{12} = a_{12},\, l_{11}u_{13} = a_{13}, \ldots, l_{11}u_{1,n+1} = a_{1,n+1}$$

which gives

$$l_{i1} = a_{i1} \qquad \text{for } i = 1, 2, \ldots, n \tag{5.42a}$$

$$u_{1j} = \frac{a_{1j}}{l_{11}} \qquad \text{for } j = 2, 3, \ldots, n+1 \tag{5.42b}$$

Similarly,

$$l_{21}u_{12} + l_{22} = a_{22} \qquad l_{31}u_{12} + l_{32} = a_{32}$$

or

$$l_{22} = a_{22} - l_{21}u_{12} \qquad l_{32} = a_{32} - l_{31}u_{12} \tag{5.42c}$$

Also,

$$l_{21}u_{13} + l_{22}u_{23} = a_{23} \qquad l_{21}u_{14} + l_{22}u_{24} = a_{24}$$

or

$$u_{23} = \frac{a_{23} - l_{21}u_{13}}{l_{22}} \qquad u_{24} = \frac{a_{24} - l_{21}u_{14}}{l_{22}} \tag{5.42d}$$

Therefore, we may proceed as given above to obtain all the elements of the matrices L and U. General equations may also be developed for l_{ij} and u_{ij} as follows:

$$l_{i1} = a_{i1} \qquad \text{for } i = 1, 2, \ldots, n \tag{5.43a}$$

$$u_{1j} = \frac{a_{1j}}{l_{11}} \qquad \text{for } j = 2, 3, \ldots, n+1 \tag{5.43b}$$

$$l_{ij} = a_{ij} - \sum_{m=1}^{j-1} l_{im}u_{mj} \qquad \text{for} \begin{bmatrix} j = 2, 3, \ldots, n \\ i = j, j+1, \ldots, n \end{bmatrix} \tag{5.43c}$$

$$u_{ij} = \frac{a_{ij} - \sum_{m=1}^{i-1} l_{im}u_{mj}}{l_{ii}} \qquad \text{for} \begin{bmatrix} i = 2, 3, \ldots, n \\ j = i+1, i+2, \ldots, n+1 \end{bmatrix} \tag{5.43d}$$

We can easily verify that these general equations yield the elements of the two matrices, given by Eqs. (5.42), by employing the corresponding values of i and j. Also, note from the above relations that if a_{11} and l_{ii} are nonzero, the two matrices exist and can be uniquely determined. As shown earlier, the diagonal elements l_{ii} are all nonzero if matrix A is nonsingular.

It is evident from the recursive formulas given for l_{ij} and u_{ij}, in Eq. (5.43), that the first column of L and the first row of U are determined first. Then the second column of L is determined from the third equation, and the second row of U from the fourth equation. We then proceed to the third column of L and the third row of U, and continue this process until both matrices are determined. The unknowns x_1, $x_2, \ldots, x_n$ are determined by back-substitution, as before, from

$$x_n = u_{n,n+1}$$

$$x_i = u_{i,n+1} - \sum_{j=i+1}^{n} u_{ij}x_j \qquad \text{for } i = n-1, n-2, \ldots, 1 \tag{5.44}$$

The main advantage of compact methods, such as Crout's method outlined above, is that a smaller number of arithmetic operations are needed, as compared to those for the Gaussian and the Gauss-Jordan elimination methods. Besides requiring less computing time, it also results in smaller round-off error. The algebra for determining l_{ij} and u_{ij} may be carried out in double precision for greater accuracy and then rounded off to single precision to reduce computer storage. This limited use of double precision is not possible in regular elimination methods, which would then require all operations and storage to be done in double precision. Since several elements in the L and U matrices are 1 or 0, considerable reduction in computer storage may be accomplished by storing both the matrices in the storage locations for the original augmented matrix. Then both l_{ij} and u_{ij} are termed a_{ij}, and the general equations may be suitably modified. The first equation, Eq. (5.43a), is automatically satisfied. For the other elements, the old a_{ij} values are replaced by new ones, as computation proceeds. Figure 5.2 shows the algorithm for Crout's method, in terms of a flow chart.

Other methods based on matrix decomposition and factorization have been developed. For symmetric matrices, which often arise in many engineering problems such as those related to the analysis of structures, an important method for factorization is Cholesky's method. This method is based on finding a lower triangular matrix L such that

$$A = LL^T \tag{5.45}$$

where L^T is the transpose of the matrix L. This factorization is possible for matrices that are symmetric and positive definite, a necessary and usually sufficient condition for which is that the eigenvalues of the matrix (Section 5.7) be positive. These properties of matrices are discussed in most books on matrices, such as Bronson (1970) and Reiner (1971). Once L has been determined, one obtains the solution x_i by computing the first unknown x_1 directly from the resulting linear equation and the remaining by substitution of the computed values of the preceding unknowns into the reduced equations, with increasing i. The Cholesky decomposition requires about $(1/6)n^3$ operations, instead of $(1/3)n^3$ needed for the Gaussian elimination.

Thus, matrix decomposition methods for solving systems of linear equations are very efficient. However, the computer programming is generally much more involved than the elimination methods, such as Gaussian or Gauss-Jordan elimination.

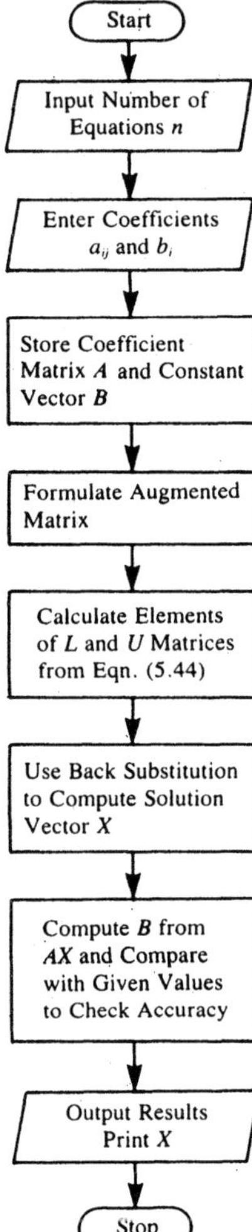

Figure 5.2 Flow chart for the use of Crout's method for solving a system of linear equations.

Consequently, engineers frequently use available programs in engineering applications, rather than develop the necessary software, for methods such as those discussed here. However, following the approach discussed earlier for elimination methods, one may also write computer programs for these methods without too much difficulty. For further details on these methods, refer to Carnahan et al. (1969), Atkinson (1978), and James et al. (1985), listed in the References at the end of this book.

5.5 NUMERICAL SOLUTION OF LINEAR SYSTEMS BY MATRIX INVERSION

In the methods discussed so far, we solved the system of linear equations, given by $AX = B$, by applying various elimination procedures directly to the given system, without finding the inverse A^{-1} of the coefficient matrix. However, if $\text{Det}(A) \neq 0$, the inverse A^{-1} exists, and the solution of the set of linear equations may be obtained as

$$X = A^{-1}B \tag{5.5}$$

If a given set of equations is to be solved, it is generally advantageous to solve the system directly, without computing the inverse matrix A^{-1}. However, as mentioned earlier, the matrix itself may be needed in the problem in order for us to study the behavior of the mathematical or physical system. Also, if several sets of equations with the same coefficient matrix A but different constant vectors B are to be solved, it is often more efficient to compute A^{-1} and to employ it with the different constant vectors B to obtain the corresponding solutions from Eq. (5.5). Another important consideration is that many computer systems have programs available for matrix inversion. These prepared programs may often be employed to solve systems of linear equations. In BASIC, for instance, the MAT statement may be used with many minicomputers and, sometimes, with microcomputers too, if the relevant software is available, to perform matrix operations. The use of programs available in the computer library was discussed in Chapter 2. As pointed out earlier, these programs may be used advantageously for practical engineering applications, if proper care is taken to ensure the applicability of the method and the accuracy of the numerical results.

5.5.1 Computational Procedure

In view of the above discussion, it is evident that matrix inversion, as an intermediate step in solving linear systems, may be desirable in some cases. To invert a square matrix, such as the coefficient matrix A, we use the following definition of the inverse A^{-1}:

$$AA^{-1} = I \tag{5.46}$$

where I is the identity or unit matrix. Therefore, the inverse may be obtained by solving the equation

$$AX = I \tag{5.47}$$

where X assumes the role of the column vector of unknowns and I assumes that of the constant vector. This equation may be solved by applying methods such as Gaussian elimination and Gauss-Jordan elimination. Then the matrix of the unknowns yields the inverse A^{-1}. As mentioned in Section 5.3, matrix inversion by Gaussian elimination requires about $(4/3)n^3$ arithmetic operations, while only $(1/3)n^3$ are needed for solving a set of linear equations. Gauss-Jordan elimination is particularly suitable for matrix inversion, since it transforms the matrix A into the identity matrix I, which also constitutes the right-hand side of Eq. (5.46), and the inverse A^{-1} is obtained directly. This method is outlined below.

The augmented matrix C for Eq. (5.47) is obtained as follows:

$$C = \begin{bmatrix} a_{11} & a_{12} & \cdots & a_{1n} & 1 & 0 & \cdots & \cdots & 0 \\ a_{21} & a_{22} & \cdots & a_{2n} & 0 & 1 & 0 & \cdots & 0 \\ \vdots & & & & & & & & \vdots \\ a_{n1} & a_{n2} & \cdots & a_{nn} & 0 & 0 & \cdots & 0 & 1 \end{bmatrix} \tag{5.48}$$

Now, if Gauss-Jordan elimination is applied to this matrix, using Eq. (5.32), until the a's are replaced by the elements of an identity matrix, the identity matrix in the augmented matrix above is transformed into the inverse A^{-1}. For illustration, let us consider the system of equations given by Eq. (5.27). The augmented matrix C for matrix inversion is

$$C = \begin{bmatrix} 2 & 1 & 3 & 1 & 0 & 0 \\ 3 & -4 & 4 & 0 & 1 & 0 \\ 1 & 4 & -1 & 0 & 0 & 1 \end{bmatrix} \tag{5.49}$$

As before, the first row is divided by the pivot element, which is 2. Then it is multiplied by 3 and subtracted from the second row to yield the second row of the reduced matrix. The new third row is similarly obtained by subtraction of the normalized first row from the third row. This process is continued, using Eq. (5.32), to yield the following matrices during reduction:

$$\begin{bmatrix} 1 & 1/2 & 3/2 & 1/2 & 0 & 0 \\ 0 & -11/2 & -1/2 & -3/2 & 1 & 0 \\ 0 & 7/2 & -5/2 & -1/2 & 0 & 1 \end{bmatrix}, \begin{bmatrix} 1 & 0 & 16/11 & 4/11 & 1/11 & 0 \\ 0 & 1 & 1/11 & 3/11 & -2/11 & 0 \\ 0 & 0 & -31/11 & -16/11 & 7/11 & 1 \end{bmatrix},$$

$$\text{and} \quad \begin{bmatrix} 1 & 0 & 0 & -12/31 & 13/31 & 16/31 \\ 0 & 1 & 0 & 7/31 & -5/31 & 1/31 \\ 0 & 0 & 1 & 16/31 & -7/31 & -11/31 \end{bmatrix}$$

As discussed before, we may employ partial pivoting, or row interchange, to avoid a zero pivot element or to increase the accuracy, by considering the pivot row and all the rows below it at each step and exchanging the rows to employ one with the largest pivot element as the pivot row for the elimination process.

Once the coefficient matrix has been transformed into I, the original identity matrix should be A^{-1}. Therefore,

$$A^{-1} = \begin{bmatrix} -12/31 & 13/31 & 16/31 \\ 7/31 & -5/31 & 1/31 \\ 16/31 & -7/31 & -11/31 \end{bmatrix} \tag{5.50}$$

We can easily verify that the above is true by multiplying this matrix by the original matrix A:

$$\begin{bmatrix} 2 & 1 & 3 \\ 3 & -4 & 4 \\ 1 & 4 & -1 \end{bmatrix} \begin{bmatrix} -12/31 & 13/31 & 16/31 \\ 7/31 & -5/31 & 1/31 \\ 16/31 & -7/31 & -11/31 \end{bmatrix} = \begin{bmatrix} 1 & 0 & 0 \\ 0 & 1 & 0 \\ 0 & 0 & 1 \end{bmatrix} = I \qquad (5.51)$$

The solution vector X may now be obtained by applying Eq. (5.5). Therefore,

$$X = A^{-1}B = [1,1,2] \qquad (5.52)$$

Similarly, the solution of sets of equations with the same A but different B may be obtained easily once A^{-1} has been determined. For instance, if B is given as [11, 10, 4], X is computed as

$$X = A^{-1}B = [2,1,2] \qquad (5.53)$$

5.5.2 Additional Considerations

Partial pivoting is generally incorporated in the program for matrix inversion. Besides avoiding problems with a zero or relatively small pivot element, it improves the accuracy of the computational results obtained. Complete pivoting, with both row and column interchanges, may also be employed to obtain the largest pivot element at each step and thus increase the accuracy. Another improvement in matrix inversion by Gauss-Jordan elimination is obtained by storing the inverted matrix in the same location as the original matrix. The identity matrix is seldom stored, although its transformed columns, which finally give the inverse, are stored in place of the columns in the coefficient matrix, that have been reduced to the diagonal form by Gauss-Jordan elimination. Most computer programs, available on large computers, for matrix inversion incorporate these features for accuracy and reduction in storage. For large systems of equations, the saving in storage is particularly important.

If the storage-saving feature, outlined above, is employed, the general equations for matrix inversion may be obtained from Eq. (5.32). Since the transformed elements of the identity matrix are stored in place of the diagonalized columns, as the inversion proceeds, we obtain

$$\begin{aligned} a'_{mj} &= \frac{a_{mj}}{a_{mm}} && \text{for } j = 1, 2, \ldots, n \quad \text{and} \quad j \neq m \\ a'_{ij} &= a_{ij} - a'_{mj}a_{im} && \text{for} \begin{bmatrix} i = 1, 2, \ldots, n \quad \text{and} \quad i \neq m \\ j = 1, 2, \ldots, n \quad \text{for each } i \quad \text{and} \quad j \neq m \end{bmatrix} \\ a'_{mm} &= \frac{1}{a_{mm}} \\ a'_{im} &= -a_{im}a'_{mm} && \text{for } i = 1, 2, \ldots, n \quad \text{and} \quad i \neq m \end{aligned} \qquad (5.54)$$

where the prime denotes the new elements which replace the old ones after each cycle. The first two equations are the same as those given earlier in Eq. (5.32). The last two are obtained from the transformation of an appended identity matrix, whose elements are zero everywhere except at the diagonal, where they are unity. If partial pivoting is used, without storing the appended matrix, the matrix obtained after the reduction process must be reordered, in the same sequence as the row interchanges, in order to obtain the inverse of the original matrix. However, each row interchange during the inversion corresponds to a column interchange in the identity matrix. Therefore, in the reordering of the final matrix, column interchanges are performed corresponding to each row interchange in the computation process and in the reverse sequence, starting with the last row interchange; see James et al. (1985).

The solution of linear equations by matrix inversion is not advantageous if only a few sets of equations with the same A but different B are to be solved. However, if a large number of such sets are to be solved, matrix inversion may be used advantageously. Such problems do arise in engineering problems. In the solution of partial differential equations by finite difference methods, for instance, large systems of linear equations differing only in the constant vector B, which results from the boundary conditions in the problem, arise. In such cases, matrix inversion may be employed to obtain the results for different boundary conditions. Also, A^{-1} is needed in the analysis of some engineering systems, such as robotics and dynamic systems. In addition, many available programs for the solution of linear systems are based on matrix inversion. Thus, A^{-1} is often computed as an intermediate step. The following example illustrates the use of matrix inversion in the solution of a system of linear equations by appending an identity matrix to the coefficient matrix A and using Gauss-Jordan elimination.

Example 5.4

Write a program for solving a system of linear algebraic equations by matrix inversion. Using this program, solve the equations obtained in Example 5.3 for the six currents in the electrical network of Fig. 5.3.1(a).

Solution

The Gauss-Jordan elimination method may be used for inverting the coefficient matrix A in the system of equations $AX = B$. Then the unknown X is obtained from the inverse of the matrix A^{-1} as $X = A^{-1}B$. The augmented matrix consists of the coefficient matrix A with an identity matrix appended to it. Gauss-Jordan elimination is applied to the coefficient matrix so that it is reduced to an identity matrix. When this is accomplished, the original identity matrix is transformed into the inverse of the matrix A^{-1}, since this amounts to solving the equation $AY = I$, where Y is the unknown matrix inverse.

Figure 5.4.1(a) gives the computer program for solving a system of equations, containing up to ten equations, by matrix inversion. The augmented matrix is read as input data, and Gauss-Jordan elimination is performed on the augmented matrix to reduce the $N \times N$ coefficient matrix to an identity matrix, where N is the number of

```
C       SOLUTION OF A SYSTEM OF LINEAR EQUATIONS BY MATRIX INVERSION
C
C
C       A(I,J) REPRESENTS THE COEFFICIENTS OF THE AUGMENTED MATRIX
C       OF N ROWS AND (2N+1) COLUMNS, WHERE N IS THE NUMBER OF
C       EQUATIONS, X(I) ARE THE UNKNOWNS AND K IS THE PIVOT ROW.
C       THE AUGMENTED MATRIX CONSISTS OF N COLUMNS OF THE
C       COEFFICIENTS OF THE MATRIX OF THE GIVEN SYSTEM, ANOTHER
C       N COLUMNS OF THE IDENTITY MATRIX AND THE LAST COLUMN OF THE
C       CONSTANTS ON THE RIGHT-HAND SIDE OF THE EQUATIONS
C
C
C       ENTER INITIAL PARAMETERS
C
        DIMENSION A(10,21),X(10)
        N=6
        M=N+1
        M1=2*N
        M2=M1+1
C
C       ENTER COEFFICIENTS OF AUGMENTED MATRIX
C
        READ(5,*)((A(I,J),J=1,M2),I=1,N)
        N1=N-1
        DO 7 K=1,N
        K1=K+1
        K2=K
C
C       FIND THE ROW WITH THE LARGEST PIVOT ELEMENT
C
        B0=ABS(A(K,K))
        DO 1 I=K,N
        B1=ABS(A(I,K))
        IF((B0-B1) .LT. 0.0) THEN
        B0=B1
        K2=I
        END IF
  1     CONTINUE
C
C       CHECK IF ROW INTERCHANGE IS NEEDED
C
        IF((K2-K) .NE. 0) THEN
C
C       INTERCHANGE ROW TO OBTAIN LARGEST PIVOT ELEMENT
C
        DO 2 J=K,M1
        C=A(K2,J)
        A(K2,J)=A(K,J)
  2     A(K,J)=C
        END IF
C
C       APPLY GAUSS-JORDAN ELIMINATION FOR MATRIX INVERSION
C
        DO 7 I=1,N
        IF (I .NE. K) THEN
```

Figure 5.4.1 Program for the solution of the system of linear equations, obtained in Example 5.3, by matrix inversion. Also shown are the computed inverse of the matrix and the solution to the equations.

```
      DO 4 J=K1,M1
4     A(I,J)=A(I,J)-A(I,K)*A(K,J)/A(K,K)
      A(I,K)=0.0
      GO TO 7
      END IF
      DO 6 J=K1,M1
6     A(K,J)=A(K,J)/A(K,K)
      A(K,K)=1.0
7     CONTINUE
      WRITE(6,8)
8     FORMAT(2X,'THE INVERSE OF THE MATRIX'/)
      DO 9 I=1,N
9     WRITE(6,10)A(I,M),A(I,M+1),A(I,M+2),A(I,M+3),A(I,M+4),
     $ A(I,M+5)
10    FORMAT(6F10.4)
C
C     FIND UNKNOWNS FROM COMPUTED MATRIX INVERSE
C
      DO 12 I=1,N
      X(I)=0.0
      DO 11 J=M,M1
11    X(I)=X(I)+A(I,J)*A(J-N,M2)
12    CONTINUE
      WRITE(6,13)
13    FORMAT(/2X,'THE SOLUTION TO THE EQUATIONS'/)
      DO 14 I=1,N
14    WRITE(6,15)I,X(I)
15    FORMAT(4X,'X(',I1,')=',F10.5)
      STOP
      END
```

(a)

NUMERICAL RESULTS

```
THE INVERSE OF THE MATRIX

  0.1243   -0.0081    0.1622    0.0311    0.0324    0.0892
  0.3432    0.3689    0.1216    0.0858    0.0243   -0.0581
 -0.5324    0.3608    0.2838    0.1169    0.0568    0.0311
  0.2405   -0.3635   -0.2297    0.0601   -0.0459   -0.0014
  0.1027   -0.2676    0.3514    0.0257    0.0703   -0.0568
  0.2270   -0.2757   -0.4865    0.0568    0.1027    0.0324

THE SOLUTION TO THE EQUATIONS

  X(1)=   0.05135
  X(2)=   0.66351
  X(3)=   0.71486
  X(4)=   0.96892
  X(5)=  -0.30541
  X(6)=  -0.25405
```

(b)

Figure 5.4.1 Continued

equations. A change in the dimension statement and the printout formats would allow the solution of a system of a larger number of equations. The program in Fig. 5.4.1(a) is written for $N = 6$, corresponding to the system given by Eqs. (5.3.1) through (5.3.6). The inverse of the matrix is printed and the unknowns are computed from the equation $X = A^{-1}B$. For convenience, the constant vector B is placed at the last column in the augmented matrix. This column is left unaltered in the application of the elimination process, so that it may be used to supply the vector B needed for computing the unknowns. The computed values, listed in Fig. 5.4.1(b), are almost identical to those obtained in Example 5.3. The matrix A may be multiplied with its inverse to check whether the identity matrix is obtained. We did so in this problem to check the accuracy, and we found that AA^{-1} was very close to the identity matrix I.

As mentioned above, matrix inversion is particularly useful if several systems of equations with the same coefficient matrix A but different constant vectors B are to be solved. Also, most computer libraries contain programs for matrix inversion. These programs may often be used effectively to solve systems of equations. In BASIC, for instance, several computers have the MAT programs for the manipulation of matrices. If these are available on a given computer, a system of equations may be solved by employing the following simple program:

```
10  DIM X(6), B(6), A(6,6), C(6,6)
20  MAT READ B
30  MAT READ A
40  MAT C = INV (A)
50  MAT X = C * B
60  MAT PRINT X;
```

where C is the inverse of the coefficient matrix A.

5.6 ITERATIVE METHODS

In the preceding sections, we discussed direct methods for solving a set of simultaneous linear equations. These methods are appropriate for a small number of equations, typically fewer than 20, if the coefficient matrix A is dense. The main limitation arises from the round-off error, which is incurred in each computation and affects the overall accuracy of the solution. The tridiagonal system is a special case for which the computation involved is much smaller than that for a general matrix. Thus, for the tridiagonal case, many more equations may be solved while the desired accuracy level is preserved. Direct methods provide the solution in a finite number of steps, and, except for the round-off error, the solution is exact. However, for a large number of equations, typically of the order of several hundred, iterative methods, which start with an assumed solution and iterate to the desired solution of the system of equations, within a specified convergence criterion, are often more efficient.

Large sets of linear equations are generally sparse, and iterative methods, which consider only the nonzero coefficients in the computation, use this sparseness advantageously. Moreover, the round-off error after each iteration simply results in a less accurate input for the next iteration. Therefore, the resulting round-off error in the numerical solution is only what arises in the computation for the final iteration. The error does not accumulate as in direct methods. However, the solution is not exact but is obtained to an arbitrary, specified, convergence criterion.

5.6.1 Basic Approach

Let us consider the set of linear equations given by Eq. (5.2). These equations may be rewritten, by solving for the unknowns x_i, as follows:

$$\begin{aligned} x_1 &= \frac{b_1 - a_{12}x_2 - a_{13}x_3 - \cdots - a_{1n}x_n}{a_{11}} \\ x_2 &= \frac{b_2 - a_{21}x_1 - a_{23}x_3 - \cdots - a_{2n}x_n}{a_{22}} \\ &\vdots \\ x_n &= \frac{b_n - a_{n1}x_1 - a_{n2}x_2 - \cdots - a_{n,n-1}x_{n-1}}{a_{nn}} \end{aligned} \tag{5.55}$$

This system may be written more concisely as

$$x_i = \frac{b_i - \sum_{j=1, j\neq i}^{n} a_{ij}x_j}{a_{ii}} \qquad \text{for } i = 1, 2, \ldots, n \tag{5.56}$$

We need initial guesses for the unknowns to start the iterative process in the above equations. If $x_1^{(0)}, x_2^{(0)}, \ldots, x_i^{(0)}, \ldots, x_n^{(0)}$ are taken as the initial values, the value of x_1 after the first iteration, $x_1^{(1)}$, is obtained from

$$x_1^{(1)} = \frac{b_1 - a_{12}x_2^{(0)} - a_{13}x_3^{(0)} - \cdots - a_{1n}x_n^{(0)}}{a_{11}}$$

Similarly,

$$x_i^{(1)} = \frac{b_i - \sum_{j=1, j\neq i}^{n} a_{ij}x_j^{(0)}}{a_{ii}} \qquad \text{for } i = 1, 2, \ldots, n \tag{5.57}$$

The values obtained after the first iteration are then used for the next iteration. Thus, this iterative process may be written as

$$x_i^{(l+1)} = \frac{b_i - \sum_{j=1, j\neq i}^{n} a_{ij}x_j^{(l)}}{a_{ii}} \qquad \text{for } i = 1, 2, \ldots, n \tag{5.58}$$

where the superscript indicates the number of the iteration. This equation is also often written as

$$x_i^{(l+1)} = F_i[x_1^{(l)}, x_2^{(l)}, \ldots, x_{i-1}^{(l)}, x_{i+1}^{(l)}, \ldots, x_n^{(l)}] \tag{5.59}$$

where the function F_i is obtained from Eq. (5.58) and represents the relationship between an unknown x_i and the other unknowns. The value of the unknown x_i after l iterations, $x_i^{(l)}$, does not appear on the right-hand side for linear equations. However, for nonlinear equations, $x_i^{(l)}$ is often present within the parentheses for F_i. Even for linear equations, a term containing $x_i^{(l)}$ may be added in order to alter the convergence characteristics, as discussed later.

5.6.2 Jacobi and Gauss-Seidel Methods

The formulation for iteration given in Eq. (5.59) is known as the *Jacobi iterative method.* To compute the values for a given iteration step, it employs the values from the previous iteration. Therefore, all the values are computed, using previous values, before any unknown is updated. This implies that computer storage is needed for the present iteration as well as for the previous one. A considerable improvement in the rate of convergence and in the storage requirements can be obtained by replacing the values from the previous iteration by new ones as soon as they are computed. Then only the values of the latest iteration are stored, and each iterative computation of the unknown employs the most recent values of the other unknowns. This computational scheme, known as the *Gauss-Seidel method,* is used extensively for solving large systems of equations that frequently arise in the numerical solution of differential equations.

Let us consider the use of the Gauss-Seidel method for computing the iterative values of the unknowns, starting with x_1 and then successively obtaining x_2, $x_3, \ldots, x_n$. Then the second iteration for, say, x_3, is obtained from

$$x_3^{(2)} = \frac{b_3 - a_{31}x_1^{(2)} - a_{32}x_2^{(2)} - a_{34}x_4^{(1)} - \cdots - a_{3n}x_n^{(1)}}{a_{33}}$$

Here, the values of x_1 and x_2 are known after the second iteration, and the others are known only after the first iteration. Similarly, the $(l+1)$th iteration for x_i may be written as

$$x_i^{(l+1)} = \frac{b_i - \sum_{j=1}^{i-1} a_{ij}x_j^{(l+1)} - \sum_{j=i+1}^{n} a_{ij}x_j^{(l)}}{a_{ii}} \qquad \text{for } i = 1, 2, \ldots, n \tag{5.60}$$

or

$$x_i^{(l+1)} = F_i[x_1^{(l+1)}, x_2^{(l+1)}, \ldots, x_{i-1}^{(l+1)}, x_{i+1}^{(l)}, \ldots, x_n^{(l)}] \tag{5.61}$$

This formulation, therefore, assumes that the computation of the unknowns x_i starts with x_1 and proceeds with increasing i until all the values are obtained for a given iteration. The Gauss-Seidel method repeatedly calculates the unknowns, replacing the values from the previous iteration by new ones and thus requiring only one computer storage space for each unknown. Programming is also simplified since the most recent value of each unknown is always employed in the computations. This iterative process will converge to the solution vector if the equations have certain characteristics, as discussed below. A better initial guess of the unknowns will also lead to faster convergence, if the process is convergent.

5.6.3 Convergence

The iterative computation of the unknowns is terminated when a specified convergence criterion is satisfied. Generally, if the change in the value of each unknown from one iteration to the next is less than a given small quantity ε, convergence is assumed to have been achieved. The convergence criterion ε may be applied to the physical value of the unknown or to its normalized value. Therefore, the condition for convergence may be written as

$$|x_i^{(l+1)} - x_i^{(l)}| \leqslant \varepsilon \qquad \text{for } i = 1, 2, \ldots, n \tag{5.62a}$$

or

$$\left|\frac{x_i^{(l+1)} - x_i^{(l)}}{x_i^{(l)}}\right| \leqslant \varepsilon \qquad \text{for } i = 1, 2, \ldots, n \tag{5.62b}$$

The second form of the convergence criterion is more appropriate if an estimate of the magnitude of the unknown x_i is not available and none of the unknowns is expected to be zero. The choice of ε is arbitrary and may often be taken as around 10^{-4} for the second form of the criterion, [Eq. (5.62b)], which specifies the maximum fractional change in each unknown from one iteration to the next. However, as discussed in Chapter 2, the dependence of the solution on the convergence criterion must be studied by varying ε so that the numerical solution obtained is essentially independent of the value chosen. The convergence criterion may also be applied to a few important unknowns, instead of all x_i, in order to reduce the computing time. Similarly, it may be applied to the sum of the absolute values or of the squares of the changes in all the unknowns between two successive iterations. With large systems of equations, such alternative forms of the convergence criterion are often employed to save computer time.

The conditions for convergence of the iterative process have been analyzed for the Jacobi and the Gauss-Seidel methods and presented in terms of the nature of the coefficient matrix A. Both of these methods have good convergence characteristics for diagonally dominant systems, that is, for systems in which each diagonal element a_{ii} is larger, in absolute value, than the sum of the magnitudes of the other elements in the row. Thus, if

$$|a_{ii}| > \sum_{j=1, j\neq i}^{n} |a_{ij}| \tag{5.63}$$

the system is said to be *diagonally dominant*, and convergence is guaranteed for linear systems. However, convergence is generally obtained with much weaker diagonal dominance. These methods are particularly useful in the solution of large systems of linear equations that arise in the numerical solution of partial differential equations by finite difference or finite element methods. The equations obtained in these cases usually have diagonal dominance, and the above iterative methods are convergent. With some modifications, these methods may also be employed for nonlinear equations, as discussed in Section 5.8.

5.6.4 An Example

In order to illustrate the Gauss-Seidel method, let us consider the following set of linear equations:

$$\begin{aligned} 5x + y + 2z &= 17 \\ x + 3y + z &= 8 \\ 2x + y + 6z &= 23 \end{aligned} \tag{5.64}$$

This system is diagonally dominant, since the dominant coefficient in each equation is the diagonal element, which is also larger than the sum of the absolute values of the other coefficients. Therefore, the Gauss-Seidel iteration is convergent for these equations. The above equations are rewritten as

$$x = \frac{17 - y - 2z}{5} \qquad y = \frac{8 - x - z}{3} \qquad z = \frac{23 - 2x - y}{6}$$

For the Gauss-Seidel method, the most recent values of x, y, and z are to be used in the iteration. If the starting values are arbitrarily chosen as $x = 1$, $y = 1$, and $z = 1$, the values for the first iteration are computed, by rounding off to three decimal digits, as follows:

$$x^{(1)} = \frac{17 - 1 - 2}{5} = 2.8$$

$$y^{(1)} = \frac{8 - 2.8 - 1}{3} = 1.4$$

$$z^{(1)} = \frac{23 - 5.6 - 1.4}{6} = 2.667$$

Similarly, the next four iterations are obtained as

$$x^{(2)} = 2.053 \qquad y^{(2)} = 1.093 \qquad z^{(2)} = 2.967$$

$$x^{(3)} = 1.995 \qquad y^{(3)} = 1.013 \qquad z^{(3)} = 3.000$$

$$x^{(4)} = 1.997 \qquad y^{(4)} = 1.001 \qquad z^{(4)} = 3.001$$

$$x^{(5)} = 2.000 \qquad y^{(5)} = 1.000 \qquad z^{(5)} = 3.000$$

The exact solution of the above system of equations is $x = 2$, $y = 1$, and $z = 3$. Therefore, the iterative procedure converges rapidly to yield a solution that is within

0.2% of the exact solution in only four iterations. The process will terminate after four iterations if ε in Eq. (5.62b) is taken as 0.02, and after five iterations if it is chosen as 0.002. More iterations may be needed for a still smaller value of ε, since changes in the fourth and higher decimal places may occur from one iteration to the next. If the Jacobi method is applied to the above system, the rate of convergence is much slower. Therefore, the Jacobi method is seldom used on traditional computers and is considered largely in order to study the convergence characteristics of other iterative methods in terms of those of the Jacobi method.

5.6.5 Relaxation Methods

19

The convergence characteristics of the Gauss-Seidel method can often be considerably improved by the use of point relaxation, which is given by

$$x_i^{(l+1)} = \omega[x_i^{(l+1)}]_{GS} + (1-\omega)x_i^{(l)} \tag{5.65}$$

where ω is a constant in the range $0 < \omega < 2$ and $[x_i^{(l+1)}]_{GS}$ is the value of x_i obtained for the $(l+1)$th iteration by using the Gauss-Seidel iteration equation, Eq. (5.60). For $\omega > 2$, the process is divergent. If $0 < \omega < 1$, the iterative scheme is known as *successive under-relaxation* (SUR), and if $1 < \omega < 2$, the scheme is termed *successive over-relaxation* (SOR). In the former case, the value for the $(l+1)$th iteration is a weighted average of the value from the previous iteration and that obtained by the use of the Gauss-Seidel method for the present iteration. In successive over-relaxation, the change in x_i from one iteration to the next, in the Gauss-Seidel scheme, is multiplied by a factor between 1.0 and 2.0 to accelerate convergence. With the optimum relaxation factor, ω_{opt}, the convergence is much faster than that for Gauss-Seidel.

The relaxation method may be written, using Eq. (5.60), as

$$x_i^{(l+1)} = \frac{\omega\left[b_i - \sum_{j=1}^{i-1} a_{ij}x_j^{(l+1)} - \sum_{j=i+1}^{n} a_{ij}x_j^{(l)}\right]}{a_{ii}} + (1-\omega)x_i^{(l)} \qquad \text{for } i = 1, 2, \ldots, n \tag{5.66}$$

It is obvious that the Gauss-Seidel method is obtained for $\omega = 1$. In this case, $x_i^{(l)}$ drops out from the right-hand side. Successive under-relaxation is generally used for nonlinear equations and for systems that result in a divergent Gauss-Seidel iteration. Successive over-relaxation is widely used for accelerating the convergence in linear systems. However, the determination of an optimum value of the relaxation factor ω is often difficult and is generally done by trial and error. For some systems, it may be available from earlier studies or from analysis. If several similar systems are to be solved, it is generally worthwhile to obtain the optimum value of ω by trying various values, over the given range, and then use it in the computations. For further details on the use of point relaxation in engineering applications, advanced books on the numerical solution of differential equations, such as Ferziger (1981) and Jaluria and Torrance (1986), may be consulted. The following example also illustrates the use of

Gauss-Seidel and successive over-relaxation methods for solving a system of linear equations.

Example 5.5

Solve the problem discussed in Example 5.2 by means of the Gauss-Seidel iterative procedure. Then modify the computer program to solve the problem by the successive over-relaxation method. Vary the relaxation factor ω to study the effect of its value on the number of iterations needed for convergence.

Solution

The system of equations to be solved is

$$-T_{i+1} + [2 + G(\Delta x)^2]T_i - T_{i-1} = 0 \qquad \text{for } i = 1, 2, \ldots, 29 \tag{5.5.1}$$

with

$$T_0 = T_{30} = 100 \tag{5.5.2}$$

The given system is rewritten as

$$T_i = \frac{T_{i+1} + T_{i-1}}{2 + S} \qquad \text{for } i = 1, 2, \ldots, 29 \tag{5.5.3}$$

where

$$S = G(\Delta x)^2 = (0.071)^2(1.0)^2 \tag{5.5.4}$$

Since T_0 and T_{30} are given as 100°C, the equations for T_1 and T_{29} become

$$T_1 = \frac{T_2 + 100}{2 + S} \tag{5.5.5}$$

$$T_{29} = \frac{100 + T_{28}}{2 + S} \tag{5.5.6}$$

Therefore, the resulting system of equations consists of Eqs. (5.5.5) and (5.5.6), and the equations from Eq. (5.5.3) for $i = 2, 3, \ldots, 28$. We solve these 29 equations by the Gauss-Seidel method to obtain the required temperature distribution.

The initial guess, or starting temperature distribution, is taken as $T(I) = 0$, for $I = 1, 2, \ldots, 29$, where $T(I)$ denotes T_i in the program shown in Fig. 5.5.1(a). Using Eqs. (5.5.3), (5.5.5), and (5.5.6), we compute the temperatures for the next iteration, and we compare the new values with the previous values to check for convergence. It is demanded that the absolute value of the difference between the two be less than the convergence criterion ε, or, for convergence,

$$|(TO)_i - T_i| \leqslant \varepsilon \qquad \text{for } i = 1, 2, \ldots, 29 \tag{5.5.7}$$

where T represents the new values and TO the previous ones. If this difference for any value of i is greater than ε, which is denoted by EPS in the program, the iterative

```
C     GAUSS-SEIDEL METHOD FOR SOLVING A SYSTEM OF LINEAR EQUATIONS
C
C
C     T(I) REPRESENTS THE TEMPERATURE DIFFERENCES FROM THE AMBIENT
C     TEMPERATURE, TO(I) DENOTES THE TEMPERATURE DIFFERENCES AFTER
C     THE PREVIOUS ITERATION, TT IS THE ACTUAL TEMPERATURE,S1 IS A
C     CONSTANT DEFINED IN THE PROBLEM AND N IS THE NUMBER OF
C     EQUATIONS
C
C
C     ENTER VALUES OF RELEVANT PARAMETERS
C
      DIMENSION T(30),TO(30)
      S1=(0.071**2)*(1.0**2)+2.0
      N=29
      NN=N-1
      EPS=0.1
      DO 10 K=1,5
C
C     INPUT STARTING VALUES
C
      J=0
      DO 1 I=1,N
  1   T(I)=0.0
C
C     STORE COMPUTED VALUES AFTER EACH ITERATION
C
  2   DO 3 I=1,N
  3   TO(I)=T(I)
C
C     COMPUTE THE END VALUES T(1) AND T(N)
C
      T(1)=(T(2)+100.0)/S1
      T(N)=(100.0+T(N-1))/S1
C
C     COMPUTE INTERMEDIATE VALUES
C
      DO 4 I=2,NN
  4   T(I)=(T(I+1)+T(I-1))/S1
C
C     CHECK FOR CONVERGENCE
C
      J=J+1
      DO 5 I=1,N
      IF(ABS(TO(I)-T(I)) .GT. EPS) GO TO 2
  5   CONTINUE
      WRITE(6,6)EPS
  6   FORMAT(//2X,'EPS=',F10.5)
      WRITE(6,7)J
  7   FORMAT(/2X,'NUMBER OF ITERATIONS=',I4/)
C
C     COMPUTE ACTUAL TEMPERATURES
C
      DO 8 I=1,N
      TT=T(I)+20.0
  8   WRITE(6,9)I,TT
  9   FORMAT(2X,'T(',I2,')=',F12.4)
  10  EPS=EPS/10.0
      STOP
      END
```

(a)

Figure 5.5.1 Computer program for the solution of the problem of Examples 5.2 and 5.5 by the Gauss-Seidel iterative method, Also shown are the computed temperatures at two values, 10^{-4} and 10^{-5}, of the convergence parameter EPS.

NUMERICAL RESULTS

EPS= 0.00010		EPS= 0.00001	
NUMBER OF ITERATIONS= 600		NUMBER OF ITERATIONS= 766	
T(1)=	114.6572	T(1)=	114.6578
T(2)=	109.7915	T(2)=	109.7928
T(3)=	105.3784	T(3)=	105.3803
T(4)=	101.3957	T(4)=	101.3981
T(5)=	97.8234	T(5)=	97.8263
T(6)=	94.6433	T(6)=	94.6467
T(7)=	91.8396	T(7)=	91.8434
T(8)=	89.3979	T(8)=	89.4022
T(9)=	87.3062	T(9)=	87.3109
T(10)=	85.5538	T(10)=	85.5588
T(11)=	84.1320	T(11)=	84.1371
T(12)=	83.0334	T(12)=	83.0388
T(13)=	82.2527	T(13)=	82.2581
T(14)=	81.7859	T(14)=	81.7914
T(15)=	81.6306	T(15)=	81.6360
T(16)=	81.7860	T(16)=	81.7914
T(17)=	82.2529	T(17)=	82.2581
T(18)=	83.0337	T(18)=	83.0388
T(19)=	84.1323	T(19)=	84.1371
T(20)=	85.5543	T(20)=	85.5588
T(21)=	87.3067	T(21)=	87.3109
T(22)=	89.3984	T(22)=	89.4022
T(23)=	91.8400	T(23)=	91.8434
T(24)=	94.6438	T(24)=	94.6467
T(25)=	97.8238	T(25)=	97.8263
T(26)=	101.3961	T(26)=	101.3981
T(27)=	105.3788	T(27)=	105.3803
T(28)=	109.7917	T(28)=	109.7928
T(29)=	114.6573	T(29)=	114.6578

(b)

Figure 5.5.1 Continued

process is repeated, taking the computed new values as the starting values for the next iteration. Once convergence has been achieved, we obtain the physical temperature TT by adding the ambient temperature of 20°C to the temperature difference T.

The resulting distribution is shown in Fig. 5.5.1(b) for two values of EPS, 10^{-4} and 10^{-5}. Only a small difference in the computed values is observed in going from the larger to the smaller value. The total number of iterations increases from 600 to 766. A comparison with the results obtained in Example 5.2 for this tridiagonal system also indicates a difference of less than 0.005% in the temperatures for the smaller value of EPS. The program for the Gauss-Seidel method is clearly much simpler than that for Gaussian elimination in Example 5.1. If the system is tridiagonal, Gaussian elimination is preferable, since it takes less computer time and is generally more accurate. However, the Gauss-Seidel method is advantageous to use when the coefficient matrix is sparse, although not tridiagonal. It can also be extended to systems of nonlinear equations, as discussed later in the text.

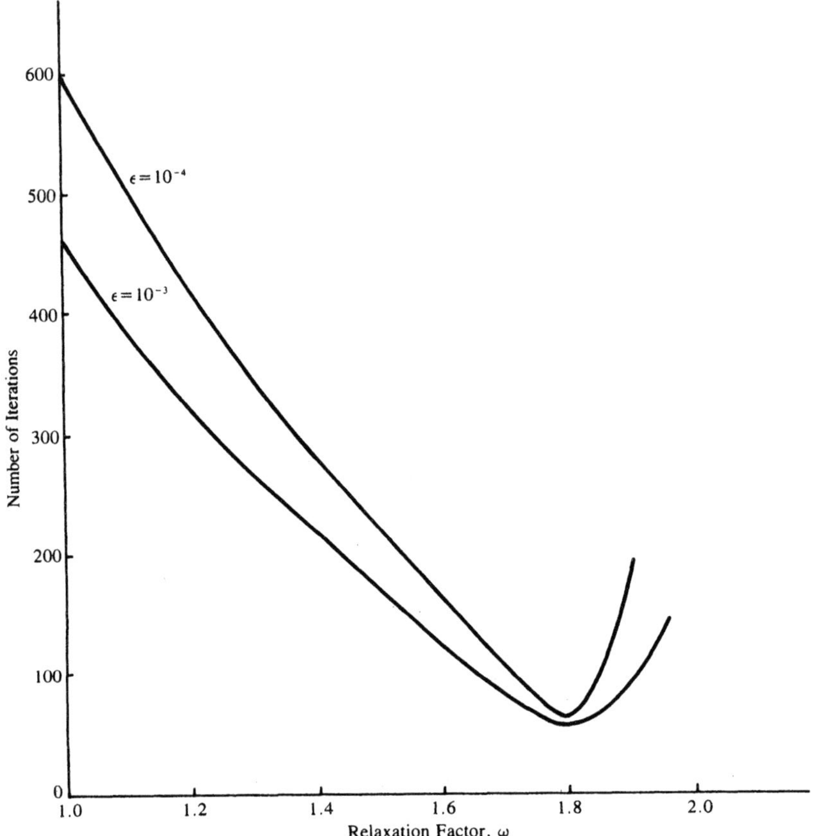

Figure 5.5.2 Variation of the number of iterations needed for convergence, in the solution of Example 5.5 by the SOR method, with the relaxation factor ω.

The given computer program can easily be modified to apply the successive over-relaxation method. The relaxation factor ω must be entered as a parameter, and the recursion formula becomes

$$T_i^{(l+1)} = \omega\left[\frac{T_{i+1}^{(l)} + T_{i-1}^{(l+1)}}{2+S}\right] + (1-\omega)T_i^{(l)} \qquad \text{for } i = 1, 2, \ldots, 29 \qquad (5.5.8)$$

This equation replaces the one given in the computer program on the basis of Eq. (5.5.3). The modified program was employed for ω varying from 1.0 to 2.0. Figure 5.5.2 shows the dependence of the number of iterations on the relaxation factor ω, at $\varepsilon = 10^{-3}$ and 10^{-4}. Note that the number of iterations at the optimum value, ω_{opt}, is almost one-tenth that for Gauss-Seidel iteration. Clearly, SOR is a very efficient method if the optimum value of the relaxation factor is known.

5.7 HOMOGENEOUS LINEAR EQUATIONS

In many problems of engineering interest, such as those encountered in vibrating systems, stability analysis, and certain electrical circuits, the constant vector B in the system of linear equations is zero, giving rise to a set of equations of the form $AX = 0$. The system of equations is then said to be *homogeneous*. A trivial solution, $X = 0$, exists for this system. However, nontrivial solutions may be obtained only if the determinant of the coefficient matrix A is zero, that is, Det $A = 0$. This occurs when all the equations of the set are not linearly independent, and one or more equations may be obtained from a linear combination of the others. In considering simultaneous nonhomogeneous linear equations, we noted from Cramer's rule that unique solutions may be obtained only if the determinant, Det A, is nonzero. However, in simultaneous homogeneous equations, the numerators in the solution by Cramer's rule, given in Eq. (5.4), are all zero, since the constant vector B is 0. Therefore, nontrivial solutions may exist only if the denominator, which is Det A, is also zero. However, unique values of the unknowns $x_1, x_2, \ldots, x_n$ are not obtained in this case, since the solution vector X when multiplied by an arbitrary constant will also satisfy the system of homogeneous equations, $AX = 0$. Therefore, the desired solution establishes relationships between the unknowns, and the number of dependent equations in the set gives the number of unknowns that must be arbitrarily chosen to obtain the rest.

5.7.1 The Eigenvalue Problem

An important class of problems involving homogeneous equations is the eigenvalue problem, which is of considerable interest in engineering applications. Such problems occur, for instance, in the analysis of structures for critical buckling loads, in stress analysis for determining the principal normal stresses, and in the natural vibration of systems to determine the frequencies and the vibrational modes. The matrix equation for an eigenvalue problem is

$$(A - \lambda I)X = 0 \tag{5.67}$$

or

$$AX = \lambda X \tag{5.68}$$

where A is a known $n \times n$ matrix, X is the solution vector, and λ is an unknown constant. Nontrivial solutions to the above system of equations are obtained only for certain values of λ. These values are known as *eigenvalues* of the coefficient matrix A, and the solution vectors X corresponding to these eigenvalues are called the *eigenvectors*, which can be determined only to within a multiplicative constant. In a vibrating system, consisting of masses and springs, as shown in Fig. 5.3, the eigenvalues are the squares of the natural frequencies of vibration, and the eigenvectors give the displacements of the masses. This problem is discussed in detail later.

From Cramer's rule, it is evident that nontrivial solutions may be obtained only if

$$\text{Det}(A - \lambda I) = 0 \tag{5.69}$$

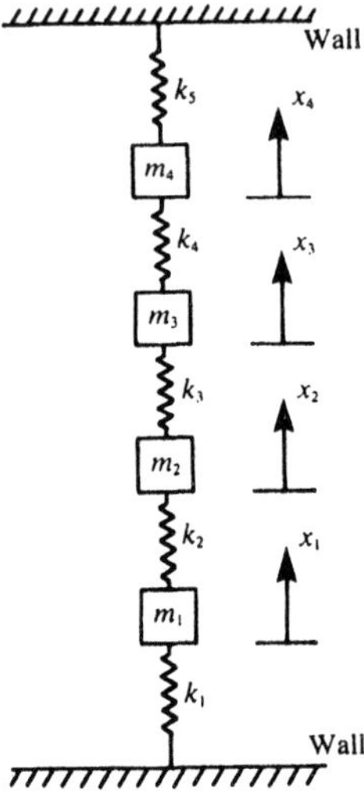

Figure 5.3 A vibrating system consisting of four masses, denoted by $m_1, \ldots, m_4$, and of five springs, whose spring constants are denoted by $k_1, \ldots, k_5$. The displacements are denoted by x_1, x_2, x_3, and x_4, giving a four-degrees-of-freedom system.

This may be written as

$$\text{Det}(A - \lambda I) = \begin{bmatrix} (a_{11} - \lambda) & a_{12} & a_{13} & \cdots & a_{1n} \\ a_{21} & (a_{22} - \lambda) & a_{23} & \cdots & a_{2n} \\ \vdots & \vdots & & \vdots & a_{n1} \\ a_{n1} & a_{n2} & a_{n3} & \cdots & (a_{nn} - \lambda) \end{bmatrix} = 0 \quad (5.70)$$

Expansion of this determinant results in a polynomial of order n in λ. This polynomial, which is known as the *characteristic polynomial* of matrix A, may be solved by the methods discussed in Chapter 4 to obtain the eigenvalues; see Examples 4.8 and 4.9. Once the eigenvalues have been obtained, one can determine the eigenvector corresponding to each value by substituting the value in the given equations. If there is only one linearly dependent equation in the set of equations, one must assume the value of one unknown to obtain the corresponding values of the remaining unknowns. Similarly, if there are two dependent equations, the values of two unknowns need to be assumed, and so on. In many engineering problems, only one dependent equation arises, and, therefore, the value of only one unknown must be chosen. Textbooks on linear algebra, such as Anton (1984) and Williams (1984), may be consulted for further details on eigenvalue problems.

The above procedure of expanding the determinant to obtain the characteristic polynomial, which may then be solved for the eigenvalues, is computationally practical only for a small number of equations, typically up to four, and for sparse matrices. For somewhat larger systems, generally of the order of ten equations, one may obtain the characteristic polynomial by using the methods developed by Leverrier and by Faddeev; see Carnahan et al. (1969). Both of these methods are quite similar and generate the coefficients $\sigma_1, \sigma_2, \ldots, \sigma_n$ of the characteristic polynomial

$$f(\lambda) = (-1)^n[\lambda^n - \sigma_1\lambda^{n-1} + \sigma_2\lambda^{n-2} + \cdots + (-1)^n\sigma_n] \quad (5.71)$$

where the σ's are given in terms of the trace of a sequence of matrices, starting with A. The trace of a matrix is the sum of its diagonal terms. Therefore,

$$\text{trace}\,(A) = \sum_{i=1}^{n} a_{ii}$$

For further details, see Carnahan et al. (1969). Once the characteristic polynomial has been obtained, the root-finding methods of Chapter 4, such as Bairstow's or Graeffe's methods, may be used to find the eigenvalues.

Since a polynomial with all real coefficients can have complex roots, complex eigenvalues may be obtained. However, in many physical problems, the coefficient matrix A is symmetric. It can be shown that all the eigenvalues of a symmetric matrix are real, which substantially simplifies the computational procedure. The solution of the eigenvalue problem for symmetric matrices is of considerable interest in engineering problems and is discussed in detail later. It may be pointed out here that even though the characteristic polynomial may be generated, by the methods of Leverrier and Faddeev, for systems containing as many as 25 or 30 equations, the solution of the polynomial is generally very involved, and the other methods outlined here are preferred to the root solving procedures of Chapter 4.

The eigenvectors may be obtained by substitution of the eigenvalues, one at a time, into the given system of equations. The equations thus obtained may be solved by the use of the Gauss-Jordan method. The method is applied to the matrix $(A - \lambda I)$, as outlined in Section 5.3, and the process carried out until the reduced matrix is such that a further application of the method is not possible due to all possible pivot elements being zero. If the system contains only one dependent equation, then the process stops with only the last column left to be reduced. The other columns contain only zeros and one. At this stage, only the independent equations are left, and, if an arbitrary value is given to one unknown, the other unknowns may be computed from the resulting nonhomogeneous equations. Similarly, if two dependent equations are present in the given set, the Gauss-Jordan method yields two unreduced columns. This requires choosing arbitrary values for two unknowns to obtain two linearly independent eigenvectors. Column interchanges, besides row interchanges, are frequently employed in the process to avoid taking a pivot element that is zero. If a column interchange is employed, the components of the eigenvector corresponding to these columns must also be interchanged. Example 5.6 discusses a physical problem and the use of Gauss-Jordan elimination for determining the eigenvectors.

Example 5.6

For the natural vibration of the three masses, m, $2m$, and m, connected by the four springs shown in Fig. 5.6.1, determine the characteristic polynomial, the eigenvalues, which correspond to the natural frequencies of vibration, and the eigenvectors, which give the amplitudes of motion of the masses. The spring constants for the four springs are k, $2k$, $2k$, and k. The displacements of the three masses are defined by the coordinates x_1, x_2, and x_3, respectively, as shown. Take $k/m = 1.0$.

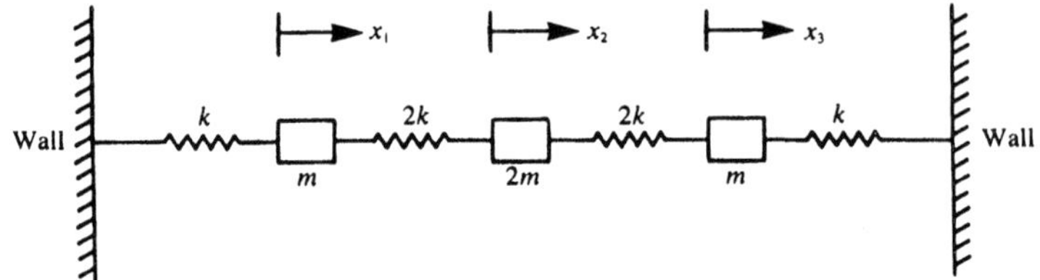

Figure 5.6.1 The vibrating mass and spring system considered in Example 5.6.

Solution

The extension in the first spring, from the left, is x_1, and that in the second spring is $(x_2 - x_1)$. Since the inward directed force due to the extension is given by a product of the spring constant and the extension, the net force acting on the first mass in the positive x_1 direction is $[2k(x_2 - x_1) - kx_1]$. Therefore, from Newton's second law,

$$m\ddot{x}_1 = 2k(x_2 - x_1) - kx_1$$

or

$$m\ddot{x}_1 + kx_1 + 2k(x_1 - x_2) = 0 \tag{5.6.1}$$

where $\ddot{x}_1$ is the second derivative of x_1 with respect to time τ and is, thus, the acceleration of the mass. Similarly, for the other masses,

$$2m\ddot{x}_2 + 2k(x_2 - x_1) + 2k(x_2 - x_3) = 0 \tag{5.6.2}$$

$$m\ddot{x}_3 + 2k(x_3 - x_2) + kx_3 = 0 \tag{5.6.3}$$

From the theory of vibrations, the solution to the above equations may be taken as

$$\begin{aligned} x_1 &= X_1 \sin \omega\tau \\ x_2 &= X_2 \sin \omega\tau \\ x_3 &= X_3 \sin \omega\tau \end{aligned} \tag{5.6.4}$$

where X_1, X_2, and X_3 are the amplitudes of motion and ω is the natural frequency in radians/second. If these equations are substituted in the equations of motion, Eqs. (5.6.1) through (5.6.3), we obtain the following system of linear homogeneous equations for $k/m = 1.0$:

$$\begin{aligned} (3 - \omega^2)X_1 - 2X_2 &= 0 \\ -X_1 + (2 - \omega^2)X_2 - X_3 &= 0 \\ -2X_2 + (3 - \omega^2)X_3 &= 0 \end{aligned} \tag{5.6.5}$$

This system may be written in matrix form as

$$AX = \lambda X \tag{5.6.6}$$

where

$$\lambda = \omega^2 \tag{5.6.7}$$

and

$$A = \begin{bmatrix} 3 & -2 & 0 \\ -1 & 2 & -1 \\ 0 & -2 & 3 \end{bmatrix} \tag{5.6.8}$$

A nontrivial solution of Eq. (5.6.5) can be obtained only if the determinant of the coefficient matrix $(A - \lambda)$ is zero. Therefore,

$$\text{Det}(A - \lambda) = \begin{bmatrix} 3-\lambda & -2 & 0 \\ -1 & 2-\lambda & -1 \\ 0 & -2 & 3-\lambda \end{bmatrix} = 0 \tag{5.6.9}$$

We obtain the characteristic polynomial for this eigenvalue problem by expanding this determinant as

$$\lambda^3 + 8\lambda^2 - 17\lambda + 6 = 0 \tag{5.6.10}$$

The roots of this polynomial equation may be determined by the root-solving methods given in Chapter 4. Using the search method to obtain the approximate location of the roots, followed by the Newton-Raphson method, we determine the eigenvalues, in s^{-2}, as follows:

$$\lambda_1 = 0.43845 \qquad \lambda_2 = 3.0 \qquad \lambda_3 = 4.56155 \tag{5.6.11}$$

To determine the eigenvectors corresponding to these eigenvalues, we substitute each eigenvalue in Eq. (5.6.5) and obtain the ratios of the amplitudes. Then, if one of the amplitudes is taken as 1.0, the others may be determined. From symmetry, $X_1 = X_3$ and both may be taken as 1.0. Then X_2 is determined for the three eigenvalues given in Eq. (5.6.11) as, respectively,

$$X_2 = 1.28078 \qquad X_2 = 0 \qquad X_2 = -0.78078 \tag{5.6.12}$$

Let us again consider the equations obtained for λ_1:

$$\begin{aligned} 2.56155X_1 - 2X_2 + 0.X_3 &= 0 \\ -X_1 + 1.56155X_2 - X_3 &= 0 \\ 0.X_1 - 2X_2 + 2.56155X_3 &= 0 \end{aligned} \tag{5.6.13}$$

If Gauss-Jordan elimination is applied to the coefficient matrix of these equations, we obtain, after the first two steps,

$$\begin{bmatrix} 1 & -0.78078 & 0 \\ 0 & 0.78078 & -1 \\ 0 & -2 & 2.56155 \end{bmatrix} \quad \text{and} \quad \begin{bmatrix} 1 & 0 & -1 \\ 0 & 1 & -1.28077 \\ 0 & 0 & 0 \end{bmatrix}$$

Therefore, further application of the Gauss-Jordan method is not possible. At this stage, two independent algebraic equations, containing the unknowns X_1, X_2, and X_3, are obtained from the first two rows of the reduced matrix. Therefore, if X_1 is taken as 1.0, X_3 is obtained as 1.0 and X_2 as 1.28077.

In many engineering problems, the set of n homogeneous equations contains $(n - 1)$ independent equations for determining the n components of the eigenvector. An application of Gauss-Jordan elimination, with partial and complete pivoting, if necessary, reduces the system to a set of independent equations at the stage where further reduction is not possible. If two dependent equations are present, one must assume two components in order to determine the rest, and so on. Therefore, small systems of equations may be solved by root-solving methods, followed by Gauss-Jordan elimination to determine the eigenvectors. Example 5.6 also illustrates the solution of homogeneous ordinary differential equations, as considered again in Chapter 8.

5.7.2 The Power Method*

The power method is a frequently employed iterative procedure for determining the eigenvalues and the corresponding eigenvectors, particularly if the largest or the smallest eigenvalue is of interest. Intermediate eigenvalues may also be determined by gradually eliminating the eigenvalues already found. However, the round-off errors accumulate, leading to lower accuracy, and the process becomes more involved as the intermediate eigenvalues are successively determined. Therefore, the method is well suited mainly for finding the largest and the smallest eigenvalues. It has the advantages of a simpler computational procedure, as compared to several other methods discussed later, and of providing the eigenvector along with the eigenvalue.

Largest Eigenvalue. Let us first consider the iterative power method for finding the largest eigenvalue of the system given by

$$AX = \lambda X \tag{5.68}$$

The method starts with an initial estimate of the eigenvector, denoted as $X^{(0)}$. Usually, all the elements of the initial vector are taken as 1 unless a better estimate is available. The vector $X^{(0)}$ is multiplied by the coefficient matrix A to obtain the vector $AX^{(0)}$. This resulting vector is normalized by dividing each of its elements by any one element, generally chosen as the largest element for accuracy or as the first element for convenience. The normalized vector is denoted as $X^{(1)}$. If the difference between the new vector $X^{(1)}$ and the old one $X^{(0)}$ is less than a chosen convergence criterion, the process is terminated. Otherwise, $X^{(1)}$ becomes the starting vector for the next iteration, and the process is repeated. When convergence has been achieved, the normalizing factor is taken as the largest component of the vector X. Then this is the largest eigenvalue λ_1, and the normalized vector is the corresponding eigenvector. The convergence of the method depends on the initial vector $X^{(0)}$ and on the ratio r of the two largest eigenvalues. Convergence is slower if the two are close to each other in magnitude, that is, if r is close to unity. The dominance ratio r is defined as

$$r = \frac{|\lambda_2|}{|\lambda_1|} \tag{5.72}$$

where λ_1 is the largest eigenvalue in magnitude and λ_2 is the next largest. Although the power method is particularly suitable for symmetric matrices, since the eigenvalues are all real in this case, it can also be used for nonsymmetric matrices. The convergence of the method to the largest eigenvalue may be proved mathematically for symmetric matrices.

Smallest Eigenvalue. In several engineering problems, the smallest eigenvalue is of particular interest. For instance, designers are interested in the lowest frequency of the natural vibration of civil engineering structures so that they can design the structures to avoid certain externally induced vibrations. The power method may be used to determine the smallest eigenvalue and the corresponding eigenvector by premultiplication of the original system, Eq. (5.68), by the inverse A^{-1} of the coefficient matrix. Therefore,

$$A^{-1}AX = \lambda A^{-1}X$$

Now, $A^{-1}A = I$, and if both sides are divided by λ, the result is

$$HX = \frac{1}{\lambda}X \tag{5.73}$$

where the inverse matrix A^{-1} is denoted as H. Therefore, if the power method is applied to Eq. (5.73), the largest eigenvalue obtained will be $1/\lambda$, which arises from the matrix H. This largest value of $1/\lambda$ corresponds to the smallest eigenvalue in magnitude. However, one must determine the inverse H of the matrix A in order to apply this method. Therefore, the procedure may not be practical for very large matrices.

Intermediate Eigenvalues. Several procedures are available for obtaining the intermediate eigenvalues, lying between the smallest and the largest eigenvalues. Most of these methods gradually remove the known eigenvalues from the problem so that the method converges to the next eigenvalue. However, this procedure, known as *deflation*, is suitable if only a few eigenvalues are needed, since the growth of round-off error often severely limits the accuracy of the results. If the largest eigenvalue λ_1 and the corresponding eigenvector X_1 have been found for the given matrix A, a new matrix A may be formed in terms of the transpose X_1^T of the matrix, as

$$A_1 = A - \frac{\lambda_1(X_1X_1^T)}{X_1^TX_1} \tag{5.74}$$

It can be shown that $X_1^TX_1$ is a scalar and equal to the sum of the squares of the components of the X_1 eigenvector. The matrix $X_1X_1^T$ is a symmetric matrix of the same dimension as A. It can also be shown that A_1 has the same eigenvalues and eigenvectors as A, except for λ_1 which is replaced by zero. Therefore, if the power method is applied to A_1, it will converge to the second largest eigenvalue λ_2 and the corresponding eigenvector X_2. Similarly, the next largest eigenvalue λ_3 and the associated eigenvector X_3 may be obtained by applying the power method to a matrix A_2, given as

$$A_2 = A_1 - \frac{\lambda_2(X_2X_2^T)}{X_2^TX_2} \tag{5.75}$$

The power method can be applied quite easily on the computer, particularly if only the largest eigenvalue is desired. For other eigenvalues, the techniques outlined above may be applied, although the method is rarely used if more than a few eigenvalues are needed. The rate of convergence can sometimes be accelerated by the addition of a constant to each diagonal element of the matrix. This shifts all the eigenvalues by a constant value and may change the dominance ratio favorably to accelerate convergence. However, the suitable amount of shift must be obtained by trial and error. Over-relaxation or under-relaxation, similar to that discussed in Section 5.6, may also be used to achieve faster convergence. Again, the optimum value of ω for the fastest convergence must often be obtained by trying several values. Example 5.7 discusses the solution of an eigenvalue problem by the power method.

Example 5.7

For the vibrating system considered in Example 5.6, obtain the largest eigenvalue and the corresponding eigenvector, using the iterative power method.

Solution

The given system of equations is written as

$$AX = \lambda X \tag{5.68}$$

where λ and A are given by Eqs. (5.6.7) and (5.6.8). To use the power method, an initial guess for the eigenvector is taken as (1,1,1), and the coefficient matrix A is multiplied with this vector, employing the formula for matrix multiplication. The resulting vector is normalized with the largest component, which gives the first approximation to the largest eigenvalue of the system. The new vector is compared with the starting vector. If the absolute value of the difference is larger than the convergence criterion ε, the new vector is taken as the starting vector for the next iteration. The process is continued until

$$|X_i - (XO)_i| \leqslant \varepsilon \qquad \text{for } i = 1, 2, 3 \tag{5.7.1}$$

where X_i is the ith component of the eigenvector after the present iteration and $(XO)_i$ is that after the previous iteration. At convergence, the normalizing factor is the largest eigenvalue, and the computed vector the desired eigenvector.

```
C       THE POWER METHOD FOR DETERMINING THE LARGEST EIGENVALUE
C       AND THE CORRESPONDING EIGENVECTOR
C
C
C       A(I,J) REPRESENTS THE ELEMENTS OF THE COEFFICIENT MATRIX,
C       X(I), OR Y(I), ARE THE COMPONENTS OF THE EIGENVECTOR, N IS
C       THE NUMBER OF EQUATIONS, XO(I) THE EIGENVECTOR AFTER THE
C       THE PREVIOUS ITERATION, XL THE DIVISOR THAT YIELDS THE
C       LARGEST EIGENVALUE AT CONVERGENCE AND EPS THE CONVERGENCE
C       CRITERION APPLIED TO THE EIGENVECTOR
C
C
        DIMENSION A(10,10),X(10),XO(10),Y(10)
C
C       INPUT OF COEFFICIENTS AND STARTING VALUES
C
        READ(5,*)N
        READ(5,*)((A(I,J),J=1,N),I=1,N)
        EPS=0.0001
        DO 1 I=1,N
  1     X(I)=1.0
  2     DO 4 I=1,N
        S=0.0
C
C       STORE VALUES FROM THE PREVIOUS ITERATION
C
        XO(I)=X(I)
C
C       COMPUTE NEW EIGENVECTOR
C
        DO 3 J=1,N
  3     S=S+A(I,J)*X(J)
  4     Y(I)=S
        K=1
C
C       FIND LARGEST COMPONENT OF EIGENVECTOR
C
        DO 5 I=2,N
        IF(Y(K) .LT. Y(I)) THEN
        K=I
        END IF
  5     CONTINUE
        XL=Y(K)
        DO 6 I=1,N
  6     X(I)=Y(I)
C
C       DIVIDE ALL COMPONENTS BY THE LARGEST VALUE
C
        DO 7 I=1,N
  7     X(I)=X(I)/XL
        DO 8 I=1,N
C
C       CHECK FOR CONVERGENCE
C
```

Figure 5.7.1 Computer program for determining the largest eigenvalue and the corresponding eigenvector for the vibrating system of Example 5.6 by the power method, along with the results obtained.

```
      IF(ABS(X(I)-XO(I)) .GT. EPS) GO TO 2
8     CONTINUE
      WRITE(6,9)XL
9     FORMAT(2X,'THE LARGEST EIGENVALUE IS =',F10.5//)
      WRITE(6,10)
10    FORMAT(2X,'THE EIGENVECTOR IS:')
      DO 11 I=1,N
11    WRITE(6,12)I,X(I)
12    FORMAT(2X,'X(',I1,')=',F10.5)
      STOP
      END
```

(a)

```
RESULTS
THE LARGEST EIGENVALUE IS =   4.56134


THE EIGENVECTOR IS:
X(1)=   1.00000
X(2)=  -0.78077
X(3)=   1.00000
```

(b)

Figure 5.7.1 Continued

The computer program for this problem, considering up to ten equations, is given in Fig. 5.7.1(a). The convergence criterion ε is denoted by EPS and is taken as 10^{-4}. Smaller values of ε were also considered, and a negligible difference in the solution was obtained. Similarly, other starting values were tried, and convergence was found to occur with essentially the same results for different values close to (1,1,1). Convergence was not achieved if values very far from these were taken, as expected. The converged results are shown in Fig. 5.7.1(b), and the values are very close to those obtained analytically in Example 5.6. For systems containing more than ten equations, the dimension statement would need to be changed. The printout format also needs to be changed to incorporate more than 1 digit in the representation of the eigenvector components if ten or more equations are considered.

5.7.3 Other Methods*

There are several other methods that are available for the solution of eigenvalue problems. Among the most important of these is Householder's method, used in conjunction with the QL algorithm. This approach is applicable only to symmetric matrices. Householder's method is used to convert an $n \times n$ symmetric matrix into a symmetric tridiagonal matrix. This form is convenient for matrix decompositions and transformations, since the number of operations for each such manipulation varies as n, rather than as n^3, which applies for the full matrix. Once the tridiagonal form has been obtained, several techniques are available for finding the eigenvalues. The LR algorithm of Rutishauser (1958) decomposes the original symmetric matrix into a product of lower triangular and upper triangular matrices. The QR algorithm of Francis (1962) decomposes the matrix into a product of an orthogonal matrix and an upper triangular matrix. Using similarity transformations, which preserve the eigenvalues of the original matrix, the first algorithm converges to a lower triangular matrix, and the second to an upper triangular matrix, with the desired eigenvalues in decreasing order of magnitude on the main diagonal. The decomposition in the LR algorithm may be done very efficiently by the use of Choleski decomposition, outlined in Section 5.4, if the matrix A is positive definite, which requires that all the eigenvalues be positive. The QR algorithm is generally more stable than the LR method and is often preferred. See Carnahan et al. (1969) and Hornbeck (1975) for details.

The QL method decomposes the matrix into the product of an orthogonal matrix and a lower triangular matrix. The method is particularly suited for tridiagonal matrices, such as those produced by Householder's method. The eventual result of decomposition and transformation is a diagonal matrix, with the eigenvalues on the diagonal. In some engineering problems, such as those concerned with the solution of the differential equations that govern the stresses in a structure, the matrix that arises is tridiagonal in form, and the above algorithms may be used efficiently, without the need of transformation by Householder's method.

Nonsymmetric matrices are also of interest in engineering problems, and the methods discussed in the preceding subsections may be employed for these. Also, the above approach for symmetric matrices may be modified to obtain the eigenvalues of an unsymmetric matrix. The application of Householder's method leads to an upper triangular form with an additional band of elements adjacent to the main diagonal. This form, known as the *Hessenberg form*, may also be obtained by elimination methods. The LR or the QR algorithms may then be employed to obtain the eigenvalues. As mentioned earlier, unsymmetric matrices may have complex eigenvalues. The power method can be modified to deal with complex eigenvalues. For further details, see the treatment of eigenvalue problems by Wilkinson (1965) and Hornbeck (1975).

Before leaving this section, we mention the Jacobi method, which is a classic, although inefficient, method for finding all the eigenvalues and eigenvectors of a symmetric matrix by the use of orthogonal transformations, which preserve the symmetry as well as the eigenvalues. If Q denotes an orthogonal matrix, we wish to use a transformation of the form Q^TAQ to reduce the elements in the ith row and the jth column of the matrix to zero. However, the reduction of one element to zero often introduces nonzero elements at positions that have been previously converted to zero. The process is, therefore, an infinite one, and the matrix eventually tends toward a diagonal form. At convergence, the eigenvalues are obtained from the diagonal elements. Eigenvectors may also be obtained along with the eigenvalues by application of the reduction procedure to an identity matrix along with the given matrix. The columns of the resulting modified matrix are then the desired eigenvectors. Various modifications of the Jacobi method, such as the *threshold* method, which eliminates elements larger than a given threshold value, have been developed. The Jacobi method is not efficient in terms of computing time, but it is often available in computer libraries and is frequently used because it is applicable to a wide variety of eigenvalue problems.

5.8 SOLUTION OF SIMULTANEOUS NONLINEAR EQUATIONS

So far, we have discussed the solution of a system of linear equations, considering both homogeneous and nonhomogeneous equations. However, in engineering problems, we are frequently faced with nonlinear equations, for which no direct methods are available and iterative procedures must be used. The solution of single, nonlinear algebraic equations, in order to find the roots, was discussed in Chapter 4. We are concerned here with the solution of a system of nonlinear equations. Such systems arise in a wide variety of problems. Thermal radiation from a heated body, for instance, varies as T^4, where T is the surface temperature. Material properties often vary nonlinearly with temperature and pressure. The forces acting on a moving particle or fluid often have a nonlinear relationship with velocity. Similarly, transcendental equations arise in many circumstances. The iterative methods outlined for linear equations are generally modified and employed for nonlinear equations. However, the methods discussed in Chapter 4 may also be considered for solving a system of nonlinear equations. An important method, which is used extensively for solving small sets of equations, is the Newton-Raphson method.

5.8.1 Newton-Raphson Method

The Newton-Raphson method is based on a Taylor series expansion of the functions $f_1, f_2, \ldots, f_n$ of Eq. (5.1). The function f_1 may be expanded in a Taylor series about $(x_1, x_2, \ldots, x_n)$, which represents an approximation to the solution. If only the first-order terms are retained and if the exact solution $(\bar{x}_1, \bar{x}_2, \ldots, \bar{x}_n)$ is substituted for the unknowns, we obtain the relation

$$f_1(\bar{x}_1, \bar{x}_2, \ldots, \bar{x}_n) \simeq f_1(x_1, x_2, \ldots, x_n) + \left(\frac{\partial f_1}{\partial x_1}\right)_{x_i}(\bar{x}_1 - x_1) + \left(\frac{\partial f_2}{\partial x_2}\right)_{x_i}(\bar{x}_2 - x_2) + \cdots + \left(\frac{\partial f_n}{\partial x_n}\right)_{x_i}(\bar{x}_n - x_n) \qquad (5.76)$$

where the partial derivatives are evaluated at $(x_1, x_2, \ldots, x_n)$. The exact solution is not known, but Eq. (5.76) provides a method for improving the approximation to the solution. If the other functions, $f_2, f_3, \ldots, f_n$, are similarly expanded about $(x_1, x_2, \ldots, x_n)$ and the exact solution substituted for the unknowns, a set of linear equations is obtained. Since $(\bar{x}_1, \bar{x}_2, \ldots, \bar{x}_n)$ is the solution vector, the functions $f_i(\bar{x}_1, \bar{x}_2, \ldots, \bar{x}_n)$, for $i = 1, 2, \ldots, n$, are all zero. The unknowns are Δx_i, where $\Delta x_i = x_i' - x_i$, x_i' being the next approximation, and may be computed from the following:

$$\begin{bmatrix} \dfrac{\partial f_1}{\partial x_1} & \dfrac{\partial f_1}{\partial x_2} & \cdots & \dfrac{\partial f_1}{\partial x_n} \\ \dfrac{\partial f_2}{\partial x_1} & \dfrac{\partial f_2}{\partial x_2} & \cdots & \dfrac{\partial f_2}{\partial x_n} \\ \vdots & \vdots & & \vdots \\ \dfrac{\partial f_n}{\partial x_1} & \dfrac{\partial f_n}{\partial x_2} & \cdots & \dfrac{\partial f_n}{\partial x_n} \end{bmatrix} \begin{bmatrix} \Delta x_1 \\ \Delta x_2 \\ \vdots \\ \Delta x_n \end{bmatrix} = \begin{bmatrix} -f_1 \\ -f_2 \\ \vdots \\ -f_n \end{bmatrix} \tag{5.77}$$

The functions and the derivatives are all evaluated at the approximate solution $(x_1, x_2, \ldots, x_n)$. Since only linear terms were retained in the Taylor series expansion, the exact solution is not obtained by solving this system of equations. However, the next approximation to the solution may be obtained as

$$\begin{aligned} x_1^{(l+1)} &= x_1^{(l)} + \Delta x_1^{(l)} \\ x_2^{(l+1)} &= x_2^{(l)} + \Delta x_2^{(l)} \\ &\vdots \\ x_n^{(l+1)} &= x_n^{(l)} + \Delta x_n^{(l)} \end{aligned} \tag{5.78}$$

where the superscript (l) represents the values after a given iteration and $(l + 1)$ those after the next iteration.

Equations (5.78) provide an iterative method for solving the system of nonlinear equations given by Eq. (5.1). An initial guess of the values of the unknowns is taken, and the functions $f_1, f_2, \ldots, f_n$ and the derivatives, needed in Eq. (5.77), are evaluated at these x values, denoted as $x_1^{(0)}, x_2^{(0)}, \ldots, x_n^{(0)}$. The set of linear equations is solved to obtain Δx_i, which is then used to obtain the next iteration $x_i^{(1)}$ from Eq. (5.78). The process is continued until all the f's are close to zero or the unknowns do not change from one iteration to the next, within a specified convergence criterion.

The method may diverge if the initial guess is too far off from the exact solution. In physical problems, some prior information is often available on the nature of the functions and on the expected solution. This information may be used advantageously in choosing the initial values. However, if no information is available, several trials, with different starting values, may be needed before the process converges. The partial derivatives are generally computed numerically, since the functions may be quite involved. This method is extensively employed in the numerical simulation of engineering systems. It is also used as a correction scheme for the solution of boundary-value problems in ordinary differential equations, as discussed in Chapter 8. Because of the computational effort required for the evaluation of the derivatives, the Newton-Raphson method is generally used when the system consists of only a relatively small number of nonlinear equations, typically less than ten. Such problems are often encountered in physical and chemical processes of interest to engineering applications. Other iterative methods, such as those based on the Jacobi and the Gauss-Seidel methods, are more suitable for large systems and are discussed below.

5.8.2 Modified Jacobi and Gauss-Seidel Methods

The system of nonlinear equations may be considered to be of the form given by Eq. (5.1). These equations are rewritten, by solving for the unknowns $x_1, x_2, \ldots, x_n$, as follows:

$$x_i = F_i(x_1, x_2, \ldots, x_i, \ldots, x_n) \qquad \text{for } i = 1, 2, \ldots, n \tag{5.79}$$

Therefore, the unknown x_i is retained on the right-hand side for nonlinear equations, since it would not, in general, be possible to solve for x_i in terms of just the other unknowns because of the nonlinearity in x_i. The nonlinearity may arise because transcendental equations are involved, because products of the unknowns are present in the equations, or because x_i appears as x_i^n, where $n \neq 1$. Also, if $\bar{x}_1, \bar{x}_2, \ldots, \bar{x}_n$ represent the solution of the given system of equations, the rearrangement of Eq. (5.1) gives, at the solution,

$$\bar{x}_i = F_i(\bar{x}_1, \bar{x}_2, \ldots, \bar{x}_i, \ldots, \bar{x}_n) \qquad \text{for } i = 1, 2, \ldots, n \tag{5.80}$$

We may now develop an iterative procedure for solving the given set of nonlinear equations. Similar to the Jacobi method for linear equations, the recursion relation may be written from Eq. (5.79) as

$$x_i^{(l+1)} = F_i[x_1^{(l)}, x_2^{(l)}, \ldots, x_i^{(l)}, \ldots, x_n^{(l)}] \qquad \text{for } i = 1, 2, \ldots, n \tag{5.81}$$

where all the unknowns are computed for the $(l + 1)$th iteration using the known values from the previous iteration. We may also replace the unknowns by the new values as soon as they are computed. This procedure is similar to the Gauss-Seidel method for linear equations and is given by

$$x_i^{(l+1)} = F_i[x_1^{(l+1)}, x_2^{(l+1)}, \ldots, x_{i-1}^{(l+1)}, x_i^{(l)}, \ldots, x_n^{(l)}] \qquad \text{for } i = 1, 2, \ldots, n \tag{5.82}$$

where $x_1^{(l+1)}$ is computed first followed by $x_2^{(l+1)}$, and so on for increasing i. Therefore, the most recently computed values of the unknowns are used for evaluating the function F_i. The formulation for this method is similar to that for the successive substitution method discussed in Chapter 4. Therefore, this method is also sometimes known as the successive substitution method for solving a system of nonlinear equations.

5.8.3 Convergence

The convergence characteristics of nonlinear equations are not as well established as those for linear equations. A general theory for the iterative solution of nonlinear equations is not available, although the behavior of certain special sets of equations has been studied in detail. The equations may also yield multiple solutions, and one would then need information on the physical aspects of the problem in order to choose the correct solution. However, the solution of equations that characterize a physical problem usually results in only one physically realistic solution, and this solution is the one that is obtained most easily. The other solutions may be physically unacceptable and are usually not readily obtained when the system of equations is solved by the above methods. We may use relaxation to alter the convergence characteristics of the iterative process. However, it is often difficult to predict the resulting behavior, since over-relaxation may even slow the convergence in some cases for nonlinear equations and accelerate it in others. Successive under-relaxation is often useful in obtaining convergence for nonlinear systems. Convergence is strongly dependent on the nature of the equations, and very often several trials, with different starting values, are needed before convergence is achieved.

Example 5.8

In the ammonia production system shown in Fig. 5.8.1, a mixture of 90 moles/s of nitrogen, 270 moles/s of hydrogen, and 0.9 mole/s of argon, which is present as an impurity, enters the chemical plant and is mixed with the residual mixture crossing a bleed valve. In the chemical reactor, a fraction of the entering mixture combines to give ammonia, which is removed by condensation. A bleed of 23.5 moles/s of the mixture is employed to avoid a buildup of argon, which adversely affects the reaction. The conversion efficiency of the reactor is $0.57e^{-0.0155F_1}$, where F_1 is the argon flow rate in moles/s. This efficiency represents the fraction of the mixture that is converted to ammonia. When mass conservation is applied to the process, the following equations are obtained:

$$F_1 = \frac{0.9}{(1 - B)} \tag{5.8.1}$$

$$P = 1 - 0.57e^{-0.0155F_1} \tag{5.8.2}$$

$$F_2 = \frac{90}{(1 - B \times P)} \tag{5.8.3}$$

$$B = 1 - \frac{23.5}{4F_2P + F_1} \tag{5.8.4}$$

where F_1 is the flow rate of argon entering the reaction chamber, F_2 is the flow rate of nitrogen, and B and P are parameters defined above. Solve this system of nonlinear equations by the successive substitution method to obtain the flow rates and the amount of ammonia produced. Such equations often arise in the analysis of chemical reactors, and the following approach may be frequently applied.

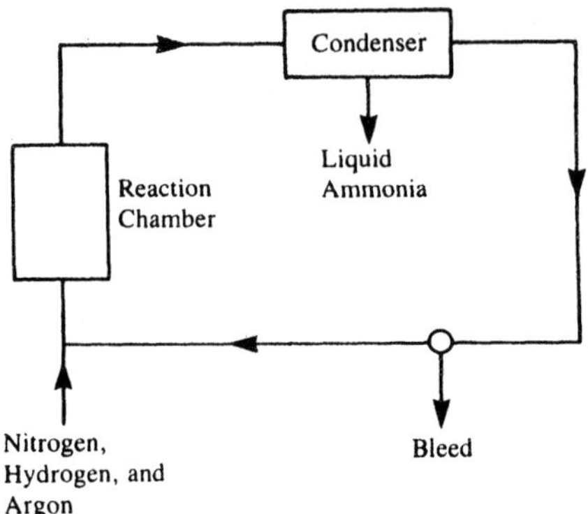

Figure 5.8.1 The ammonia production system considered in Example 5.8.

Solution

The four unknowns are taken as F_1, P, F_2, and B. The starting value of B is arbitrarily chosen as 0.1, and the other quantities are computed. The total flow rate of the mixture entering the plant is denoted by C, where $C = F_1 + 4F_2$, since the flow rate of nitrogen is F_2 moles/s and that of hydrogen is $3F_2$ moles/s. The amount of ammonia produced is denoted by D and is given by

$$D = 2F_2 \times 0.57e^{-0.0155F_1} \tag{5.8.5}$$

since each mole of nitrogen gives two moles of ammonia, as seen from the chemical equation

$$N_2 + 3H_2 = 2NH_3 \tag{5.8.6}$$

The convergence criterion is applied to the total flow rate C as follows:

$$|C - CO| \leqslant \varepsilon \tag{5.8.7}$$

where CO is the flow rate after the previous iteration and ε is the convergence criterion.

The physical problem discussed here is evidently quite involved. However, the computer program, shown in Fig. 5.8.2(a), is very simple. An initial guess for C is employed for use in the convergence check after the first iteration. The convergence criterion ε is denoted by EPS and is taken as 10^{-3}. It was ascertained that a still smaller value of ε did not significantly affect the numerical results. The results obtained after each iteration for the flow rate of argon, the total flow rate, and the amount of ammonia collected, in moles/s, are indicated in Fig. 5.8.2(b). The convergence is slow, partly because of the nature of the equations and partly because of the first-order convergence of this method. Convergence was not obtained for values of B very far from the chosen starting value of 0.1. Also, a change in the sequence of the computations performed resulted in divergence in some cases. Despite these problems with convergence, the successive substitution method is a much simpler method than the Newton-Raphson method, for nonlinear equations, and is widely used in engineering applications.

```
C     THE SUCCESSIVE SUBSTITUTION METHOD FOR NONLINEAR EQUATIONS
C
C
C     HERE F1 IS THE FLOW RATE OF ARGON IN MOLES/S, F2 IS THE FLOW
C     RATE OF NITROGEN, C IS THE TOTAL FLOW RATE, B AND P ARE
C     THE PARAMETERS DEFINED IN THE PROBLEM, D IS THE AMOUNT OF
C     AMMONIA COLLECTED IN MOLES/S, CO IS THE VALUE OF C AFTER
C     THE PREVIOUS ITERATION AND EPS IS THE CONVERGENCE CRITERION
C     APPLIED TO THE TOTAL FLOW RATE C
C
C
C     INPUT OF STARTING VALUES
C
      EPS=0.0001
      B=0.1
      C=180.0
 1    CO=C
C
C     COMPUTATION OF THE UNKNOWN QUANTITIES
C
      F1=0.9/(1.0-B)
      P=1.0-0.57*EXP(-0.0155*F1)
      F2=90.0/(1.0-B*P)
      B=1.0-23.5/(4.0*F2*P+F1)
      C=F1+4.0*F2
      D=0.57*EXP(-0.0155*F1)*2.0*F2
      WRITE(6,2)F1,C,D
C
C     CONVERGENCE CHECK
C
      IF(ABS(C-CO) .LE. EPS) THEN
 2    FORMAT(2X,'ARGON:',F12.5,4X,'FLOW:',F12.5,4X,'NH3:',F12.5)
      ELSE
      GO TO 1
      END IF
      STOP
      END
```

(a)

RESULTS

```
ARGON:     1.00000    FLOW:   377.52020    NH3:   105.65780
ARGON:     6.36528    FLOW:   621.94090    NH3:   158.95630
ARGON:    11.64363    FLOW:   708.79150    NH3:   165.87840
ARGON:    14.43962    FLOW:   749.70330    NH3:   167.52770
ARGON:    15.88013    FLOW:   770.51630    NH3:   168.14510
ARGON:    16.62993    FLOW:   781.34420    NH3:   168.42190
ARGON:    17.02342    FLOW:   787.03170    NH3:   168.55670
ARGON:    17.23091    FLOW:   790.03320    NH3:   168.62510
ARGON:    17.34062    FLOW:   791.62100    NH3:   168.66060
ARGON:    17.39871    FLOW:   792.46190    NH3:   168.67910
ARGON:    17.42950    FLOW:   792.90770    NH3:   168.68890
ARGON:    17.44583    FLOW:   793.14420    NH3:   168.69410
ARGON:    17.45448    FLOW:   793.26920    NH3:   168.69680
ARGON:    17.45906    FLOW:   793.33560    NH3:   168.69830
ARGON:    17.46149    FLOW:   793.37080    NH3:   168.69900
ARGON:    17.46278    FLOW:   793.38960    NH3:   168.69950
ARGON:    17.46347    FLOW:   793.39960    NH3:   168.69970
ARGON:    17.46382    FLOW:   793.40450    NH3:   168.69980
ARGON:    17.46400    FLOW:   793.40720    NH3:   168.69990
ARGON:    17.46411    FLOW:   793.40890    NH3:   168.69990
ARGON:    17.46417    FLOW:   793.40990    NH3:   168.69990
ARGON:    17.46422    FLOW:   793.41040    NH3:   168.69990
ARGON:    17.46423    FLOW:   793.41060    NH3:   168.60000
ARGON:    17.46423    FLOW:   793.41060    NH3:   168.69990
```

(b)

Figure 5.8.2 Computer program for the solution of the system of nonlinear algebraic equations obtained in Example 5.8 by the successive substitution method. Also shown is the convergence of the computed results.

Example 5.9

For the physical problem described in Example 4.7, employ the Newton-Raphson method to solve the system of nonlinear equations, given by Eq. (4.7.4), to obtain the flow rate R and the pressure P.

Solution

The system of equations that govern the flow through a duct due to a fan, as given in Example 4.7, are

$$R = \left(\frac{P - 80}{10.5}\right)^{3/5} \tag{5.9.1}$$

$$P = \left(\frac{15 - R}{75 \times 10^{-6}}\right)^{1/2} \tag{5.9.2}$$

To use the Newton-Raphson method, take the two functions that are to be reduced to zero as

$$R_1 = \left(\frac{P - 80}{10.5}\right)^{3/5} - R = R_1(P,R) \tag{5.9.3}$$

$$P_1 = \left(\frac{15 - R}{75 \times 10^{-6}}\right)^{1/2} - P = P_1(P,R) \tag{5.9.4}$$

The initial guesses for R and P and the convergence criterion ε are inputs to the program, which is shown in BASIC in Fig. 5.9.1(a). The convergence criterion is applied to a parameter B, where

$$B = R_1^2 + P_1^2 \tag{5.9.5}$$

The four partial derivatives, $\partial R_1/\partial R$, $\partial R_1/\partial P$, $\partial P_1/\partial R$, and $\partial P_1/\partial P$, are computed at the starting values of R and P, employing analytical differentiation of the functions R_1 and P_1. These derivatives are denoted by RR, RP, PR, and PP, respectively. The increments in R and P, DR and DP, respectively, for the next iteration are then obtained from Eq. (5.77), which gives

$$\frac{\partial R_1}{\partial R}\Delta R + \frac{\partial R_1}{\partial P}\Delta P = -R_1 \tag{5.9.6}$$

$$\frac{\partial P_1}{\partial R}\Delta R + \frac{\partial P_1}{\partial P}\Delta P = -P_1 \tag{5.9.7}$$

The new values of R and P are determined from Eq. (5.78). The iterative process is repeated until convergence is achieved, as indicated by $B < \varepsilon$. Figure 5.9.1(b) shows the results for the starting values of R and P taken as 2 and 100, respectively, and ε, or EPS, taken as 10^{-4}. For the same convergence criterion, the flow rate was obtained as 6.7320255 m^3/s in Example 4.7, indicating a close agreement with the present results.

```
100 REM  NEWTON-RAPHSON METHOD FOR A SYSTEM OF NONLINEAR EQUATIONS
110 REM
120 REM  R AND P ARE THE TWO INDEPENDENT VARIABLES, R1 AND P1 ARE
130 REM  THE FUNCTIONS THAT ARE TO BE REDUCED TO ZERO FOR FINDING THE
140 REM  ROOTS, RR, RP, PR AND PP ARE THE PARTIAL DERIVATIVES OF THE
150 REM  FUNCTION, R1 OR P1, DENOTED BY THE FIRST LETTER, WITH RESPECT
160 REM  TO THE VARIABLE DENOTED BY THE SECOND LETTER, DR AND DP ARE THE
170 REM  INCREMENTS IN R AND P, RESPECTIVELY, AND EPS IS THE CONVERGENCE
180 REM  CRITERION APPLIED TO THE QUANTITY B DEFINED IN THE PROGRAM
190 REM
200 REM
210 REM  ENTER STARTING VALUES
220    INPUT "R=?   ",R
230    INPUT "P=?   ",P
240    INPUT "EPS=? ",EPS
250    R1=((P-80)/10.5)^.6 - R
260    P1=((15-R)*(10^6)/75)^.5 - P
270 REM
280 REM  CHECK FOR CONVERGENCE
290    B=R1^2 + P1^2
300    IF B < EPS THEN 490
310 REM
320 REM  COMPUTE PARTIAL DERIVATIVES
330    RR=-1
340    RP=3/(5*(10.5^.6)*((P-80)^.4))
350    PR=-1/(2*((7.500001E-05)^.5)*((15-R)^.5))
360    PP=-1
370 REM
380 REM  DETERMINE INCREMENTS FOR THE NEXT ITERATION
390    D=RR*PP - RP*PR
400    DR=(-R1*PP + P1*RP)/D
410    DP=(-P1*RR + R1*PR)/D
420 REM
430 REM  CALCULATE VALUES OF R AND P FOR THE NEXT ITERATION
440    R=R+DR
450    P=P+DP
460    PRINT "R=";R,"P=";P
470    IF R > 15 THEN 520
480    GOTO 250
490    PRINT
500    PRINT "THE REQUIRED SOLUTION IS:"
510    PRINT "THE FLOW RATE=";R,"      PRESSURE=";P
520    END
```

(a)

RESULTS

```
RUN
R=?   2
P=?   100
EPS=? 0.0001
R= 9.873579    P= 290.2549
R= 6.86436     P= 338.1764
R= 6.732573    P= 332.0232
R= 6.732087    P= 332.0222

THE REQUIRED SOLUTION IS:
THE FLOW RATE= 6.732087           PRESSURE= 332.0222
```

(b)

Figure 5.9.1 Program in BASIC for the solution of the system of nonlinear equations, from Example 5.9, by the Newton-Raphson method. The results obtained are also shown.

Convergence is found to occur for a wide range of starting values. The flow rate is known to be less than $15\,m^3/s$, as seen from Eq. (5.9.4), and this condition is incorporated in the program to terminate the computation if $R > 15$. Also note that convergence is very rapid, despite the considerable difference between the starting and the converged values. The convergence was much slower with the successive substitution method, as illustrated by Example 4.7. Due to its superior convergence characteristics, the Newton-Raphson method is the preferred method if the number of equations is small.

5.9 SUMMARY

The solution of simultaneous linear and nonlinear algebraic equations is considered in this chapter. Several methods are discussed, and their advantages over other methods and their applicability to the various systems of equations that arise in engineering problems are indicated. The choice of the method for the solution of a given system depends on whether the equations are linear or nonlinear and on whether they are homogeneous or nonhomogeneous. It also depends on the number of equations to be solved and the nature of the equations, particularly the sparseness of the coefficient matrix. The selection of the method for a given situation is also often influenced by the need to determine other quantities, such as the inverse or the determinant of the coefficient matrix, besides the unknown vector X. Similarly, several systems of equations with the same coefficient matrix but different constant vectors may have to be solved. This additional consideration is often an important factor in the selection of the method. In several cases, the available software in the computer library may also make a given method more attractive than the others. This last consideration is of particular importance in engineering applications, where interest often lies in obtaining the solution to a wide variety of systems of equations.

For linear, nonhomogeneous equations, the methods discussed here include Gaussian elimination, Gauss-Jordan elimination, Crout's method, matrix inversion, and iterative methods. Gaussian elimination is the simplest direct method, in terms of computer programming, and is appropriate for a small number of equations, typically of the order of 20 or less, because of the round-off error. However, if the system is tridiagonal, this method may be used advantageously for several hundred equations without significant loss of accuracy due to round-off error. In the finite difference solution of partial differential equations, tridiagonal systems are often obtained, and Gaussian elimination (TDMA) is the preferred method. If many systems that have the same coefficient matrix A but different constant vector B are to be solved, Gauss-Jordan elimination may be used advantageously, since all systems are solved in one elimination process. Similarly, matrix inversion determines A^{-1}, which is the same for all the systems, and the solution X is obtained simply by multiplication of the inverse with the constant vector, that is, $X = A^{-1}B$. Many computer systems have programs available for matrix inversion, and these may be used for solving the given set of equations. In BASIC, the MAT C = INV(A) statement will yield the inverse of the matrix A on computers that have the MAT instructions available. Since this software is available on many minicomputers and also on some microcomputers, matrix inversion is frequently employed to solve systems of equations in engineering applications. Crout's method and other compact methods based on decomposition of the matrix A are generally more efficient than elimination methods. The round-off error is also less since the number of operations is smaller than that for Gaussian elimination. Computer programming is somewhat more involved. However, because of their advantage over elimination methods in computing time, compact methods have become quite popular in recent years.

Iterative methods, such as Jacobi, Gauss-Seidel, and relaxation methods, are particularly suitable for large systems of equations, typically of order 100 or larger, and for sparse coefficient matrices. The round-off error in the solution is due only to the error that arises in the final iteration. If the optimum value of the relaxation factor ω is known, the successive over-relaxation (SOR) method generally requires less computing time than most direct methods. The optimum value ω_{opt} is usually not known, and several values may have to be tried to determine it numerically. The Jacobi method is seldom used, since it requires greater computer storage and computational effort, on traditional computers, than the Gauss-Seidel method, which remains a popular choice, along with SOR, for solving large systems of equations. These methods, with some modifications, are also suitable for nonlinear equations, which generally cannot be solved by direct methods. Relaxation may also be used in this case, but the effect of relaxation on convergence is often unpredictable in nonlinear equations. Therefore, the modified Gauss-Seidel method is generally used. For a small number of equations, typically less than ten, the Newton-Raphson method is preferable since its convergence is more rapid.

The solution of linear, homogeneous equations requires methods quite different from those for nonhomogeneous equations. The eigenvalue problem is discussed in detail. For a small number of equations, the characteristic polynomial may be solved to obtain the eigenvalues, and the Gauss-Jordan method may be applied to determine the corresponding eigenvectors. The power method, which is an iterative method for determining the largest eigenvalue and the related eigenvector, may be used for moderately sized systems. The smallest eigenvalue can also be determined easily. However, intermediate eigenvalues can be obtained accurately if only a few of them are desired, since the round-off error accumulates as these are successively determined. The Jacobi method is a classical iterative method that yields all the eigenvalues and the eigenvectors of symmetric matrices by obtaining the matrix in a diagonal form. This method is frequently employed, even though it is quite inefficient, because the program is available on many computers. The most efficient method for large matrices is Householder's method used in conjunction with the QL algorithm. The former converts a symmetric matrix into a tridiagonal form. Various methods, such as the QL algorithm, are available for extracting the eigenvalues from a tridiagonal matrix by the use of decompositions and transformations.

Systems of algebraic equations arise in many engineering applications and also in the solution of differential equations by finite difference or finite element methods. Therefore, the methods presented in this chapter are important in many practical circumstances and will also be employed in subsequent chapters.

PROBLEMS

5.1. Compare the Gaussian elimination and Gauss-Jordan elimination methods for solving a system of linear equations. Which one is more accurate? Which one is more efficient? Indicate the advantages, if any, of the latter method over the former.

5.2. Draw the flow chart for solving a system of linear equations by Gaussian elimination.

5.3. Consider the electrical network shown and write down the algebraic equations for determining the four loop currents indicated. Solve this system of linear equations by Gaussian elimination.

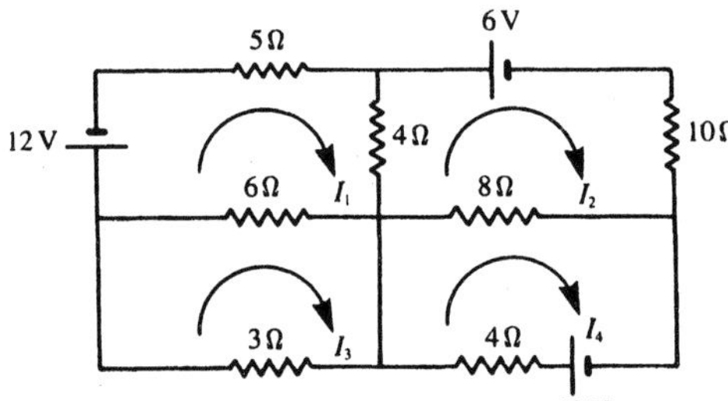

Problem 5.3

5.4. (a) If the 3 Ω resistance in the network for Prob. 5.3 is replaced by an open circuit, resulting in only three loops, compute the three loop currents by the Gaussian elimination method. (b) If the currents through the remaining six resistances are denoted by $I_1, I_2, \ldots,$ I_6, write down the six linear equations that govern these currents. Compute their values using Gaussian elimination, and compare the results with those for the loop currents in the first part of the problem.

5.5. A third-order polynomial of the form $y = Ax^3 + Bx^2 + Cx + D$ is to be fitted to four time-velocity data points. At time $x = 0, 1, 2,$ and 3 seconds, the velocity is measured as 7, 14, 29, and 58 m/s. Using Gauss-Jordan elimination, find the curve that passes through these points.

5.6. A fourth-order polynomial passes through the five points for which the independent and dependent variables, x and y, respectively, are given as $(-2,37)$, $(-1,7)$, $(0,5)$, $(1,13)$, and $(2,61)$. Find the polynomial by any suitable method. Here, x represents the spatial location and y the species concentration in a chemical reactor.

5.7. Six data points are obtained in the calibration of a velocity-measuring device. At velocities of 0, 0.2, 0.4, 0.6, 0.8, and 1.0 m/s, the voltage signals from the instrument are obtained as, respectively, 1.2, 1.74, 2.63, 3.99, 6.04, and 9.1 volts. Find the fifth-order polynomial that passes through these points.

5.8. For the physical problem described in Example 5.2, take the temperature at the left face as 200°C and that at the right face as 20°C. Write down the resulting tridiagonal system of equations, and obtain the solution by Gaussian elimination.

5.9. Derive the recursion formulas for solving a tridiagonal system of equations by Gaussian elimination.

5.10. Write a computer program to compute the magnitude of an $n \times n$ determinant by Gaussian elimination, where n can be up to 10. Using this program, compute the magnitudes of the determinants of the following matrices that arise in the dynamic analysis of structures:

$$\begin{bmatrix} 1 & 2 & -1 & 0 \\ 3 & 2 & 1 & 0 \\ 2 & 1 & 1 & 3 \\ 0 & 1 & 2 & 0 \end{bmatrix} \quad \text{and} \quad \begin{bmatrix} 2 & 0 & 3 & 1 \\ 3 & 1 & 2 & 2 \\ -1 & 0 & -2 & 4 \\ 3 & 2 & 1 & -2 \end{bmatrix}$$

5.11. A system of linear equations is given as follows:

$$\begin{aligned} x_1 - 2x_2 + 2x_3 + 3x_4 &= 5 \\ x_1 + 5x_2 - 3x_3 - 5x_4 &= 13 \\ -x_1 + x_2 + 3x_3 - 4x_4 &= -6 \\ 2x_1 + 3x_2 - x_3 - 2x_4 &= 18 \end{aligned}$$

Determine whether all the equations in this set are independent.

5.12. Solve the following system of linear equations by the Gauss-Jordan method, to indicate the basic procedure involved:

$$\begin{aligned} 2x_1 + 2x_2 + 3x_3 &= 15 \\ x_1 + 3x_2 + x_3 &= 10 \\ 3x_1 - x_2 + 2x_3 &= 7 \end{aligned}$$

5.13. Using the Gauss-Jordan method with partial pivoting, invert the matrices given in Prob. 5.10. Also, multiply the inverse A^{-1} with the corresponding matrix A, for the two cases, and compare the values obtained with the identity matrix. Comment on the difference, if any, between the two.

5.14. Find the currents in the three resistances located on the three sides of the triangle in the network shown. The voltages shown at the ends are the positive voltages imposed at these points. Use matrix inversion to solve this set of equations.

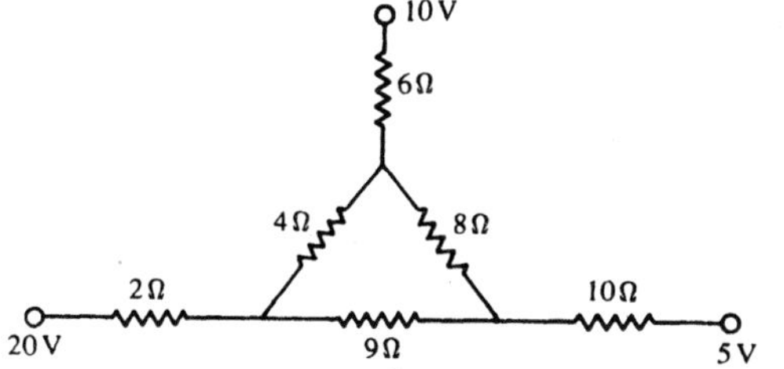

Problem 5.14

5.15. For the electrical network shown on page 228, obtain the governing set of linear equations using the voltage drops across the resistances as unknowns. Use Crout's method to solve this system of equations. Also solve the system by Gaussian elimination, and compare the accuracy obtained by the two methods.

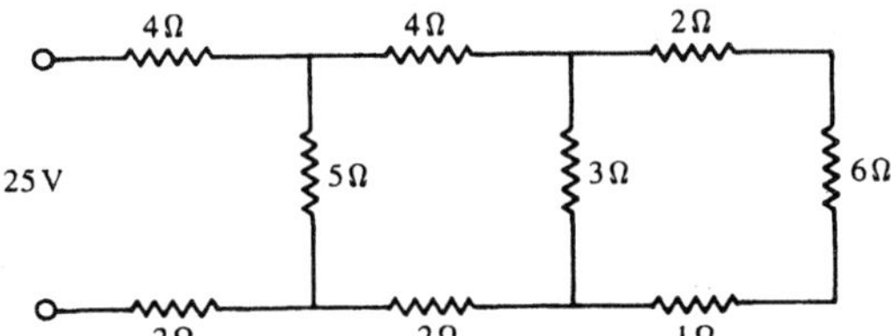

Problem 5.15

5.16. Solve the following system of linear equations using Gaussian elimination and also Gauss-Jordan elimination, both being employed with and without partial pivoting:

$$x_1 + 2x_2 + 4x_3 = 18$$

$$2x_1 + 3x_2 - 5x_3 = -18$$

$$4x_1 - x_2 - x_3 = -14$$

Compare the results obtained with each other and with the analytical solution. Does pivoting improve the accuracy of the results?

5.17. An industrial organization produces four items, x_1, x_2, x_3, and x_4. A portion of the amount produced for each is used in the manufacture of other items, and the net product is sold. The balance between the output and the production rate, resulting from the various inputs, gives rise to the following four equations, corresponding to the four items:

$$2x_1 + x_2 + 0.x_3 + 6x_4 = 64$$

$$5x_1 + 2x_2 + 0.x_3 + 0.x_4 = 37$$

$$0.x_1 + 7x_2 + 2x_3 + 2x_4 = 66$$

$$0.x_1 + 0.x_2 + 8x_3 + 9x_4 = 104$$

Using Gauss-Jordan elimination, with pivoting, solve this set of equations. Also, use Crout's method and compare the results obtained with those from the former method. Comment on the difference between the two methods.

5.18. Compute the magnitude of the determinant of the coefficient matrix of the following system of equations to determine whether it is a singular matrix.

$$3x + 7y - 6z = -15$$

$$x - 3y + 2z = 27$$

$$9x + 5y - 6z = 51$$

What can you say about the given system on the basis of the computed value? How will you solve such a system?

5.19. Using the power method, determine the largest eigenvalue and the corresponding eigenvector of the following matrices.

$$\begin{bmatrix} 8 & 4 & 5 \\ 6 & 10 & 3 \\ 8 & 6 & 20 \end{bmatrix} \quad \text{and} \quad \begin{bmatrix} 10 & 5 & 8 \\ 5 & 40 & 6 \\ 8 & 10 & 20 \end{bmatrix}$$

Vary the convergence criterion ε, applied to the eigenvector, from 10^{-4} to 10^{-1}, and study the resulting effect on the number of iterations needed for convergence, starting with an initial guess of the eigenvector as (1,1,1).

5.20. Solve the following equations, giving at least three complete steps, by the Gauss-Seidel method:

$$2x + 8y + 3z = 27$$
$$x + 3y + 5z = 22$$
$$6x + y + 2z = 14$$

Do you expect the numerical scheme to converge? Justify your answer.

5.21. Consider the following diagonally dominant system of equations:

$$8x + y + 2z = 29$$
$$x + 9y + z = 34$$
$$2x + 3y + 7z = 48$$

Using the Jacobi method and also the Gauss-Seidel method, solve this system. Compare the number of iterations needed for convergence in the two cases, if the convergence criterion ε is taken as 10^{-3} and applied to the computed value of the unknowns.

5.22. Find the optimum value ω_{opt} for fastest convergence, if the successive over-relaxation (SOR) method is employed for the system of equations given in Prob. 5.21. Plot the number of iterations to convergence against the relaxation factor ω and discuss your findings.

5.23. The temperature distribution in the square region shown is governed by the Laplace equation (Chapter 9). The finite difference approximation to this equation yields a system of algebraic equations given by

$$T_{i,j} = \frac{T_{i+1,j} + T_{i-1,j} + T_{i,j+1} + T_{i,j-1}}{4}$$

where i is the number of the row and j is the column in which a grid point is located. For the nine points shown, obtain nine linear equations for determining the temperatures at these points. Solve this system by the Gauss-Seidel method. Note that the temperatures at the boundaries are given and are used in the equations for all the temperatures, except for the one at position number 5.

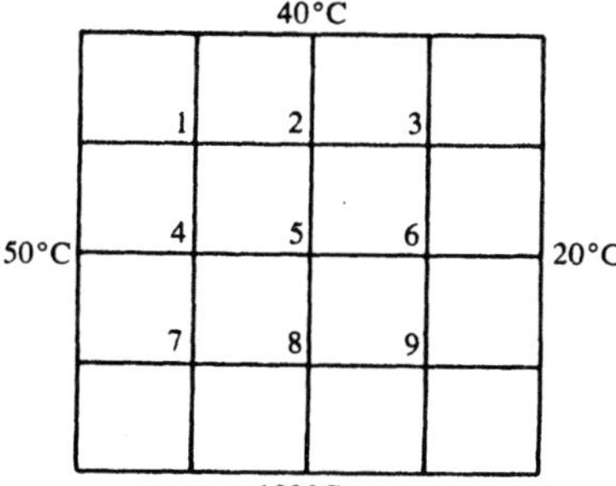

Problem 5.23

5.24. Solve Prob. 5.23 if the bottom surface is at a temperature of 1.0, while the others are at 0.0.

5.25. Using the Gauss-Seidel iterative method, solve the following system of equations:

$$\begin{aligned} 8x_1 - x_2 - 2x_3 + 3x_4 &= 22 \\ 2x_1 + 6x_2 - x_3 + x_4 &= 12 \\ x_1 + 2x_2 + 10x_3 + 2x_4 &= 20 \\ 3x_1 - 3x_2 + 2x_3 + 9x_4 &= 32 \end{aligned}$$

Vary the convergence criterion, which is applied to the computed results, and discuss the effect of this variation on the number of iterations and on the numerical solution.

5.26. Find the optimum value ω_{opt} of the relaxation factor for solving the system of equations in Prob. 5.25 by the SOR method, and compare the number of iterations to convergence with that for the Gauss-Seidel method.

5.27. As done for Example 5.6, obtain the system of linear homogeneous equations that govern the vibration of the three-mass system shown. Using the power method, determine the largest eigenvalue and the corresponding eigenvector.

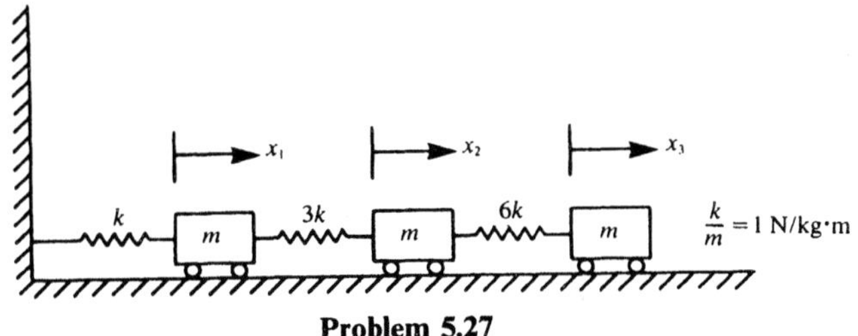

Problem 5.27

5.28. Obtain the characteristic polynomial for Prob. 5.27 and find all the eigenvalues by root solving. Determine the eigenvectors by the Gauss-Jordan method, as outlined in Example 5.6.

5.29. For the vibrating system shown on page 231, use the power method to obtain the largest and the smallest eigenvalues and the corresponding eigenvectors. Neglect the effect of gravity.

5.30. The forces acting on a body give rise to stresses in the material. At a given point in the material, the state of stress is given by the matrix

$$\begin{bmatrix} 8 & 3 & 6 \\ 5 & 10 & 2 \\ 6 & 7 & 20 \end{bmatrix} \times 10^6 \text{ N/m}^2$$

The largest principal stress, which determines the failure of the material, is the largest eigenvalue of the stress matrix. Using the iterative power method, find the largest eigenvalue and the corresponding eigenvector.

5.31. Obtain all the eigenvalues of the stress matrix

$$\begin{bmatrix} 12 & 6 & 8 \\ 8 & 40 & 7 \\ 6 & 12 & 20 \end{bmatrix} \times 10^6 \text{ N/m}^2$$

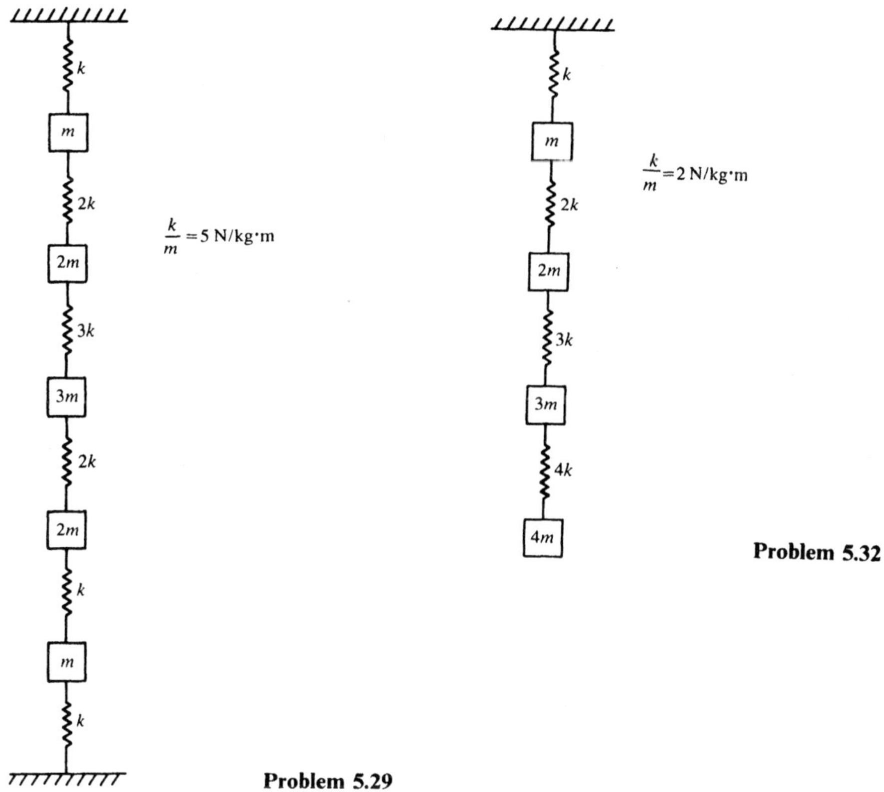

Problem 5.29

Problem 5.32

and compare the largest value obtained with the elements of the given matrix. Is the result physically expected?

5.32. For the four-mass vibrating system shown, obtain the governing algebraic equations, neglecting gravitational effects. Determine the largest eigenvalue and the corresponding eigenvector, using the power method. Also obtain the smallest eigenvalue and compare it with the computed largest eigenvalue. Comment on the physical significance of the difference.

5.33. Water flows through two parallel pipe networks, each of which contains a pump to provide the necessary pressure difference Δp. The water flow rates through the two circuits are Q_1 and Q_2, the total flow rate being Q. Therefore,

$$Q = Q_1 + Q_2$$

The characteristics of two pumps are given in terms of the relationship between the pressure difference and the flow rate as

$$\Delta p = 550 - 10Q_1^2$$

$$\Delta p = 700 - 15Q_2^2$$

Also, the pressure difference may be computed from changes in elevation and friction in the pipes to give

$$\Delta p = 68 + 8Q^2$$

Using the successive substitution method, solve this system of nonlinear equations to obtain the flow rates and the pressure difference.

5.34. Solve Prob. 5.33 by the Newton-Raphson method, and discuss the difference in the convergence characteristics and in the programming from the successive substitution method.

5.35. Solve the following by using Newton's method:

$$x^3 + 3y^2 = 21$$

$$x^2 + 2y + 2 = 0$$

Show two complete cycles of iteration to locate the root for $x > 0$.

5.36. Consider the physical problem discussed in Example 4.4. The problem may be posed in terms of the single equation given earlier or in terms of the two equations

$$\theta = 70 - 70 \exp\left[-\frac{1000}{21(5 + 20w)}\right]$$

$$250 = 4.2w\theta$$

Solve this system of nonlinear equations by the Newton-Raphson method, and compare the results with those obtained earlier in Example 4.4.

5.37. Solve the following set of nonlinear equations, which govern the flow rates in a network of four pipes, by the Gauss-Seidel iterative method:

$$7a + b^2 + c + d = 3.7$$

$$a^2 + 8b + 3c - d = 4.9$$

$$2a - 2b + 5c + d^2 = 8.8$$

$$a - b + c^2 + 14d = 18.2$$

Because of the physical nature of the problem considered, a, b, c, and d are all real and may be positive or negative. A negative value indicates flow in a direction opposite to that assumed in the analysis.

5.38. Solve the problem discussed in Example 5.9 by the successive substitution method, and compare the convergence characteristics with those for the Newton-Raphson method used in the given example.

5.39. Alternating current electrical circuits are generally solved by the use of complex variables, since the sinusoidal variation can be represented conveniently by complex quantities. The impedance for a resistor is simply its resistance, whereas the impedances for inductors and capacitors are functions of the frequency ω. For an inductor, the impedance is $i\omega L$, where $i = \sqrt{-1}$, and for a capacitor it is $-i/\omega C$, where L is the inductance in henries and C the capacitance in farads. For the circuit shown, the ac power source is 15 volts with a frequency of ω. The phase angle is arbitrarily taken as zero. Considering the two loop currents I_1 and I_2 and using Kirchhoff's law, obtain two algebraic equations for the

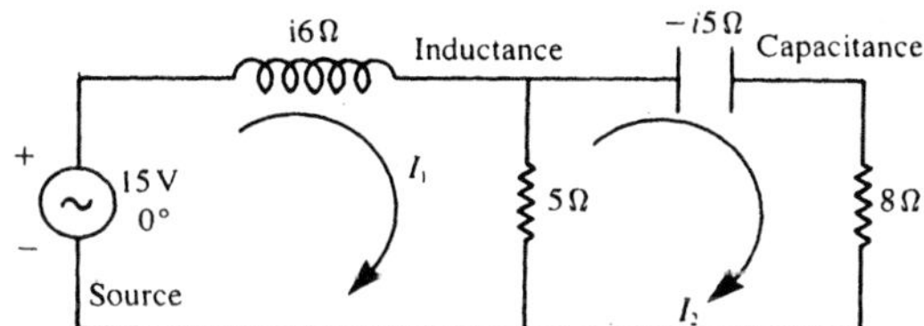

Problem 5.39

currents. Then taking the currents, voltages, and impedances as complex, separate the real and imaginary parts to obtain four linear equations. Solve these equations to obtain the loop currents.

5.40. Following the procedure outlined in Prob. 5.39, obtain the linear equations, with complex coefficients, for the ac electrical circuit shown. Again, obtain the corresponding linear equations with real coefficients by separating the real and imaginary parts. Solve these equations to obtain the magnitude and phase angle of the currents.

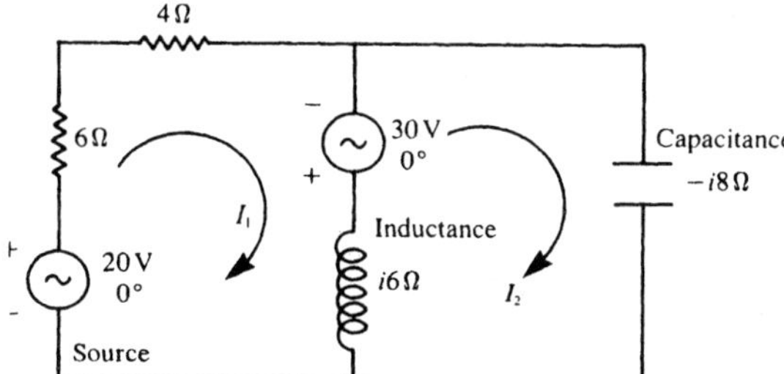

Problem 5.40

5.41. Under what conditions would the use of under-relaxation in an iterative scheme, such as Gauss-Seidel, be valuable in altering the convergence characteristics of the method? Give examples.

5.42. What would be the result of using over-relaxation with the Jacobi method?

5.43. Discuss some physical circumstances for which the iterative methods will be more advantageous to use than the direct methods. Justify your answer.

6

Numerical Curve Fitting and Interpolation

6.1 INTRODUCTION

A problem of considerable interest in engineering applications is that of representing data at a set of discrete points by means of a smooth and continuous function. Experimental and numerical studies generally yield results at a finite number of data points, which are often tabulated. However, a much more useful representation is by means of a smooth curve that passes through or as close as possible to the data points. Then the equation of the curve can be employed to obtain values at intermediate points where tabulated results are not available. Also, in the numerical simulation of engineering processes and systems, it is more convenient to use a curve fit of the available data on the characteristics of the components, such as pumps and blowers, rather than tabulated results, to obtain the values needed.

Curve fitting is needed in a wide variety of engineering problems. The property data for materials are generally available at discrete values of the independent variable, such as pressure, temperature, and concentration. Curve fitting yields a function $f(x)$, where x is the independent variable and $f(x)$ is a material property, such as density, specific heat, equilibrium constant for a chemical reaction, and electrical resistance. Then this function $f(x)$ may be used to obtain the desired material property at arbitrary values of the independent variable over a given range. Curve fitting of property data is needed in many diverse engineering applications, such as the simulation of power plants, refrigeration systems, chemical reactors, environmental processes, electronic systems, and building structures. Similarly, experimental data on many processes of engineering interest, such as wind speed at various heights above the surface of a lake, velocity of a moving body as a function of time, the electrical current in an electronic circuit as a function of the input voltage, and the deflection of a structure under a changing load, are generally obtained in terms of a continuous function $f(x)$, which can be subsequently employed in the analysis and design of relevant engineering processes and systems. The calibration curves for measuring devices, such as pressure transducers and flow meters, are similarly obtained from data taken at discrete points.

There are two basic approaches to curve fitting. The first one involves determining a curve that passes through every given data point and is known as an *exact fit*; see Fig. 6.1(a). Therefore, at the data points, the curve obtained yields values that are identical to the given data. An exact fit is appropriate if the data have a high level of accuracy, as is often the case for numerical simulations and for material property data. The number of parameters in the approximating curve, which is

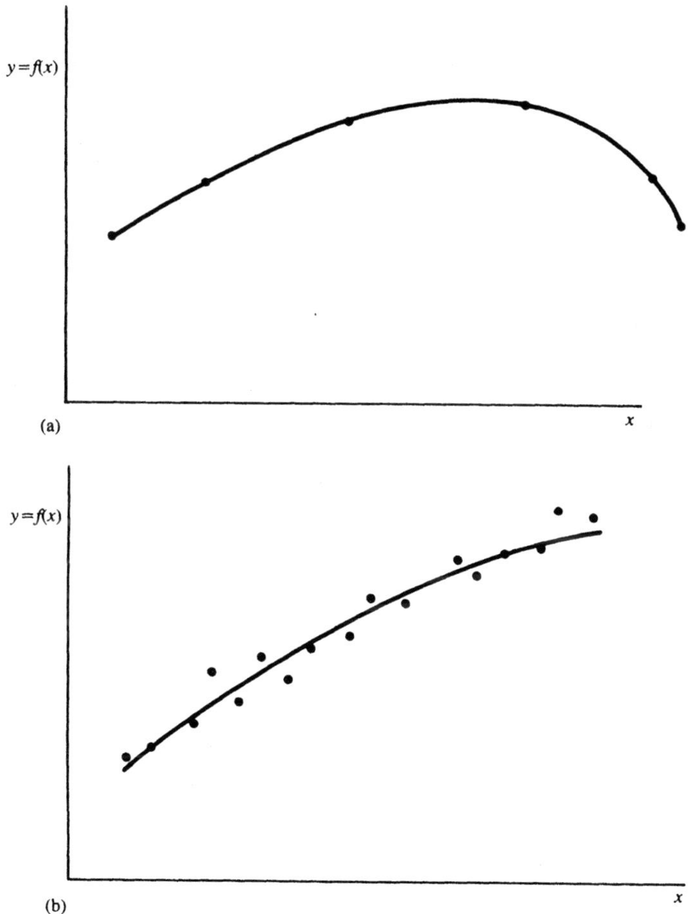

Figure 6.1 Curve fitting to given sets of data points. (a) An exact fit and (b) a best fit.

generally taken as a polynomial, must be equal to the number of data points. If the latter is large, the determination of the parameters becomes quite involved. Also, the curve obtained is not very convenient to use and is often ill-conditioned. Thus, if a large number of data points are available, the second approach, known as the *best fit*, is more appropriate. In this case, the curve does not pass through every data point. However, the difference between the values given by the approximating curve and the given data is minimized, so that the error in obtaining the values from the curve is small; see Fig. 6.1(b). The number of parameters in the curve is typically much smaller than the number of data points, and simple functions, such as linear and exponential forms, are frequently used for curve fitting. This approach is also suitable if the error in the data is significant, so that the fitted curve need not pass through each data point and a best fit is more appropriate.

Interpolation is employed to determine the dependent variable $y = f(x)$ at intermediate values between the given data points, and extrapolation is used for finding $f(x)$ outside the range of the given data, in x. Both are extensively used in engineering applications and also in the development of numerical procedures for differentiation, integration, root solving, and the solution of differential equations. The use of interpolation and extrapolation in differentiation and in root solving was demonstrated in Chapters 3 and 4. The applications to other problems in numerical analysis will be outlined in the following chapters. The basic approach involves fitting an exact curve to a finite number of discrete points and then applying the desired mathematical operation, such as differentiation or integration, to the smooth function obtained.

There are several numerical methods that may be employed for determining the interpolating curve from a given set of data points. Besides the direct evaluation of the parameters of an interpolating polynomial by substituting the given data and solving the resulting set of linear equations, as outlined in Chapter 5, interpolation with Lagrange polynomials and Newton's divided-difference polynomials is also discussed here. Splines, which fit subsets of the data with lower-order polynomials, such as a cubic, are also important in interpolation and are presented in this chapter.

The choice of the function $f(x)$ to obtain a best fit to a given data set is also an important consideration. Although polynomials, particularly straight lines which lead to linear regression, are very often employed for curve fitting, other forms, such as exponential and sinusoidal functions, are also used. As mentioned in Chapter 1, the physical nature of the given problem may often be employed to choose the appropriate form of the function $f(x)$ for a best fit. Periodic processes, such as those encountered in natural phenomena, are usually fitted with sinusoidal functions, as shown in Fig. 1.5. Similarly, in chemical reactions where the rate of change of concentration is proportional to the concentration at any time, the concentration varies exponentially, and, therefore, if measurements are taken in such processes, exponential functions are employed for curve fitting. Calibration curves for devices, such as those for measuring pressure and velocity, are generally obtained as polynomials by using curve-fitting techniques for a best fit. A few examples of curve fitting are shown in Fig. 6.2. The use of the physical or chemical background of the

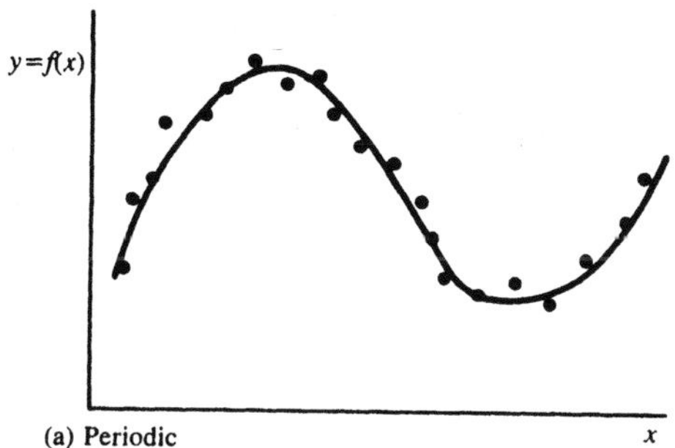

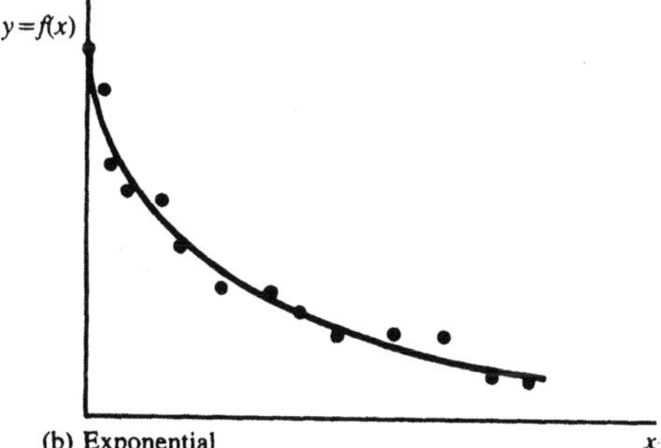

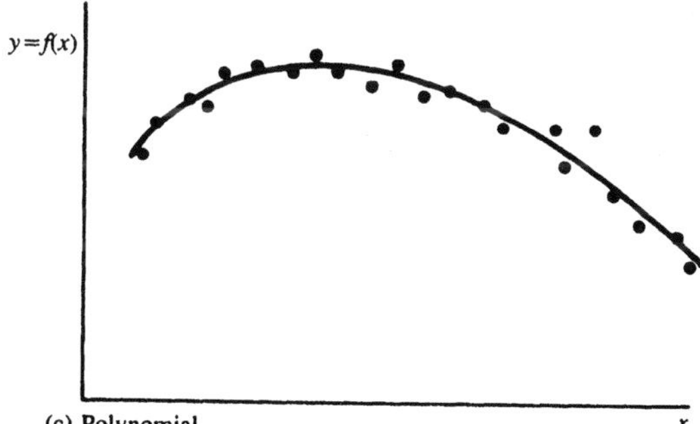

Figure 6.2 A few examples of curve fitting, employing different forms of the function $f(x)$ for a best fit.

given problem or of any prior information on the variation of the dependent variable helps in the choice of the most appropriate form of the curve for a best fit. A proper choice of the function $f(x)$ for curve fitting not only reduces the number of parameters to be determined numerically but also yields the desired result in a simple and useful form.

There are several numerical methods for obtaining the best fit to a given data set. The most widely used technique is based on the minimization of the sum of the squares of the differences between the actual data and the values obtained from the best fit. This method, known as the *least-squares fit*, is discussed in detail and applied to different forms of the chosen function $f(x)$ for curve fitting. Other methods are also outlined. Also considered is the circumstance where the dependent variable is a function of more than one independent variable, say, $f(x,y)$. Such problems are of interest, for instance, in chemical reactors and power plants, where the fluid properties depend on two or more independent variables, for instance, pressure, temperature, and concentration.

In this chapter, the numerical methods for obtaining an exact fit to tabulated data at discrete points are considered first. Various interpolation formulations and techniques are discussed. Interpolation with splines, particularly cubic splines, is outlined. The use of the least-squares method for obtaining a best fit with simple polynomials is discussed in detail. Other forms of the curve for a best fit are also considered. Finally, functions of more than one independent variable are considered, and the corresponding numerical curve-fitting procedures are presented.

6.2 EXACT FIT AND INTERPOLATION

An exact fit of tabulated data is frequently obtained in engineering applications, using polynomials in most cases. One chooses a general form of the approximating polynomial and substitutes the given data points in the equation for the polynomial in order to evaluate the parameters in the chosen curve. Therefore, the approximating polynomial passes through each data point and yields the exact value, as the given data, at these points. Since the polynomial is exact at the given data points, it is known as an *exact fit*. Once the approximating curve has been found, one can employ it to determine the values of the dependent variable y, which is a function $f(x)$ of the independent variable x, at arbitrary values of x not included in the tabulated data. If the chosen value of x lies within the range covered by the given data, the function $f(x)$ is found at an intermediate value of x, and the process is known as *interpolation*. If x lies beyond the range of the data, the process is called *extrapolation*.

Both interpolation and extrapolation are widely used in engineering and in numerical analysis, as shown in Chapters 3 and 4 and later in Chapters 7 and 8. Extrapolation is employed less frequently than interpolation, since there are uncertainties associated with evaluating the function at x values beyond the range for which data are available. If the function is known to be well behaved beyond the range of data, extrapolation may be used. Otherwise, substantial error may arise in the extrapolated value. However, interpolation is routinely used for a wide variety of engineering problems. Measurements and numerical simulations are carried out at a finite number of discrete data points, and curve fitting is used to obtain values at intermediate points. Several examples of engineering problems where this approach is used were given in the preceding section. As mentioned earlier, an exact fit is appropriate if the given data are very accurate and if the number of data points is relatively small, typically less than ten. Precise measurements of material properties, such as density, reflectivity, specific heat, and electrical resistivity, and results from numerical investigations are often obtained with a very small amount of error. Then one may use the numerical methods for an exact fit to determine an approximating polynomial, which may be subsequently used for computing the dependent variable at arbitrary values of the independent variable.

6.2.1 Exact Fit with an *n*th-Order Polynomial

A polynomial of degree n can be devised to exactly fit $(n + 1)$ data points. The general form of the polynomial may be taken as

$$y = f(x) = a_0 + a_1 x + a_2 x^2 + \cdots + a_n x^n \tag{6.1}$$

where y is the dependent variable, x is the independent variable, and the a's are constants. Two available data points are adequate to describe a first-degree, or linear, equation. Similarly, three data points are needed for a second-degree, or quadratic, equation and four points for a third-degree, or cubic, equation as shown in Fig. 6.3. The available data may be denoted as (x_i, y_i) for $i = 0, 1, 2, \ldots, n$, where y_i is the value of the function y at $x = x_i$. Then these values may be substituted in the chosen general form of the polynomial, Eq. (6.1), to yield

$$y_i = a_0 + a_1 x_i + a_2 x_i^2 + \cdots + a_n x_i^n \qquad \text{for } i = 0, 1, 2, \ldots, n \tag{6.2}$$

Since x_i and y_i are known for the given $(n + 1)$ points, Eq. (6.2) yields $(n + 1)$ linear equations for the unknown constants a_0 to a_n, as i is varied from 0 to n.

A numerical solution of this linear system, employing the methods given in the preceding chapter, will give these constants, and will thus determine the polynomial that exactly expresses the given data points. Example 5.1 demonstrated the solution of such a linear system for determining a fifth-order polynomial that provides an exact fit to the given six data points on the specific volume of steam. The matrix equation that represents the system of equations yielded by Eq. (6.2) is

$$\begin{pmatrix} 1 & x_0 & x_0^2 & \cdots & x_0^n \\ 1 & x_1 & x_1^2 & \cdots & x_1^n \\ \vdots & \vdots & \vdots & & \vdots \\ 1 & x_n & x_n^2 & \cdots & x_n^n \end{pmatrix} \begin{pmatrix} a_0 \\ a_1 \\ \vdots \\ a_n \end{pmatrix} = \begin{pmatrix} y_0 \\ y_1 \\ \vdots \\ y_n \end{pmatrix} \tag{6.3}$$

The determinant of the above coefficient matrix is known as the *Vandermond determinant.* It is nonzero unless a point is duplicated, that is, $x_i = x_j$ for $i \neq j$. Therefore, as shown in Chapter 5, a unique solution may be obtained for $a_0, a_1, \ldots, a_n$ from Eq. (6.3), giving a unique polynomial that yields the exact value of the dependent variable at the given data points.

The approach outlined above is fairly simple and can be used for an arbitrary distribution of data points. However, as mentioned earlier, it is appropriate for relatively small sets of data and for cases where the given data are very accurate. An

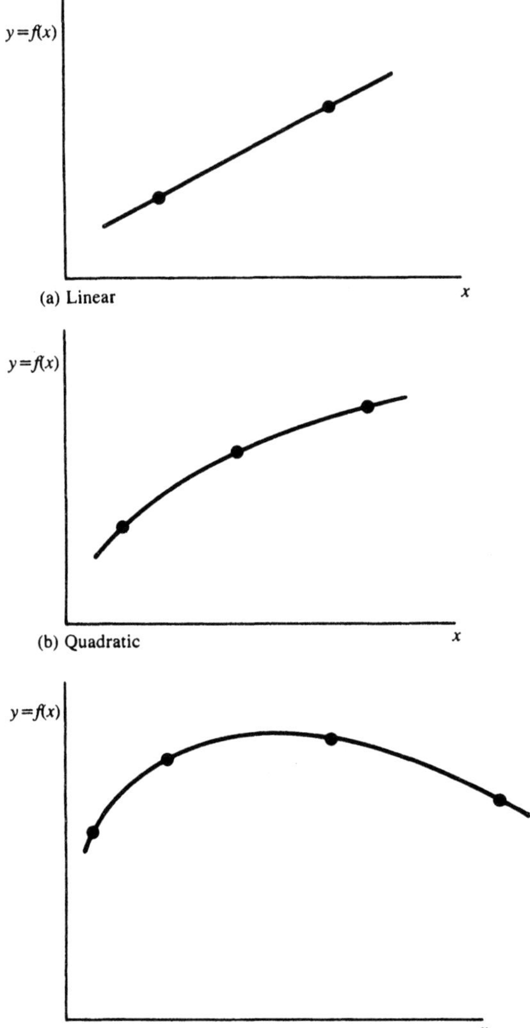

Figure 6.3 Exact fits to given data, using polynomials of first, second, or third order.

exact fit is generally employed if the number of data points is typically less than ten. For a larger number of points, higher-order polynomials are needed. The coefficients of the polynomial may then be quite small, particularly for the higher-order terms and if the independent variable attains large values. An example of such a representation is the variation of a material property, such as electrical resistivity ρ with temperature T, given by the polynomial

$$\rho(T) = a_0 + a_1 T + a_2 T^2 + \cdots + a_{19} T^{19} + a_{20} T^{20}$$

where T could vary from, say, 20°C to 300°C. Then the value of the coefficients, particularly the higher-order ones such as a_{19} and a_{20}, will be very small, giving rise to accuracy problems in interpolation. The polynomial may also be ill-conditioned, so that small changes in T result in large changes in $\rho(T)$. Then even the round-off error is magnified to yield inaccurate results from the use of such a polynomial for interpolation.

One method of avoiding these problems is to use a polynomial of lower order, based on a corresponding smaller data set chosen from the given data. Sometimes, normalization of the independent variable, say, by defining a new variable $\tilde{T}$ where $\tilde{T} = T/20$ in the above example, reduces the range of variation of the independent variable and thus avoids very small values of the coefficients. This approach was employed in Example 5.1. The normalizing characteristic quantity, such as 20°C in the above example or 10°C in Example 5.1, may often be chosen arbitrarily or on the basis of physical reasoning to reduce the range of the independent variable to a desired level.

6.2.2 Uniformly Spaced Independent Variable

In certain cases, the values of the function y are given at uniformly spaced values of the independent variable x. Such a circumstance arises, for instance, in numerical calculations and experimental studies where the values of x may be chosen arbitrarily and are taken as equally spaced for convenience. This is particularly true of tabulated data for material properties.

If the independent variable is uniformly spaced, as shown in Fig. 6.4, the determination of the polynomial $f(x)$ which exactly fits the given data can be simplified by choosing the following alternate form of the polynomial, instead of the general form given by Eq. (6.1):

$$y - y_0 = a_1\left[\frac{n}{L}(x - x_0)\right] + a_2\left[\frac{n}{L}(x - x_0)\right]^2 + \cdots + a_n\left[\frac{n}{L}(x - x_0)\right]^n \tag{6.4}$$

where L is the range, $x_n - x_0$, of x; n is the order of the polynomial; and the a's are the coefficients to be determined. Thus, the number of data points is $(n + 1)$, with (x_0, y_0), $(x_1, y_1), \ldots, (x_i, y_i), \ldots, (x_n, y_n)$ representing the given data that are to be numerically curve fitted with a polynomial. Here,

$$\frac{L}{n} = x_1 - x_0 = \frac{x_2 - x_0}{2} = \frac{x_3 - x_0}{3}, \ldots$$

since the values of x are equally spaced. If the data points are successively substituted into Eq. (6.4), we obtain the following in terms of the Δy's shown in Fig. 6.4:

$$\begin{aligned}
\Delta y_1 &= y_1 - y_0 = a_1 + a_2 + a_3 + \cdots + a_n \\
\Delta y_2 &= y_2 - y_0 = 2a_1 + 2^2 a_2 + 2^3 a_3 + \cdots + 2^n a_n \\
\Delta y_3 &= y_3 - y_0 = 3a_1 + 3^2 a_2 + 3^3 a_3 + \cdots + 3^n a_n \\
&\vdots \\
\Delta y_n &= y_n - y_0 = na_1 + n^2 a_2 + n^3 a_3 + \cdots + n^n a_n
\end{aligned} \tag{6.5}$$

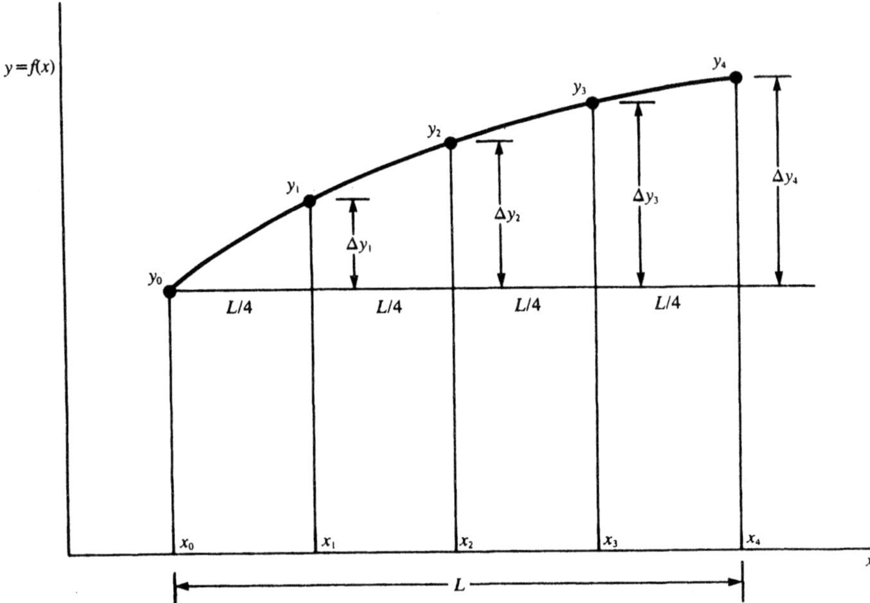

Figure 6.4 Exact fit to uniformly spaced data.

TABLE 6.1 Constants for Polynomials of the Form Given by Eq. (6.4) for Data in Which the Independent Variable Is Uniformly Spaced.

Polynomial	a_4	a_3	a_2	a_1
Fourth degree, $n=4$	$\frac{1}{24}(\Delta y_4 - 4\Delta y_3 + 6\Delta y_2 - 4\Delta y_1)$	$\frac{1}{6}(3\Delta y_1 + \Delta y_3 - 3\Delta y_2) - 6a_4$	$\frac{1}{2}(\Delta y_2 - 2\Delta y_1) - 3a_3 - 7a_4$	$\Delta y_1 - a_2 - a_3 - a_4$
Cubic, $n=3$		$\frac{1}{6}(3\Delta y_1 + \Delta y_3 - 3\Delta y_2)$	$\frac{1}{2}(\Delta y_2 - 2\Delta y_1) - 3a_3$	$\Delta y_1 - a_2 - a_3$
Quadratic, $n=2$			$\frac{1}{2}(\Delta y_2 - 2\Delta y_1)$	$\Delta y_1 - a_2$
Linear, $n=1$				Δy_1

Equations (6.5) can be solved for $a_1, a_2, \ldots, a_n$ in terms of $\Delta y_1, \Delta y_2, \ldots, \Delta y_n$ for chosen values of n to obtain the constants for an nth-order polynomial. Table 6.1 gives the constants for polynomials of the first four degrees, that is, $n = 1$ through $n = 4$. Then, for a given set of $(n + 1)$ data points, the degree of the polynomial is n, and the values of the constants are obtained simply by substituting the values of the differences $\Delta y_1, \Delta y_2, \ldots, \Delta y_n$ into the expressions given in the table. If several data sets are to be considered, the expressions for the coefficients of the polynomials, such as those given in Table 6.1, may be stored in the computer and the constants computed as each data set is entered. If only one or two sets of data are to be curve fitted or if higher-order polynomials are to be used, it would generally be easier to simply solve Eq. (6.5) for the constants, using the methods given in Chapter 5 for systems of linear algebraic equations. However, programs based on this approach for curve fitting are frequently available on computers, and the expressions for the coefficients are already stored, so that the given data, at uniformly spaced values of the independent variable, are entered and the program yields the constants of an approximating polynomial of the form given by Eq. (6.4). Interpolated values are then obtained from the resulting polynomial, as shown in the following example.

Example 6.1

In a fluid flow system, which experimentally simulates the flow generated in a room due to a fire, the flow rate F is measured at several values of a governing parameter R, known as the *Richardson number*, where R depends on the heat input by the fire, room dimensions, and so on. The data obtained are as follows:

R	0.025	0.05	0.1	0.2	0.3	0.4	0.5
F	1.4198	2.548	4.2	5.978	6.908	7.613	7.799

Employing the last five data points, obtain an exact fit and compute the interpolated values at $R = 0.25$ and 0.35. Also, obtain the flow rates at $R = 0$, 0.025, and 0.05 by extrapolation. Compare the last two extrapolated values with the given data.

Solution

Since the five data points to be considered for an exact fit, with a fourth-order polynomial, are uniformly spaced, the approach given in Section 6.2.2 may be employed. The independent variable is R, and the dependent variable is F. Denoting the data points as (R_0,F_0), $(R_1,F_1), \ldots, (R_4,F_4)$, we may compute the ΔF's as

$$
\begin{aligned}
\Delta F_1 &= F_1 - F_0 = 5.978 - 4.2 \\
\Delta F_2 &= F_2 - F_0 = 6.908 - 4.2 \\
\Delta F_3 &= F_3 - F_0 = 7.613 - 4.2 \\
\Delta F_4 &= F_4 - F_0 = 7.799 - 4.2
\end{aligned}
\qquad (6.1.1)
$$

where $R_0 = 0.1$ and $F_0 = 4.2$. Also, the uniform spacing is $R_1 - R_0 = R_2 - R_1 =$

```
RO=  ? 0.1
FO=  ? 4.2
F1=  ? 5.978
F2=  ? 6.908
F3=  ? 7.613
F4=  ? 7.799

THE COEFFICIENTS OF THE POLYNOMIAL ARE:

A1= 26.38918          A2=-115.5793
A3= 333.0837          A4=-382.0837


ENTER THE VALUE OF R AT WHICH INTERPOLATION IS DESIRED
R=  ? 0.0
R= 0           F= .0339973
ENTER THE VALUE OF R AT WHICH INTERPOLATION IS DESIRED
R=  ? 0.025
R= .025        F= 1.418069
ENTER THE VALUE OF R AT WHICH INTERPOLATION IS DESIRED
R=  ? 0.05
R= .05         F= 2.547569
ENTER THE VALUE OF R AT WHICH INTERPOLATION IS DESIRED
R=  ? 0.25
R= .25         F= 6.488571
ENTER THE VALUE OF R AT WHICH INTERPOLATION IS DESIRED
R=  ? 0.35
R= .35         F= 7.285508
ENTER THE VALUE OF R AT WHICH INTERPOLATION IS DESIRED
R=  ? 0.4
R= .4          F= 7.613002
ENTER THE VALUE OF R AT WHICH INTERPOLATION IS DESIRED
R=  ? 0.5
R= .5          F= 7.799
```

Figure 6.1.1 Coefficients of the polynomial obtained for an exact fit in Example 6.1, along with the interpolated results for the dependent variable F at several values of the independent variable R.

$R_3 - R_2 = R_4 - R_3 = 0.1$. The total range L is 0.4, and the degree n of the polynomial is 4. Therefore, from Eq. (6.4), the polynomial to be determined may be written as follows:

$$F - F_0 = a_1\left(\frac{R-R_0}{0.1}\right) + a_2\left(\frac{R-R_0}{0.1}\right)^2 + a_3\left(\frac{R-R_0}{0.1}\right)^3 + a_4\left(\frac{R-R_0}{0.1}\right)^4 \quad (6.1.2)$$

This equation gives

$$F = F_0 + A_1(R - R_0) + A_2(R - R_0)^2 + A_3(R - R_0)^3 + A_4(R - R_0)^4 \quad (6.1.3)$$

where $A_1 = a_1/0.1$, $A_2 = a_2/(0.1)^2$, and so on.

A simple computer program may be written to compute the differences from Eq. (6.1.1) and the coefficients in Eq. (6.1.2) from Table 6.1. These coefficients are then converted to the coefficients of Eq. (6.1.3) for convenience in the application of the polynomial for interpolation. Figure 6.1.1 presents the results from such a program. The data points are entered interactively, and the coefficients of the polynomial in Eq. (6.1.3) are computed. This polynomial is then employed to compute the interpolated or extrapolated values of the flow rate F at several values of R, entered interactively into the program.

Note from Fig. 6.1.1 that the polynomial obtained yields the exact values at the given data points, as expected. The extrapolated values at $R = 0.025$ and 0.05 are close to those obtained experimentally and given in the problem. The value at $R = 0$ is expected to be zero on physical grounds. However, the extrapolated value is nonzero, although it is fairly small, being equal to 0.034. Thus, over the range of R considered here, $0 \leqslant R \leqslant 0.5$, the computed polynomial yields good accuracy. For values of R larger than 0.5, extrapolation does not give satisfactory results, since F is obtained as decreasing with increasing R, which is contrary to the behavior expected for the physical problem considered. However, interpolation in the range $0.1 \leqslant R \leqslant 0.5$ yields physically realistic results, being exact at the data points employed for deriving the polynomial.

6.3 LAGRANGE INTERPOLATION

A method that is widely used for obtaining an exact fit to a given data set is Lagrange interpolation. It is based on the use of a special form of the interpolating polynomial, known as the *Lagrange polynomial.* For a quadratic function, it is written as

$$y = f(x) = a_0(x - x_1)(x - x_2) + a_1(x - x_0)(x - x_2) + a_2(x - x_0)(x - x_1) \quad (6.6)$$

where (x_0,y_0), (x_1,y_1), and (x_2,y_2) are the three given data points and a_0, a_1, and a_2 are the constants to be determined from these points. It can easily be seen that Eq. (6.6) represents a second-order polynomial, which can also be rewritten in the form given by Eq. (6.1). Substituting the given data points successively in Eq. (6.6), we can easily determine the constants as follows:

$$\begin{aligned} a_0 &= \frac{y_0}{(x_0 - x_1)(x_0 - x_2)} \\ a_1 &= \frac{y_1}{(x_1 - x_0)(x_1 - x_2)} \\ a_2 &= \frac{y_2}{(x_2 - x_0)(x_2 - x_1)} \end{aligned} \quad (6.7)$$

Similarly, if $(n + 1)$ data points are given, an nth-order Lagrange polynomial may be written by taking n factors in each term, instead of two taken for the quadratic function of Eq. (6.6). Thus, an nth-order Lagrange polynomial is written as

$$\begin{aligned} y = f(x) = {} & a_0(x - x_1)(x - x_2)\cdots(x - x_n) + a_1(x - x_0)(x - x_2)\cdots(x - x_n) \\ & + \cdots + a_n(x - x_0)(x - x_1)\cdots(x - x_{n-1}) \end{aligned} \quad (6.8)$$

The coefficients a_i, where i varies from 0 to n, can be determined by substitution of the $(n + 1)$ data points into Eq. (6.8) to obtain expressions such as those given by Eq. (6.7). The resulting interpolating polynomial is

$$y = f(x) = \sum_{i=0}^{n} y_i \prod_{\substack{j=0 \\ j \neq i}}^{n} \left(\frac{x - x_j}{x_i - x_j} \right) \tag{6.9}$$

where the product sign $\prod$ denotes multiplication of the n factors obtained by varying j from 0 to n, excluding $j = i$, for the quantity within the parentheses. Thus, for example, a third-order Lagrange polynomial is obtained from Eq. (6.9) as

$$y = \frac{(x - x_1)(x - x_2)(x - x_3)}{(x_0 - x_1)(x_0 - x_2)(x_0 - x_3)} y_0 + \frac{(x - x_0)(x - x_2)(x - x_3)}{(x_1 - x_0(x_1 - x_2)(x_1 - x_3)} y_1$$
$$+ \frac{(x - x_0)(x - x_1)(x - x_3)}{(x_2 - x_0)(x_2 - x_1)(x_2 - x_3)} y_2 + \frac{(x - x_0)(x - x_1)(x - x_2)}{(x_3 - x_0)(x_3 - x_1)(x_3 - x_2)} y_3$$

Lagrange interpolation is applicable to an arbitrary distribution of the independent variable x. The determination of the coefficients of the polynomial does not require the solution of a system of equations, as was the case for the methods discussed in the preceding section. The interpolating polynomial, Eq. (6.9), can easily be entered and the necessary calculations performed on a computer for obtaining the desired exact fit to the given data. The programming is quite simple, as illustrated in Example 6.2. Because of the applicability of the method to arbitrary distributions of data points and the ease with which it may be applied, Lagrange interpolation is widely employed for engineering applications. Programs available on many computers for interpolation are also frequently based on this method.

Example 6.2

The deflection of a structure under loading is measured at five different values of the force applied X, in kilonewtons (kN). The deflection Y is in centimeters, and the data are given as follows:

X (*kN*)	0.5	1.0	1.5	2.0	2.5
Y (cm)	3.0	3.9	5.2	7.3	10.5

Employing Lagrange interpolation, compute the deflection at the intermediate load values of 0.75, 1.25, 1.8, and 2.2 kN. Also obtain the extrapolated values at 0 and 3.0 kN. Such problems are of interest in civil engineering, though many more data points are generally obtained, requiring higher-order polynomials for an exact fit or resorting to a best fit.

Solution

The Lagrange polynomial to be computed is given by Eq. (6.9). Since five data points are given, a fourth-order polynomial can be derived to exactly fit the given points. The

```
C               LAGRANGE INTERPOLATION
C
C     X IS THE INDEPENDENT VARIABLE AND Y THE DEPENDENT VARIABLE,
C     WITH X(I) AND Y(I) REPRESENTING THE GIVEN DATA POINTS.
C     N IS THE NUMBER OF DATA POINTS, XL THE VALUE OF X AT WHICH
C     INTERPOLATION IS DESIRED AND YL THE CORRESPONDING COMPUTED
C     VALUE OF Y AT X=XL. A(I) REPRESENTS THE COEFFICIENTS OF
C     THE LAGRANGE POLYNOMIAL AND M IS THE NUMBER OF POINTS
C     AT WHICH INTERPOLATED VALUES ARE NEEDED.
C
C
      DIMENSION X(10),Y(10),A(10)
C
C     ENTER THE GIVEN DATA
C
      READ(5,*)N
      READ(5,*)M
      READ(5,*)(X(I),I=1,N)
      READ(5,*)(Y(I),I=1,N)
      WRITE(6,10)
   10 FORMAT(2X,'THE VALUES FROM LAGRANGE INTERPOLATION ARE:'//)
      DO 6 K=1,M
      READ(5,*)XL
C
C     COMPUTE THE COEFFICIENTS OF THE LAGRANGE POLYNOMIAL
C
      DO 2 J=1,N
      A(J)=Y(J)
      DO 1 I=1,N
      IF(I .NE. J) THEN
      A(J)=A(J)/(X(J)-X(I))
      END IF
    1 CONTINUE
    2 CONTINUE
C
C     CALCULATE THE INTERPOLATED VALUE OF THE DEPENDENT VARIABLE
C
      YL=0.0
      DO 4 J=1,N
      S=1.0
      DO 3 I=1,N
      IF(I .NE. J) THEN
      S=S*(XL-X(I))
      END IF
    3 CONTINUE
    4 YL=YL+S*A(J)
C
C     PRINT THE CALCULATED RESULTS
C
      WRITE(6,5)XL,YL
    5 FORMAT(2X,'XL=',F9.4,4X,'YL=',F9.4)
    6 CONTINUE
      WRITE(6,7)
    7 FORMAT(//2X,'COEFFICIENTS OF THE LAGRANGE POLYNOMIAL ARE:')
      DO 9 I=1,N
      WRITE(6,8)I,A(I)
    8 FORMAT(/4X,'A(',I1,')=',F9.4)
    9 CONTINUE
      STOP
      END
```

Figure 6.2.1 Computer program in FORTRAN 77 for Lagrange interpolation.

coefficients a_i of the polynomial in Eq. (6.8) are given by the product

$$\prod_{\substack{j=0 \\ j \neq i}}^{n} \left(\frac{Y_i}{X_i - X_j} \right)$$

where X_i and Y_i are the values of the independent and dependent variables respectively, at the data points. The interpolated value of Y at a given X is then obtained from Eq. (6.8).

We can easily write a computer program to calculate the coefficients of the polynomial and then to use these to determine the corresponding interpolated value of Y. Figure 6.2.1 shows the program in FORTRAN 77 for Lagrange interpolation. The number of data points N is read, along with the given data. Also read is the number of intermediate points M at which interpolated or extrapolated values of the dependent variable are desired. The values of the independent variable at which interpolation/extrapolation is needed are denoted by XL and are read from the data appended to the main program. The coefficients A(I) of the Lagrange polynomial are calculated, and then the interpolated/extrapolated value of the dependent variable, denoted by YL, is obtained from Eq. (6.8). The values of XL are sequentially changed according to the given input, and the corresponding values of YL are computed. Finally, the calculated results are printed in tabular form, as shown in Figure 6.2.2, along with the coefficients A(I) of the Lagrange polynomial.

```
THE VALUES FROM LAGRANGE INTERPOLATION ARE:

XL=   0.5000     YL=    3.0000
XL=   1.0000     YL=    3.9000
XL=   1.5000     YL=    5.2000
XL=   2.0000     YL=    7.3000
XL=   2.5000     YL=   10.5000
XL=   0.0000     YL=    2.0000
XL=   0.7500     YL=    3.4289
XL=   1.2500     YL=    4.4727
XL=   1.8000     YL=    6.3426
XL=   2.2000     YL=    8.4346
XL=   3.0000     YL=   15.0000


COEFFICIENTS OF THE LAGRANGE POLYNOMIAL ARE:

  A(1)=    2.0000

  A(2)=  -10.4000

  A(3)=   20.8000

  A(4)=  -19.4667

  A(5)=    7.0000
```

Figure 6.2.2 Computed interpolated values from Lagrange interpolation and coefficients of the Lagrange polynomial for Example 6.2.

Note that, as expected, the calculated values of the dependent variable are exact at the data points employed for obtaining the Lagrange polynomial. The interpolated values at XL = 0.75, 1.25, 1.8, and 2.2 kN are found to be within the expected range. Also, a deflection of 2.0 cm is obtained at zero load, indicating the deflection due only to the weight of the structure. The extrapolated value at XL = 3.0 kN is 15.0 cm, which is qualitatively satisfactory, since the deflection increases with load. However, both values at XL = 0 and 3.0 are beyond the range of the given data and their accuracy is not known. Thus, these values must be used with caution, unless validation from further experimentation is obtained.

6.4 NEWTON'S DIVIDED-DIFFERENCE INTERPOLATING POLYNOMIAL

An extensively used form of the polynomial for interpolation is Newton's divided-difference polynomial. It can be used for an arbitrary distribution of data points, although simplified formulas result for uniformly spaced points and form the basis for several interpolation schemes, such as forward, backward, and central Newton-Gregory formulas, as outlined later in this section.

6.4.1 General Formulas

First-order, or linear, interpolation is the simplest form of interpolation and is obtained by drawing a straight line connecting two data points, as sketched in Fig. 6.5. The value of the function $f(x)$ at a given value of the independent variable x can be obtained from the interpolating straight line. Thus, from geometry,

$$\frac{f(x) - f(x_0)}{x - x_0} = \frac{f(x_1) - f(x_0)}{x_1 - x_0}$$

or

$$\begin{aligned} f(x) &= f(x_0) + \frac{f(x_1) - f(x_0)}{x_1 - x_0}(x - x_0) \\ &= c_0 + c_1(x - x_0) \end{aligned} \tag{6.10}$$

where c_0 and c_1 are coefficients of the interpolating polynomial. Here, c_1 represents a finite divided-difference approximation of the first derivative, as given by Eq. (3.11). Only two coefficients are needed here because the interpolating polynomial is a straight line.

In a similar way, a second-order, or quadratic, interpolation may be considered. On the basis of Eq. (6.10), the general form of the polynomial is taken as

$$y = f(x) = c_0 + c_1(x - x_0) + c_2(x - x_0)(x - x_1) \tag{6.11}$$

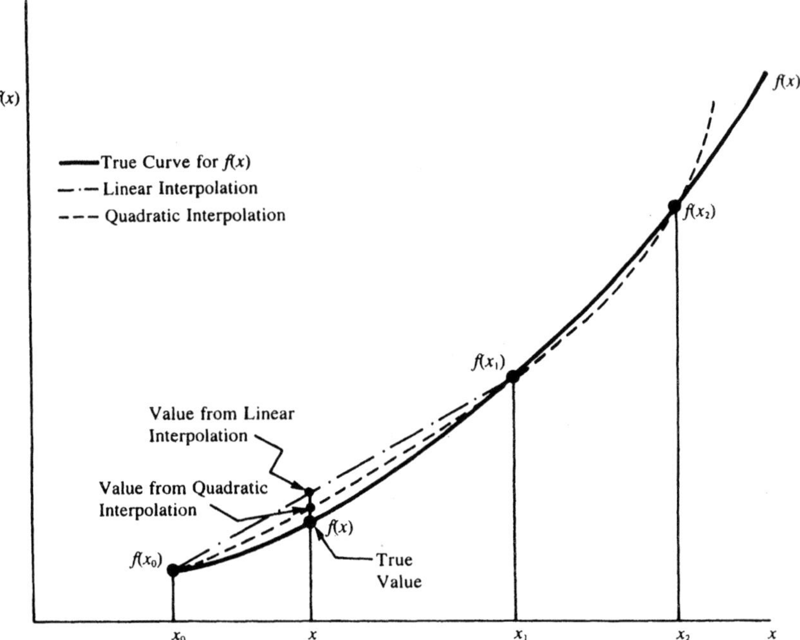

Figure 6.5 Interpolation with a straight line and a second-order polynomial for the derivation of Newton's divided-difference method.

Three data points are needed to determine the coefficients c_0, c_1, and c_2. Employing the first point, denoted by (x_0, y_0), we obtain c_0 as

$$c_0 = f(x_0) \tag{6.12}$$

Similarly, the second point, (x_1, y_1), yields

$$c_1 = \frac{f(x_1) - c_0}{x_1 - x_0} = \frac{f(x_1) - f(x_0)}{x_1 - x_0} \tag{6.13}$$

We obtain the third coefficient c_2 by substituting the third point, (x_2, y_2), in Eq. (6.11). Employing the results given by Eqs. (6.12) and (6.13), we obtain

$$c_2 = \frac{f(x_2) - f(x_0)}{(x_2 - x_0)(x_2 - x_1)} - \frac{f(x_1) - f(x_0)}{(x_1 - x_0)(x_2 - x_1)}$$

$$= \frac{\dfrac{f(x_2) - f(x_1)}{x_2 - x_1} - \dfrac{f(x_1) - f(x_0)}{x_1 - x_0}}{x_2 - x_0} \tag{6.14}$$

Therefore, from the above equations, c_0 and c_1 for the quadratic interpolation are identical to those for the linear interpolation. The third term on the right-hand side of Eq. (6.11) improves the interpolation by introducing curvature, as shown graphically in Fig. 6.5. The coefficient c_2 is similar to the finite difference representation of the second derivative; see Eq. (3.15). The general form of the polynomial, Eq. (6.11), is similar to the Taylor series expansion, presented in Section 3.2. Also, the first-order divided difference, Eq. (6.13), can be used to determine the second-order divided difference, Eq. (6.14). These features allow the development of a recursive formula for determining the coefficients of Newton's interpolating polynomials of arbitrary order.

From the above discussion, the general form for an nth-order Newton's polynomial may be written as follows:

$$\begin{aligned} y = f(x) &= c_0 + c_1(x - x_0) + c_2(x - x_0)(x - x_1) + \cdots \\ &\quad + c_n(x - x_0)(x - x_1)\cdots(x - x_{n-1}) \end{aligned} \tag{6.15}$$

To determine the $(n + 1)$ coefficients, $c_0, c_1, \ldots, c_n$, in the nth-order polynomial, we need $(n + 1)$ data points. When these points, denoted by $(x_0, y_0), (x_1 y_1), \ldots, (x_n, y_n)$, are substituted in the general form of the polynomial, the coefficients are given by the equations

$$\begin{aligned} c_0 &= f(x_0) \\ c_1 &= \frac{f(x_1) - f(x_0)}{x_1 - x_0} = F(x_1, x_0) \\ c_2 &= \frac{F(x_2, x_1) - F(x_1, x_0)}{x_2 - x_0} = F(x_2, x_1, x_0) \\ &\vdots \\ c_n &= F(x_n, x_{n-1}, \ldots, x_1, x_0) \end{aligned} \tag{6.16}$$

where the function F denotes finite divided differences. Therefore,

$$\begin{aligned} F(x_i, x_j) &= \frac{f(x_i) - f(x_j)}{x_i - x_j} \\ F(x_i, x_j, x_k) &= \frac{F(x_i, x_j) - F(x_j, x_k)}{x_i - x_k} \\ F(x_n, x_{n-1}, \ldots, x_1, x_0) &= \frac{F(x_n, x_{n-1}, \ldots, x_2, x_1) - F(x_{n-1}, x_{n-2}, \ldots, x_1, x_0)}{x_n - x_0} \end{aligned} \tag{6.17}$$

Note from the above expressions that a recursive formula may be written to determine the coefficients. Therefore, the problem is well suited for digital computation. The higher-order differences are determined from the lower-order differences. Therefore, we evaluate the coefficients by starting with c_0 and successively calculating c_1, c_2, c_3, and so on, up to c_n. Once the coefficients have been determined, the interpolating polynomial is obtained from Eq. (6.15), which may also be written as

$$y = f(x) = f(x_0) + (x - x_0)F(x_1, x_0) + (x - x_0)(x - x_1)F(x_2, x_1, x_0) + \cdots + (x - x_0)(x - x_1)\cdots(x - x_{n-1})F(x_n, x_{n-1}, \ldots, x_1, x_0) \quad (6.18)$$

The general form of Newton's interpolating polynomial is similar to the Taylor series expansion, since terms representing higher-order derivatives are successively added to improve the accuracy of the representation. As given by Eq. (3.7), the remainder term R_n in a Taylor series expansion is

$$R_n = \frac{d^{n+1}f}{dx^{n+1}}(\xi)\frac{(x_{i+1} - x_i)^{n+1}}{(n+1)!} \qquad \text{where } x_i < \xi < x_{i+1} \quad (3.7)$$

The derivative is evaluated at a point ξ which lies in the interval from x_i to x_{i+1}. Similarly, for an nth-order Newton's interpolating polynomial, the expression for the remainder and, thus, for the error is

$$R_n = \frac{d^{n+1}f}{dx^{n+1}}(\xi)\frac{(x - x_0)(x - x_1)\ldots(x - x_n)}{(n+1)!} \qquad \text{where } x_0 < \xi < x_n \quad (6.19)$$

Since the function $f(x)$ and its derivatives are not known, in general, the $(n + 1)$th derivative may be replaced by the corresponding finite divided difference. Thus, R_n may be written as

$$R_n = [F(x, x_n, x_{n-1}, \ldots, x_1, x_0)](x - x_0)(x - x_1)\ldots(x - x_n) \quad (6.20)$$

This expression can be used to estimate the error if an additional data point (x_{n+1}, y_{n+1}) is available as follows:

$$R_n = [F(x_{n+1}, x_n, \ldots, x_1, x_0)](x - x_0)(x - x_1)\ldots(x - x_n) \quad (6.21)$$

Since the additional data point is generally not available, the interpolating polynomial itself may be used to obtain an additional point, and the error determined from Eq. (6.21). Example 6.3 illustrates the use of Newton's method for interpolation.

6.4.2 Uniformly Spaced Data

Several simplified formulas can be derived from the above results if the data are given at equally spaced values of the independent variable x. If Δx is the interval between the data, the values of x are given by

$$x_i = x_0 + i\,\Delta x \qquad \text{for } i = 1, 2, \ldots, n \quad (6.22)$$

Such uniformly spaced data are obtained, for instance, from numerical simulations of engineering systems, tables of material properties, and experimental studies in which the independent variable is taken at uniformly spaced intervals for convenience. Then the coefficients c_0, c_1, and c_2 are given by

$$
\begin{aligned}
c_0 &= f(x_0) \\
c_1 &= \frac{f(x_1) - f(x_0)}{\Delta x} = \frac{\Delta f_0}{\Delta x} \\
c_2 &= \frac{f(x_2) - 2f(x_1) + f(x_0)}{2(\Delta x)^2} = \frac{\Delta^2 f_0}{2!(\Delta x)^2}
\end{aligned}
\tag{6.23}
$$

where Δf_0 is known as the first forward difference and $\Delta^2 f_0$ as the second forward difference at $x = x_0$. These constitute the numerator of the forward finite difference approximations of the first and second derivatives to $O(\Delta x)$; see Fig. 3.5. Therefore, in general, the coefficient c_n of the polynomial is given by

$$c_n = F(x_n, x_{n-1}, \ldots, x_1, x_0) = \frac{\Delta^n f_0}{n!(\Delta x)^n} \tag{6.24}$$

From Eq. (6.18), Newton's interpolating polynomial can be written for equally spaced data as follows:

$$
\begin{aligned}
y = f(x) = f(x_0) &+ \frac{\Delta f_0}{\Delta x}(x - x_0) + \frac{\Delta^2 f_0}{2!(\Delta x)^2}(x - x_0)(x - x_0 - \Delta x) + \cdots \\
&+ \frac{\Delta^n f_0}{n!(\Delta x)^n}(x - x_0)(x - x_0 - \Delta x)\ldots[x - x_0 - (n-1)\Delta x)] + R_n
\end{aligned}
\tag{6.25}
$$

where the remainder R_n is the same as that given by Eq. (6.19). The above interpolating polynomial is known as the *Newton-Gregory forward interpolation formula*. One can generate a forward difference table by taking forward differences at each x, then taking differences of the differences, and so on. An example of such a forward difference table is shown in Table 6.2(a). The general formula for computing these differences at $x = x_i$ is

$$\Delta^n f_i = \Delta^{n-1} f_{i+1} - \Delta^{n-1} f_i \tag{6.26}$$

Then these differences may be substituted into Eq. (6.25) to yield the interpolating polynomial. The subscript gives the location, in x, where the difference is evaluated, and the superscript indicates the order of the difference. The lowest order differences Δf_i are given by $n = 1$. Also, $n = 0$ corresponds to the values of the function f_i.

In a similar way, a backward difference polynomial, known as the *Newton-Gregory backward interpolation polynomial*, may be derived. The backward differences at $x = x_n$ are denoted by ∇f_n, $\nabla^2 f_n$, and so on, and are obtained from Fig. 3.6 or a backward difference table generated in a manner analogous to that for forward differences; see Table 6.2(b). The corresponding interpolating polynomial is written as follows:

$$
\begin{aligned}
y = f(x) = f(x_0) &+ \frac{\nabla f_n}{\Delta x}(x - x_n) + \frac{\nabla^2 f_n}{2!(\Delta x)^2}(x - x_n)(x - x_n + \Delta x) + \cdots \\
&+ \frac{\nabla^n f_n}{n!(\Delta x)^n}(x - x_n)(x - x_n + \Delta x)\ldots[x - x_n + (n-1)\Delta x] + R_n
\end{aligned}
\tag{6.27}
$$

TABLE 6.2 Examples of Difference Tables for Computing the Interpolation Polynomials, Using Divided Differences, for Uniformly Spaced Data. (a) Forward Differences and (b) Backward Differences.

x	$f(x)$	Δf	$\Delta^2 f$	$\Delta^3 f$	$\Delta^4 f$	$\Delta^5 f$	$\Delta^6 f$
1	−4	3	5	2	1	1	1
2	−1	8	7	3	2	2	
3	7	15	10	5	4		
4	22	25	15	9			
5	47	40	24				
6	87	64					
7	151						

(a)

x	$f(x)$	∇f	$\nabla^2 f$	$\nabla^3 f$	$\nabla^4 f$	$\nabla^5 f$	$\nabla^6 f$
1	−4						
2	−1	3					
3	7	8	5				
4	22	15	7	2			
5	47	25	10	3	1		
6	87	40	15	5	2	1	
7	151	64	24	9	4	2	1

(b)

where $\Delta x = x_0 - x_1 = x_2 - x_1$, and so on, and R_n is the remainder which can be derived in a manner similar to that given earlier for the forward difference formulation. Also, the general formula for the backward difference at $x = x_i$ is

$$\nabla^n f_i = \nabla^{n-1} f_i - \nabla^{n-1} f_{i-1} \tag{6.28}$$

where the subscripts denote the value of x at which the difference is obtained and the superscript gives the order of the difference.

Thus, if the data are given at equally spaced values of the independent variable, the above simplified formulas may be employed. The corresponding forward, or backward, differences are generated, using Eq. (6.26) or Eq. (6.28), and the desired value of the function $f(x)$ is determined from the interpolating polynomial, for a given value of x. The choice of the formula, forward or backward, for interpolation depends on the value of x, in relation to the given data points, at which $f(x)$ is to be determined. Thus, if x is close to x_0, the forward difference formula is more appropriate than the backward formula. Similarly, if x is close to x_n, the backward difference form is used. Several central difference formulas have also been derived to accommodate interpolation in the region near the middle of the distribution of the data points. Consult Carnahan et al. (1969) and Hornbeck (1975), listed in the References, for the relevant formulas.

6.4.3 Extrapolation

The process of estimating the function $f(x)$ at a point x which lies beyond the range of the given data points is known as *extrapolation*. However, the most accurate estimation for $f(x)$ is generally obtained when x is close to the middle of the range. Also, the behavior of the function beyond the given data is not known. Thus, the estimated value of $f(x)$ could be in considerable error. Because of the element of uncertainty involved, values of the function obtained by extrapolation must be treated with caution. If any information is available on the nature of the function, and on its behavior beyond the range of the given data, one must consider the extrapolated values in terms of this information to judge their validity and accuracy.

Extrapolation is frequently needed in engineering applications. We are all familiar with predictions of weather, future trends in economic parameters, expected output from engineering systems, demand for engineering products, and so on. Extrapolation is, therefore, necessary for future planning of engineering resources and output. It is also often needed for the control and design of systems and processes. If the data are available at discrete, evenly spaced points, the Newton-Gregory forward or backward formula, as appropriate, may be used for extrapolation. Lagrange interpolation or Newton's divided-difference polynomials may be employed for an arbitrary distribution of data points. The procedures for extrapolation are similar to those for interpolation. However, since estimations are being made for points beyond the range of the given data, an element of uncertainty arises, and extreme care must be exercised in the use of the values obtained.

Example 6.3

Solve the problem given in Example 6.1 by Newton's divided-difference interpolation, employing the data over the range $0.05 \leqslant R \leqslant 0.4$. Use polynomials of increasing order and compute the remainder term in each case.

Solution

The data points to be considered for deriving the interpolating polynomial are as follows:

X	0.05	0.1	0.2	0.3	0.4
Y	2.548	4.2	5.978	6.908	7.613

where X is the independent variable and Y the dependent variable. We use different symbols here, as compared to those in Example 6.1, in order to derive a generalized solution procedure, based on Newton's divided-difference polynomials. Since five data points are employed for an exact fit, a fourth-order polynomial of the form given by Eq. (6.15) may be derived.

Figure 6.3.1 gives the program in BASIC for computing the interpolating

```
100 REM      NEWTON'S DIVIDED-DIFFERENCE INTERPOLATING POLYNOMIAL
110 REM
120 REM      X IS THE INDEPENDENT VARIABLE, Y THE DEPENDENT VARIABLE,
130 REM      XP THE GIVEN VALUE OF X AT WHICH INTERPOLATION IS DESIRED,
140 REM      N THE NUMBER OF DATA POINTS, F(I,J) A MATRIX WHOSE FIRST
150 REM      COLUMN CONSISTS OF THE GIVEN VALUES OF THE DEPENDENT VARIABLE
160 REM      AND WHOSE FIRST ROW CONSISTS OF THE COEFFICIENTS OF THE
170 REM      INTERPOLATING POLYNOMIAL, AND R IS THE REMAINDER
180 REM
190 REM
200 REM      ENTER INPUT QUANTITIES
210 REM
220      DIM X(11),F(11,11)
230      READ N
240      PRINT "THE NUMBER OF DATA POINTS= ";N
250      FOR J=1 TO N
260      READ X(J),F(J,1)
270      NEXT J
280 REM
290 REM      COMPUTE THE DIVIDED DIFFERENCES
300 REM
310      FOR K=1 TO (N-1)
320      L=K+1
330      FOR M=1 TO (N-K)
340      F(M,L)=(F(M+1,K)-F(M,K))/(X(M+K)-X(M))
350      NEXT M
360      NEXT K
370 REM
380 REM      PRINT THE CALCULATED COEFFICIENTS OF THE POLYNOMIAL
390 REM
400      FOR I=1 TO N
410      PRINT "C(";I;")=";F(1,I)
420      NEXT I
430 REM
440 REM      ENTER VALUE OF X AT WHICH INTERPOLATION IS NEEDED
450 REM      AND COMPUTE THE INTERPOLATED VALUE OF Y
460 REM
465      PRINT
470      INPUT "XP=   ";XP
480      Y=0
490      B=1
500      FOR I=1 TO N
510      Y=Y+F(1,I)*B
520      PRINT "INTERPOLATED VALUE OF Y=";Y
530 REM
540 REM      CALCULATE THE REMAINDER TERM
550 REM
560      B=B*(XP-X(I))
570      R=B*F(1,I+1)
580      IF I >= N THEN 605
590      PRINT "REMAINDER TERM=";R
600      NEXT I
605      PRINT
610      PRINT "IF YOU WANT INTERPOLATION AT ADDITIONAL POINTS, TYPE 1"
620      INPUT MORE
630      IF MORE = 1 THEN 465
640      DATA 5
650      DATA  0.05,2.548,0.1,4.2,0.2,5.978,0.3,6.908,0.4,7.613
660      END
```

Figure 6.3.1 Computer program in BASIC for deriving Newton's divided-difference interpolating polynomial and employing it for computing interpolated values of the dependent variable Y and the remainder term.

```
THE NUMBER OF DATA POINTS=  5
C( 1 )= 2.548
C( 2 )= 33.03999
C( 3 )=-101.7333
C( 4 )= 237.333
C( 5 )=-381.4272

XP=  ? 0.0
INTERPOLATED VALUE OF Y= 2.548
REMAINDER TERM=-1.652
INTERPOLATED VALUE OF Y= .8960004
REMAINDER TERM=-.5086664
INTERPOLATED VALUE OF Y= .3873341
REMAINDER TERM=-.237333
INTERPOLATED VALUE OF Y= .1500011
REMAINDER TERM=-.1144282
INTERPOLATED VALUE OF Y= 3.557293E-02

IF YOU WANT INTERPOLATION AT ADDITIONAL POINTS, TYPE 1
? 1

XP=  ? 0.025
INTERPOLATED VALUE OF Y= 2.548
REMAINDER TERM=-.8259999
INTERPOLATED VALUE OF Y= 1.722
REMAINDER TERM=-.1907499
INTERPOLATED VALUE OF Y= 1.53125
REMAINDER TERM=-7.787488E-02
INTERPOLATED VALUE OF Y= 1.453376
REMAINDER TERM=-3.441785E-02
INTERPOLATED VALUE OF Y= 1.418958

IF YOU WANT INTERPOLATION AT ADDITIONAL POINTS, TYPE 1
? 1

XP=  ? 0.2
INTERPOLATED VALUE OF Y= 2.548
REMAINDER TERM= 4.956
INTERPOLATED VALUE OF Y= 7.504
REMAINDER TERM=-1.525999
INTERPOLATED VALUE OF Y= 5.978001
REMAINDER TERM= 0
INTERPOLATED VALUE OF Y= 5.978001
REMAINDER TERM= 0
INTERPOLATED VALUE OF Y= 5.978001

IF YOU WANT INTERPOLATION AT ADDITIONAL POINTS, TYPE 1
? 1

XP=  ? 0.25
INTERPOLATED VALUE OF Y= 2.548
REMAINDER TERM= 6.607999
INTERPOLATED VALUE OF Y= 9.155999
REMAINDER TERM=-3.051998
INTERPOLATED VALUE OF Y= 6.104001
REMAINDER TERM= .3559995
INTERPOLATED VALUE OF Y= 6.460001
REMAINDER TERM= 2.860705E-02
INTERPOLATED VALUE OF Y= 6.488608
```

Figure 6.3.2 Numerical results obtained from Newton's divided-difference method for the problem considered in Example 6.3.

```
IF YOU WANT INTERPOLATION AT ADDITIONAL POINTS, TYPE 1
? 1

XP=  ? 0.35
INTERPOLATED VALUE OF Y= 2.548
REMAINDER TERM= 9.911997
INTERPOLATED VALUE OF Y= 12.46
REMAINDER TERM=-7.629995
INTERPOLATED VALUE OF Y= 4.830003
REMAINDER TERM= 2.669996
INTERPOLATED VALUE OF Y= 7.499998
REMAINDER TERM=-.2145527
INTERPOLATED VALUE OF Y= 7.285445

IF YOU WANT INTERPOLATION AT ADDITIONAL POINTS, TYPE 1
? 1

XP=  ? 0.4
INTERPOLATED VALUE OF Y= 2.548
REMAINDER TERM= 11.564
INTERPOLATED VALUE OF Y= 14.112
REMAINDER TERM=-10.68199
INTERPOLATED VALUE OF Y= 3.430004
REMAINDER TERM= 4.983993
INTERPOLATED VALUE OF Y= 8.413998
REMAINDER TERM=-.8009971
INTERPOLATED VALUE OF Y= 7.613001

IF YOU WANT INTERPOLATION AT ADDITIONAL POINTS, TYPE 1
? 1

XP=  ? 0.5
INTERPOLATED VALUE OF Y= 2.548
REMAINDER TERM= 14.868
INTERPOLATED VALUE OF Y= 17.416
REMAINDER TERM=-18.31199
INTERPOLATED VALUE OF Y=-.8959904
REMAINDER TERM= 12.81598
INTERPOLATED VALUE OF Y= 11.91999
REMAINDER TERM=-4.119414
INTERPOLATED VALUE OF Y= 7.800576

IF YOU WANT INTERPOLATION AT ADDITIONAL POINTS, TYPE 1
? 2
```

Figure 6.3.2 Continued

polynomial. The number of data points and the corresponding data are obtained from the data statements in the program. We calculate the divided differences from the formulas given in Eq. (6.17) and use them to determine the coefficients C(I) of the divided-difference polynomial. A matrix F(I,J) is used to store the divided differences. The first column of this matrix consists of the given values of Y at the five data points, and the first row contains the coefficients of the polynomial. The value of X at which interpolation is desired is denoted by XP and is entered interactively by the user. The interpolated value is obtained by means of Eq. (6.18). The remainder term **R** is also calculated, employing Eq. (6.21) and the computed value of the corresponding higher-order divided difference. The various symbols used are defined in the program, and the important steps in the computation are indicated.

The numerical results obtained are presented in Fig. 6.3.2. The number of data points employed N is printed, along with the computed values of the coefficients C(I), I = 1 to I = N. The value XP of the independent variable at which interpolation is sought is entered interactively. The program computes the interpolated value of the dependent variable Y, employing zeroth-, first-, second-, third-, and fourth-order approximations. These approximations refer to the first term, the first two terms, the first three terms, and so on, in Eq. (6.18). Thus, the last, or fourth-order, approximation is the most accurate one. This is also shown by the presented results since the remainder term decreases as the order of the approximation increases. The remainder term for the last approximation involves an additional point and is not computed here. For the other approximations, we compute the remainder term from Eq. (6.21) by simply using the next-order divided difference, which is known from earlier calculations.

Note again that the interpolating polynomial yields the exact value of the dependent variable at the given data points, as expected. Also, the interpolated values at X = 0.25 and 0.35 are close to those obtained earlier in Example 6.1. The extrapolated values at X = 0.025 and 0.5 agree closely with the experimental data, and the value at X = 0 with that obtained earlier. Thus, Newton's method may be used as an alternative to the procedure outlined in Example 6.1. However, this method does not require uniformly spaced data points, as needed for the method employed in Example 6.1. Also, the method yields the remainder term which reflects the increase in the accuracy of the interpolation as the order of the approximation is increased. The program is more involved than Lagrange interpolation, as given in Example 6.2. However, this method has the important advantages of ease in employing varying orders of approximation and ease of evaluating the accuracy by means of the remainder term. Both Lagrange and Newton's divided-difference interpolation are widely used in engineering applications.

6.5 NUMERICAL INTERPOLATION WITH SPLINES*

In the preceding sections, we considered several methods and forms of interpolating functions for an exact fit to a given data set. In many engineering problems, such as calibration of measuring instrumentation, numerical simulation of systems, and measurement of material properties, the available data points are relatively few, the function $f(x)$ is reasonably well behaved, and the accuracy level is very high, so that these techniques for an exact fit are appropriate. However, there are several cases where an alternative approach, which is based on curve fitting of small subsets of data points with lower-order polynomials, provides a better representation of the data. Such interpolating polynomials that are employed to yield a piecewise exact fit to the data are known as *spline functions*. The basic concept is based on the drafting technique of using a thin, flexible strip, known as a *spline*, to draw a smooth curve through a given distribution of points. Although the interpolating polynomial may be linear, quadratic, cubic, or of some other order, the cubic spline function is the most widely used one and is discussed here. Splines are advantageous to use when the conventional interpolation methods, such as those discussed in the previous sections, yield polynomials of high order and the interpolating curve is of wiggly or oscillating character, as shown in Fig. 6.6. In such cases, spline interpolation often yields a better

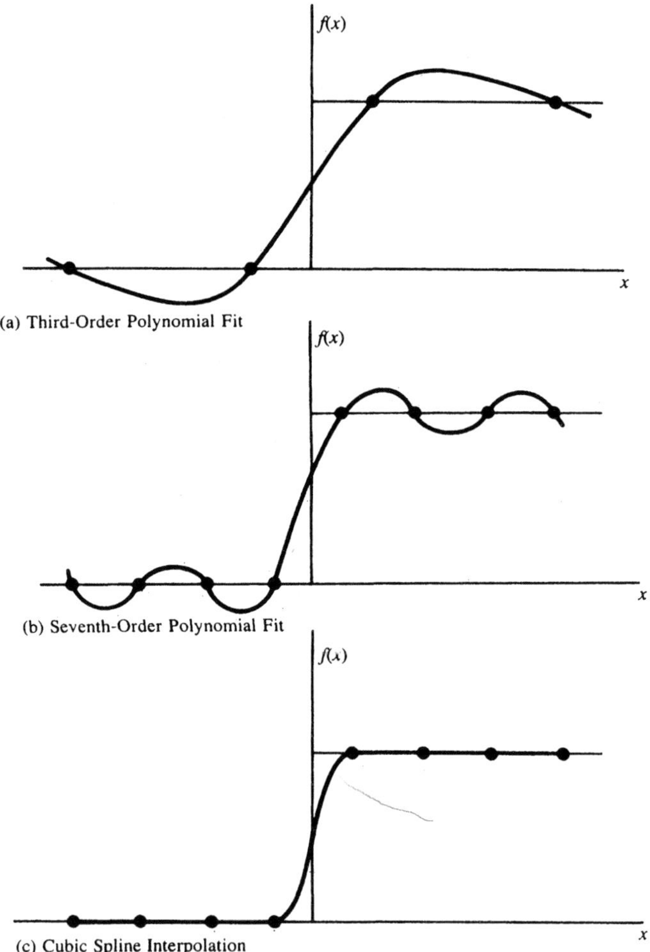

Figure 6.6 Interpolation with single polynomials over the entire range and with piecewise cubic splines for a step change in the dependent variable.

approximation. This approach is particularly valuable in the interpolation of relatively accurate material property data over wide ranges of the independent variable such as temperature and pressure.

Let us consider two arbitrary points x_i and x_{i+1} at which the function $f(x)$ is given. The general form of the cubic that passes through these points and provides the interpolation function between the two may be taken as

$$f_i(x) = a_0 + a_1 x + a_2 x^2 + a_3 x^3 \qquad \text{for } x_i \leqslant x \leqslant x_{i+1} \tag{6.29}$$

There are four unknown constants in this polynomial. Since the curve passes through the two points x_i and x_{i+1}, two conditions that must be used are

$$f_i(x_i) = a_0 + a_1 x_i + a_2 x_i^2 + a_3 x_i^3 \tag{6.30}$$

$$f_i(x_{i+1}) = a_0 + a_1 x_{i+1} + a_2 x_{i+1}^2 + a_3 x_{i+1}^3 \tag{6.31}$$

The remaining two conditions may be chosen arbitrarily to obtain a smooth transition from one cubic distribution to the adjacent ones. An effective choice is the continuity of the first and second derivatives at the two points. Thus, the slope and curvature of $f_i(x)$ match those of $f_{i-1}(x)$ at $x = x_i$ and those of $f_{i+1}(x)$ at $x = x_{i=+1}$. A special treatment will be needed at the end points of the given data, as discussed later.

Since the second derivative of a cubic is a straight line over each interval, as shown in Fig. 6.7, a first-order Lagrange interpolation may be derived from Section 6.3 to represent the second derivative, over the interval $x_i \leqslant x \leqslant x_{i+1}$, as follows:

$$f_i''(x) = f''(x_i)\frac{x_{i+1} - x}{x_{i+1} - x_i} + f''(x_{i+1})\frac{x - x_i}{x_{i+1} - x_i} \tag{6.32}$$

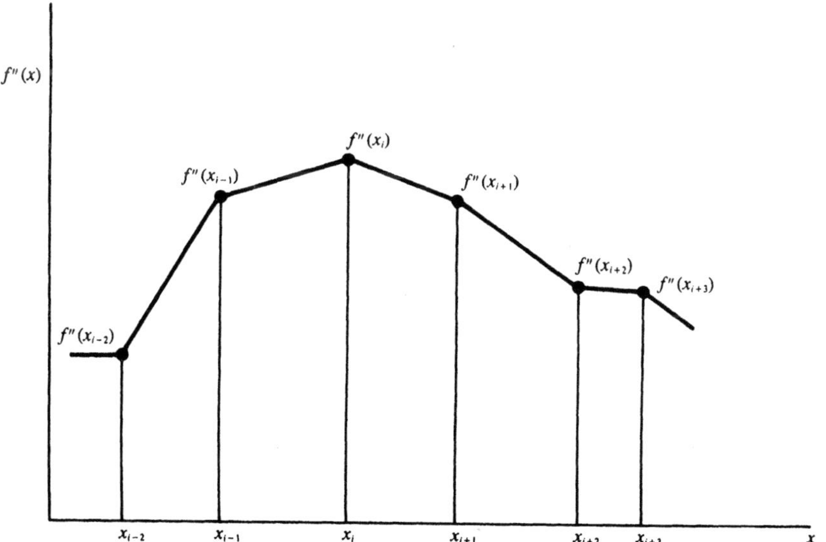

Figure 6.7 Variation of the second derivative $f''(x)$ over the subintervals that constitute the given range of data for spline interpolation.

Integrating this equation twice and applying Eqs. (6.30) and (6.31) to determine the constants that arise, we obtain the cubic $f_i(x)$ over $x_i \leqslant x \leqslant x_{i+1}$:

$$f_i(x) = f''(x_i)\frac{(x_{i+1} - x)^3}{6\,\Delta x_i} + f''(x_{i+1})\frac{(x - x_i)^3}{6\,\Delta x_i} + \left[\frac{f(x_i)}{\Delta x_i} - \frac{\Delta x_i}{6} f''(x_i)\right](x_{i+1} - x)$$
$$+ \left[\frac{f(x_{i+1})}{\Delta x_i} - \frac{\Delta x_i}{6} f''(x_{i+1})\right](x - x_i) \qquad \text{where } \Delta x_i = x_{i+1} - x_i \tag{6.33}$$

Equation (6.33) yields the interpolating cubic distributions over each of the subintervals in the range $x_0 \leqslant x \leqslant x_n$. We determine the second derivatives in Eq. (6.33) by using the matching condition for the first derivative, that is,

$$f_i'(x_i) = f_{i-1}'(x_i) \tag{6.34}$$

Now, $f_i(x)$ may be differentiated and x set equal to x_i to obtain the derivative at the left-hand limit of interval i. Similarly, $f_{i-1}(x)$ is differentiated and x set equal to x_i to yield the derivative at the right-hand limit of interval $(i - 1)$. The two results obtained are equated to give a set of linear simultaneous equations of the form

$$\Delta x_{i-1} f''(x_{i-1}) + 2(x_{i+1} - x_{i-1}) f''(x_i) + x_i f''(x_{i+1})$$
$$= 6\left[\frac{f(x_{i+1}) - f(x_i)}{\Delta x_i} - \frac{f(x_i) - f(x_{i-1})}{\Delta x_{i-1}}\right] \qquad \text{where } \Delta x_{i-1} = x_i - x_{i-1} \tag{6.35}$$

Here, the fact that $f''(x_i)$ is the same when x_i is approached from either side, as shown in Fig. 6.7, has also been used. This condition may be stated as

$$f_i''(x_i) = f_{i-1}''(x_i)$$

For $(n + 1)$ data points, represented by the values x_i of the independent variable, where $i = 0, 1, 2, \ldots, n$, the number of intervals is n. Therefore, n cubics are generated by Eq. (6.33) for spline interpolation. However, there are $(n + 1)$ unknown second derivatives in the equations for the n cubics. Equation (6.35), when written for all the interior points, that is, for $i = 1, 2, 3, \ldots, (n - 1)$, yields $(n - 1)$ equations for the evaluation of the second derivative. Since only $f''(x_{i-1})$, $f''(x_i)$, and $f''(x_{i+1})$ appear as unknowns in Eq. (6.35), the system of equations is tridiagonal and may easily be solved by Gaussian elimination, as illustrated in Example 5.2. However, we still need two additional conditions at the end points of the data set, that is, for $i = 0$ and $i = n$, in order to determine $f''(x_0)$ and $f''(x_n)$. These conditions are usually taken as

$$f''(x_0) = 0 \quad \text{and} \quad f''(x_n) = 0 \tag{6.37}$$

Thus, the analogous elastic strip for drawing a curve through the given points is allowed to assume a natural, unconstrained straight line beyond the given range of data points. This spline, known as a *natural cubic spline*, is the one most frequently employed for an arbitrary data set.

Several other approximations for the end conditions have been employed for different types of data; see Ferziger (1981). For data that are expected to lie on a periodic curve, the end conditions are often taken as $f''(x_0) = f''(x_{n-1})$ and $f''(x_1) = f''(x_n)$, representing the repetitive nature of the curve. Another frequently used set of conditions is $f''(x_0) = f''(x_1)$ and $f''(x_{n-1}) = f''(x_n)$, which makes f'' constant in the intervals at the end. It also makes $f(x)$ quadratic in these intervals. Other choices for the end conditions are also possible. However, the natural spline is the most commonly employed interpolating spline. Equation (6.33) gives the cubic equation for each interval, and Eq. (6.35), along with Eq. (6.37) or one of the other end conditions chosen, gives the tridiagonal system for obtaining the $(n + 1)$ second derivatives.

Spline interpolation has become quite important in recent years for a wide variety of engineering applications. Measurements of material properties, such as density, reflectivity, modulus of elasticity, mass diffusion coefficients, absorption coefficient, and specific heat, and the results from numerical simulations of engineering systems frequently give rise to large sets of very accurate data. Although a best fit may be used in some cases, as discussed in the next section, an exact fit is more appropriate since the results are known to be accurate, and a curve that passes through all the data points is desirable. However, because of the large number of points, a polynomial of fairly high order will generally be needed to fit all the points. Since this leads to ill-conditioning and computational difficulties, as mentioned earlier, a piecewise interpolating polynomial of lower order will be more valuable. Spline interpolation is, therefore, the commonly used approach for such data sets. The natural cubic spline is the most popular choice, although linear, quadratic, and other types of splines can also be used. For further details on splines, see Ahlberg et al. (1967). The following example illustrates the use of spline interpolation in a problem of practical interest.

Example 6.4

Thermocouple junctions of dissimilar metals and alloys are extensively used in engineering applications for temperature measurement. A voltage difference V is generated between two junctions at different temperatures. Calibration tables, which give the voltage V in millivolts (mV) for one junction at 0°C and the other at temperature T in °C, are available in the literature for several types of thermocouple junctions. The values for a Chromel-Alumel thermocouple, which is generally known as type K thermocouple and consists of nickel-chromium and nickel-aluminum alloys, are given as follows:

T (°C)	10	20	30	40	50	60	70	80	90
V (mV)	0.397	0.798	1.203	1.611	2.022	2.436	2.85	3.266	3.681

T (°C)	100	110	120	130	140	150
V (mV)	4.095	4.508	4.919	5.327	5.733	6.137

```
C                             SPLINE INTERPOLATION
C
C         V IS THE INDEPENDENT VARIABLE, T THE DEPENDENT VARIABLE, M THE NUMBER
C         OF DATA POINTS, T2 THE SECOND DERIVATIVE OF THE DEPENDENT VARIABLE, VP
C         THE VALUE OF V AT WHICH THE INTERPOLATED VALUE TP IS DESIRED AND
C         V(I),T(I) REPRESENT THE VALUES AT THE DATA POINTS.
C
C
          DIMENSION V(15),T(15),T2(15)
C
C         ENTER INPUT VARIABLES AND DATA
C
          PRINT *,'ENTER THE NUMBER OF DATA POINTS'
          READ *,M
          OPEN(UNIT=11,FILE='V.DAT')
          OPEN(UNIT=12,FILE='T.DAT')
          READ (11,*) (V(I),I=1,M)
          READ (12,*) (T(I),I=1,M)
          CLOSE(UNIT=11)
          CLOSE(UNIT=12)
C
C         CALL SUBROUTINE TO COMPUTE THE SECOND DERIVATIVE T2
C
          CALL DERIVATIVE(M,V,T,T2)
C
C         SPECIFY VALUE OF V FOR INTERPOLATION
C
    2     PRINT *,'ENTER THE VALUE OF V FOR INTERPOLATION'
          READ *,VP
C
C         CALL SUBROUTINE TO USE SPLINE INTERPOLATION
C
          CALL SPLINE(M,V,T,T2,VP,TP)

C
C         OUTPUT RESULTS
C
          WRITE(1,4)VP,TP
    4     FORMAT(2X,'VOLTAGE V=',F9.5,4X,'TEMPERATURE T=',F9.5//)
          PRINT *,'IF YOU WANT ADDITIONAL INTERPOLATION , TYPE 1'
          READ *,MORE
          IF (MORE .EQ. 1) GO TO 2
          STOP
          END
C
C
C                        SUBROUTINE DERIVATIVE
C
C         THIS SUBROUTINE CALCULATES THE SECOND DERIVATIVE VALUES T2 NEEDED
C         FOR A CUBIC SPLINE INTERPOLATION. A,B AND C ARE THE ELEMENTS IN
C         EACH ROW OF THE TRIDIAGONAL MATRIX AND D REPRESENTS THE CONSTANTS
C         ON THE RIGHT-HAND SIDE OF THE EQUATIONS THAT YIELD THE T2 VALUES.
C
C
          SUBROUTINE DERIVATIVE(M,V,T,T2)
          DIMENSION V(15),T(15),T2(15),A(15),B(15),C(15),D(15)
C
C         COMPUTE THE ELEMENTS OF THE TRIDIAGONAL MATRIX
```

Figure 6.4.1 Computer program in FORTRAN 77 for spline interpolation.

```
C
        C(1)=V(2)-V(1)
        DO 1 I=2,M-1
        A(I)=V(I)-V(I-1)
        B(I)=2.0*(V(I+1)-V(I-1))
        C(I)=V(I+1)-V(I)
    1   D(I)=6.0*((T(I+1)-T(I))/C(I) - (T(I)-T(I-1))/A(I))
C
C      SOLVE THE TRIDIAGONAL SYSTEM FOR THE SECOND DERIVATIVE
C
        DO 2 I=3,M-1
        B(I)=B(I)-A(I)*C(I-1)/B(I-1)
    2   D(I)=D(I)-A(I)*D(I-1)/B(I-1)
        T2(1)=0.0
        T2(M)=0.0
        T2(M-1)=D(M-1)/B(M-1)
        DO 3 I=2,M-2
        IN=M-I
    3   T2(IN)=(D(IN)-C(IN)*T2(IN+1))/B(IN)
        RETURN
        END
C
C
C                          SUBROUTINE SPLINE
C
C      THIS SUBROUTINE OBTAINS THE RELEVANT CUBIC SPLINE AND COMPUTES THE
C      DESIRED INTERPOLATED VALUE OF THE DEPENDENT VARIABLE
C
        SUBROUTINE SPLINE(M,V,T,T2,VP,TP)
        DIMENSION V(15),T(15),T2(15)
C
C      DETERMINE THE INTERVAL IN WHICH VP LIES
C
        DO 1 I=1,M-1
        IF (VP .LE. V(I+1)) THEN
        S1=V(I+1)-V(I)
        S2=VP-V(I)
        S3=V(I+1)-VP
C
C      COMPUTE THE INTERPOLATED VALUE FROM THE CUBIC SPLINE
C
        TP=T2(I)*S3*(S3**2/S1-S1)/6.0 + T2(I+1)*S2*(S2**2/S1-S1)/6.0
     $     +T(I)*S3/S1 + T(I+1)*S2/S1
        GO TO 2
        END IF
    1   CONTINUE
    2   RETURN
        END
```

Figure 6.4.1 Continued

Employing cubic spine interpolation, obtain the temperatures if the voltage output values are 1.0, 3.0, 4.343, 5.855, 6.0, and 6.097 mV. Compare the results obtained at 4.343, 5.855, and 6.097 mV with those given in the literature as 106°C, 143°C, and 149°C, respectively.

Solution

This problem is well suited for spline interpolation, since the tabulated values are very accurate and since the large number of data points makes an exact fit with a single polynomial difficult to apply and also inaccurate, as outlined in the text. The voltage V is measured in engineering processes, and the temperature T is obtained by interpolation from the calibration data. Thus, V is taken as the independent variable and T as the independent variable.

Figure 6.4.1 presents the computer program in FORTRAN for spline inter-

```
 ENTER THE NUMBER OF DATA POINTS
15
 ENTER THE VALUE OF V FOR INTERPOLATION
1.0
  VOLTAGE V=  1.00000    TEMPERATURE T= 24.99982

 IF YOU WANT ADDITIONAL INTERPOLATION , TYPE 1
1
 ENTER THE VALUE OF V FOR INTERPOLATION
3.0
  VOLTAGE V=  3.00000    TEMPERATURE T= 73.60907

 IF YOU WANT ADDITIONAL INTERPOLATION , TYPE 1
1
 ENTER THE VALUE OF V FOR INTERPOLATION
4.343
  VOLTAGE V=  4.34300    TEMPERATURE T=106.00078

 IF YOU WANT ADDITIONAL INTERPOLATION , TYPE 1
1
 ENTER THE VALUE OF V FOR INTERPOLATION
5.855
  VOLTAGE V=  5.85500    TEMPERATURE T=143.01590

 IF YOU WANT ADDITIONAL INTERPOLATION , TYPE 1
1
 ENTER THE VALUE OF V FOR INTERPOLATION
6.0
  VOLTAGE V=  6.00000    TEMPERATURE T=146.60565

 IF YOU WANT ADDITIONAL INTERPOLATION , TYPE 1
1
 ENTER THE VALUE OF V FOR INTERPOLATION
6.097
  VOLTAGE V=  6.09700    TEMPERATURE T=149.00882

 IF YOU WANT ADDITIONAL INTERPOLATION , TYPE 1
2
```

Figure 6.4.2 Computed results from spline interpolation for the problem considered in Example 6.4.

polation. The number of data points is entered interactively by the user, and the program reads the relevant data from files V.DAT and T.DAT, for voltage and temperature, respectively, stored in the computer. Two subroutines, DERIVATIVE and SPLINE, are employed in the program. The former generates the tridiagonal matrix for the second derivative $f''(x_i)$, denoted by T2 in the program. The elements of the matrix are obtained from Eqs. (6.35) and (6.37). The Thomas algorithm, derived in Example 5.2, is employed to obtain the values of the second derivative needed for the cubic spline given by Eq. (6.33). The voltage at which interpolation is desired is denoted by VP and is entered into the main program by the user. The subroutine SPLINE determines the interval in which VP lies, derives the relevant cubic spline using Eq. (6.33), and computes the interpolated value TP of the temperature. The main program prints the results and inquires whether interpolation at another value of V is needed. Thus, interpolated results may be obtained at the desired values of the output voltage V.

The numerical results obtained are presented in Fig. 6.4.2. The interpolated temperatures for V = 1.0, 3.0, and 6.0 mV are found to be physically realistic and to lie in the appropriate subintervals of the given data. Also, the values for V = 4.343, 5.855, and 6.097 are very close to those given in the literature as 106, 143, and 149°C, respectively, lending strong support to the accuracy of the interpolated results obtained. In fact, several additional values of V were considered, and the results were found to be very accurate. Thus, cubic spline interpolation may be used satisfactorily for this problem and other similar ones, such as material property data. An exact fit with a single polynomial is generally not appropriate for these cases since a large number of data points are given. Also, a best fit is not suitable since the data are generally very accurate. For these reasons, spline interpolation is extensively used in engineering systems and processes to generate interpolated values for material properties, characteristics of components, calibration, and for several other applications.

6.6 METHOD OF LEAST SQUARES FOR A BEST FIT

In the preceding sections, we discussed interpolation with approximating functions that pass through each given data point. Such an exact fit is appropriate if the given data are of a high level of accuracy. If the number of points is relatively small, a single polynomial approximation may be employed for interpolation. If a large number of points are given, spline interpolation, which yields lower-order polynomials such as cubics to fit small subsets of the data, can be used to piecewise approximate the data for obtaining values of the dependent variable at intermediate points. However, the data obtained in many engineering applications have a significant amount of associated error. Experimental data, for instance, would generally have some noise, or error, whose magnitude would depend on the instrumentation and the arrangement employed for the measurements. In such cases, a polynomial interpolation that demands that the approximating curve pass through each data point is not satisfactory.

A better approach is to derive a function that provides a best fit to the given data by somehow minimizing the difference between the given values of the dependent variable and those obtained from the approximating curve. Figure 6.8 shows a few circumstances where a best fit will be much more satisfactory than an exact fit. Because of the error associated with the data, it is not necessary for the approximating

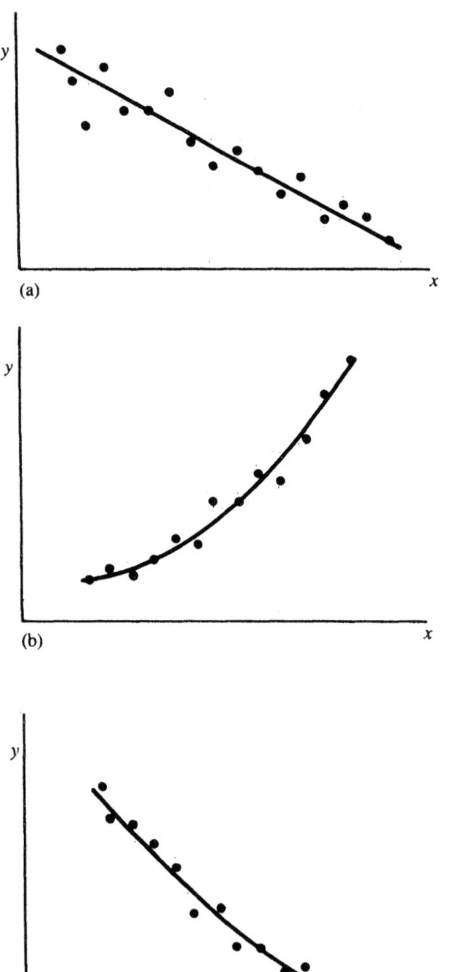

Figure 6.8 Data distributions for which a best fit is more appropriate than an exact fit.

curve to match each data point. A curve that adequately represents the general trend of the data, without necessarily passing through each point, will be useful in characterizing the data and deriving correlating equations for quantitatively describing the physical or chemical process under consideration. Such correlations are extremely important in engineering applications and are often the desired output from an experimental study. The measurements on the deflection of a building structure due to the flow of water, for instance, can be used to yield a best fit that then can be employed in the design of such structures for locating them in streams and in the sea. Similarly, measurements of the velocities of accelerating automobiles can be used to derive correlating equations that characterize the dependence of the acceleration on various parameters, such as shape, weight, and fuel mixture. Heat and mass transfer from surfaces are often measured for different geometries and flow conditions. The results obtained are then curve fitted to yield correlations that can be used for future analysis and design of similar processes and systems.

6.6.1 Basic Considerations

Several criteria can be used to derive the curve that best fits the data. If the approximating function is denoted by $f(x)$ and the given data points by (x_i, y_i), where y is the dependent variable, x is the independent variable, and $i = 1, 2, \ldots, n$, the error e_i at $x = x_i$ is given by

$$e_i = y_i - f(x_i) \tag{6.38}$$

One method for obtaining a best fit to the data is to minimize the sum of these individual errors, that is, minimize $\sum_{i=1}^{n} e_i$. However, this approach is not satisfactory since this criterion allows the errors to cancel out and thus does not yield a unique curve. Moreover, the curve may not represent the general trend of the data at all. If the sum of the absolute values of the errors, $\sum_{i=1}^{n} |e_i|$, is minimized, the result is better; but, again, a unique best fit is generally not obtained. Another approach that may be used is the minimax criterion, which minimizes the maximum error, $(e_i)_{max}$, for the data points. However, this method is heavily influenced by a single point that may have large error. Although unsuitable for obtaining a best fit in most engineering problems, this approach is often appropriate for fitting a simple function to a much more complicated one, as outlined by Carnahan et al. (1969).

The most commonly used approach for a best fit is the method of least squares. In this method, the sum S of the squares of the errors is minimized. The expression for S is

$$S = \sum_{i=1}^{n} e_i^2 = \sum_{i=1}^{n} [y_i - f(x_i)]^2 \tag{6.39}$$

This approach generally yields a unique curve that provides a good representation of the given data, if the approximating function is properly chosen. As outlined in Section 1.4, one must employ the basic nature of the problem under consideration in choosing the form of the approximating function. Thus, for the measurements of the average daily temperature over the year, a sinusoidal function will yield a good best fit to the data; see Fig. 1.5. Similarly, in most experimental studies of engineering systems and processes, the expected trends are known from the physical or chemical nature of the problem, allowing one to choose an appropriate function for curve fitting.

Let us consider, as an example, the measurement of a physical variable, which may be, say, the length, weight, or density of a given material. If n measurements are taken, the results will generally differ because of the experimental error involved. Let us denote these measurements as $l_1, l_2, \ldots, l_n$. If L denotes the desired best fit to these measurements, then

$$S = (l_1 - L)^2 + (l_2 - L)^2 + \cdots + (l_n - L)^2 \tag{6.40}$$

To minimize this sum S of the squares of the deviations, we differentiate S twice to obtain

$$\frac{dS}{dL} = -2(l_1 - L) - 2(l_2 - L) - \cdots - 2(l_n - L) = -2\left[\sum_{i=1}^{n} l_i - nL\right] \tag{6.41}$$

$$\frac{d^2S}{dL^2} = 2n \tag{6.42}$$

Since n is positive, the value of L for which $dS/dL = 0$ gives a minimum value of the sum S. From Eq. (6.41), this value is obtained as

$$L = \frac{\sum_{i=1}^{n} l_i}{n} \tag{6.43}$$

Therefore, if the sum S is minimized, the value of the quantity L is simply the arithmetic mean of the measurements. One will expect this value to be the best representation of the data if the measurements are all taken with equal care and are thus of comparable accuracy. This example provides a physical basis for the method of least squares and may easily be extended to a function $f(x)$, using the consideration of a single unknown variable L given above.

6.6.2 Linear Regression

The procedure of obtaining a best fit to a given data set is often known as *regression*. Let us first consider fitting a straight line to a set of data points denoted by (x_1, y_1), $(x_2, y_2), \ldots, (x_n, y_n)$, where x is the independent variable and y the dependent variable. Although engineering applications usually lead to nonlinear functions, there are several circumstances where a linear variation closely approximates the measurements. Moreover, exponential and power-law forms, which are very frequently encountered in practical problems, can often be reduced to linear variations, as illustrated later in this section. Consequently, linear regression is very important in a wide variety of engineering applications, particularly in the derivation of correlating equations from experimental data.

The equation of the straight line for curve fitting may be taken as

$$f(x) = a + bx \tag{6.44}$$

where a and b are the coefficients that must be determined from the given set of n data points. Thus, a and b are to be chosen such that the sum S of the squares of the deviations of the data points from the values obtained from the equation of the straight line, Eq. (6.44), is a minimum. This implies that

$$S = \sum_{i=1}^{n} (y_i - a - bx_i)^2 \rightarrow \text{minimum} \tag{6.45}$$

The minimum occurs when the partial derivatives of S with respect to a and b are both zero. Thus,

$$\frac{\partial S}{\partial a} = \sum_{i=1}^{n} -2(y_i - a - bx_i) = 0 \tag{6.46}$$

$$\frac{\partial S}{\partial b} = \sum_{i=1}^{n} -2(y_i - a - bx_i)x_i = 0 \tag{6.47}$$

These equations may be simplified and expressed as

$$\sum y_i - \sum a - \sum bx_i = 0$$

$$\sum y_i x_i - \sum ax_i - \sum bx_i^2 = 0$$

which may be written for the unknowns a and b as

$$na + b\sum x_i = \sum y_i \tag{6.48}$$

$$a\sum x_i + b\sum x_i^2 = \sum x_i y_i \tag{6.49}$$

where the summations are all from $i = 1$ to $i = n$.

Equations (6.48) and (6.49) are linear in the unknowns and may be solved simultaneously to yield the desired values of a and b. Using Cramer's rule, we obtain a and b in terms of the relevant determinants as follows:

$$a = \frac{\begin{vmatrix} \sum y_i & \sum x_i \\ \sum x_i y_i & \sum x_i^2 \end{vmatrix}}{\begin{vmatrix} n & \sum x_i \\ \sum x_i & \sum x_i^2 \end{vmatrix}} \tag{6.50}$$

$$b = \frac{\begin{vmatrix} n & \sum y_i \\ \sum x_i & \sum x_i y_i \end{vmatrix}}{\begin{vmatrix} n & \sum x_i \\ \sum x_i & \sum x_i^2 \end{vmatrix}} \tag{6.51}$$

where the vertical bars indicate magnitude of the determinant. We may employ the given set of n data points to compute $\sum x_i$, $\sum y_i$, $\sum x_i^2$, and $\sum x_i y_i$. Then we use these values to calculate the determinants in Eqs. (6.50) and (6.51). These then yield the coefficients a and b for the straight line, Eq. (6.44), that provides a best fit to the given data.

To quantify the accuracy with which the computed straight line fits the given data, we compute the sum of the squares of the deviations of the data from the mean to represent the spread before regression is applied. Denoting this sum by S_m and the mean by $\bar{y}$, we have

$$S_m = \sum_{i=1}^{n} (y_i - \bar{y})^2$$

The spread in the data that remains after regression is indicated by S, where

$$S = \sum_{i=1}^{n} (y_i - a - bx_i)^2$$

Therefore, the extent of improvement due to curve fitting by a straight line is indicated by

$$r^2 = \frac{S_m - S}{S_m} \tag{6.52}$$

where r is known as the *correlation coefficient*. A good correlation for linear regression is indicated by a high value of r, the maximum of which is 1.0. However, the given data should also be plotted along with the computed curve, in order to determine, qualitatively, how good a representation of the data is provided by the fit. Equation (6.52) can also be used for higher-order polynomials and nonpolynomial forms of the function for a best fit, as outlined later in this section. See Draper and Smith (1981) for further details on the application of regression analysis.

6.6.3 Best Fit with a Polynomial

Linear regression yields a straight line that provides a best fit to a given data set. It is simple to apply, since only two unknown coefficients, a and b in Eq. (6.44), are to be determined. In many cases, particularly if the range of the independent variable is relatively small, a straight line provides a fairly good representation of the data. Also, as outlined in the next section, certain nonlinear forms, such as exponentials, may be transformed to yield linear variations. However, the data may have a definite trend that is poorly represented by a straight line. An example of such a situation is shown in Fig. 6.9, which illustrates that a straight line is not a satisfactory choice for curve fitting in this case. A polynomial, such as a parabola or a cubic, will be more appropriate.

In order to obtain a best fit to the given data, let us consider an mth-order polynomial, given as

$$f(x) = c_0 + c_1x + c_2x^2 + \cdots + c_mx^m \tag{6.53}$$

Then the sum S of the squares of the deviations of the data from the curve is given by

$$S = \sum_{i=1}^{n} (y_i - c_0 - c_1x_i - c_2x_i^2 - \cdots - c_mx_i^m)^2 \tag{6.54}$$

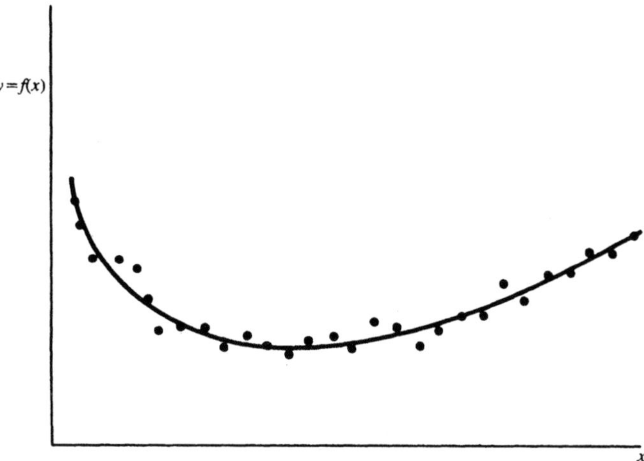

Figure 6.9 A polynomial best fit to given data.

We determine the coefficients $c_0, c_1, \ldots, c_m$ by extending the procedure outlined in the preceding section for linear regression. Therefore, S is differentiated with respect to each of the coefficients, and the partial derivatives are set equal to zero in order to minimize S. This gives

$$\begin{aligned}\frac{\partial S}{\partial c_0} &= -2\sum(y_i - c_0 - c_1x_i - c_2x_i^2 - \cdots - c_mx_i^m) = 0\\ \frac{\partial S}{\partial c_1} &= -2\sum x_i(y_i - c_0 - c_1x_i - c_2x_i^2 - \cdots - c_mx_i^m) = 0\\ &\vdots\\ \frac{\partial S}{\partial c_m} &= -2\sum x_i^m(y_i - c_0 - c_1x_i - c_2x_i^2 - \cdots - c_mx_i^m) = 0\end{aligned} \qquad (6.55)$$

Equations (6.55) may be simplified and rearranged to yield the following system of $(m + 1)$ linear equations for the $(m + 1)$ unknowns $c_0, c_1, \ldots, c_m$:

$$\begin{aligned}nc_0 + c_1\sum x_i + c_2\sum x_i^2 + \cdots + c_m\sum x_i^m &= \sum y_i\\ c_0\sum x_i + c_1\sum x_i^2 + c_2\sum x_i^3 + \cdots + c_m\sum x_i^{m+1} &= \sum x_iy_i\\ &\vdots\\ c_0\sum x_i^m + c_1\sum x_i^{m+1} + c_2\sum x_i^{m+2} + \cdots + c_m\sum x_i^{2m} &= \sum x_i^my_i\end{aligned} \qquad (6.56)$$

where all the summations are from $i = 1$ to $i = n$. It can easily be verified that the equations for linear regression, Eqs. (6.48) and (6.49), are obtained for a first-order polynomial, $m = 1$. The methods given in Chapter 5 may be employed to solve the above system of equations, which are linear in the unknown coefficients $c_0, c_1, \ldots, c_m$.

Curve fitting with polynomials is generally restricted to small values of the order m of the polynomial, in order to avoid extensive calculations for the determination of the coefficients and to obtain simple correlating curves that approximate the general trends of the data. Typical values of m range from 1 to 4, the appropriate value being chosen on the basis of the accuracy and spread of the data, as well as the number of data points. For a relatively large spread of the data, a lower-order polynomial fit will generally be more appropriate. Computation is involved in evaluating the summations in Eq. (6.56) and then solving this system of equations, which may be recast in matrix notation. For a second-order polynomial, for instance, we have

$$\begin{pmatrix} n & \sum x_i & \sum x_i^2 \\ \sum x_i & \sum x_i^2 & \sum x_i^3 \\ \sum x_i^2 & \sum x_i^3 & \sum x_i^4 \end{pmatrix} \begin{pmatrix} c_0 \\ c_1 \\ c_2 \end{pmatrix} = \begin{pmatrix} \sum y_i \\ \sum x_i y_i \\ \sum x_i^2 y_i \end{pmatrix} \tag{6.57}$$

Various elimination and matrix inversion or decomposition methods, given in Chapter 5, may be employed for solving Eq. (6.56) or (6.57) for the coefficients. Gaussian elimination is the most popular choice because of the small number of equations to be solved in most cases. The correlation coefficient r may again be determined from Eq. (6.52) to evaluate how good a fit is given by the resulting polynomial.

6.6.4 Nonpolynomial Forms

The method of least squares is not restricted to polynomials for curve fitting and may easily be applied to various other forms that contain constant coefficients. An example of a physical situation where such a form is more appropriate than a polynomial is the periodic variation in ambient temperature considered in Chapter 1; see Fig. 1.5. Equations (1.10) through (1.12) give some of the sinusoidal functions that may be employed for curve fitting. Considering the function given in Eq. (1.11), for example, we obtain

$$f(x) = A \sin \omega x + B \cos \omega x \tag{6.58}$$

and

$$S = \sum_{i=1}^{n} (y_i - A \sin \omega x_i - B \cos \omega x_i)^2 \tag{6.59}$$

where the sum S is to be minimized for a best fit. Thus,

$$\frac{\partial S}{\partial A} = \sum_{i=1}^{n} -2(y_i - A \sin \omega x_i - B \cos \omega x_i)\sin \omega x_i = 0$$

$$\frac{\partial S}{\partial B} = \sum_{i=1}^{n} -2(y_i - A \sin \omega x_i - B \cos \omega x_i)\cos \omega x_i = 0$$

This gives the equations

$$A\sum(\sin \omega x_i)^2 + B\sum(\sin \omega x_i \cos \omega x_i) = \sum y_i \sin \omega x_i \tag{6.60}$$

$$A\sum(\sin \omega x_i \cos \omega x_i) + B\sum(\cos \omega x_i)^2 = \sum y_i \cos \omega x_i \tag{6.61}$$

which can be easily solved for A and B.

Nonpolynomial forms are important in a wide variety of engineering problems. If the function chosen for curve fitting has constant coefficients, such as A and B in Eq. (6.58), the method of least squares can be easily applied. However, if the constants do not appear as coefficients, for example, the constant a in Eq. (1.10), a straightforward application of the method is not possible. Therefore, the nonpolynomial forms employed for the curve fitting of various types of engineering data are chosen such that the constants to be determined appear only as coefficients.

Besides periodic processes, an example of which is considered above, several engineering applications involve power-law and exponential variations, some of which can be linearized as outlined below. The example given by Eqs. (1.1) and (1.2), for instance, concerns an exponential variation. Similarly, processes that approach a constant magnitude at large values of the independent variable x can often be represented by polynomials with negative exponents, for example,

$$y = c_0 + c_1 x^{-1} + c_2 x^{-2} \tag{6.62}$$

Processes where such an equation may be applicable are the charging of a capacitor in an electrical circuit, the free fall of an object under gravity to attain a terminal velocity, and the dissolution of salt in a given amount of liquid until saturation occurs. However, unless the physical or chemical nature of the given data indicates the suitability of a particular nonpolynomial form, curve fitting is first explored using a polynomial, with varying orders of the polynomial, to obtain a satisfactory representation of the data; see Example 6.5.

Linearization. In several cases, a nonlinear form chosen to curve fit the given data may be linearized by suitable transformations so that linear regression may be applied. Consider, for example, the exponential variation that is commonly encountered in engineering problems, as shown in Fig. 1.2. The general form of an exponential variation may be taken as

$$f(x) = c_1 e^{c_2 x} \tag{6.63}$$

where c_1 and c_2 are constants to be determined for a best fit. In engineering applications, c_1 is generally positive, and c_2 may be positive, as in the convective heating of a metal block, or negative, as in radioactive decay and discharge of a capacitor. If the natural logarithm of Eq. (6.63) is taken, we obtain

$$\log[f(x)] = \log c_1 + c_2 x \tag{6.64}$$

Thus, $\log[f(x)]$ is a linear function of x, and linear regression may be applied using x as the independent variable and the natural logarithm of y, where $y = f(x)$, as the dependent variable. Then y_i in Eqs. (6.50) and (6.51) is replaced by $\log y_i$. Also, $a = \log c_1$ and $b = c_2$, where a and b are the coefficients for linear regression, Eq. (6.44). This approach is frequently employed for obtaining correlating equations for measured heat and mass transfer rates from bodies and surfaces under different physical and chemical conditions.

Similarly, the power-law variation given by the general form

$$f(x) = c_1 x^{c_2} \tag{6.65}$$

is frequently employed for the representation of certain engineering processes. Again, a natural logarithm of the equation is taken to yield

$$\log[f(x)] = \log c_1 + c_2 \log x \tag{6.66}$$

where the logarithm to base 10 may also be taken for convenience, instead of the natural logarithm. Again, linear regression may be applied, with $\log x$ and $\log[f(x)]$ as the independent and dependent variables, respectively, to obtain the coefficients c_1 and c_2.

Similarly, various other forms, such as

$$f(x) = \frac{c_1}{c_2 + x} \tag{6.67}$$

$$f(x) = c_1 + c_2 x^{-1} \tag{6.68}$$

$$f(x) = \frac{c_1 x}{c_2 + x} \tag{6.69}$$

may be linearized by taking the reciprocal of $f(x)$, of x, or of both as the independent and dependent variables. Thus, these equations may be rewritten as

$$Y = \left(\frac{c_2}{c_1}\right) + \left(\frac{1}{c_1}\right) x \tag{6.70}$$

$$y = c_1 + c_2 X \tag{6.71}$$

$$Y = \left(\frac{1}{c_1}\right) + \left(\frac{c_2}{c_1}\right) X \tag{6.72}$$

where $y = f(x)$, $Y = 1/f(x)$, and $X = 1/x$. Therefore, linear regression may be applied to these transformed equations to obtain the coefficients c_1 and c_2. These examples also indicate the importance of linear regression in the curve fitting of engineering data. Many processes of practical interest are governed by exponential, power-law, and other forms given above, and linear regression is employed to determine the unknown coefficients. Example 6.5 illustrates the use of the method of least squares for obtaining a best fit to a given data set.

Example 6.5

(a) In a chemical reaction, the effect of the concentration C of a catalyst on the rate R of the reaction is investigated experimentally. The measurements of C in g/m^3

and of R in g/s yield the following:

C (g/m^3)	0.1	0.2	0.5	1.0	1.2	1.8	2.0	2.6	3.5	4.0
R (g/s)	1.85	1.91	2.07	2.32	2.40	2.54	2.56	2.53	2.03	1.24

Using the method of least squares and considering polynomials up to the fifth order, obtain a best fit to these data.

(b) A small, heated metal block cools in air, and its temperature T is measured as a function of time τ to give the following data:

τ (s)	1	2	5	10	15	20	25	30
T (°C)	109.58	99.25	73.78	45.15	26.78	17.24	9.85	6.97

From physical considerations of the problem, the temperature is expected to decay exponentially, as $Ae^{-a\tau}$, where A and a are constants. Employing the program developed in Part (a), obtain a best fit to the given data and determine the constants A and a.

Solution

(a) From the data presented, we know that the reaction rate R increases with concentration C of the catalyst up to a point and then decreases. Thus, we expect that linear regression would not be satisfactory, and therefore we attempt curve fitting with polynomials. However, we can also obtain the results for linear regression from the numerical scheme by choosing the order of the polynomial as 1. In fact, we employ linear regression in Part (b), as outlined below.

Denoting the independent variable by x and the dependent variable by y, for generality, we obtain a system of linear equations, as given by Eq. (6.56). Then we solve this system to obtain the coefficients c_i of the polynomial. Figure 6.5.1 presents the computer program in FORTRAN 77 for the least-squares method for polynomial regression. The data points are represented by X(I) and Y(I), and the coefficients of the polynomial by C(I). The order of the polynomial is denoted by MP, which gives the number of coefficients to be determined as N = MP + 1. The various other symbols employed are defined in the program.

In the given program, the input data and the chosen order of the polynomial for curve fitting are read from an appended data set. The system of linear equations, given by Eq. (6.56), is then generated. The corresponding augmented matrix, with the constant vector on the right-hand side of Eq. (6.56) being stored as the $(N+1)$th column, is obtained. Gaussian elimination is employed for the solution of this system of equations. A subroutine called GAUSS is employed. This subroutine applies the Gaussian elimination algorithm to the system of equations. Thus, the coefficient matrix is reduced to an upper triangular matrix, and back-substitution is employed to

```
C     LEAST SQUARES METHOD FOR POLYNOMIAL REGRESSION
C
C
C     X IS THE INDEPENDENT VARIABLE, Y THE DEPENDENT VARIABLE AND
C     X(I), Y(I) REPRESENT THE DATA POINTS. MP IS THE ORDER OF THE
C     POLYNOMIAL, ND THE NUMBER OF DATA POINTS, N THE NUMBER OF
C     LINEAR EQUATIONS TO BE SOLVED AND M THE NUMBER OF ROWS IN THE
C     AUGMENTED MATRIX A(I,J). C(I) REPRESENTS THE COEFFICIENTS OF
C     THE POLYNOMIAL.
C
C
         DIMENSION A(10,11),C(10),X(25),Y(25)
C
C     ENTER THE INPUT DATA
C
         READ(5,*)MP
         READ(5,*)ND
         READ(5,*)(X(I),I=1,ND)
         READ(5,*)(Y(I),I=1,ND)
         N=MP+1
         M=N+1
C
C     INITIALIZE THE COEFFICIENT MATRIX
C
         DO 1 I=1,N
         DO 1 J=1,M
  1      A(I,J)=0.0
C
C     COMPUTE ELEMENTS OF THE AUGMENTED MATRIX
C
         DO 5 I=1,N
         DO 3 J=1,N
         L=I+J-2
         DO 2 K=1,ND
  2      A(I,J)=A(I,J)+X(K)**L
  3      CONTINUE
         DO 4 K=1,ND
  4      A(I,M)=A(I,M)+Y(K)*X(K)**(I-1)
  5      CONTINUE
C
C     CALL SUBROUTINE TO SOLVE THE SYSTEM OF EQUATIONS
C
         CALL GAUSS(N,A,C)
         WRITE(6,12)MP
  12     FORMAT(2X,'THE ORDER OF THE POLYNOMIAL=',I2/)
         WRITE(6,9)
  9      FORMAT(2X,'THE CONSTANTS OF THE POLYNOMIAL ARE:'/)
         DO 10 I=1,N
  10     WRITE(6,11)I,C(I)
  11     FORMAT(2X,'C(',I1,')=',F12.5)
C
C     CALCULATE THE VALUES OBTAINED FROM THE POLYNOMIAL IN ORDER
C     TO CHECK THE ACCURACY OF THE RESULTING BEST FIT
C
         WRITE(6,13)
  13     FORMAT(/2X,'THE VALUES CALCULATED FROM THE BEST FIT ARE:'/)
         DO 7 I=1,ND
         Y(I)=0.0
```

Figure 6.5.1 Computer program for polynomial regression, using the method of least squares.

```
         DO 6 J=1,N
  6      Y(I)=Y(I)+C(J)*X(I)**(J-1)
  7      WRITE(6,8)I,X(I),I,Y(I)
  8      FORMAT(2X,'X(',I2,')=',F10.4,5X,'Y(',I2,')=',F10.4)
         STOP
         END
C
C
C     SUBROUTINE GAUSS
C
C     THIS SUBROUTINE SOLVES A SYSTEM OF LINEAR EQUATIONS BY
C     REDUCING THE AUGMENTED MATRIX TO UPPER TRIANGULAR FORM
C     AND THEN EMPLOYING BACK-SUBSTITUTION
C
         SUBROUTINE GAUSS(N,A,C)
         DIMENSION A(10,11),C(10)
         N1=N-1
         M=N+1
C
C     FIND THE ROW WITH THE LARGEST PIVOT ELEMENT
C
         DO 2 K=1,N1
         K1=K+1
         K2=K
         B0=ABS(A(K,K))
         DO 3 I=K1,N
         B1=ABS(A(I,K))
         IF((B0-B1) .LT. 0.0) THEN
         B0=B1
         K2=I
         END IF
  3      CONTINUE
         IF((K2-K) .NE. 0) THEN
C
C     INTERCHANGE ROWS TO OBTAIN THE LARGEST PIVOT ELEMENT
C
         DO 5 J=K,M
         D=A(K2,J)
         A(K2,J)=A(K,J)
  5      A(K,J)=D
         END IF
         DO 2 I=K1,N
C
C     APPLY THE GAUSSIAN ELIMINATION ALGORITHM
C
         DO 6 J=K1,M
  6      A(I,J)=A(I,J)-A(I,K)*A(K,J)/A(K,K)
  2      A(I,K)=0.0
C
C     APPLY BACK SUBSTITUTION
C
         C(N)=A(N,M)/A(N,N)
         DO 7 I1=1,N1
         I=N-I1
         S=0.0
         J1=I+1
         DO 8 J=J1,N
  8      S=S+A(I,J)*C(J)
  7      C(I)=(A(I,M)-S)/A(I,I)
         RETURN
         END
```

Figure 6.5.1 Continued

determine the coefficients C(I). The subroutine GAUSS is the same as that developed earlier for Example 5.1 and employs partial pivoting for accuracy and for avoiding a zero pivot element. The computed coefficients are printed and thus a polynomial of form given by Eq. (6.53) is obtained for a best fit. The values of the dependent variable Y(I) at the given data points X(I) are calculated using this polynomial. These values are then compared with the given data to estimate the accuracy of the best fit obtained.

Figure 6.5.2 shows the computed results for polynomials of order 1, 3, and 5. A

```
THE ORDER OF THE POLYNOMIAL= 1

THE CONSTANTS OF THE POLYNOMIAL ARE:

C(1)=      2.25954
C(2)=     -0.06778

THE VALUES CALCULATED FROM THE BEST FIT ARE:

X( 1)=     0.1000      Y( 1)=     2.2528
X( 2)=     0.2000      Y( 2)=     2.2460
X( 3)=     0.5000      Y( 3)=     2.2257
X( 4)=     1.0000      Y( 4)=     2.1918
X( 5)=     1.2000      Y( 5)=     2.1782
X( 6)=     1.8000      Y( 6)=     2.1375
X( 7)=     2.0000      Y( 7)=     2.1240
X( 8)=     2.6000      Y( 8)=     2.0833
X( 9)=     3.5000      Y( 9)=     2.0223
X(10)=     4.0000      Y(10)=     1.9884


THE ORDER OF THE POLYNOMIAL= 3

THE CONSTANTS OF THE POLYNOMIAL ARE:

C(1)=      1.82437
C(2)=      0.43013
C(3)=      0.09669
C(4)=     -0.05961

THE VALUES CALCULATED FROM THE BEST FIT ARE:

X( 1)=     0.1000      Y( 1)=     1.8683
X( 2)=     0.2000      Y( 2)=     1.9138
X( 3)=     0.5000      Y( 3)=     2.0562
X( 4)=     1.0000      Y( 4)=     2.2916
X( 5)=     1.2000      Y( 5)=     2.3768
X( 6)=     1.8000      Y( 6)=     2.5643
X( 7)=     2.0000      Y( 7)=     2.5945
X( 8)=     2.6000      Y( 8)=     2.5487
X( 9)=     3.5000      Y( 9)=     1.9587
X(10)=     4.0000      Y(10)=     1.2772
```

Figure 6.5.2 Calculated results for Example 6.5(a), using polynomials of order 1, 3, and 5 for a best fit.

```
THE ORDER OF THE POLYNOMIAL= 5

THE CONSTANTS OF THE POLYNOMIAL ARE:

C(1)=       1.79415
C(2)=       0.54588
C(3)=       0.11940
C(4)=      -0.19583
C(5)=       0.06476
C(6)=      -0.00849

THE VALUES CALCULATED FROM THE BEST FIT ARE:

X( 1)=     0.1000      Y( 1)=     1.8497
X( 2)=     0.2000      Y( 2)=     1.9066
X( 3)=     0.5000      Y( 3)=     2.0762
X( 4)=     1.0000      Y( 4)=     2.3199
X( 5)=     1.2000      Y( 5)=     2.3959
X( 6)=     1.8000      Y( 6)=     2.5409
X( 7)=     2.0000      Y( 7)=     2.5613
X( 8)=     2.6000      Y( 8)=     2.5293
X( 9)=     3.5000      Y( 9)=     2.0302
X(10)=     4.0000      Y(10)=     1.2399
```

Figure 6.5.2 Continued

comparison of the results with the given data shows that linear regression is in considerable error, as expected. The third-order polynomial fit is fairly accurate, although the fifth-order regression is more accurate. However, because of the smaller computational effort required and the ease in application to engineering problems, a third-order polynomial best fit is very frequently employed in practice, rather than higher-order polynomials. These trends are illustrated more clearly in Fig. 6.5.3, where the given data points are plotted, along with some of the polynomials derived from the method of least squares. Again, note that a third-order polynomial yields a fairly accurate representation of the data.

(b) Consider regression with an exponential of the form

$$y = Ae^{-ax} \tag{6.5.1}$$

Taking natural logarithms of both sides, we obtain

$$\begin{aligned} \log y &= \log A - ax \\ &= B - ax \end{aligned} \tag{6.5.2}$$

where $B = \log A$ is a constant. Thus, we may apply linear regression to the given data, employing $\log y$ as the dependent variable and x as the independent variable. The results obtained are shown in Fig. 6.5.4. The constants for the linear best fit are C(1) and C(2), which correspond to B and $-a$ in Eq. (6.5.2). Therefore, $\log A = B = C(1)$, which gives $A = \exp[C(1)]$, and $a = -C(2)$. The resulting constants A and a are obtained from the program as 119.1 and 0.09695, respectively. The term exponent in this figure refers to $-a$ in Eq. (6.5.1).

The values of the dependent variable Y(I) are also calculated from the best fit

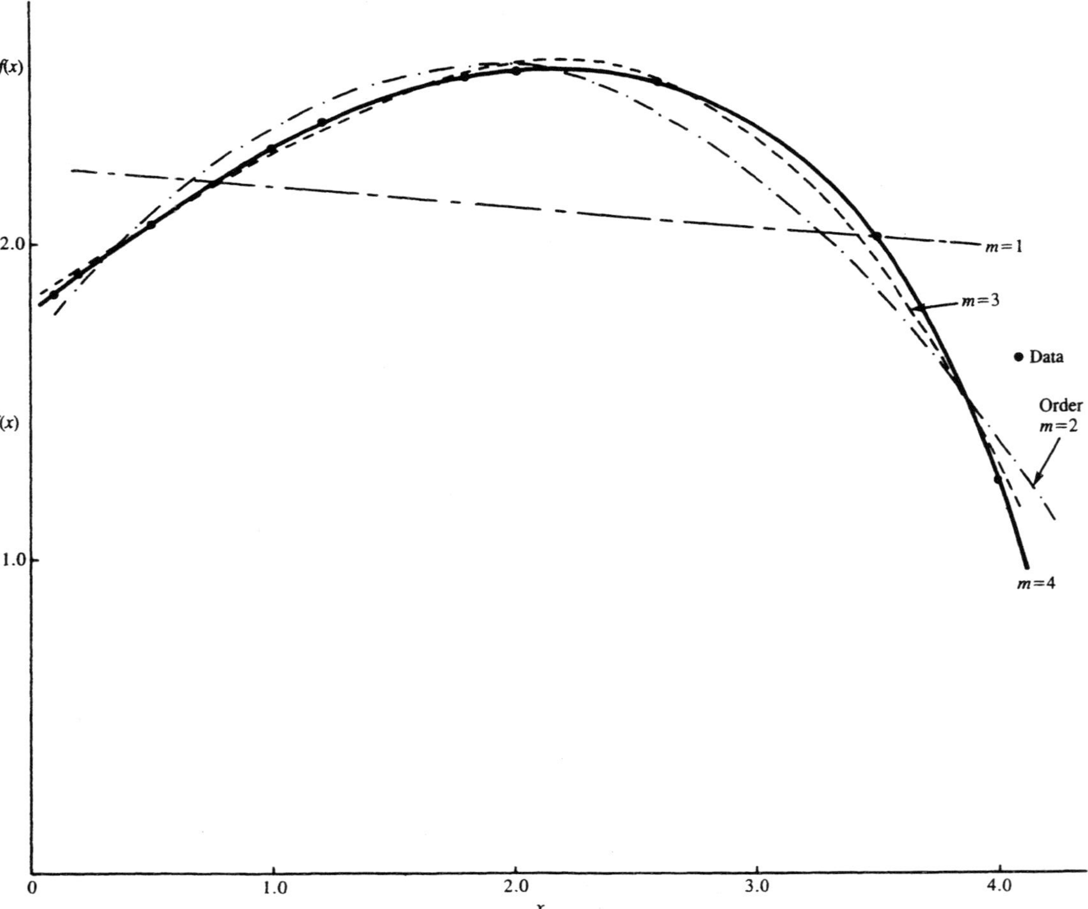

Figure 6.5.3 Comparison between the given data and the best fit obtained, using polynomials of various degrees.

```
THE CONSTANTS OF THE POLYNOMIAL ARE:

C(1)=      4.78000
C(2)=     -0.09695

CONSTANT A= 119.10     EXPONENT=-0.09695

THE VALUES CALCULATED FROM THE BEST FIT ARE:

X(1)=      1.0000      Y(1)=   108.0993
X(2)=      2.0000      Y(2)=    98.1109
X(3)=      5.0000      Y(3)=    73.3503
X(4)=     10.0000      Y(4)=    45.1726
X(5)=     15.0000      Y(5)=    27.8195
X(6)=     20.0000      Y(6)=    17.1326
X(7)=     25.0000      Y(7)=    10.5510
X(8)=     30.0000      Y(8)=     6.4978

THE CORRELATION COEFFICIENT = 0.9998
```

Figure 6.5.4 Numerical results obtained with an exponential best fit for the problem considered in Example 6.5(b).

and are found to be close to the given data. The correlation coefficient r for this problem is found to be 0.9998, which indicates a very good representation of the given data by the exponential function $y = 119.1 \exp(-0.09695x)$. Similarly, the program given in Fig. 6.5.1 may be employed for other nonpolynomial forms, as discussed in Section 6.6.3.

6.7 FUNCTION OF TWO OR MORE INDEPENDENT VARIABLES

In the preceding sections, we considered curve fitting for dependent variables that are functions of only one independent variable. However, in engineering applications, we frequently encounter functions of two or more independent variables. In many cases, interest lies in representing the dependence of such functions on only one independent variable, while the others are held constant at given values. Then, an exact or a best fit, as appropriate, may be employed, as discussed earlier, to characterize this variation. However, there are several circumstances where it is necessary to consider the variation of the dependent variable y with two or more independent variables, say, x_1, x_2, and so on. The pressure generated by a pump, for instance, depends on both the speed and the flow rate. Similarly, properties of gases, such as density, depend on the pressure as well as the temperature. Although curve fitting may be carried out with only one independent variable, taking the others at specified values and thus generating a number of curves that fit the data, it is often more convenient and desirable to seek a single function such as $f(x_1,x_2)$ that represents the dependence on all the independent variables. Since this problem is generally much more involved than curve fitting with a single independent variable, we shall consider only a few simple cases here.

6.7.1 Exact Fit

Let us consider a variable y which is a function of two independent variables x_1 and x_2. Then if an exact fit with a second-order polynomial is sought, we may employ the general equation

$$y = A + Bx_1 + Cx_1^2 \tag{6.73}$$

where the coefficients A, B, and C are functions of x_2. Again, employing second-order polynomials, we may write

$$A = a_0 + a_1x_2 + a_2x_2^2 \tag{6.74}$$

$$B = b_0 + b_1x_2 + b_2x_2^2 \tag{6.75}$$

$$C = c_0 + c_1x_2 + c_2x_2^2 \tag{6.76}$$

Equation (6.73) may be written at three different values of x_2 as follows:

$$y = A_1 + B_1x_1 + C_1x_1^2 \tag{6.77}$$

$$y = A_2 + B_2x_1 + C_2x_1^2 \tag{6.78}$$

$$y = A_3 + B_3x_1 + C_3x_1^2 \tag{6.79}$$

where (A_1, B_1, C_1) correspond to one value of x_2, (A_2, B_2, C_2) to another, and (A_3, B_3, C_3) to a third value of x_2.

The first step involves determining the coefficients in Eqs. (6.77) through (6.79) by employing three data points, in terms of y and x_1, at each value of x_2. Thus, as shown in Fig. 6.10, we need nine data points to evaluate these nine coefficients. Each curve in Fig. 6.10 is represented by a second-order polynomial, which is determined at the given value of x_2 if three data points are available for this curve. Thus, a set of three equations is solved, as discussed in Section 6.2, to obtain the coefficients A_1, B_1, and C_1 in Eq. (6.77). Similarly, the coefficients in Eqs. (6.78) and (6.79) are determined. Thus, we now have the values A_1, A_2, and A_3 for the variable A in Eq. (6.74) at three values of x_2. Using these values, we may determine the coefficients a_0, a_1, and a_2. Similarly, we determine the coefficients in Eqs. (6.75) and (6.76) using the values B_1, B_2, B_3 and C_1, C_2, C_3 at the three given values of x_2. Thus, the procedure for an exact fit is applied twice to obtain all the relevant coefficients.

The coefficients obtained from the nine data points shown in Fig. 6.10 yield an exact second-order polynomial fit to the given data. The resulting general equation is written as

$$y = (a_0 + a_1x_2 + a_2x_2^2) + (b_0 + b_1x_2 + b_2x_2^2)x_1 + (c_0 + c_1x_2 + c_2x_2^2)x_1^2 \tag{6.80}$$

Thus, the functional dependence of y on x_1 and x_2 is represented by this equation. This approach may easily be extended to higher-order polynomials and functions of

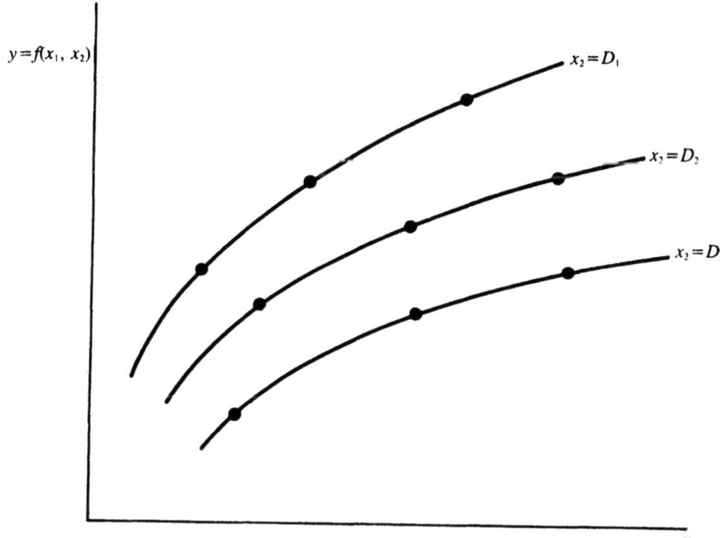

Figure 6.10 Sketch of a function, $f(x_1,x_2)$, of two independent variables x_1 and x_2, showing the nine data points needed for an exact fit with second-order polynomials.

more than two variables. However, the solution becomes more involved because of the larger number of coefficients to be determined. For instance, if third-order polynomials are employed instead of the parabolas in Eq. (6.80), we must determine sixteen coefficients, employing four data points in terms of y and x_1 at four different values of x_2. Similarly, 25 data points are needed for fourth-order polynomials. Similar increase in the complexity of the solution and of the resulting polynomial fit arises if functions of more than two independent variables are considered. Other forms of the function for the exact fit, besides that given in Eq. (6.73), may also be considered.

6.7.2 Best Fit

A best fit is often more appropriate than an exact fit for functions of two or more independent variables. Experimental data with a significant amount of error, for example, are better represented by a best fit than by a curve that passes through each data point. These considerations have been discussed earlier in relation to functions of one independent variable and apply equally well to multiple variables.

Let us first consider multiple linear regression, assuming the dependent variable y to be a linear function of x_1 and x_2 as

$$y = f(x_1,x_2) = c_0 + c_1x_1 + c_2x_2 \tag{6.81}$$

where c_0, c_1, and c_2 are constants to be computed. Employing the procedure given earlier for linear regression, we determine the sum S to be minimized as follows:

$$S = \sum_{i=1}^{n} (y_i - c_0 - c_1 x_{1,i} - c_2 x_{2,i})^2 \tag{6.82}$$

where the subscript i, which varies from 1 to n, is used to denote the n data points. Differentiating S with respect to the coefficients and setting the partial derivatives equal to zero yields the minimum value of S. Thus,

$$\frac{\partial S}{\partial c_0} = -2\sum (y_i - c_0 - c_1 x_{1,i} - c_2 x_{2,i}) = 0$$

$$\frac{\partial S}{\partial c_1} = -2\sum x_{1,i}(y_i - c_0 - c_1 x_{1,i} - c_2 x_{2,i}) = 0$$

$$\frac{\partial S}{\partial c_2} = -2\sum x_{2,i}(y_i - c_0 - c_1 x_{1,i} - c_2 x_{2,i}) = 0$$

where the summations are from $i = 1$ to $i = n$.

The above equations yield the following system of linear equations for the unknowns c_0, c_1 and c_2:

$$nc_0 + c_1 \sum x_{1,i} + c_2 \sum x_{2,i} = \sum y_i \tag{6.83}$$

$$c_0 \sum x_{1,i} + c_1 \sum (x_{1,i})^2 + c_2 \sum x_{1,i} x_{2,i} = \sum x_{1,i} y_i \tag{6.84}$$

$$c_0 \sum x_{2,i} + c_1 \sum x_{1,i} x_{2,i} + c_2 \sum (x_{2,i})^2 = \sum x_{2,i} y_i \tag{6.85}$$

These simultaneous equations may be solved for c_0, c_1, and c_2 to give the best fit, Eq. (6.81). In this case, a regression plane is obtained instead of a line, since y varies with two independent variables x_1 and x_2. The correlation coefficient is again obtained from Eq. (6.52), with appropriate change in the definition of S_m to take the dependence of y on both x_1 and x_2 into account.

By employing the following general form of the function for a best fit, we can extend the procedure outlined above for multiple regression to functions of more than two variables:

$$y = f(x_1, x_2, \ldots, x_m) = c_0 + c_1 x_1 + c_2 x_2 + \cdots + c_m x_m \tag{6.86}$$

The system of linear equations for evaluating the coefficients $c_0, c_1, \ldots, c_m$ may easily be derived as given above for the case of two independent variables. Similarly, multiple polynomial regression, with orders higher than linear, may be derived for introducing curvature into the best fit. Also, linearization of nonlinear functions, such as exponentials and power-law variations, can often be carried out, as outlined earlier for functions of a single variable x. Then multiple linear regression may be applied. Thus, if y is of the general form

$$y = c_0 x_1^{c_1} x_2^{c_2} \ldots x_m^{c_m} \tag{6.87}$$

the equation may be transformed by taking its natural logarithm to give

$$\log y = \log c_0 + c_1 \log x_1 + c_2 \log x_2 + \cdots + c_m \log x_m \tag{6.88}$$

Multiple linear regression may now be applied. Example 6.6 illustrates the use of this procedure for a practical circumstance.

Example 6.6

The flow of water in an open channel with a slight downward slope is an important circumstance in civil engineering applications. The channel is specified in terms of its hydraulic radius R, which is the cross-sectional area divided by the wetted perimeter consisting of the sides and bottom of the channel, and the slope S. The slope is given as $\tan\theta$, where θ is the angle that the bottom makes with the horizontal, considered positive for downhill flow. The volume flow rate Q in m^3/s is measured as a function of R and S for certain open channels to yield the following data:

S \ R (m)	0.5	1.0	1.5	2.0
1.5×10^{-3}	1.91	3.10	4.11	5.03
5×10^{-3}	3.48	6.66	7.51	9.19
9×10^{-3}	4.67	7.59	10.08	12.33

It is expected, from theoretical considerations, that Q varies as AR^bS^c, where A, b, and c are constants. Obtain a best fit to the given data and determine these constants.

Solution

Since the dependent variable Q is a function of two independent variables R and S, multiple regression, as outlined in Section 6.7.2, may be applied. The form of the function to be employed is

$$Q = AS^bR^c \tag{6.6.1}$$

Taking natural logarithms of both sides, we obtain

$$\log Q = \log A + b \log S + c \log R \tag{6.6.2}$$

Thus, multiple linear regression may be used with $\log S$ and $\log R$ as the independent variables and $\log Q$ as the dependent variables.

The dependent variable $\log Q$ is denoted by y, and $\log S$ and $\log R$ by x_1 and x_2, respectively. Then, the system of linear equations to be solved for the constants c_0, c_1, and c_2, in the linear function $y = c_0 + c_1x_1 + c_2x_2$, for a best fit is given by Eqs. (6.83) through (6.85). The coefficients of this system of equations may be obtained from the twelve data points given. Thus, $n = 12$, and the summations, such as $\sum x_{1,i}$ and $\sum x_{1,i}y_i$, are for $i = 1$ to $i = 12$. Employing a simple program in BASIC, we obtain the following system of equations:

$$12c_0 - 66.045c_1 + 1.216c_2 = 20.575 \tag{6.6.3}$$

$$-66.045c_0 + 370.164c_1 - 6.695c_2 = -109.871 \tag{6.6.4}$$

$$1.216c_0 - 6.695c_1 + 3.376c_2 = 4.345 \tag{6.6.5}$$

The above system of linear equations was solved by Gaussian elimination, using the program given in Example 5.1, to yield

$$c_0 = 4.4235 \qquad c_1 = 0.505 \qquad c_2 = 0.6945 \tag{6.6.6}$$

From Eq. (6.6.2), $c_0 = \log A$, $c_1 = b$, and $c_2 = c$. This gives $A = \exp(c_0) = \exp(4.4235) = 83.387$. Therefore, the best fit to the given data is obtained as

$$Q = 83.387 S^{0.505} R^{0.6945} \tag{6.6.7}$$

In a similar way, other power-law and exponential variations may be treated for functions of two or more independent variables. Multiple polynomial regression may also be employed for certain circumstances, using a similar, although more complicated, approach.

6.8 SUMMARY

This chapter presents numerical methods for the curve fitting of data given at discrete points, considering both an exact fit and a best fit. In the former case, the approximating curve passes through each data point and is appropriate if the data have a high level of accuracy and a relatively small number of points are given. Various forms of the approximating function are considered, including the general equation of a polynomial, Lagrange polynomial, and Newton's divided-difference polynomials. The use of these interpolating polynomials for evaluating the function at intermediate points, where data are not available, is discussed. Lagrange interpolation is particularly useful for an arbitrary distribution of points and is widely used. If a large number of very accurate data points are given, spline interpolation, which provides a piecewise exact fit to the data, is more appropriate than a single curve, since polynomials of high order may be ill-conditioned and are also inconvenient to use in practical circumstances. The equations for cubic splines are derived. Examples are given to demonstrate the use of interpolation in engineering problems.

A best fit, which minimizes the error between the data and the approximating curve without forcing it to pass through each given data point, is extensively employed for correlating engineering data. It is more suitable than an exact fit for data that have a significant amount of associated error. Experimental data generally do have some error, and a best fit is used for representing the observed trends. This approach is generally used with lower-order polynomials, such as straight lines, parabolas, and cubics, to obtain a best fit to a large number of data points. The method of least squares is discussed in detail, considering linear regression, polynomial regression, and nonpolynomial forms. In several important engineering applications, special forms, such as exponential and power-law variations, are of interest. These forms may often be linearized by suitable transformations, and linear regression may be applied. Finally, functions of two or more variables are considered. A few simple procedures for an exact fit, as well as for a best fit, are outlined.

The choice of the form of the approximating function for curve fitting is an important consideration. Frequently, the physical or chemical nature of the problem under consideration may be employed to determine the general nature of the variation and the function chosen appropriately. If no prior information is available on the expected trends, a rough plot of the data may be used to guide the choice of the function for curve fitting. A best fit is much more extensively used in engineering problems than an exact fit, because of the presence of significant error in most available data and also because a large number of data points are often given. One may start with simple linear regression and then proceed to parabolas and cubics, in order to check whether a better representation is obtained with a higher-order function.

Lagrange interpolation is a very popular choice for an exact fit, since a system of linear equations does not have to be solved, as is the case for the general form of an nth-order polynomial. Newton's method is particularly useful if the data points are evenly spaced. Extrapolation is also employed in some cases to compute the value of the function at a point beyond the range of the given data. However, one should exercise extreme care while using extrapolated values, since the behavior of the function beyond the given range is often not known. There are also several special interpolating functions, such as Chebyshev polynomials, that are employed in the analysis of engineering systems and processes; see Hornbeck (1975). Also, there are other methods for deriving the interpolating function. One such method is Hermite interpolation which uses both the function and its derivative at a given number of data points, as outlined by Ferziger (1981).

PROBLEMS

6.1. Consider a second-order Lagrange polynomial and show that it may be recast in the general form of a second-order polynomial given by Eq. (6.1). Obtain the relationship between the coefficients of the two polynomials.

6.2. Show that Lagrange interpolation is a more efficient method for interpolation than that obtained by using the general form of an nth-order polynomial, as demonstrated in Example 5.1.

6.3. Compare the Lagrange and Newton's divided-difference interpolation methods, indicating their respective advantages over the other. Which one is expected to require less computer time for interpolation with an arbitrary distribution of data points?

6.4. The specific heat C of pure copper is given at 100, 200, 400, 600, and 800 K as 252, 356, 397, 417, and 433 J/kg·K. Employ Lagrange interpolation to compute the values at 300 K and 500 K. Also, compute the extrapolated value at 1000 K and compare it with the value of 451 given in the literature.

6.5. The density of air at 200, 300, 400, and 500 K is obtained as 1.7458, 1.1614, 0.8711, and 0.6964 kg/m^3, respectively, from tabulated property data in the literature. For this uniformly spaced data, obtain a third-order interpolating polynomial.

6.6. A car showroom has 100 cars at the beginning of a week, and the number left after each day is tabulated as follows:

Time (Days)	0	1	2	3	4	5	6
Cars Left (N)	100	75	65	52	46	39	34

We wish to extrapolate these results to predict the cars left at the end of the week. Using an exact fit, predict the number of cars left in the showroom after seven days. Comment on the result obtained.

6.7. Use a second-order and also a third-order polynomial regression for Prob. 6.6. Compare the results obtained with that obtained earlier with an exact fit, and comment on the difference. Which method would you expect to yield a more dependable prediction? Discuss.

6.8. The force F on a structure due to winds is measured as a function of wind speed V. The results at speeds of 5, 10, 15, 20, and 25 m/s are obtained as 36.2, 52.5, 85.6, 150.0, and 210.9 newtons. Obtain a fourth-order interpolating polynomial that provides an exact fit to these data points.

6.9. The future worth FW of a given sum of money R after n years is $R(1 + x)^n$, where x is the interest rate per unit amount, say \$1.00, compounded annually. Therefore, the future worth ratio FW/R gives the future worth per unit deposit and may be determined at interest rates of 8%, 10%, 12%, and 15% for 15 years as 3.172, 4.177, 5.474, and 8.137. Employing Newton's divided-difference interpolation method, compute the corresponding values at 9% and 12.5% interest rates. Also give the resulting future worth for a deposit of \$5000 at these rates.

6.10. The voltage v applied across an electrical circuit is varied, and the resulting current i measured. For v values of 1, 2, 3.5, 5, and 6 volts, the current is 1.5, 1.8, 2.6, 3.0, and 3.5 amperes. Use Newton's divided-difference method to obtain the electrical current at $v = 4$ and 5.5 volts.

6.11. An important fluid property is the kinematic viscosity which determines the viscous, or frictional, forces acting in a flow. The kinematic viscosity of air multiplied by 10^6 is given at 350, 450, 500, 550, and 650 K as 20.92, 32.39, 38.79, 45.57, and 60.21 m^2/s, respectively. Using any suitable interpolation method, compute the intermediate values at 400 K and 600 K. Compare the results obtained with the values given in the literature as 26.41×10^{-6} m^2/s and 52.69×10^{-6} m^2/s, respectively.

6.12. From the data given in Prob. 6.9, determine the interest rate if the future worth ratio FW/R is 6.5.

6.13. The calibration table for a copper-constantan thermocouple which is employed for temperature measurement gives the temperature T in °C for different values of the voltage output V in millivolts (mV). Using interpolation with a cubic spline for the following data, compute the temperatures corresponding to thermocouple outputs of 0.9 mV and 1.75 mV:

T (°C)	10	20	30	40	50	60	70	80
V (mV)	0.391	0.789	1.196	1.611	2.035	2.467	2.908	3.357

6.14. Using the data in Prob. 6.13 with Lagrange interpolation, calculate the voltage output at $T = 65°C$. Employ a fourth-order polynomial and choose appropriate data points.

6.15. The transport rate $\dot{m}$ of a chemical species at a porous surface is measured as a function of the difference in concentration ΔC between the surface and the ambient medium. The results obtained are as follows:

ΔC (kg/m^3)	0.1	0.3	0.4	0.5	0.7	0.9	1.0
$\dot{m}$ (kg/s)	2.53	3.33	3.58	3.78	4.12	4.38	4.5

A power-law variation of the form $\dot{m} = A(\Delta C)^a$ is expected to govern this mass transfer process. Obtain a best fit to the given data by the method of least squares and determine the constants A and a.

6.16. The concentration of salt decreases in a container because of mass transfer at the surface. The concentration C is measured as a function of time τ to yield

τ (s)	0.1	0.2	0.3	0.5	1.0	2.0	4.0	4.5	5.0
C (kg/m^3)	83.3	81.7	80.0	76.9	69.6	57.0	38.2	34.6	31.3

An exponential variation of the form $C = Be^{-b\tau}$ is expected on physical grounds. Obtain a best fit to the data, and determine the constants B and b.

6.17. The temperature T, pressure p (in kilopascals), and specific volume v, which is inverse of density, for saturated steam are obtained from tabulated data in the literature as follows:

T (°C)	10	20	30	40	50	60	70	80	90
p (kPa)	1.23	2.34	4.25	7.38	12.35	19.94	31.19	47.39	70.13
v (m^3/kg)	106.4	57.79	32.90	19.52	12.03	7.67	5.04	3.41	2.36

Obtain a best fit, with a third-order polynomial, to the T-v data. Using the polynomial obtained, compute the specific volumes at 55°C and 75°C. Also calculate the value at 100°C, and compare it with the tabulated value of 1.673 m^3/kg.

6.18. Using the data given in Prob. 6.17, obtain a best fit to the specific volume dependence on pressure. Consider both second- and third-order polynomials. Using the polynomials obtained, calculate the specific volumes at 5.0 and 25.0 kilopascals.

6.19. The pressure-temperature relationship for saturated steam is suggested to be of the form $\log p = C + D/T$, where C and D are constants and log represents the natural logarithm. Using linear regression with the data in Prob. 6.17, determine the values of these constants. Is the given functional dependence of p on T a satisfactory representation?

6.20. The acceleration of certain objects is studied in an experimental test track for automobiles. The distance traveled by an object L is measured as a function of time τ to yield the following:

τ (s)	0.1	0.2	0.5	1.0	1.5	1.8	2.0	3.0
L (m)	0.26	0.55	1.56	3.90	7.41	10.28	12.6	30.9

Obtain a best fit to this data, considering first-, second, and third-order polynomials. Using these polynomials, calculate the values of the dependent variable L at the time intervals employed for the given data to evaluate the accuracy of the polynomial representations.

6.21. In Prob. 6.20, calculate the correlation coefficient to estimate the improvement in the representation of the data by means of curve fitting.

6.22. For the experimental data given in Example 6.1, obtain a best fit, using second- and third-order polynomials. Compare the interpolated values obtained by the exact fit in Example 6.1 with those obtained from the best fit. Comment on the difference.

6.23. Solve the problem given in Example 6.2 by Newton's divided-difference method, and compare the interpolated results obtained with those given earlier.

6.24. The temperature T of a small copper sphere cooling in air is measured as a function of time τ to yield the following:

τ (s)	0.2	0.6	1.0	1.8	2.0	3.0	5.0	6.0	8.0
T (°C)	146.0	129.5	114.8	90.3	85.1	63.0	34.6	25.6	14.1

An exponential temperature decrease is expected from theoretical considerations. Using linear regression, obtain the exponent c and the constant C, where $T = Ce^{-c\tau}$ represents the variation.

6.25. The temperature of a furnace wall is expected to vary sinusoidally with a time period of one day, because of the daily start-up and shutdown. The measured temperatures at several time intervals τ, where τ is measured from midnight, are given as follows:

τ (h)	2	3	5	8	10	15	18	22	24
T (°C)	86.5	97.7	104.0	101.7	92.5	62.3	55.0	67.5	80.0

Obtain a best fit to these data, using the method of least squares and assuming a sinusoidal variation of the form $A \sin(2\pi\tau/24) + B \cos(2\pi\tau/24) + C$, where A, B, and C are constants to be determined.

6.26. Calculate the correlation coefficients for the various polynomials considered for a least-squares best fit in Example 6.5, and discuss the trends indicated by the results obtained.

6.27. Derive an expression for the correlation coefficient corresponding to a third-order polynomial best fit to a data set represented by (x_i, y_i), $i = 1$ to $i = n$. Discuss the physical implications of this coefficient. Can the correlation coefficient be related to the accuracy of the best fit obtained?

6.28. Consider an equation of the form $y = \sin(\pi a x) + Ax^b$, where A, a, and b are constants. Can the method of least squares be applied to this equation for a given set of data points? Discuss.

6.29. Outline a procedure for obtaining a best fit with a power-law function of the form $y = a + bx^n$, where a, b, and n are constants.

6.30. Consider a functional dependence of y on the independent variable x of the form given by Eq. (6.62). Using this equation, outline a procedure for deriving a best fit to given data.

6.31. Six data points generated by a polynomial are given. Outline a method for finding the order of the polynomial. Also apply your method to y values of 3.61, 5.38, 11.0, 18.34, 28.63, and 35.0 corresponding to x values of 0.2, 0.5, 1.0, 1.4, 1.8, and 2.0, respectively, where x is the independent variable and y the dependent variable.

6.32. Five data points are given, with one of them in considerable error. How will you find this point, using the interpolation methods discussed in the text? Consider, as an example, the following data set:

x	0.25	0.75	1.25	2.5	3.0
y	2.80	4.60	5.75	7.94	6.5

where x is the independent variable and y the dependent variable.

6.33. The decay of the electrical current I in an electronic circuit is measured as a function of time τ, following the opening of a switch. The data obtained are given as follows:

τ (s)	0.5	1.0	1.5	2.5	3.5	5.0	6.5	9.0	9.5
I (amperes)	13.2	10.1	8.7	6.9	6.3	5.1	4.7	4.2	4.0

From theoretical considerations, the current is expected to follow a variation of the form $A\tau^{-a}$, where A and a are constants. Obtain a best fit to the given data, and determine the values of these constants.

6.34. The flow rate Q in circular pipes is measured as a function of the pressure difference Δp and diameter D. The resulting data for the flow rate in m^3/s are given as follows:

Δp (atm) \ D (m)	0.3	0.5	1.0	1.4
0.5	0.13	0.43	2.1	4.55
0.9	0.25	0.81	4.0	8.69
1.2	0.34	1.12	5.5	11.92
1.8	0.54	1.74	8.59	18.63

Using the method of least squares, obtain a best fit for the flow rate as a function of the two independent variables D and Δp. It is expected that Q varies as $BD^a \Delta p^b$, where B, a, and b are constants to be determined.

7

Numerical Integration

7.1 INTRODUCTION

A problem that frequently arises in engineering applications is that of integration of a given function $f(x)$ over a specified range of the independent variable x. In many cases, the function $f(x)$ is continuous, finite, and well behaved over the range of integration $a \leqslant x \leqslant b$, where a and b are constants. Then, the integral I where

$$I = \int_a^b f(x)\,dx \tag{7.1}$$

may often be determined by using available mathematical techniques. The results for common elementary functions such as $\sin x$, $\cos x$, e^x, x^2, $1/x$, and so on, are well known, and those for many more complicated functions are given in integral tables. Analytical, or closed-form, expressions for integrals, whenever available, are of considerable value since they are exact, that is, without the errors that inevitably arise in numerical methods. Moreover, they are of general applicability, so that the effect of varying the physical parameters, associated with the problem, on the integral may easily be investigated. In addition, analytical results can be employed in the evaluation of a numerical integration scheme for correctness and accuracy.

In engineering problems, the function $f(x)$ is often too complicated to be integrated analytically. The limits of integration may be infinite, and the function $f(x)$ itself may be discontinuous or infinite at some point. Also, the function may be available only at certain discrete points, say, from an experimental study or from the numerical solution of a differential equation. In this last circumstance, curve fitting, as discussed in the preceding chapter, may sometimes yield a function $f(x)$ that can be integrated analytically. Otherwise, numerical integration is necessary. Similarly, for the various other circumstances mentioned above, analytical methods may not be available or may be too difficult to be applied, making it essential to use numerical integration.

Integration, which is also often called *quadrature*, basically refers to the area between the curve of $f(x)$ versus x and the x axis, from $x = a$ to $x = b$, as shown graphically in Fig. 7.1. As expected, the integral I is positive if the area above the x axis is larger than that below it. This graphical representation of the integral $I = \int_a^b f(x)\,dx$ will frequently be referred to in the development of formulas for numerical integration.

In this chapter, various methods for the numerical integration of a continuous or discretized function $f(x)$ are presented. The most common approach is based on replacing the function $f(x)$ or the tabulated data with a simple polynomial that can be easily integrated. This approach gives rise to the Newton-Cotes formulas, the simplest one being obtained when the function $f(x)$ is taken as constant over the various segments into which the given range $a \leqslant x \leqslant b$ is divided. The most commonly used Newton-Cotes formulas are the trapezoidal, Simpson's one-third, and Simpson's three-eighths rules, which are based, respectively, on linear, parabolic, and cubic, or third-order, polynomial approximations to the functions. Although these formulas are derived for continuous functions, their application to evenly and unequally spaced data is also discussed, since experimental and numerical results are generally available at such discrete values of the independent variable x.

We also determine the truncation errors in these formulas to evaluate the resulting accuracy. As the number of segments n into which the region is divided is increased, or the step size Δx is reduced, the truncation error decreases, so that the

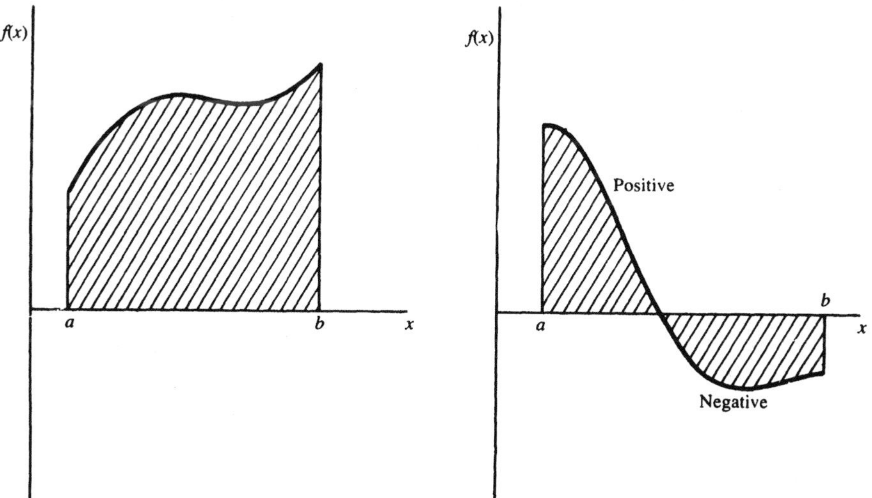

Figure 7.1 Graphical representation of the integral of a function $f(x)$ over x, between the limits $x = a$ and $x = b$, as the area between the curve and the x axis.

numerical value of the integral approaches the exact value. However, as Δx is reduced to very small values, the computational effort and the round-off error increase substantially, as discussed in Section 2.3, resulting in an increase in the total error with a further reduction in Δx. Thus, even though the mathematical definition of integration demands that $\Delta x \to 0$, a lower limit on Δx is imposed by the round-off error in numerical integration. These considerations are again discussed later in this chapter.

In many engineering applications, an accuracy higher than that provided by the relatively simple trapezoidal and Simpson's rules is demanded from numerical integration. Various methods for improving the accuracy, such as Richardson's extrapolation and higher-order integration formulas, are discussed. Romberg integration, which provides very high accuracy, without an associated substantial increase in the computational effort and the round-off error, as encountered at very small segment size Δx, is of particular importance in such applications and is discussed in detail.

Also considered in this chapter is Gauss quadrature, which is particularly suitable for cases where the evaluation of the integrand $f(x)$ is involved and thus is time consuming. Adaptive methods, which increase the accuracy of the computation by focusing on intervals in which the inaccuracy is larger than that in other intervals, are also outlined. Finally, improper integrals, in which the integrand becomes infinite at some point or the limits of integration are infinite, are discussed, and some of the techniques that may be employed for computing the integral are presented.

Before proceeding to the various methods for numerical integration, we will present a few examples of engineering problems in which numerical integration is often needed, in order to provide a physical background for the discussion to follow. In electrical engineering, the root mean square (RMS) value of an electrical current $I(\tau)$, which varies periodically with time τ, is given by

$$I_{\mathrm{RMS}} = \frac{1}{\tau_c} \sqrt{\int_0^{\tau_c} I^2(\tau)\, d\tau} \tag{7.2a}$$

where τ_c is the time for one cycle. Numerical integration is generally needed for an arbitrary periodic variation of $I(\tau)$. Periodic processes are also encountered in natural phenomena, such as the daily and yearly variations of environmental temperatures, and numerical integration is employed to compute the resulting transport of mass and energy, say, from the surface of a lake. The integral of the current $I(\tau)$ entering a capacitor, $\int_0^\tau I(\tau)\, d\tau$, gives the stored charge $Q(\tau)$. Thus, the voltage $V(\tau)$ across the capacitor, due to the current in a given electrical circuit containing the capacitor, may be determined, as shown in Example 7.1. A similar integral arises in civil engineering for water storage in a reservoir due to the inflow minus the outflow, both of which are time dependent. The variation with time is often very complicated, or the values are known only at certain discrete data points, making it necessary to use numerical integration.

Integration is very important in radiative heat transfer where integrals over surfaces, volumes, the wavelength interval of the radiation, and the total angle of the incident radiation are needed to compute the energy transport rates. In most practical cases, these integrals are too complicated to be solved by analytical methods. The integral of the emissive power of a blackbody, Eq. (3.3.1), over wavelength ranging from zero to infinity is one such example. The mass or energy transfer from a surface is frequently obtained from an integral of the transport rate, given as a time-dependent mass or heat transfer flux, per unit area and time. Although some simple problems may be solved analytically, most practical circumstances require numerical integration. Such problems often arise in chemical reactors and manufacturing processes.

The volume flow rate in a circular tube, Q, is obtained by an integral of the velocity distribution $V(r)$ as follows:

$$Q = \int_0^R V(r) 2\pi r \, dr \tag{7.2b}$$

where R is the radius of the tube and r the radial distance from the axis. In most practical cases, $V(r)$ is a complicated function or is available only at discrete data points, thus requiring numerical integration for the computation of Q. In dynamic systems, the work done W is related to the force $F(x)$ and distance x as

$$W = \int_{x_1}^{x_2} F(x) \, dx \tag{7.2c}$$

where x_1 and x_2 are the initial and final positions. Again, for an arbitrary functional dependence $F(x)$, numerical integration is needed.

The few examples outlined above indicate the importance of numerical integration in many diverse engineering fields. Some relatively simple integral expressions are also given. However, many more complicated forms are often encountered in engineering. For example, multiple integrals commonly arise in radiation due to integration over several independent variables. Improper integrals, due to the integrand becoming singular or the integration limits becoming infinite, are also often of interest. Many of these cases are considered in this chapter. We now proceed to the derivation of some of the commonly used formulas for numerical integration.

7.2 RECTANGULAR AND TRAPEZOIDAL RULES FOR INTEGRATION

The most commonly used schemes for numerical integration are the Newton-Cotes formulas, which are based on the approximation of a complicated function $f(x)$, or of tabulated data, with a simple polynomial that can be integrated easily. Thus, the integral I is written as

$$I = \int_a^b f(x) \, dx \simeq \int_a^b P_m(x) \, dx \tag{7.3}$$

where $P_m(x)$ is an mth-order polynomial of the form

$$P_m(x) = p_0 + p_1x + p_2x^2 + \cdots + p_mx^m \tag{7.4}$$

The p's are constants that we determine by choosing an interpolating polynomial that yields the same values of the dependent variable as the given function $f(x)$ at a finite number of points, as done in Section 6.2.1. However, the replacement of $f(x)$ by $P_m(x)$ is done piecewise over each of the n intervals into which the total range of x is subdivided, as discussed below. The general approach to the derivation of the Newton-Cotes formulas is based on Lagrange interpolation, discussed in the preceding chapter. However, the first few approximations may be derived by simple direct methods, based on the graphical interpretation of integration.

The first step in the numerical integration of a function $f(x)$ is the division of the integration range $a \leqslant x \leqslant b$ into a finite number n of intervals or strips, as shown in Fig. 7.2. If Δx is the width of each interval, then $\Delta x = (b - a)/n$. The largest value of Δx is $(b - a)$, which is obtained when the entire range of integration is taken as a single interval. The independent variable x varies from $x = a$ to $x = b$ in steps of Δx, so that x may be written as

$$x_i = a + i\,\Delta x \qquad \text{where } i = 0, 1, 2, \ldots, (n-1), n \tag{7.5}$$

Thus, $x_0 = a$, $x_n = b$, and x_i represents the value at an intermediate grid point, as shown in Fig. 7.2. The corresponding ordinates are denoted by f_0, f_1, $f_2, \ldots,$ $f_i, \ldots, f_n$. The interpolating polynomial, Eq. (7.4), is now applied piecewise to the function or data over these intervals of constant width. More than one segment will be needed for a polynomial of order higher than that for a straight line, $m = 1$, in order to provide the necessary number of points for the determination of all the coefficients of the chosen interpolating polynomial.

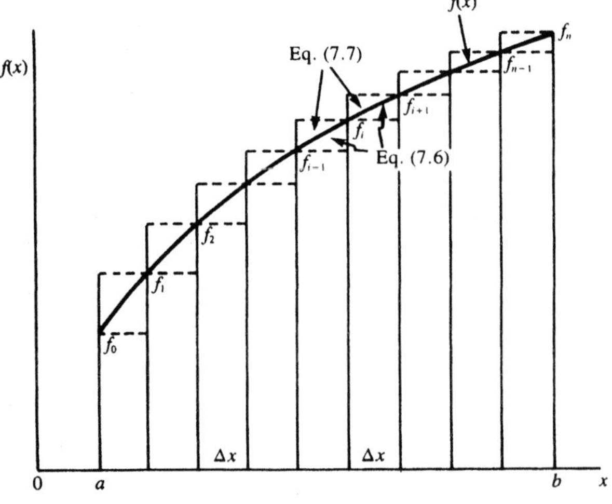

Figure 7.2 Approximation of an integral by a finite number of rectangular strips or segments.

7.2.1 The Rectangular Rule

The simplest approximation to the function $f(x)$ is a zeroth-order polynomial, that is, a constant value over each interval. Then the function $f(x)$ is approximated as a constant, at f_i or f_{i+1}, over the interval $x_i \leqslant x \leqslant x_{i+1}$. Thus, the area under the curve in this interval is taken as $f_i\,\Delta x$, or $f_{i+1}\,\Delta x$. For an increasing function, as sketched in Fig. 7.2, the approximation of the function as f_i over the interval underestimates the actual area under the curve, and the approximation as f_{i+1} overestimates the integral. Similarly, for a decreasing function, the former approximation provides an upper bound for the integral, and the latter approximation a lower bound.

Therefore, the given integral I is approximated, in the rectangular numerical integration scheme, by

$$I = \int_a^b f(x)\,dx \simeq \sum_{i=0}^{n-1} f_i\,\Delta x \tag{7.6}$$

or

$$I = \int_a^b f(x)\,dx \simeq \sum_{i=0}^{n-1} f_{i+1}\,\Delta x \tag{7.7}$$

The first formulation sums the ordinates at the beginning of each interval and multiplies the sum with the step size Δx, to give the numerical approximation to the integral. The second formulation sums the ordinates at the end of each interval and approximates the integral by the product of this sum with Δx. As shown in Fig. 7.2 and as mentioned above, the two formulations provide the upper and lower bounds for the given integral if the function $f(x)$ is a monotonically increasing or decreasing function of x. It must also be noted that the difference between the integrals from the two formulations is simply $|f_n - f_0|\,\Delta x$, that is, the product of Δx and the difference between the two end ordinates.

The rectangular rule yields the exact value of the integral only if $f(x)$ is a constant. For an arbitrary function, the truncation error is generally very large and the method is seldom used. However, this discussion serves to illustrate the basic concepts involved in numerical integration. It will be shown later that the truncation error in this scheme is $O[(\Delta x)^2]$ per step, resulting in a total truncation error, over the entire range of integration, of $O(\Delta x)$. Therefore, it is a first-order scheme, and more accurate methods are generally used in engineering problems.

7.2.2 The Trapezoidal Rule

The next order approximation of the function $f(x)$ is by means of a first-order polynomial, which implies that the function is replaced by a straight line over each interval, as sketched in Fig. 7.3. Then the area under the curve in each element or interval is replaced by that of a trapezoid. If the areas of these trapezoids are denoted

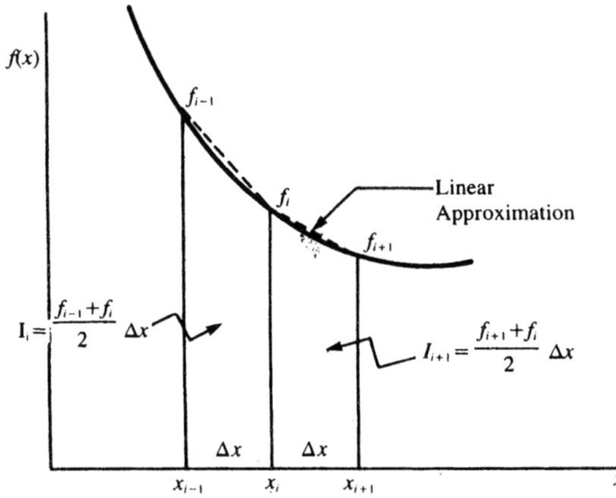

Figure 7.3 Approximation of the given function by straight lines over each of the segments, in which the integration domain is subdivided, for the trapezoidal rule.

by $I_1, I_2, \ldots, I_n$, as indicated in Fig. 7.3, then

$$\begin{aligned} I_1 &= \left(\frac{f_0 + f_1}{2}\right)\Delta x \\ I_2 &= \left(\frac{f_1 + f_2}{2}\right)\Delta x \\ &\vdots \\ I_i &= \left(\frac{f_{i-1} + f_i}{2}\right)\Delta x \\ &\vdots \\ I_n &= \left(\frac{f_{n-1} + f_n}{2}\right)\Delta x \end{aligned} \tag{7.8}$$

Therefore, the integral I may be approximated by

$$I = \int_a^b f(x)\,dx \simeq \frac{\Delta x}{2}(f_0 + 2f_1 + 2f_2 + \cdots + 2f_{n-1} + f_n) \tag{7.9}$$

It can easily be shown that the result obtained by this method is simply the average of the results from the two formulations of the rectangular rule, given by Eqs. (7.6) and (7.7).

The trapezoidal rule for numerical integration is extensively used in engineering applications. It is fairly simple to program, as illustrated by Example 7.1. It also imposes no constraints on the choice of the number of intervals n. Simpson's one-third rule, which is discussed later in this chapter, for instance, requires n to be even. Since each interval can be treated separately by the trapezoidal rule, as given in Eq. (7.8), the method can easily be extended to numerical integration with intervals of unequal width. This is of particular relevance in the integration of a function that is given at a finite number of data points, as is the case in several engineering applications. As shown below, the truncation error, per step, in the trapezoidal method of integration is $O[(\Delta x)^3]$. This results in a total truncation error, over the entire range of integration, of $O[(\Delta x)^2]$, making it a second-order method. One can considerably improve the accuracy of the numerical results obtained by determining the truncation error and incorporating it into the solution. This approach is based on Richardson extrapolation and leads to Romberg integration, discussed in Section 7.4.

7.2.3 Truncation Error

In order to derive the truncation errors associated with the rectangular and trapezoidal rules for numerical integration, let us define a function $y(x)$ as

$$y(x) = \int_a^x f(x)\,dx \tag{7.10}$$

so that $y(x)$ is the integral of the function $f(x)$ from $x = a$ to x, as shown graphically in Fig. 7.4. Also, from Eq. (7.10),

$$y'(x) = f(x) \qquad y''(x) = f'(x) \qquad y'''(x) = f''(x) \qquad \ldots \tag{7.11}$$

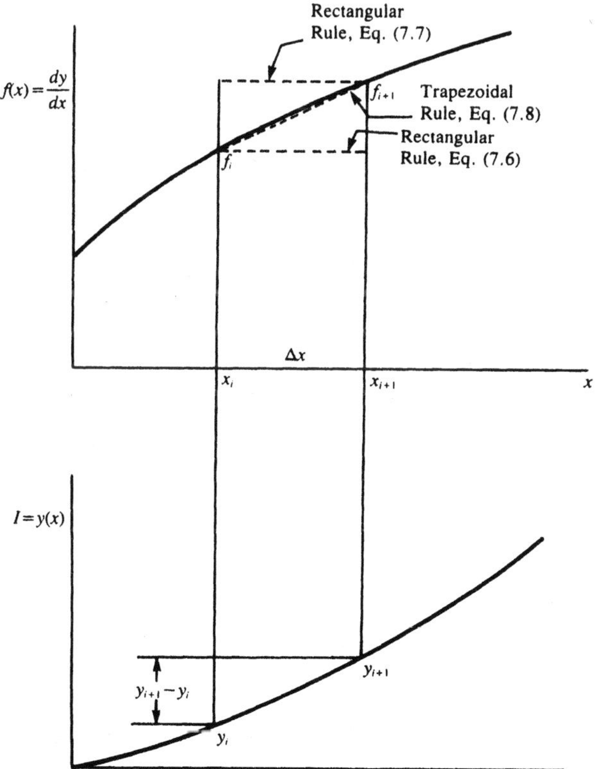

Figure 7.4 Sketch of a function and its integral for the estimation of truncation error in numerical integration by the rectangular and trapezoidal rules.

Since $y(x)$ represents the integral of the function, the exact integral over the range $x_i \leqslant x \leqslant x_{i+1}$ is $y(x_{i+1}) - y(x_i)$. With $y(x_i)$ denoted by y_i, the exact area under the curve in the given interval is, therefore, $y_{i+1} - y_i$.

Assuming both $y(x)$ and $f(x)$ to be continuous and smooth over the interval $x_i \leqslant x \leqslant x_{i+1}$, we may expand y_{i+1} and f_{i+1} in a Taylor series about $x = x_i$. Thus, if the derivatives are denoted by primes, the expansion for y_{i+1} is given by

$$y_{i+1} = y_i + \Delta x y_i' + \frac{(\Delta x)^2}{2!} y_i'' + \frac{(\Delta x)^3}{3!} y_i''' + O[(\Delta x)^4]$$

This formula gives the exact integral over the interval $x_i \leqslant x \leqslant x_{i+1}$ as

$$\begin{aligned} y_{i+1} - y_i &= \Delta x y_i' + \frac{(\Delta x)^2}{2!} y_i'' + \frac{(\Delta x)^3}{3!} y_i''' + O[(\Delta x)^4] \\ &= \Delta x f_i + \frac{(\Delta x)^2}{2!} f_i' + \frac{(\Delta x)^3}{3!} f_i'' + O[(\Delta x)^4] \end{aligned} \tag{7.12}$$

using the relations in Eq. (7.11). Similarly, y_i may be expanded in a Taylor series about $x = x_{i+1}$ as follows:

$$y_i = y_{i+1} - \Delta x f_{i+1} + \frac{(\Delta x)^2}{2!} f_{i+1}' - \frac{(\Delta x)^3}{3!} f_{i+1}'' + O[(\Delta x)^4] \tag{7.13}$$

Rectangular Rule. In the rectangular rule, the integral over the interval $x_i \leqslant x \leqslant x_{i+1}$ is approximated by $f_i \Delta x$ in the first formulation, Eq. (7.6), and by $f_{i+1} \Delta x$ in the second formulation, Eq. (7.7). The exact integral is $y_{i+1} - y_i$. Therefore, from Eq. (7.12), the truncation error (TE) in the first formulation of the rectangular rule is

$$\begin{aligned} \text{TE} &= \underbrace{(y_{i+1} - y_i)}_{\substack{\text{Exact} \\ \text{Value}}} - \underbrace{f_i \Delta x}_{\substack{\text{Numerical} \\ \text{Approximation}}} \\ &= \frac{(\Delta x)^2}{2} f_i' + O[(\Delta x)^3] \end{aligned}$$

This implies that the leading term of the truncation error associated with this step is $[(\Delta x)^2/2] f_i'$. Similarly, from Eq. (7.13), the truncation error in the second formulation of the rectangular rule is obtained as follows:

$$\begin{aligned} \text{TE} &= \underbrace{(y_{i+1} - y_i)}_{\substack{\text{Exact} \\ \text{Value}}} - \underbrace{f_{i+1} \Delta x}_{\substack{\text{Numerical} \\ \text{Approximation}}} \\ &= -\frac{(\Delta x)^2}{2} f_{i+1}' + O[(\Delta x)^3] \end{aligned}$$

Using the remainder theorem, discussed in Chapter 3, we write the error per step in the two formulations, respectively, as follows:

$$\text{TE/step} = \frac{(\Delta x)^2}{2} f'(\xi) \quad \text{and} \quad -\frac{(\Delta x)^2}{2} f'(\xi) \qquad \text{where } x_i < \xi < x_{i+1} \tag{7.14}$$

Trapezoidal Rule. Before proceeding to the total error, over the entire range of integration, let us first determine the error associated with the trapezoidal rule. Expanding the function $f(x)$ in a Taylor series about $x = x_i$, we obtain

$$f_{i+1} = f_i + \Delta x f_i' + \frac{(\Delta x)^2}{2!} f_i'' + \frac{(\Delta x)^3}{3!} f_i''' + O[(\Delta x)^4]$$

Therefore,

$$f_i' = \frac{f_{i+1} - f_i}{\Delta x} - \frac{\Delta x}{2} f_i'' - \frac{(\Delta x)^2}{6} f_i''' + O[(\Delta x)^3] \tag{7.15}$$

This formula is simply the forward difference approximation for f_i', along with the associated truncation error, as derived in Chapter 3. Substituting this expression into Eq. (7.12), we obtain

$$\begin{aligned} y_{i+1} - y_i &= \Delta x f_i + \frac{(\Delta x)^2}{2}\left[\frac{f_{i+1} - f_i}{\Delta x} - \frac{\Delta x}{2} f_i'' - \frac{(\Delta x)^2}{6} f_i''' - \cdots\right] \\ &\quad + \frac{(\Delta x)^3}{3!} f_i'' + O[(\Delta x)^4] \qquad (7.16) \\ &= \Delta x \frac{f_{i+1} + f_i}{2} - \frac{1}{12}(\Delta x)^3 f_i'' + O[(\Delta x)^4] \end{aligned}$$

Since $(y_{i+1} - y_i)$ is the exact area under the curve and $\Delta x(f_{i+1} + f_i)/2$ the trapezoidal area in the interval considered, the truncation error per step is

$$\text{TE/step} = -\frac{1}{12}(\Delta x)^3 f_i'' + O[(\Delta x)^4]$$

Again, using the remainder theorem, we write the truncation error for integration over the interval $x_i \leqslant x \leqslant x_{i+1}$ as

$$\text{TE/step} = -\frac{1}{12}(\Delta x)^3 f''(\xi) \qquad \text{where } x_i < \xi < x_{i+1} \tag{7.17}$$

This expression gives the truncation error in the $(i+1)$th strip, or subinterval; see Fig. 7.2. Therefore, the truncation error per step in the trapezoidal rule is $O[(\Delta x)^3]$. Since the error is zero if $f'' = 0$, the method is exact only for a linear function.

Total Error. The total error in integrating the function $f(x)$ over the entire interval $a \leqslant x \leqslant b$ is obtained by the summation of the errors over n subintervals. Therefore, the total truncation error E for the trapezoidal rule is

$$E = \sum_{i=0}^{n-1}\left[-\frac{1}{12}(\Delta x)^3 f''(\xi_i)\right] \qquad \text{where } x_i < \xi_i < x_{i+1} \tag{7.18}$$

The maximum total error may be estimated from this expression by employing the largest value of f'' in each subinterval. However, the second derivative f'' may not be easy to evaluate in many practical circumstances. A more useful alternative expression for the total error is obtained by defining an arithmetic mean f''_{av} of the values of $f''(\xi_i)$ in the n strips. Then

$$\sum_{i=0}^{n-1} f''(\xi_i) = nf''_{av} \tag{7.19}$$

Therefore, the total truncation error E may be expressed in terms of f''_{av} as follows:

$$\begin{aligned} E &= -\frac{1}{12}(\Delta x)^3 nf''_{av} = -\frac{1}{12}(\Delta x)^3 \frac{b-a}{\Delta x} f''_{av} \\ &= -\frac{1}{12}(\Delta x)^2(b-a)f''_{av} \end{aligned} \tag{7.20}$$

Assuming f''_{av} to remain essentially constant as the step size Δx is varied, we write the total truncation error as

$$E \simeq S_T(\Delta x)^2 = O[(\Delta x)^2] \tag{7.21}$$

where $S_T = -(b-a)f''_{av}/12$ is assumed to be a constant, as indicated by the approximation ($\simeq$) sign. Thus, the trapezoidal rule is a second-order method.

Proceeding in a similar manner for the rectangular rule, we can show that the total truncation errors, for the two formulations of Eqs. (7.6) and (7.7), are, respectively,

$$E = \frac{\Delta x}{2}(b-a)f'_{av} \quad \text{and} \quad -\frac{\Delta x}{2}(b-a)f'_{av} \tag{7.22}$$

where f'_{av} is the average of the $f'(\xi_i)$ values in the n strips. This expression may again be written as

$$E \simeq S_R(\Delta x) = O(\Delta x) \tag{7.23}$$

where $S_R = -(b-a)f'_{av}/2$ is again assumed to be a constant. Therefore, the rectangular rule is a first-order method. Both the trapezoidal and the rectangular rules for numerical integration are quite simple to program. The difference between the two lies only in the incorporation of the ordinates at the ends of the total range in the summation of the ordinates for the numerical scheme. The rectangular rule uses only the ordinate at $x = a$ in the first formulation and at $x = b$ in the second formulation, whereas the trapezoidal rule uses the average of the two. The ordinates in the interior region are summed in all three cases; see Eqs. (7.6), (7.7), and (7.9). Since the trapezoidal rule is more accurate than the rectangular rule, there is no reason to use the latter. In fact, the trapezoidal rule is among the most widely used schemes for numerical integration in problems of engineering interest because of its simplicity.

Accuracy. As shown in the expressions for the truncation errors in the rectangular and trapezoidal integration methods, the error decreases as the step size Δx is decreased, that is, as the number of segments n is increased. This behavior is expected, as discussed in detail in Section 2.3. Thus, the accuracy of the numerical results can be improved by decreasing Δx, a process generally known as *grid refinement*. However, as Δx is decreased, the number of segments increases and so does the computational effort. This results in an increase in the round-off error. Therefore, the total error, which includes the truncation and round-off errors, is reduced by decreasing Δx to a certain point, beyond which the round-off error becomes substantial and the total error increases with decreasing Δx; see Fig. 2.7. All of these considerations were discussed in Section 2.3 and are repeated here to emphasize the importance of numerical errors and the need to vary the grid size, Δx, keeping it larger than the constraint imposed by the round-off error, to ensure that the numerical solution is essentially independent of the value chosen. The above expressions for the truncation error will again be employed in Section 7.4 for developing schemes of higher accuracy.

Example 7.1

A capacitor in an electrical circuit is initially at zero charge. At time τ of 1 s, a switch is closed, and a time-dependent electric current $I(\tau)$ charges up the capacitor. The current is given as

$$I(\tau) = 4(1 - e^{-0.5})e^{-0.5(\tau - 1)}(1 - e^{-\tau})$$

Using the trapezoidal rule for numerical integration, compute the charge Q and the voltage V across the capacitor as functions of time up to $\tau = 20$ s. The capacitance C of the capacitor is 0.025 farad.

Solution

The charge Q stored by the capacitor is given by the integral

$$Q(\tau) = \int_1^\tau I(\tau')\,d\tau' \tag{7.1.1}$$

where τ is the time at which the charge is to be determined and τ' is simply a dummy variable. The voltage across the capacitor is given by

$$V(\tau) = \frac{Q(\tau)}{C} \tag{7.1.2}$$

Here, it is assumed that the charge and thus the voltage across the capacitor are zero at $\tau = 1$ s, as given in the problem. Therefore, this problem requires the application of the trapezoidal rule for evaluating the integral

$$Q(\tau) = \int_1^\tau 4(1 - e^{-0.5})e^{-0.5(\tau' - 1)}(1 - e^{-\tau'})\,d\tau' \tag{7.1.3}$$

Figure 7.1.1 presents the computer program in FORTRAN 77 for solving this problem. The current $I(\tau)$ is represented by the function F(X) or F(T), where T is time. The lower limit of the integral is specified as TMIN = 1.0. The upper limit TMAX,

```
C                TRAPEZOIDAL RULE FOR NUMERICAL INTEGRATION
C
C       F(X) IS THE FUNCTION TO BE INTEGRATED AND REPRESENTS THE ELECTRIC
C       CURRENT AS A FUNCTION OF TIME T IN SECONDS, V IS THE VOLTAGE, Q
C       IS THE ELECTRICAL CHARGE IN COULOMBS, C IS THE CAPACITANCE IN
C       FARADS, DT IS THE TIME STEP, N IS THE NUMBER OF SUBDIVISIONS,
C       AND TMIN AND TMAX ARE THE MINIMUM AND MAXIMUM VALUES OF T.
C
        IMPLICIT REAL (A-H,O-Z)
C
C       DEFINE FUNCTION TO BE INTEGRATED
C
        F(X)=4.0*(1.0-EXP(-0.5))*(EXP(-0.5*(X-1.0)))*(1.0-EXP(-X))
C
C       ENTER INPUT VALUES
C
        TMIN=1.0
        C=0.025
        PRINT *,'ENTER THE STEP SIZE DT'
        READ *,DT
        DO 6 J=1,6
        WRITE(1,7)DT
  7     FORMAT(//5X,'STEP SIZE DT=',F7.5)
        WRITE(1,1)
  1     FORMAT(/6X,'TIME T',19X,'CHARGE Q',17X,'VOLTAGE V')
        WRITE(1,2)
  2     FORMAT(5X,8('-'),17X,10('-'),15X,11('-')/)
C
C       VARY TIME AT WHICH CHARGE IS TO BE COMPUTED
C
        DO 3 TMAX=2.0,20.0,2.0
        N=(TMAX-TMIN)/DT
C
C       COMPUTE SUM OF INTERIOR ORDINATES FOR TRAPEZOIDAL RULE
C
        SUM=0.0
        T=TMIN+DT
        DO 4 I=1,N-1
        SUM=SUM+F(T)
        T=T+DT
  4     CONTINUE
C
C       APPLY TRAPEZOIDAL RULE
C
        Q=(DT/2.0)*(F(TMIN)+2.0*SUM+F(TMAX))
        V=Q/C
        WRITE(1,5)TMAX,Q,V
  5     FORMAT(5X,F5.2,21X,E9.4,16X,E9.4)
  3     CONTINUE
  6     DT=DT/2
        STOP
        END
```

Figure 7.1.1 Computer program in FORTRAN 77 for computing the integral of the time-dependent current $I(\tau)$, given in Example 7.1, using the trapezoidal rule.

```
100 REM     TRAPEZOIDAL RULE FOR NUMERICAL INTEGRATION
110 REM
120 REM     DEFINE FUNCTION TO BE INTEGRATED
130 REM
140   DEF FNA(X)=4*(1-EXP(-.5))*(EXP(-.5*(X-1)))*(1-EXP(-X))
150 REM
160 REM     ENTER INPUT VALUES
170 REM
180   TMIN=1
190   C=.025
200   PRINT "ENTER THE STEP SIZE DT"
210   INPUT DT
220 REM
230 REM     VARY TIME AT WHICH CHARGE IS TO BE COMPUTED
240 REM
250   FOR TMAX=2 TO 20 STEP 2
260   N=(TMAX-TMIN)/DT
270 REM
280 REM     COMPUTE SUM OF INTERIOR ORDINATES FOR TRAPEZOIDAL RULE
290 REM
300   SUM=0
310   T=TMIN+DT
320   FOR I=1 TO (N-1)
330   SUM=SUM+FNA(T)
340   T=T+DT
350   NEXT I
360 REM
370 REM     APPLY TRAPEZOIDAL RULE
380 REM
390   Q=(DT/2)*(FNA(TMIN)+2*SUM+FNA(TMAX))
400   V=Q/C
410   PRINT "TIME=";TMAX,"CHARGE=";Q,"VOLTAGE=";V
420   NEXT TMAX
430   END
```

(*a*)

RESULTS

```
ENTER THE STEP SIZE DT
? 0.125
TIME= 2        CHARGE= .9381998          VOLTAGE= 37.52799
TIME= 4        CHARGE= 2.063364          VOLTAGE= 82.53455
TIME= 6        CHARGE= 2.503398          VOLTAGE= 100.1359
TIME= 8        CHARGE= 2.666577          VOLTAGE= 106.6631
TIME= 10       CHARGE= 2.726671          VOLTAGE= 109.0668
TIME= 12       CHARGE= 2.748782          VOLTAGE= 109.9513
TIME= 14       CHARGE= 2.756916          VOLTAGE= 110.2766
TIME= 16       CHARGE= 2.759908          VOLTAGE= 110.3963
TIME= 18       CHARGE= 2.761009          VOLTAGE= 110.4404
TIME= 20       CHARGE= 2.761414          VOLTAGE= 110.4566
```

(*b*)

Figure 7.1.2 Computer program in BASIC for Example 7.1, along with the results at $\Delta\tau = 0.125$ s.

which represents τ in Eq. (7.1.1), is varied from 2 s to 20 s, in increments of 2 s each. The step size, or the segment width, $\Delta\tau$ is denoted by DT and is entered interactively. The number of subdivisions N is calculated for each time level τ at which the charge Q and voltage V are to be determined. The sum of the ordinates in the interior region of the integration domain is computed, and the trapezoidal rule is applied to yield the numerical value of the integral in Eq. (7.1.3). This gives the electrical charge Q at time

TABLE 7.1.1 Numerical Results Obtained for the Charge Q and the Voltage V, as Functions of time τ, for Example 7.1, at Several Values of the time step $\Delta\tau$.

STEP SIZE DT=0.25000

TIME T	CHARGE Q	VOLTAGE V
2.00	.9368E+00	.3747E+02
4.00	.2062E+01	.8250E+02
6.00	.2503E+01	.1001E+03
8.00	.2666E+01	.1066E+03
10.00	.2726E+01	.1091E+03
12.00	.2748E+01	.1099E+03
14.00	.2757E+01	.1103E+03
16.00	.2760E+01	.1104E+03
18.00	.2761E+01	.1104E+03
20.00	.2761E+01	.1104E+03

STEP SIZE DT=0.12500

TIME T	CHARGE Q	VOLTAGE V
2.00	.9382E+00	.3753E+02
4.00	.2063E+01	.8253E+02
6.00	.2503E+01	.1001E+03
8.00	.2667E+01	.1067E+03
10.00	.2727E+01	.1091E+03
12.00	.2749E+01	.1100E+03
14.00	.2757E+01	.1103E+03
16.00	.2760E+01	.1104E+03
18.00	.2761E+01	.1104E+03
20.00	.2761E+01	.1105E+03

STEP SIZE DT=0.06250

TIME T	CHARGE Q	VOLTAGE V
2.00	.9386E+00	.3754E+02
4.00	.2064E+01	.8254E+02
6.00	.2504E+01	.1001E+03
8.00	.2667E+01	.1067E+03
10.00	.2727E+01	.1091E+03
12.00	.2749E+01	.1100E+03
14.00	.2757E+01	.1103E+03
16.00	.2760E+01	.1104E+03
18.00	.2761E+01	.1104E+03
20.00	.2761E+01	.1105E+03

τ. The voltage V is then computed from Eq. (7.1.2). Figure 7.1.2 gives the corresponding program in BASIC, and, as before, the similarities and differences between the two programming languages are demonstrated.

Table 7.1.1 presents some of the numerical results obtained in tabular form, with time τ being varied from 2 s to 20 s. The step size $\Delta\tau$ was also varied, starting with 2 s and then successively halving it. The results remain essentially unchanged as $\Delta\tau$ is decreased from 0.125 s to 0.0625 s, indicating the former to be adequate for this computation. The corresponding results from the program in BASIC are shown in Fig. 7.1.2 at $\Delta\tau = 0.125$ s, indicating close agreement with those in Table 7.1.1. The results at $\Delta\tau = 2$ s were found to be in considerable error. The effect of the segment size $\Delta\tau$ is shown more clearly in Fig. 7.1.3 by a plot of the charge Q at $\tau = 2$, 4, and 8 s

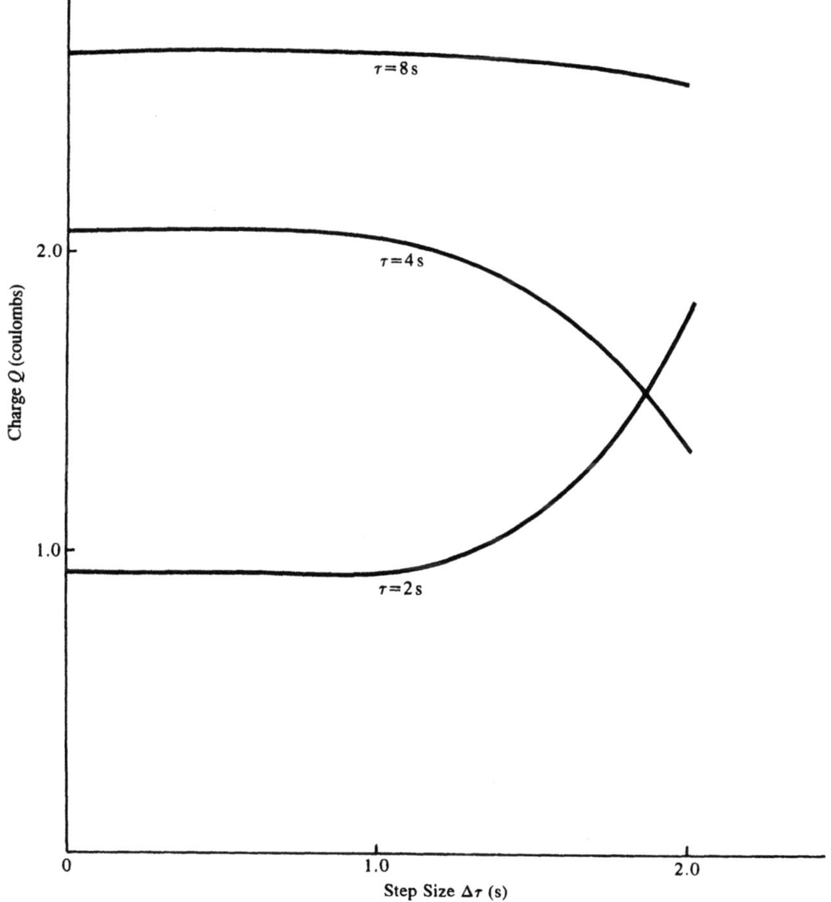

Figure 7.1.3 Variation of the computed capacitor charge Q, at $\tau = 2$, 4, and 8 seconds, with the time step $\Delta\tau$ employed.

versus $\Delta\tau$. Clearly, the accuracy is improved as $\Delta\tau$ is decreased, over the range considered, due to the decrease in the truncation error. Also, the effect of $\Delta\tau$ on the results is smaller at large values of τ. This behavior is expected from the function being integrated, since the integrand approaches zero at large time, giving a constant value of the integral as τ becomes large.

Also note from the results presented in Fig. 7.1.3 that the charge Q approaches a constant value of 2.761 coulombs as time increases beyond about 18 s. This indicates that the capacitor is fully charged by this time. A constant voltage difference of 110.5 volts across the capacitor is also attained. The electrical current asymptotically approaches zero, as expected from the form of the given function $I(\tau)$. Thus, as $\tau \to \infty$, the charge Q and voltage V approach constant values, indicating a finite constant value of the integral as the upper integration limit approaches infinity. The integral for $\tau \to \infty$ may also be evaluated analytically to yield Q as 2.761756, which agrees closely with the numerical result obtained.

7.3 SIMPSON'S RULES FOR NUMERICAL INTEGRATION

One can improve the accuracy with which an integral is computed by increasing the number of segments n into which the range of integration is divided, constrained by the round-off error which becomes significant as the step size Δx is reduced to very small values, or by employing higher-order polynomials $P_m(x)$ to approximate the function $f(x)$. The trapezoidal rule uses a straight line to approximate the function over each segment. Simpson's one-third rule, usually referred to as simply *Simpson's rule*, uses second-order polynomials, that is, parabolas, to approximate the function. One connects successive groups of three points on the $f(x)$ versus x curve with parabolas to determine the area under the curve over the interval defined by these points. Similarly, a third-order polynomial, $m = 3$, requires four points on the curve for the approximation of the function and leads to what is known as *Simpson's three-eighths rule*. Since each segment of the integration domain is associated with only two points on the curve, as shown in Fig. 7.2, Simpson's one-third rule requires a minimum of two segments and an even number of segments n into which the total range of integration is subdivided. Simpson's three-eighths rule requires a minimum of three segments, and, if it is used in conjunction with the one-third rule, n may be odd or even.

7.3.1 Simpson's One-Third Rule

The function $f(x)$ in the integral

$$I = \int_a^b f(x)\,dx \tag{7.1}$$

is replaced by a second-order polynomial, or a parabola, for numerical integration by

Simpson's rule. Three points on the curve of $f(x)$ versus x are needed to determine this parabola. Consider the three points (x_{i-1}, f_{i-1}), (x_i, f_i), and (x_{i+1}, f_{i+1}), as shown in Fig. 7.5. A parabola that passes through these three points may be found and the area under the curve of $f(x)$ approximated by that under the parabola. Two segments, each of width Δx, are involved in this computation, since three points are needed to define the parabola.

We may employ the various methods of interpolation discussed in Chapter 6 to determine the parabola passing through the three given points. Lagrange interpolation provides the general method for deriving Newton-Cotes formulas. However, because only a second-order polynomial, $m = 2$, is under consideration here, we may simply employ the general form of the equation for a parabola and determine the coefficients by substituting the coordinates for the three points into this equation, as done in Section 6.2.1. Therefore, the second-order polynomial may be taken as

$$P_2(x) = Ax^2 + Bx + C \tag{7.24}$$

Since this parabola passes through the three points being considered, as shown in Fig. 7.5, the constants A, B, and C can be determined from

$$f_{i-1} = A(-\Delta x)^2 + B(-\Delta x) + C$$

$$f_i = C$$

$$f_{i+1} = A(\Delta x)^2 + B(\Delta x) + C$$

where x_i has been taken at the origin, $x = 0$, to simplify the calculation. Such a choice does not affect the generality of the derivation. From the above equations,

$$A = \frac{f_{i-1} - 2f_i + f_{i+1}}{2(\Delta x)^2} \qquad B = \frac{f_{i+1} - f_{i-1}}{2\,\Delta x} \qquad C = f_i$$

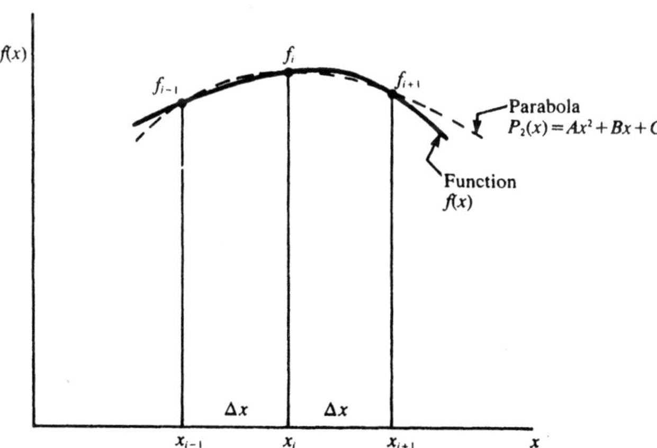

Figure 7.5 Replacement of the function $f(x)$ over the width of two segments by a parabola for the derivation of Simpson's one-third rule.

The area under the polynomial of Eq. (7.24) is denoted by I_p and is given by

$$I_p = \int_{-\Delta x}^{\Delta x} (Ax^2 + Bx + C)\,dx = \left[A\frac{x^3}{3} + B\frac{x^2}{2} + Cx\right]_{-\Delta x}^{\Delta x} = \frac{2}{3}A(\Delta x)^3 + 2C(\Delta x)$$

Therefore, the area under the curve in the two segments is approximated by

$$I_p = \left[\frac{f_{i-1} - 2f_i + f_{i+1}}{3} + 2f_i\right]\Delta x = \left(\frac{f_{i-1} + 4f_i + f_{i+1}}{3}\right)\Delta x \tag{7.25}$$

where the expressions for A and C, given above, have been substituted. This formula is known as *Simpson's rule*, or as *Simpson's one-third rule*, because the step size Δx is divided by 3 in the formula. The use of the one-third in the terminology distinguishes this method from a similar one, derived later, in which Δx is multiplied by 3/8, instead of 1/3, and which is known as *Simpson's three-eighths rule.*

We may use the expression in Eq. (7.25) for the integral over two segments to determine the integral over the entire range $a \leqslant x \leqslant b$, which is divided into n segments of equal width Δx. Here, n must be even in order to consider groups of two segments for the application of Eq. (7.25); see Fig. 7.6. Then the total integral I is approximated by the following:

$$I = \int_a^b f(x)\,dx \simeq \sum_{j=1}^{n/2} (I_p)_j$$

where $(I_p)_j$ is the integral given by Eq. (7.25) for the jth group of two segments. Thus, i in Eq. (7.25) is given by $i = 2j - 1$. Therefore,

$$I \simeq \frac{\Delta x}{3}\left(f_0 + 4\sum_{i=1,3,5,\ldots}^{n-1} f_i + 2\sum_{i=2,4,6,\ldots}^{n-2} f_i + f_n\right) \tag{7.26}$$

It is shown later in this section that the truncation error per step in numerical integration by Simpson's rule is $O[(\Delta x)^5]$, which results in a total error of $O[(\Delta x)^4]$.

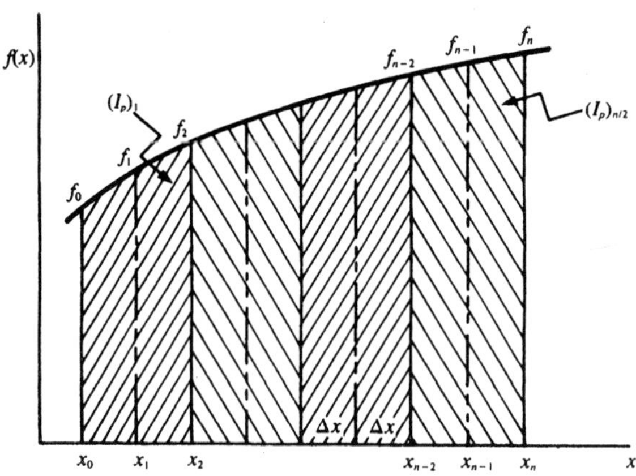

Figure 7.6 Application of Simpson's rule, with an even number of strips, for the numerical integration of a function $f(x)$ over the range $x = a$ to $x = b$.

Therefore, this is a fourth-order method and is much more accurate than the trapezoidal rule for an arbitrary function $f(x)$. If the function being integrated is a polynomial of order zero, one, or two, Simpson's rule is exact, that is, there is no truncation error. In fact, it is exact even for third-order polynomials, or cubics, since the leading term in the truncation error contains only the fourth derivative f'''', all terms containing the lower derivatives having canceled out. Computer programming for Simpson's rule is more involved than that for the trapezoidal rule; compare Examples 7.1 and 7.2. However, because of its much higher accuracy level, Simpson's rule is widely used in engineering applications, where accuracy is usually important. Example 7.2 demonstrates the use of this method in a problem of practical interest.

7.3.2 Simpson's Three-Eighths Rule

If a third-order polynomial, $m = 3$, is employed to approximate the integrand $f(x)$ by requiring that it pass through four points on the curve of $f(x)$, Simpson's three-eighths rule is obtained. A minimum of three segments are needed in order to provide the four points for the determination of the polynomial, as shown in Fig. 7.7. The general equation for the polynomial is taken as

$$P_3(x) = Ax^3 + Bx^2 + Cx + D \tag{7.27}$$

In a manner similar to that given above for the derivation of the one-third rule, the polynomial is determined by substitution of the coordinates of the four points through which it passes. The integral over the three segments is approximated by the corresponding integral I_p of the polynomial $P_3(x)$. The resulting expression for I_p is

$$I_p = \frac{3}{8}\Delta x(f_{i-1} + 3f_i + 3f_{i+1} + f_{i+2}) \tag{7.28}$$

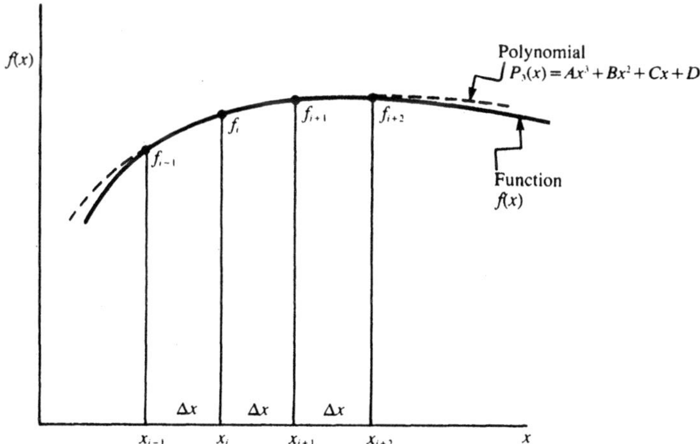

Figure 7.7 Replacement of the function $f(x)$ over the width of three segments by a third-order polynomial for the derivation of Simpson's three-eighths rule.

For application of this method over the entire range of integration, n must be a multiple of 3. Then the integral I is approximated by

$$I = \int_a^b f(x)\,dx \simeq \sum_{j=1}^{n/3} (I_p)_j$$

where $(I_p)_j$ is the integral for the jth group of three segments, giving i in Eq. (7.28) as $i = 3j - 2$.

$$I = \frac{3}{8}\Delta x\left[f_0 + 3\sum_{i=1,4,7,\ldots}^{n-2}(f_i + f_{i+1}) + 2\sum_{i=3,6,9,\ldots}^{n-3} f_i + f_n\right] \tag{7.29}$$

It is shown below that the truncation error in numerical integration by Simpson's three-eighths rule is of the same order as that for the one-third rule. Because of this and the requirement that the number of segments n must be a multiple of 3, the three-eighths rule is seldom used by itself. Simpson's one-third rule is easier to program, and the constraint on n is only that it must be even. However, if the two methods are used together, no such constraint on n is needed. If n is even, Simpson's one-third rule is employed for numerical integration over the entire region. If n is odd, one can use Simpson's three-eighths rule, for instance, to compute the area under the curve in the first three segments and the one-third rule for the remaining even number of segments. Thus, a combination of the two methods provides fourth-order accuracy in the numerical results, without restricting the number of segments that may be employed, except that $n \geqslant 2$.

7.3.3 Truncation Errors

The derivation of the truncation errors associated with Simpson's rules for numerical integration follows the procedure presented for the trapezoidal rule. Thus, $y(x)$ represents the integral $\int_a^x f(x)\,dx$, and $y'(x) = f(x)$, $y''(x) = f'(x)$, and so on. The exact area under the curve of $f(x)$ over the two segments shown in Fig. 7.8 is $y_{i+1} - y_{i-1}$. Expanding y_{i+1} and y_{i-1} in Taylor series about y_i, we obtain

$$y_{i+1} = y_i + \Delta x y_i' + \frac{(\Delta x)^2}{2!}y_i'' + \frac{(\Delta x)^3}{3!}y_i''' + \frac{(\Delta x)^4}{4!}y_i'''' + \frac{(\Delta x)^5}{5!}y_i''''' + \cdots$$

$$y_{i-1} = y_i - \Delta x y_i' + \frac{(\Delta x)^2}{2!}y_i'' - \frac{(\Delta x)^3}{3!}y_i''' + \frac{(\Delta x)^4}{4!}y_i'''' - \frac{(\Delta x)^5}{5!}y_i''''' + \cdots$$

Therefore,

$$y_{i+1} - y_{i-1} = 2\,\Delta x f_i + \frac{(\Delta x)^3}{3}f_i'' + \frac{(\Delta x)^5}{60}f_i'''' + O[(\Delta x)^7] \tag{7.30}$$

where the relationships between $y(x)$ and $f(x)$, from Eq. (7.11), have been used. We need a finite difference approximation for f_i'' to obtain the expression for Simpson's

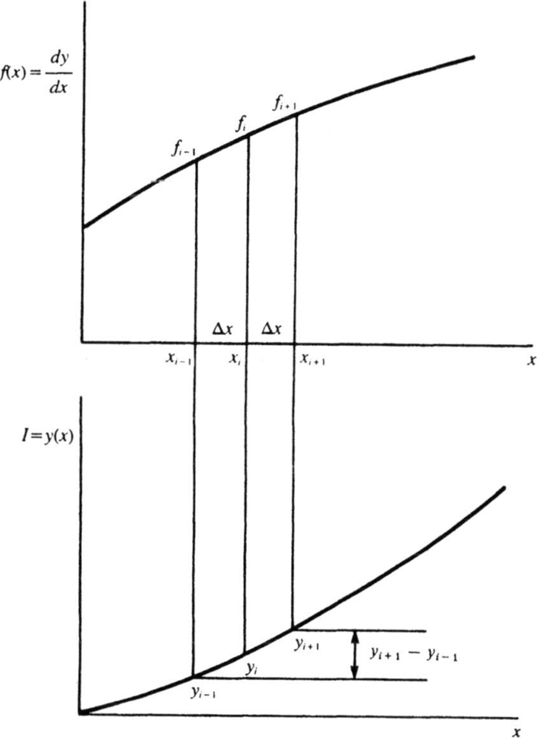

Figure 7.8 Sketch of a function $y'(x)$ and its integral $y(x)$ for the estimation of truncation error in Simpson's rule for numerical integration.

one-third rule on the right-hand side of Eq. (7.30). From Section 3.4.2.,

$$f_i'' = \frac{f_{i+1} - 2f_i + f_{i-1}}{(\Delta x)^2} - \frac{1}{12}(\Delta x)^2 f_i'''' + O[(\Delta x)^4] \tag{7.31}$$

Substituting this expression for f_i'' into Eq. (7.30), we obtain the resulting equation:

$$\underbrace{y_{i+1} - y_{i-1}}_{\text{Exact Value}} = \underbrace{\frac{\Delta x}{3}(f_{i+1} + 4f_i + f_{i-1})}_{\text{Simpson's One - Third Rule}} - \underbrace{\frac{1}{90}(\Delta x)^5 f_i'''' + O[(\Delta x)^7]}_{\text{Truncation Error}}$$

From the remainder theorem, the truncation error per step (TE/step) of Simpson's one-third rule for numerical integration is

$$\text{TE/step} = -\frac{1}{90}(\Delta x)^5 f''''(\xi) \qquad \text{where } x_{i-1} < \xi < x_{i+1} \tag{7.32}$$

Therefore, the error will be zero if $f'''' = 0$. This implies that the method is exact for a third-order polynomial. We obtain the total error E by summing the errors in all the steps:

$$E = \sum_{i=1}^{n/2} \left[-\frac{1}{90}(\Delta x)^5 f''''(\xi_i) \right] \qquad \text{where } x_{2i-2} < \xi_i < x_{2i}$$

Again, defining an arithmetic mean f''''_{av} of the values of f'''' in the $n/2$ subintervals, each of width $2\Delta x$, we may write the total truncation error as

$$E = -\frac{1}{90}(\Delta x)^5 \frac{n}{2} f''''_{av} = -\frac{1}{90}(\Delta x)^5 \frac{b-a}{2\,\Delta x} f''''_{av}$$
$$E = -\frac{1}{180}(\Delta x)^4 (b-a) f''''_{av} \tag{7.33}$$

If f''''_{av} is assumed to remain essentially unchanged as Δx is varied, the total error E may be expressed as

$$E \simeq S_s(\Delta x)^4 = O[(\Delta x)^4] \tag{7.34}$$

where $S_s = -(b-a)f''''_{av}/180$ and is assumed to be a constant. This indicates that Simpson's one-third rule is fourth-order accurate. On the basis of this expression for the truncation error, higher-order accuracy may be obtained by the use of Richardson extrapolation, as discussed in the next section.

The truncation error for Simpson's three-eighths rule may be derived in a similar way. It can be shown that the truncation error per step in the expression given by Eq. (7.28) is

$$\text{TE/step} = -\frac{3}{80}(\Delta x)^5 f''''(\xi) \qquad \text{where } x_{i-1} < \xi < x_{i+2}$$

Proceeding as before, we can show that the total error E is

$$E = -\frac{1}{80}(\Delta x)^4 (b-a) f''''_{av} \simeq S_{ST}(\Delta x)^4$$

where f''''_{av} is the arithmetic mean of the values of f'''' in the $n/3$ subintervals, each of width $3\Delta x$. Therefore, the method is also fourth-order accurate, and the total error is somewhat larger than that for the one-third rule for a given step size Δx. However, if the total integration region is divided into three segments for the application of the three-eighths rule and into two segments for the one-third rule, the former yields more accurate results, because the step size is $(b-a)/3$ in the first case and $(b-a)/2$ in the second case. The smaller step size for the application of the three-eighths rule then yields a smaller total truncation error.

Example 7.2

The velocity profile in the turbulent flow of a fluid in a smooth circular pipe may be represented by the empirical power-law equation

$$U(x) = 5\left(1 - \frac{x}{R}\right)^{1/7} \tag{7.2.1}$$

where $U(x)$ is the axial velocity in the pipe, in m/s; x is the radial distance from the axis, in meters; and R is the radius of the pipe. The total volume flow rate in the pipe is then given by the integral $\int_0^R U(x) 2\pi x\,dx$. Using Simpson's one-third rule, for $R = 0.1$ m, compute this integral as accurately as possible. Also determine the average velocity.

Solution

The integral to be evaluated numerically is

$$I = \int_0^{0.1} 5\left(1 - \frac{x}{0.1}\right)^{1/7} 2\pi x\,dx \tag{7.2.2}$$

Since x and R are in meters and the velocity U is in m/s, this integral will yield the volume flow rate in m^3/s. We obtain the average velocity V_{av} by dividing the flow rate by the area of cross section of the pipe. Therefore,

$$V_{av} = \frac{I}{\pi R^2} = \frac{I}{\pi (0.1)^2}\ \text{m/s} \tag{7.2.3}$$

Figure 7.2.1 presents the computer program in FORTRAN 77 for the numerical integration of the given function $2\pi x U(x)$ by Simpson's one-third rule. The function to be integrated is defined as FUNC(Y), the limits of integration are given, and the number of subdivisions to be employed is entered interactively. The subroutine SIMPSN computes the integral I by Simpson's one-third rule and yields the resulting flow rate, which is denoted by FLOW. The average velocity is then obtained by the use of Eq. (7.2.3).

The numerical results obtained from the numerical scheme are presented in Fig. 7.2.2. The computed flow rate and the average velocity are shown for the number of subdivisions N ranging from 10 to 5120. Note that N = 320 is quite adequate for this problem. Again, at much larger values of N, the round-off error is expected to become significant and to increase the total error, acting against the decrease in the truncation error, as N is increased. Although N is successively doubled a chosen number of times in this program, a better approach would be to continuously monitor the effect of increasing N on the numerical value of the integral. If this effect is smaller than a chosen convergence criterion, then the computation may be terminated. In the program shown in Fig. 7.2.1, N is increased to values much larger than necessary, in order to determine if the effect of round-off error becomes significant at the larger values. The results shown indicate that the trends are pretty much as expected, and the round-off error is small over the range of N considered.

The ratio of the average velocity V_{av} to the velocity V_{max} at the axis can be shown analytically to be given by the expression

$$\frac{V_{av}}{V_{max}} = \frac{2n^2}{(n+1)(2n+1)} \tag{7.2.4}$$

where $1/n$ is the exponent in Eq. (7.2.1). In the present case, $n = 7$ and $V_{max} = 5$ m/s,

```
C          SIMPSON'S ONE-THIRD RULE FOR NUMERICAL INTEGRATION
C
C      U(Y) IS THE VELOCITY DISTRIBUTION, FUNC(Y) THE FUNCTION TO BE
C      INTEGRATED, X THE RADIAL DISTANCE, XMIN AND XMAX THE MINIMUM
C      AND MAXIMUM VALUES OF X, DX THE SEGMENT SIZE, N THE NUMBER OF
C      SUBDIVISIONS, VAVG THE AVERAGE VELOCITY, AND FLOW THE REQUIRED
C      INTEGRAL WHICH GIVES THE FLOW RATE.
C
       IMPLICIT REAL (A-H,O-Z)
C
C      DEFINE FUNCTION TO BE INTEGRATED
C
       U(Y)=5.0*((1.0-Y/0.1)**(1.0/7.0))
       FUNC(Y)=2.0*3.14159*Y*U(Y)
C
C      ENTER INPUT QUANTITIES
C
       XMIN=0.0
       XMAX=0.1
       PRINT *,'ENTER THE NUMBER OF SUBDIVISIONS N'
       READ *,N
       DO 4 J=1,10
C
C      CALL SUBROUTINE FOR SIMPSON'S ONE-THIRD RULE
C
       CALL SIMPSN(FUNC,XMIN,XMAX,N,FLOW)
       VAVG=FLOW/(3.14159*0.1*0.1)
       WRITE(1,1)FLOW,VAVG,N
  1    FORMAT(/5X,'FLOW RATE =',E12.4,5X,'AVG. VEL.=',E12.4,5X,'N=',I5)
       N=2*N
  4    CONTINUE
       STOP
       END
C
C
C
       SUBROUTINE SIMPSN(FUNC,XMIN,XMAX,N,FLOW)
C
C      THIS SUBROUTINE COMPUTES THE INTEGRAL BY SIMPSON'S
C      ONE-THIRD RULE FOR NUMERICAL INTEGRATION
C
       IMPLICIT REAL (A-H,O-Z)
       SUM=0.0
       X=XMIN
       DX=(XMAX-XMIN)/N
       DO 2 I=1,N-1
       X=X+DX
       S=I
       IF((S/2.0 - I/2) .GT. 0.1)THEN
       SUM=SUM+4.0*FUNC(X)
       ELSE
       SUM=SUM+2.0*FUNC(X)
       END IF
  2    CONTINUE
C
C      APPLY SIMPSON'S RULE TO COMPUTE THE INTEGRAL
C
       FLOW=(DX/3.0)*(FUNC(XMIN)+SUM+FUNC(XMAX))
       RETURN
       END
```

Figure 7.2.1 Computer program for the numerical integration of the function given in Example 7.2 by Simpson's rule to determine the flow rate and the average velocity in the pipe.

```
 ENTER THE NUMBER OF SUBDIVISIONS N
10
        FLOW RATE =  0.1230E+00     AVG. VEL.=  0.3917E+01     N=   10
        FLOW RATE =  0.1259E+00     AVG. VEL.=  0.4008E+01     N=   20
        FLOW RATE =  0.1272E+00     AVG. VEL.=  0.4049E+01     N=   40
        FLOW RATE =  0.1278E+00     AVG. VEL.=  0.4068E+01     N=   80
        FLOW RATE =  0.1281E+00     AVG. VEL.=  0.4076E+01     N=  160
        FLOW RATE =  0.1282E+00     AVG. VEL.=  0.4080E+01     N=  320
        FLOW RATE =  0.1282E+00     AVG. VEL.=  0.4082E+01     N=  640
        FLOW RATE =  0.1283E+00     AVG. VEL.=  0.4083E+01     N= 1280
        FLOW RATE =  0.1283E+00     AVG. VEL.=  0.4083E+01     N= 2560
        FLOW RATE =  0.1283E+00     AVG. VEL.=  0.4083E+01     N= 5120
```

Figure 7.2.2 Numerical results on the flow rate and the average velocity, obtained in Example 7.2, for various values of the number of subdivisions N.

being the velocity at the axis, $y = 0$. This gives $V_{av} = 0.8167 V_{max} = 0.40835$ m/s. This analytical value agrees closely with the numerical result obtained, lending support to the accuracy of the numerical scheme.

7.4 HIGHER-ACCURACY METHODS

Accuracy is generally important in engineering applications. In the dynamics of bodies, for instance, the integration of the force over distance gives the change in the kinetic energy, an accurate determination of which is necessary to study the damping and accelerating characteristics of the body for a suitable control system. Similarly, the integral of mass transfer rate over time yields the total mass transfer from a chemical reactor. An accurate evaluation of this integral is needed for supplying the required inflow of material into the system. Because of the high level of accuracy generally needed in engineering problems, methods have been developed for improving the accuracy of the numerical results obtained from integration formulas such as those discussed in the preceding sections. Higher-order Newton-Cotes formulas may also be employed for obtaining greater accuracy in numerical integration. This section discusses several of these higher-accuracy methods.

7.4.1 Richardson Extrapolation

Richardson extrapolation, which is also called *deferred approach to the limit*, is a numerical method for improving the accuracy of the results obtained from a given numerical scheme, provided an estimate of the total discretization error is available.

Although the technique is applied to numerical integration here, it can be used for a wide variety of numerical problems, such as the solution of differential equations by finite difference methods. Let us first consider integration by the trapezoidal rule. The total truncation error is given by Eq. (7.21) as $E \simeq S_T(\Delta x)^2$. Then, if I is the exact integral, and I_1 and I_2 are the numerical values of the integral obtained with step sizes Δx_1 and Δx_2, we may write, using TE to represent the total discretization error,

$$I \simeq I_1 + S_T(\Delta x_1)^2 \tag{7.35a}$$

and

$$I \simeq I_2 + S_T(\Delta x_2)^2 \tag{7.35b}$$

From these equations, the constant S_T may be estimated as follows:

$$S_T \simeq \frac{I_2 - I_1}{(\Delta x_1)^2 - (\Delta x_2)^2} \tag{7.36}$$

Then, from Eq. (7.35b),

$$I \simeq I_2 + \frac{I_2 - I_1}{(\Delta x_1/\Delta x_2)^2 - 1} \tag{7.37}$$

Equation (7.37) does not yield the exact value of the integral I, since the expression for the truncation error is only an approximate one and since TE is employed instead of discretization error. However, an improved estimate for I is obtained from Eq. (7.37). The second term in this equation represents the truncation error for integration with a step size of Δx_2. If Δx_2 is taken as half of Δx_1, that is, $\Delta x_1/\Delta x_2 = 2$, then

$$I \simeq I_2 + \frac{I_2 - I_1}{3} = \frac{4I_2 - I_1}{3} \tag{7.38}$$

It can be easily shown that this expression for the integral is identical to that obtained from Simpson's one-third rule with a step size of Δx_2. Therefore, the integral is obtained to fourth-order accuracy.

Similarly, if Simpson's one-third rule is considered, the total truncation error is $S_s(\Delta x)^4$. As before, we may write

$$I \simeq I_1 + S_s(\Delta x_1)^4 \tag{7.39a}$$

and

$$I \simeq I_2 + S_s(\Delta x_2)^4 \tag{7.39b}$$

where, again, I is the exact integral and I_1 and I_2 are the numerical values from Simpson's rule for step sizes Δx_1 and Δx_2, respectively. Then

$$S_s \simeq \frac{I_2 - I_1}{(\Delta x_1)^4 - (\Delta x_2)^4} \tag{7.40}$$

and

$$I \simeq I_2 + \frac{I_2 - I_1}{(\Delta x_1/\Delta x_2)^4 - 1} \tag{7.41}$$

Equation (7.41) yields an improved estimate of the integral. Since truncation error of $O[(\Delta x)^4]$ has been eliminated, this expression gives the results to a sixth-order accuracy. If $\Delta x_2 = \Delta x_1/2$, a more accurate approximation to the integral is obtained from

$$I \simeq I_2 + \frac{I_2 - I_1}{15} = \frac{16I_2 - I_1}{15} \tag{7.42}$$

Note that the accuracy of the numerical results obtained from the trapezoidal rule or from Simpson's rule can be substantially improved by computing the integral twice, with two different step sizes, and using Eq. (7.37) or Eq. (7.41). Generally, numerical integration is carried out with a chosen step size, which is then halved to yield the second estimate I_2 of the integral. Then Eq. (7.38) or Eq. (7.42) yields the improved estimate. This method does not require any major change in the computer program since the numerical scheme is simply applied twice. The computational effort is essentially doubled. However, because of the considerable improvement in accuracy and the simplicity of its application, Richardson extrapolation is frequently used with numerical schemes for integration and also with various other numerical schemes, such as those for solving ordinary and partial differential equations by finite difference methods.

7.4.2 Romberg Integration

Richardson extrapolation substantially improves the accuracy of the numerical results by eliminating the leading term in the truncation error. Thus, one may obtain fourth-order accuracy by applying the trapezoidal rule twice, with different step sizes, and using Richardson extrapolation to determine the improved value of the integral, as given above. One may apply this technique in succession to eliminate still higher-order terms in the truncation error. This leads to an efficient method, known as *Romberg integration*, which is widely used for obtaining numerical results of high accuracy.

The truncation error in the trapezoidal rule for numerical integration was obtained in terms of the dominant term by the use of the remainder theorem. However, it can be shown (Ralston, 1965 and Davis and Rabinowitz, 1967) that if the higher-order terms are included, the error E in the trapezoidal rule may be written as

$$E = A_1(\Delta x)^2 + A_2(\Delta x)^4 + A_3(\Delta x)^6 + A_4(\Delta x)^8 + \cdots \tag{7.43}$$

where the A's are constants. The leading term, of order $(\Delta x)^2$, was eliminated by the application of Richardson extrapolation in the preceding section. If the integral computed by the trapezoidal rule, with n segments, is denoted as $I_{0,n}$ and the improved value of the integral by Richardson extrapolation as $I_{1,n}$, then, from Eq. (7.38),

$$I_{1,n} = \frac{4I_{0,n} - I_{0,n/2}}{4^1 - 1} \tag{7.44}$$

where the two integrals by the trapezoidal rule are obtained with $n/2$ and n intervals, corresponding to step sizes Δx and $\Delta x/2$, respectively.

Similarly, we eliminate the second term in the series representing the total truncation error, Eq. (7.43), by applying Richardson extrapolation again. Since this term is of fourth order, the next extrapolation will be of sixth-order accuracy. Denoting this second extrapolation as $I_{2,n}$, we obtain

$$I_{2,n} = \frac{4^2 I_{1,n} - I_{1,n/2}}{4^2 - 1} \tag{7.45}$$

The process may be continued indefinitely, improving the value of the integral by successively eliminating the higher-order terms in the error. The general formula for the kth-order extrapolation is obtained as follows:

$$I_{k,n} = \frac{4^k I_{k-1,n} - I_{k-1,n/2}}{4^k - 1} \tag{7.46}$$

Thus, the value of the integral may be improved to the desired level of accuracy. Figure 7.9 shows a schematic of Romberg integration. First, $I_{0,1}$ and $I_{0,2}$ are determined from the trapezoidal rule, with one and two elements, respectively. These yield the first extrapolation $I_{1,2}$. Similarly, $I_{0,4}$ used with $I_{0,2}$ gives $I_{1,4}$. The second extrapolation, $I_{2,4}$, is computed from $I_{1,2}$ and $I_{1,4}$. This computation of the integrals by the trapezoidal rule and of the improved values by Richardson extrapolation continues until the results remain essentially unchanged from one order of extrapolation to the next. Thus, if $|I_{2,4} - I_{1,4}|$ is less than a chosen convergence criterion ε,

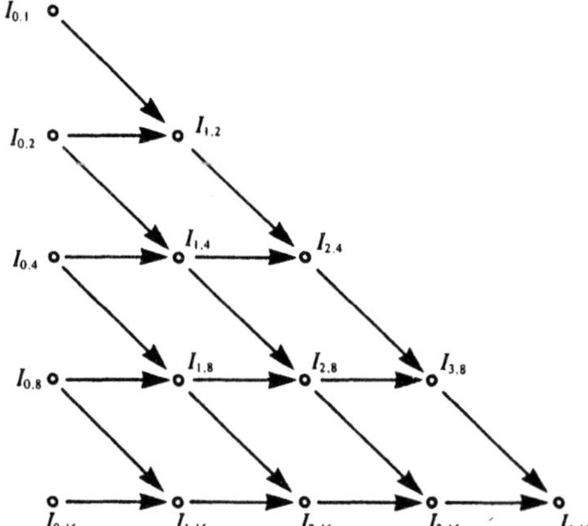

Figure 7.9 A schematic of Romberg integration, indicating the various levels of extrapolation.

the process is terminated there. If not, we employ eight elements for the trapezoidal rule to determine the improved values of the integral for eight elements. Then, we compare $I_{3,8}$ with $I_{2,8}$ for convergence. Thus, the criterion for convergence may be written as

$$|I_{\lambda,n} - I_{\lambda-1,n}| \leqslant \varepsilon \tag{7.47}$$

where λ represents the highest order of extrapolation that can be obtained with n elements.

Romberg integration can, therefore, be used to obtain results of arbitrary accuracy, as far as truncation error is concerned. However, the round-off error, as always, imposes a limitation on the accuracy that may be achieved in practice. Example 7.3 demonstrates the use of Romberg integration in the accurate evaluation of an integral.

7.4.3 Higher-Order Newton-Cotes Formulas

So far, we have considered only the zeroth-, first-, second-, and third-order Newton-Cotes formulas. In general, these formulas are quite adequate for most engineering applications. The trapezoidal and Simpson's one-third rules are extensively used. The three-eighths rule is generally used in conjunction with the one-third rule if an odd number of segments is to be employed. We derived the formulas for these cases simply by taking the general form of the polynomial, whose coefficients were determined by making this curve pass through the required number of points on the plot of $f(x)$ versus x. However, the general approach for the derivation of Newton-Cotes formulas is based on the use of Lagrange interpolation, presented in Chapter 6. Since the segments are of uniform width, the points employed in the determination of the interpolating polynomial are uniformly distributed.

Closed Newton-Cotes formulas are those for which the data points at the two ends of the integration interval are known. Thus, the rectangular, trapezoidal, and Simpson's rules are all closed formulas. Open Newton-Cotes formulas are based on integration limits that lie beyond the range of available data. Although seldom used for numerical integration, open formulas are of interest in the solution of ordinary differential equations, as illustrated in Chapter 8. Table 7.1 gives the formulas and the truncation errors per step for several Newton-Cotes closed integration schemes. Note that the accuracy improves substantially as the order of the polynomial increases from zero to two. The error for Simpson's three-eighths rule is of the same order as that for the one-third rule. In fact, the former is somewhat larger in magnitude. Therefore, the three-eighths rule is seldom used by itself and is generally used in conjunction with the one-third rule.

The accuracy again increases substantially as the order of the polynomial changes from two or three to four, the formula for which is often known as *Boole's rule.* The next-order polynomial is found to yield an accuracy of the same order as Boole's law, as observed for Simpson's rules. Because of the constraints imposed on the minimum number of points and, therefore, on the minimum and total number of

TABLE 7.1 Newton-Cotes Closed Integration Formulas, Along with the Truncation Error per Step and the Minimum Number of Segments Needed.

Minimum no. of segments	Points	Name	Formula	Truncation Error per Step
1	2	Rectangular rule	$f_i \Delta x$ or $f_{i+1} \Delta x$	$\frac{1}{2}(\Delta x)^2 f'(\xi)$ or $-\frac{1}{2}(\Delta x)^2 f'(\xi)$
1	2	Trapezoidal rule	$\frac{f_{i-1}+f_i}{2}\Delta x$	$-\frac{1}{12}(\Delta x)^3 f''(\xi)$
2	3	Simpson's one-third rule	$\frac{f_{i-1}+4f_i+f_{i+1}}{3}\Delta x$	$-\frac{1}{90}(\Delta x)^5 f''''(\xi)$
3	4	Simpson's three-eighths rule	$\frac{3(f_{i-1}+3f_i+3f_{i+1}+f_{i+2})}{8}\Delta x$	$-\frac{3}{80}(\Delta x)^5 f''''(\xi)$
4	5	Boole's rule	$\frac{2(7f_{i-2}+32f_{i-1}+12f_i+32f_{i+1}+7f_{i+2})}{45}\Delta x$	$-\frac{8}{945}(\Delta x)^7 f^{VI}(\xi)$
5	6	—	$\frac{5(19f_{i-2}+75f_{i-1}+50f_i+50f_{i+1}+75f_{i+2}+19f_{i+3})}{288}\Delta x$	$-\frac{275}{12{,}096}(\Delta x)^7 f^{VI}(\xi)$

segments that must be used for higher-order formulas, these are generally more difficult to program and are less versatile. As a consequence, the trapezoidal and Simpson's rules are the most extensively employed methods for numerical integration in engineering problems. However, Richardson extrapolation is frequently used, as in Romberg integration, to improve the accuracy of the numerical results.

Example 7.3

A mathematical function frequently encountered in the analysis of several engineering problems is the Gaussian error function, erf z, which is defined as

$$\operatorname{erf} z = \frac{2}{\sqrt{\pi}} \int_0^z e^{-x^2}\, dx \tag{7.3.1}$$

Using Romberg integration, compute the value of the error function at $z = 0.5$, 1.0, 1.5, and 2.0, with a convergence criterion parameter ε in Eq. (7.47) of 10^{-5}.

Solution

The recursion formula for computing the various orders of extrapolation in Romberg integration is given by Eq. (7.46). For convenience, the first-order approximation, or extrapolation, is taken as the trapezoidal rule, with Richardson extrapolation being the second-order extrapolation. Then $I_{1,m}$ represents the integral by the trapezoidal rule with m segments. With this change, the recursion formula for the kth-order extrapolation becomes

$$I_{k,n} = \frac{4^{k-1} I_{k-1,n} - I_{k-1,n-1}}{4^{k-1} - 1} \tag{7.3.2}$$

where $I_{k-1,n}$ represents the more accurate extrapolation of $(k-1)$th order, and $I_{k-1,n-1}$ the less accurate one, that is, at half the number of segments as the former. The recursion formula given by Eq. (7.3.2) is more convenient to use than that given earlier by Eq. (7.46), since we start with the first-order extrapolation, which is simply the trapezoidal rule, and we can successively double the number of segments, starting with 1 and keeping track of increasing accuracy by means of the subscript n.

The computer program for solving this problem is presented in Fig. 7.3.1. The given function $F(x) = (2/\sqrt{\pi})\exp(-x^2)$, which is to be integrated, is defined; the input variables, such as the convergence criterion ε, which is denoted by EPS, are entered; and the value of z at which the error function erf z is to be computed is given. Integration by the trapezoidal rule is carried out with one segment, over the range $x = 0$ to $x = z$. The segment size is successively halved, making maximum use of the segment areas already calculated. The program then computes the higher-order extrapolations, using Eq. (7.3.2). At each iterative step, corresponding to a given number of subdivisions M, the difference between the two highest possible extrapolations $I_{\lambda,n}$ and $I_{\lambda-1,n}$, as defined in Eq. (7.47) and denoted by Y in the program, is determined. Using Eq. (7.47), if this difference is less than or equal to ε, the program is

```
C                              ROMBERG INTEGRATION
C
C       F(X) IS THE FUNCTION TO BE INTEGRATED, X THE INDEPENDENT
C       VARIABLE, XMIN AND XMAX THE MINIMUM AND MAXIMUM VALUES OF X,
C       DX THE SEGMENT WIDTH, ERF(Z) THE ERROR FUNCTION AT Z, EPS
C       THE CONVERGENCE CRITERION, Y THE VALUE OF THE INTEGRAL
C       CORRESPONDING TO AN EXTRAPOLATION, M THE NUMBER OF SEGMENTS,
C       AND DIF THE DIFFERENCE BETWEEN THE RESULTS FOR THE TWO HIGHEST
C       ORDERS OF EXTRAPOLATION AT A GIVEN NUMBER OF SEGMENTS.
C
        DIMENSION Y(8,8)
C
C       DEFINE FUNCTION TO BE INTEGRATED
C
        F(X)=(2.0/SQRT(3.14159))*EXP(-X**2)
C
C       ENTER INPUT VALUES
C
        EPS=0.00001
        XMIN=0.0
        DO 7 J=1,4
        PRINT *,'ENTER THE VALUE OF Z'
        READ *,XMAX
        DX=XMAX-XMIN
C
C       FIRST ORDER (TRAPEZOIDAL RULE) CALCULATION
C
        N=1
        Y(1,1)=0.5*DX*(F(XMIN)+F(XMAX))
        WRITE(1,1)N,Y(1,1)
    1   FORMAT(2X,'NO. OF ITERATIONS=',I2,5X,'ERF(Z)=',F9.6)
    2   M=2**(N-1)
        DX=DX/2.0
        N=N+1
        Y(1,N)=0.5*Y(1,N-1)
        DO 3 K=1,M
        X=XMIN+(2.0*K-1)*DX
    3   Y(1,N)=Y(1,N)+DX*F(X)
C
C       COMPUTE HIGHER ORDER EXTRAPOLATIONS
C
        DO 4 K=2,N
    4   Y(K,N)=(4.0**(K-1)*Y(K-1,N)-Y(K-1,N-1))/(4.0**(K-1)-1.0)
        DIF=ABS(Y(N,N)-Y(N-1,N))
        WRITE(1,5)N,Y(N,N)
    5   FORMAT(2X,'NO. OF ITERATIONS=',I2,5X,'ERF(Z)=',F9.6)
C
C       APPLY CONVERGENCE CRITERION
C
        IF(DIF .GT. EPS)THEN
        IF(N .LT. 8)THEN
        GO TO 2
        ELSE
        PRINT *,'MORE THAN 8 ITERATIONS'
        END IF
        ELSE
        WRITE(1,6)XMAX,Y(N,N)
    6   FORMAT(/4X,'VALUE OF Z=',F5.3,5X,'ERF(Z)=',F9.6//)
        END IF
    7   CONTINUE
        STOP
        END
```

Figure 7.3.1 Computer program for the numerical integration of the function given in Example 7.3 by Romberg integration to compute the error function erf(z) at various values of z.

```
 ENTER THE VALUE OF Z
0.5
  NO. OF ITERATIONS= 1       ERF(Z)= 0.501791
  NO. OF ITERATIONS= 2       ERF(Z)= 0.520602
  NO. OF ITERATIONS= 3       ERF(Z)= 0.520499

    VALUE OF Z=0.500       ERF(Z)= 0.520499

 ENTER THE VALUE OF Z
1.0
  NO. OF ITERATIONS= 1       ERF(Z)= 0.771744
  NO. OF ITERATIONS= 2       ERF(Z)= 0.843103
  NO. OF ITERATIONS= 3       ERF(Z)= 0.842711
  NO. OF ITERATIONS= 4       ERF(Z)= 0.842700

    VALUE OF Z=1.000       ERF(Z)= 0.842700

 ENTER THE VALUE OF Z
1.5
  NO. OF ITERATIONS= 1       ERF(Z)= 0.935482
  NO. OF ITERATIONS= 2       ERF(Z)= 0.954759
  NO. OF ITERATIONS= 3       ERF(Z)= 0.966706
  NO. OF ITERATIONS= 4       ERF(Z)= 0.966097

    VALUE OF Z=1.500       ERF(Z)= 0.966097

 ENTER THE VALUE OF Z
2.0
  NO. OF ITERATIONS= 1       ERF(Z)= 1.149046
  NO. OF ITERATIONS= 2       ERF(Z)= 0.936492
  NO. OF ITERATIONS= 3       ERF(Z)= 0.998920
  NO. OF ITERATIONS= 4       ERF(Z)= 0.995266
  NO. OF ITERATIONS= 5       ERF(Z)= 0.995322

    VALUE OF Z=2.000       ERF(Z)= 0.995322
```

Figure 7.3.2 Numerical results obtained for Example 7.3, indicating the number of iterative steps needed for Romberg integration and the computed values of the error function.

terminated; otherwise, the segment size is halved and the computation continued. After convergence has been achieved for a given value of z, other values of z, as given in the problem, are successively entered until all the required numerical values have been obtained.

The numerical results obtained from the given computer program are shown in Fig. 7.3.2. The number of iterative steps needed in each case are shown, along with the computed value of the error function. The number of iterations needed are found to increase with z, over the range considered. The value of the error function is listed in most books on mathematical functions, and the values corresponding to $z = 0.5$, 1.0, 1.5, and 2.0 are given, respectively, as 0.5205, 0.8427, 0.9661, and 0.995322. Clearly, these values given in the literature are very close to those obtained from the present computation. Since the number of segments needed for four iterative steps is eight, the results indicate the rapid increase in accuracy as the segment width is halved. As mentioned earlier, Romberg integration is an extremely efficient and accurate method. Consequently, it is very widely employed in engineering problems.

7.5 INTEGRATION WITH SEGMENTS OF UNEQUAL WIDTH

All of the formulas for numerical integration presented so far were based on segments of equal width. This implies that the points, on the curve of the function $f(x)$, that were used for determining the interpolating polynomial were equally spaced. However, this procedure is not necessarily the most efficient one. In regions where the function varies very gradually or where its value is small, the number of points for function evaluation may be reduced without significantly affecting the accuracy of the results. The optimum distribution of points for the numerical integration of a given function may also be derived to obtain maximum accuracy with a given number of function evaluations. In these cases, the segments into which the range of integration is subdivided are not of equal width. Similarly, experimental or numerical data may be available at only specified points which may be unevenly distributed. Procedures for the numerical integration of such data are needed. This section considers various methods for improving the efficiency of numerical integration by using unequal segments and also the methods that may be employed for integrating a function whose value is given at unevenly spaced data points.

7.5.1 Unequally Spaced Data

Experimental results are often obtained at unevenly spaced values of the independent variable. In an experimental study of the displacement of a moving body as a function of time, for instance, more frequent measurements are generally taken at small times, just after the onset of motion, than at large times. Similarly, the pressure loading on a building due to the wind is measured only at discrete locations, which may not be evenly spaced. Numerical solutions also often yield results at unequally spaced data points. Thus, we are faced with the problem of integrating a function $f(x)$ which is available simply as data at arbitrary, unevenly spaced values of the independent variable.

One approach for solving this problem is to employ the curve-fitting techniques given in the preceding chapter, in order to obtain a continuous function $f(x)$. Then the range of integration may be subdivided into segments of equal width, and numerical integration may be carried out by, say, the Newton-Cotes formulas, using the ordinates obtained from the function $f(x)$ derived from curve fitting. This approach is frequently employed in engineering problems, since experimental and numerical data are often curve fitted for using the results in other computations, as discussed in Chapter 6.

The second approach employs the data as given and simply obtains the integral in each segment. The trapezoidal rule may be applied to each segment and the results summed to yield the integral I as follows:

$$I = \Delta x_1 \frac{f_0 + f_1}{2} + \Delta x_2 \frac{f_1 + f_2}{2} + \Delta x_3 \frac{f_2 + f_3}{2} + \cdots + \Delta x_n \frac{f_{n-1} + f_n}{2} \quad (7.48)$$

where $\Delta x_1, \Delta x_2, \ldots, \Delta x_n$ are the widths of the n segments that correspond to $(n + 1)$ data points, x_i represents the value of the independent variable at a given point, and f_i is the corresponding value of the function. Here, $\Delta x_1, \Delta x_2, \ldots, \Delta x_n$ are not equal, as was assumed for deriving Eq. (7.9). A computer program can be easily written to compute the integral I, using Eq. (7.48) for unequal segment widths.

If two adjacent segments are of equal width, Simpson's one-third rule may be used to obtain the integral over these segments. Thus, if Δx_{i-1} and Δx_i are equal, the integral I_s for these two subdivisions is given by

$$I_s = \left(\frac{f_{i-1} + 4f_i + f_{i+1}}{3} \right) \Delta x \quad (7.49)$$

Similarly, if three segments are of equal width, Simpson's three-eighths rule, Eq. (7.28), may be employed. Since Simpson's rules are more accurate than the trapezoidal rule, an implementation of Simpson's rule in the numerical integration, wherever possible, will increase the accuracy of the results. Thus, a program may be developed that checks the widths of adjacent segments before applying numerical integration to the data. If two consecutive segments are of equal width, Simpson's one-third rule is employed, and if three segments are of equal width, Simpson's three-eighths rule is used. If the widths of two adjacent segments are different, the trapezoidal rule is employed. Example 7.4 demonstrates an application of this procedure to unevenly spaced experimental data.

7.5.2 Adaptive Quadrature*

One can employ Romberg integration to obtain numerical results to any desired accuracy, within the constraints imposed by the round-off error. However, since segments of equal width are employed, the entire range of integration is treated uniformly. This approach is not the most efficient one if the function is slowly varying or small in magnitude in certain regions, where fewer points can be taken. Adaptive quadrature enables one to increase the number of points in regions where the accuracy is not at the desired level, while keeping fewer points in regions where satisfactory accuracy has been attained.

Several methods have been developed to achieve such an uneven distribution of points. The main idea is to focus on regions where the error is larger than the desired value. Suppose the integral

$$I = \int_a^b f(x)\, dx \quad (7.1)$$

is to be computed with accuracy ε. Then the error in each subinterval of width Δx must be less than $\varepsilon\, \Delta x/(b-a)$, so that the total error is less than ε. We start by dividing the range of integration into equal segments of width Δx. The integral over each segment is determined by use of, say, the trapezoidal rule. Then each segment is subdivided into two subintervals of width $\Delta x/2$, and the integral is computed. From Richardson's extrapolation, Eq. (7.36), an estimate of the error in these segments of width $\Delta x/2$ is $(i_2 - i_1)/3$, where i_2 is the integral with two segments of width $\Delta x/2$ each and i_1 is the integral over the segment of width Δx. If this error is less than $\varepsilon(\Delta x/2)/(b-a)$, no further subdivisions of the corresponding segments are needed, and the estimate of the integral over the segments is taken for the computation of the total integral over the entire region. If the error is larger than $\varepsilon(\Delta x/2)/(b-a)$ in certain segments, these segments are halved and the above procedure is repeated until the specified accuracy has been attained in these.

This method, therefore, allows one to systematically reduce the error in regions where it is too large, while keeping regions where satisfactory accuracy has been attained unaffected. Simpson's rule may also be used instead of the trapezoidal rule. However, since the final accuracy is prescribed, the trapezoidal rule is more appropriate because it is simpler to use. Adaptive quadrature is particularly useful for complicated functions that have a large variation over certain regions and a small variation over others. In the integration of $\exp(-50x^2)$ over $0 \leqslant x \leqslant 1$, for instance, many more subintervals are needed at small x than at large x to attain uniform accuracy over the entire region. Adaptive quadrature is a valuable method in such cases. For further details, see Forsythe et al. (1977), Ferziger (1981), and Gerald and Wheatley (1984).

Example 7.4

In an experiment on the motion of accelerating bodies, such as airplanes and rockets, the velocity V, in m/s, of a body is measured at several time intervals τ, in seconds. The data obtained are tabulated for time ranging from 0 to 2.0 seconds as follows:

τ	0.0	0.1	0.2	0.3	0.5	0.7	0.8	1.0
V	9.50	10.00	10.57	11.24	12.97	15.38	16.93	20.9

τ	1.1	1.3	1.5	1.6	1.7	1.8	2.0
V	23.41	29.74	38.17	43.33	49.21	55.88	71.90

Compute the distance traveled by the body x as a function of time τ.

Solution

The experimental data are given at unevenly distributed values of the independent variable τ. The distance traveled x is given by the integral

$$x = \int_0^{\tau} V(\tau')\, d\tau' \tag{7.4.1}$$

where τ' is a dummy variable. We shall employ the procedure outlined in Section 7.5.1

```
C                INTEGRATION WITH SEGMENTS OF UNEQUAL WIDTH
C
C       V IS THE VELOCITY, T THE TIME, X THE DISTANCE TRAVELLED, M THE
C       TOTAL NUMBER OF DATA POINTS, EPS THE ALLOWABLE DIFFERENCE BETWEEN
C       ADJACENT SEGMENT WIDTHS FOR APPLYING SIMPSON'S RULES, DT1, DT2, AND
C       DT3 THE THREE NEXT SEGMENT WIDTHS, AND I THE DATA POINT UPTO WHICH
C       THE INTEGRAL IS COMPUTED TO YIELD THE DISTANCE X.
C
        DIMENSION V(20),T(20)
C
C       ENTER INPUT DATA
C
        PRINT *,'ENTER THE NUMBER OF DATA POINTS'
        READ *,M
        OPEN(UNIT=15,FILE='TIM.DAT')
        OPEN(UNIT=16,FILE='VELO.DAT')
        READ (15,*)(T(I),I=1,M)
        READ (16,*)(V(I),I=1,M)
        CLOSE(UNIT=15)
        CLOSE(UNIT=16)
        EPS=0.0001
        I=1
        X=0.0
C
C       COMPUTE WIDTHS OF NEXT THREE SEGMENTS
C
  2     DT1=T(I+1)-T(I)
        DT2=T(I+2)-T(I+1)
        DT3=T(I+3)-T(I+2)
        DT=DT1
        IF(ABS(DT2-DT1) .GE. EPS)THEN
C
C       TRAPEZOIDAL RULE
C
        X=X+(V(I+1)+V(I))*DT/2
        I=I+1
        PRINT *,'TRAPEZOIDAL RULE'
        ELSE IF(ABS(DT3-DT2) .GE. EPS)THEN
C
C       SIMPSON'S ONE-THIRD RULE
C
        X=X+(V(I)+4.0*V(I+1)+V(I+2))*DT/3
        I=I+2
        PRINT *,'SIMPSON ONE-THIRD RULE'
        ELSE
C
C       SIMPSON'S THREE-EIGHTHS RULE
C
        X=X+(V(I)+3.0*V(I+1)+3.0*V(I+2)+V(I+3))*DT*3.0/8.0
        I=I+3
        PRINT *,'SIMPSON THREE-EIGHTHS RULE'
        END IF
C
C       OUTPUT RESULTS
C
        WRITE(1,5)I,T(I),V(I),X
  5     FORMAT(/2X,'I=',I3,4X,'TIME=',F7.4,4X,'VELOCITY=',F8.4,4X,
     $  'DISTANCE=',F8.4/)
        IF(I .LT. M) THEN
        GO TO 2
        END IF
        STOP
        END
```

Figure 7.4.1 Computer program in FORTRAN 77 for evaluating the integral given in Example 7.4, using the tabulated velocity data at unevenly distributed time intervals.

```
 ENTER THE NUMBER OF DATA POINTS
15
 SIMPSON THREE-EIGHTHS RULE

  I=  4     TIME= 0.3000     VELOCITY= 11.2400     DISTANCE=  2.9906

 SIMPSON ONE-THIRD RULE

  I=  6     TIME= 0.7000     VELOCITY= 15.3800     DISTANCE=  8.2240

 TRAPEZOIDAL RULE

  I=  7     TIME= 0.8000     VELOCITY= 16.9300     DISTANCE=  9.8395

 TRAPEZOIDAL RULE

  I=  8     TIME= 1.0000     VELOCITY= 20.9000     DISTANCE= 13.6225

 TRAPEZOIDAL RULE

  I=  9     TIME= 1.1000     VELOCITY= 23.4100     DISTANCE= 15.8380

 SIMPSON ONE-THIRD RULE

  I= 11     TIME= 1.5000     VELOCITY= 38.1700     DISTANCE= 27.8740

 SIMPSON THREE-EIGHTHS RULE

  I= 14     TIME= 1.8000     VELOCITY= 55.8800     DISTANCE= 41.8116

 TRAPEZOIDAL RULE

  I= 15     TIME= 2.0000     VELOCITY= 71.9000     DISTANCE= 54.5896
```

Figure 7.4.2 Computed distance traveled x as a function of time τ from the velocity data given in Example 7.4. The various integration schemes employed over different regions of the given data are also indicated.

for numerical integration to obtain $x(\tau)$. As discussed in the text, Simpson's three-eighths rule is employed when three adjacent segments are of equal width, and Simpson's one-third rule when only two adjacent segments are of equal width. If the size of a given subdivision is different from that of the next subdivision, the trapezoidal rule must be used.

Figure 7.4.1 shows the computer program used to evaluate the integral in Eq. (7.4.1), employing the unevenly spaced data given in the problem. The given data are stored in separate data files, VELO.DAT and TIM.DAT, which are read by the computer. In general, the complete data set may be stored and an arbitrary number of data points may be chosen for the computation. The widths of three adjacent subdivisions, starting with $\tau = 0$ and proceeding in the direction of increasing time, are determined. A small allowable difference ε, denoted by EPS in the program, is employed to determine whether the segment widths are equal. Here, ε is taken as 10^{-4}, which is an arbitrary small quantity and will not affect the numerical results for the given data set. Obviously, ε is used to avoid problems caused by round-off error, because of which two equal segment widths may be indicated as different due to differences in round-off error. If the first two segments are of different width, the trapezoidal rule is employed. If the widths are equal, the third segment is also considered, and if all three are equal in width, Simpson's three-eighths rule is used. Otherwise, Simpson's one-third rule is employed.

The numerical results obtained are shown in Fig. 7.4.2. The data point up to which the given integral is computed, the corresponding time, the velocity, and the total distance x traveled up to this point are given. The method used for numerical integration in the preceding subdivision(s) is also indicated. Thus, the three-eighths rule is employed first, followed by Simpson's one-third rule, then the trapezoidal rule, and so on. The velocity V is printed in order to ensure that the input data have been correctly read. At the end of 2 s, the total distance traveled is obtained as 54.5896 meters. This program is quite flexible and can be used for a wide variety of unevenly spaced data.

7.5.3 Gauss Quadrature*

In many engineering problems, the evaluation of the integrand $f(x)$ is very involved and time-consuming. Gauss quadrature is based on a variety of interpolating functions and gives maximum accuracy for a given number of function evaluations. However, the x locations where the function $f(x)$ is to be evaluated are adjustable. Thus, for a two-point formula, the integral

$$I = \int_{-1}^{1} f(x)\,dx \tag{7.50a}$$

is approximated by

$$I \simeq A_1 f(x_1) + A_2 f(x_2) \tag{7.50b}$$

where A_1, A_2, x_1, and x_2 are all unknowns. The integration limits are taken as -1 and 1. The integral between the finite limits a and b can be transformed into the limits ± 1 by means of the transformation

$$\xi = \frac{2x - (a + b)}{b - a} \tag{7.51}$$

Thus, the integral may be taken over the limits -1 to 1 without loss of generality. This simplifies the computation and generalizes the formulation. Example 7.5 illustrates how this transformation is carried out in practice.

Now, if we require that Eq. (7.50b) yield the integrals for constant, linear, parabolic, and cubic functions exactly, the four coefficients in the equation can be determined. Thus, employing 1, x, x^2, and x^3 as the functions, and substituting these for $f(x_1)$ and $f(x_2)$, we have

$$A_1 f(x_1) + A_2 f(x_2) = \int_{-1}^{1} 1\,dx = 2 \quad \text{or} \quad A_1 + A_2 = 2$$

$$A_1 f(x_1) + A_2 f(x_2) = \int_{-1}^{1} x\,dx = 0 \quad \text{or} \quad A_1 x_1 + A_2 x_2 = 0$$

$$A_1 f(x_1) + A_2 f(x_2) = \int_{-1}^{1} x^2\,dx = \frac{2}{3} \quad \text{or} \quad A_1 x_1^2 + A_2 x_2^2 = \frac{2}{3}$$

$$A_1 f(x_1) + A_2 f(x_2) = \int_{-1}^{1} x^3\,dx = 0 \quad \text{or} \quad A_1 x_1^3 + A_2 x_2^3 = 0$$

which gives

$$A_1 = A_2 = 1 \qquad x_1 = -\frac{1}{\sqrt{3}} \qquad x_2 = \frac{1}{\sqrt{3}}$$

Therefore,

$$I \simeq f\left(-\frac{1}{\sqrt{3}}\right) + f\left(\frac{1}{\sqrt{3}}\right) \tag{7.52}$$

which is known as the *two-point Gauss-Legendre formula*. This integral estimate is exact for polynomials up to the third order, that is, cubics, and is, therefore, of the same order of accuracy as Simpson's rule. It is interesting to note that this accuracy is achieved on the basis of only two function evaluations, at $x = \pm 1/\sqrt{3}$.

The above discussion indicates the general features and the power of Gauss quadrature. The general formula for this method may now be written as follows:

$$I \simeq A_1 f(x_1) + A_2 f(x_2) + A_3 f(x_3) + \cdots + A_n f(x_n) \tag{7.53}$$

where n is the number of points in the range $-1 \leqslant x \leqslant 1$ at which the function $f(x)$ is calculated. The integral under consideration is $\int_{-1}^{1} f(x)\,dx$, which is obtained from the integral $\int_a^b f(x)\,dx$ by means of the transformation given by Eq. (7.51). The derivation of higher-order formulas for Gauss quadrature is quite involved. Basically, orthogonal polynomials, such as Legendre, Chebyshev, Hermite, and Laguerre polynomials, are taken to represent the function over the range $-1 \leqslant x \leqslant 1$. These polynomials are generally discussed in books on advanced calculus. The locations where the function is to be evaluated are actually the n zeros of an nth-degree Legendre polynomial. Table 7.2 gives the n locations, along with the corresponding weights $A_1, A_2, \ldots, A_n$, for the Gauss-Legendre formulas, considering n up to 24, which should be adequate for most practical problems. Other commonly employed quadrature formulas are the Gauss-Chebyshev, Gauss-Laguerre, and Gauss-Hermite integration formulas, given by Abramowitz and Stegun (1964) and Stroud and Secrest (1966). Although the derivation of these formulas is complicated, the application to numerical integration is not, as illustrated in Example 7.5.

TABLE 7.2 Weighting Factors A and the Values of the Independent Variable x at Which the Function $f(x)$ Must Be Evaluated for the Gauss-Legendre Formulas, Considering up to the 24-Point Approximation.

n	$\pm x_i$	A_i
2	0.5773502692	1.0000000000
3	0.0000000000	0.8888888889
	0.7745966692	0.5555555556
4	0.3399810436	0.6521451549
	0.8611363116	0.3478548451
5	0.0000000000	0.5688888889
	0.5384693101	0.4786286705
	0.9061798459	0.2369268850
6	0.2386191861	0.4679139346
	0.6612093865	0.3607615730
	0.9324695142	0.1713244924
7	0.0000000000	0.4179591837
	0.4058451514	0.3818300505
	0.7415311856	0.2797053915
	0.9491079123	0.1294849662
8	0.1834346425	0.3626837834
	0.5255324099	0.3137066459
	0.7966664774	0.2223810345
	0.9602898565	0.1012285363
9	0.0000000000	0.3302393550
	0.3242534234	0.3123470770
	0.6133714327	0.2606106964
	0.8360311073	0.1806481607
	0.9681602395	0.0812743884
10	0.1488743390	0.2955242247
	0.4333953941	0.2692667193
	0.6794095683	0.2190863625
	0.8650633667	0.1494513492
	0.9739065285	0.0666713443
12	0.1252334085	0.2491470458
	0.3678314990	0.2334925365
	0.5873179543	0.2031674267
	0.7699026742	0.1600783285
	0.9041172564	0.1069393260
	0.9815606342	0.0471753364
16	0.0950125098	0.1894506105
	0.2816035508	0.1826034150
	0.4580167777	0.1691565194
	0.6178762444	0.1495959888
	0.7554044084	0.1246289713
	0.8656312024	0.0951585117
	0.9445750231	0.0622535239
	0.9894009350	0.0271524594
20	0.0765265211	0.1527533871
	0.2277858511	0.1491729865
	0.3737060887	0.1420961093
	0.5108670020	0.1316886384
	0.6360536807	0.1181945320
	0.7463319065	0.1019301198
	0.8391169718	0.0832767416
	0.9122344283	0.0626720483
	0.9639719273	0.0406014298
	0.9931285992	0.0176140071
24	0.0640568929	0.1279381953
	0.1911188675	0.1258374563
	0.3150426797	0.1216704729
	0.4337935076	0.1155056681
	0.5454214714	0.1074442701
	0.6480936519	0.0976186521
	0.7401241916	0.0861901615
	0.8200019860	0.0733464814
	0.8864155270	0.0592985849
	0.9382745520	0.0442774388
	0.9747285560	0.0285313886
	0.9951872200	0.0123412298

The error E in Gauss-Legendre quadrature, which is usually referred to simply as *Gauss quadrature*, is obtained for an n-point formula as

$$E \simeq \frac{2^{2n+1}(n!)^4}{(2n+1)(2n!)^3} f^{(2n)}(\xi) \qquad \text{where } -1 < \xi < 1 \tag{7.54}$$

Therefore, a polynomial of degree $(2n - 1)$ is integrated exactly, since the $(2n)$th derivative, $f^{(2n)}$, is zero in this case, resulting in zero error. This implies that if n points are employed in Gauss quadrature, the accuracy obtained is of the same order as that obtained with a polynomial of order $(2n - 1)$ in Newton-Cotes formulas. Therefore, Gauss quadrature is a powerful method that is frequently used in engineering applications. However, a systematic reduction in error, as achieved by Romberg integration, is not possible in Gauss quadrature, and one must repeat the entire integration scheme with higher-order formulas if a greater accuracy is needed. Gauss quadrature maximizes the accuracy for a given number of function evaluations and is particularly suitable for complicated functions. However, the method is not applicable to problems where tabulated data are given at arbitrary locations, since function evaluations at definite points are needed. In some cases, it may be possible to take the data at the points specified by the quadrature formula and, thus, use Gauss quadrature for numerical integration.

Example 7.5

In a civil engineering application, a vertical plate 1 m high and 1.2 m wide is positioned in a stream of flowing water in a channel. The pressure p exerted on the plate due to the flow is measured at several vertical locations x, where $x = 0$ represents the top edge of the plate. Curve fitting is employed to obtain p as a function of x. The resulting expression for pressure in newtons/(meters squared) and x in meters is

$$p(x) = 10 + 4.6x - 16.2x^2 + 8.9x^3 - 41.3x^4 + 22.6x^5 \tag{7.5.1}$$

Using Gauss quadrature with two as well as four function evaluations, compute the total force exerted on the plate due to the flow.

Solution

The resulting force on the plate F is given by the equation

$$F = \int_0^1 p(x)\, 1.2dx$$

$$= 1.2 \int_0^1 p(x)\, dx \qquad \text{newtons} \tag{7.5.2}$$

where $1.2dx$ represents the area of a differential surface element at a vertical location given by x. Since $p(x)$ is a fifth-order polynomial in x, the above integral can easily be evaluated analytically to yield 4.631667, giving the force F as 5.558 N. However, let us apply Gauss quadrature to this problem in order to demonstrate the use of this method for numerical integration.

We must first change the limits of the integration to -1 and 1 by employing Eq. (7.51). Thus, since $a = 0$ and $b = 1$, ξ is given by

$$\xi = \frac{2x - (0 + 1)}{1} = 2x - 1$$

or

$$x = 0.5\xi + 0.5 \tag{7.5.3}$$

which gives

$$dx = 0.5d\xi \tag{7.5.4}$$

Therefore, the expression for the total force F on the plate becomes

$$F = 1.2 \int_{-1}^{1} [10 + 4.6(0.5\xi + 0.5) - 16.2(0.5\xi + 0.5)^2 + 8.9(0.5\xi + 0.5)^3 - 41.3(0.5\xi + 0.5)^4 + 22.6(0.5\xi + 0.5)^4]0.5d\xi \tag{7.5.5}$$

or,

$$F = 1.2 \int_{-1}^{1} f(\xi)\, d\xi \tag{7.5.6}$$

where $f(\xi)$ is the function to be integrated, as given above.

Using the two-point Gauss-Legendre formula, F is given by

$$F = 1.2\left[f\left(-\frac{1}{\sqrt{3}}\right) + f\left(\frac{1}{\sqrt{3}}\right)\right] \tag{7.5.7}$$

Similarly, for the four-point Gauss-Legendre formula,

$$F = 1.2[A_1 f(x_1) + A_2 f(x_2) + A_3 f(x_3) + A_4 f(x_4)] \tag{7.5.8}$$

where the required x's and A's are given in Table 7.2. Two function evaluations are involved in the former case, and four in the latter. Thus, for the two-point formula,

$$f\left(-\frac{1}{\sqrt{3}}\right) = 5.129892$$

$$f\left(\frac{1}{\sqrt{3}}\right) = -0.5826686$$

which gives

$$F = 1.2(4.547223) = 5.456668 \text{ newtons}$$

Similarly, for the four-point formula,

$$F = 1.2[Af(-C) + Bf(-D) + Bf(D) + Af(C)]$$

where $A = 0.347854845$, $B = 0.652145155$, $C = 0.861136312$, and $D = 0.339981044$. This gives $F = 5.558$ newtons. Thus, the four-point formula gives a very high level of accuracy with only four evaluations of the function $f(\xi)$. Even for the two-point formula, with only two function evaluations, the error is only 1.82%. This error figure indicates the efficiency of this method and its considerable value for complicated functions frequently encountered in engineering applications.

7.6 NUMERICAL INTEGRATION OF IMPROPER INTEGRALS

In the preceding sections, the limits of integration a and b in the integral $\int_a^b f(x)\,dx$ were taken as finite, and the integrand $f(x)$ was assumed to be continuous and finite over the range $a \leqslant x \leqslant b$. However, in engineering computations, we are often faced with integrals in which either the limits of integration are infinite or the integrand is singular at some point in the range of integration. Such integrals are known as *improper integrals*, and special procedures are often required for their evaluation. Some of these integrals are discussed here. It is assumed that the integral exists and is finite. This assumption is often based on the nature of the physical quantity represented by the integral. For instance, if the integral represents the total energy lost by a given body, the integral is expected to be finite. Similarly, if an integral over time yields the total distance traveled by a body that is decelerating due to an applied force, the integral must approach a finite value as the upper limit of integration approaches infinity. The analytical methods for proving that an integral exists and is finite are generally given in most advanced calculus books.

7.6.1 Integrals with Infinite Limits

In problems of engineering interest, we often encounter integrals of the form $\int_a^\infty f(x)\,dx$, $\int_{-\infty}^b f(x)\,dx$, or $\int_{-\infty}^\infty f(x)\,dx$, where either one or both of the limits of integration are infinite. In the example of the retarding body, outlined above, the lower limit is finite, say, time $\tau = 0$, and the upper limit is infinite. The mass transfer from an infinite surface, which approximates, say, the surface of a large lake, would involve an integral over the range $-\infty \leqslant x \leqslant \infty$. Similarly, flow rates in jets and plumes often require integration from $-\infty$ to ∞, since no walls, which limit the extent of the flow, are assumed to be present. Statistical distributions, like Gaussian and Poisson distributions, also generally involve infinite limits of integration.

There are several methods by which such integrals may be evaluated. The most common and often convenient approach is to write a given integral of the form $\int_a^\infty f(x)dx$ as $\int_a^b f(x)\,dx$ and to evaluate the integral with increasing values of b, until any further increase in b results in a negligible change in the integral. This approach was demonstrated in Example 7.1, where the charge Q in the capacitor was computed by the integral $\int_0^\tau I(\tau')\,d\tau'$, $I(\tau)$ being the current, τ the time, and τ' simply a dummy variable. Thus as $\tau \to \infty$, the charge attains a finite value. We obtained this result by increasing the upper limit of integration until the charge Q remained unchanged as τ was increased further. Similarly, for integrals of the form $\int_{-\infty}^{b} f(x)\,dx$ or $\int_{-\infty}^{\infty} f(x)\,dx$, this approach may be used, suitably decreasing the lower limit and/or increasing the upper limit. Another illustration of this method of handling infinite limits of integration is given in Example 7.6.

The integrand $f(x)$ may approach zero in an asymptotic manner as $x \to \infty$. In some cases, the dominant terms at large x can be employed to simplify the function and integrate it analytically. Thus, the given integral is written as

$$\int_a^\infty f(x)\,dx = \int_a^s f(x)\,dx + \int_s^\infty \tilde{f}(x)dx \tag{7.55}$$

where s is chosen to be sufficiently large so that the function $f(x)$ may be replaced by a simpler asymptotic approximation $\tilde{f}(x)$ for $x \geqslant s$. Then the first integral on the right-hand side of Eq. (7.55) is evaluated numerically and the second integral analytically. Examples of functions for which this approach may be employed are $f(x) = 1/(e^x + e^{-2x} + 3x^{-2})$ and $f(x) = 1/(e^{5x} + 2x^3 + 1)$. At large x, the first function may be approximated as e^{-x}, and the second as e^{-5x}, both of which may be integrated analytically over $s \leqslant x < \infty$ to yield e^{-s} and $e^{-5s}/5$, respectively. Again, s may be varied until the numerical value of the total integral $\int_a^\infty f(x)\,dx$ shows a negligible change with a further increase in s. This procedure, wherever applicable, is more efficient than replacing the upper limit by a large number b and computing the integral for increasing values of b, as outlined above.

In some cases, a transformation of the independent variable may be employed to change the infinite limit of integration into a finite one. Commonly used transformations are $y = x^{-n}$ and $y = e^{-x}$, both of which give zero for the new variable y as x goes to infinity. For instance, consider the following two integrals:

$$I = \int_1^\infty \frac{x}{1 + x + x^3}\,dx \tag{7.56a}$$

$$I = \int_0^\infty \frac{1}{e^x + e^{-x}}\,dx \tag{7.56b}$$

We can transform the first integral into one with finite limits by using the transformation $y = 1/x$. This gives $dx = -dy/y^2$, and the integral becomes

$$I = \int_0^\infty \frac{x}{1 + x + x^3}\,dx = \int_1^0 \frac{\frac{1}{y}}{1 + \frac{1}{y} + \frac{1}{y^3}} \cdot -\frac{dy}{y^2} = \int_0^1 \frac{dy}{1 + y^2 + y^3} \tag{7.57a}$$

Similarly, we transform the integral in Eq. (7.56b) by employing $y = e^{-x}$ as follows:

$$I = \int_0^\infty \frac{dx}{e^x + e^{-x}} = \int_1^0 \frac{1}{y + \frac{1}{y}} \cdot -\frac{dy}{y} = \int_0^1 \frac{dy}{1 + y^2} \tag{7.57b}$$

Thus, integrals over infinite regions may be transformed into integrals over finite regions. However, in some cases, the transformed integrand may be singular at one of the limits. Then the problem with an infinite limit is replaced by one involving a singular integrand, discussed in the following subsection. In general, it is easier to handle an infinite limit of integration than an integrand that becomes singular. Therefore, one must consider whether or not a given transformation simplifies the problem.

7.6.2 Singular Integrand

Another class of improper integrals is the one in which the limits of integration are finite but the integrand is singular in the range of integration, generally at one or both limits. However, the singularity is assumed to be gentle enough for the integral to exist and be finite. Examples of such integrals are

$$\int_0^2 \frac{e^x}{x}\,dx \qquad \int_0^1 \frac{1}{\sqrt{x}}\,dx \qquad \int_0^1 \frac{x^3+1}{\sqrt{1-x^2}}\,dx \qquad \int_0^{\pi/2} \frac{d\theta}{2\sqrt{\cos\theta}}$$

In all of these cases, the integrand becomes singular at the lower or the upper limit of integration, and the integral can be shown to exist.

There are several methods for dealing with such improper integrals. Among these are integrating by parts, subtracting out the singularity, using a power series to approximate the integral near the point of singularity, and transforming the variables. Obviously, the appropriate method depends on the nature of the singularity, and all of these techniques may be considered to determine if one of them would work. Let us illustrate the use of some of these strategies to eliminate the singularity by means of examples.

The integral

$$\int_0^2 \frac{e^x}{\sqrt{x}}\,dx \tag{7.58a}$$

can be integrated by parts to yield

$$\int_0^2 \frac{e^x}{\sqrt{x}}\,dx = 2\sqrt{x}\,e^x\Big|_0^2 + \int_0^2 2\sqrt{x}\,e^x\,dx \tag{7.58b}$$

The first term on the right-hand side can easily be evaluated, and the second term is an integral that is not singular. This remaining integral can thus be computed analytically or by means of the various numerical methods presented in this chapter. The singularity in this integral can also be subtracted out as follows:

$$\int_0^2 \frac{e^x}{\sqrt{x}}\,dx = \int_0^2 \frac{1}{\sqrt{x}}\,dx + \int_0^2 \frac{e^x-1}{\sqrt{x}}\,dx \tag{7.59}$$

The first integral on the right-hand side is singular, but it can easily be integrated analytically. The second integral is nonsingular, since $(e^x - 1)/\sqrt{x}$ becomes zero at the lower limit of integration.

The singularity in the integral

$$\int_0^1 \frac{x^3 + 1}{\sqrt{1 - x^2}}\, dx \tag{7.60a}$$

can be eliminated by use of the transformation $x = \sin y$. Then $\sqrt{1 - x^2} = \cos y$, and $dx = \cos y\, dy$. The integral becomes

$$\int_0^1 \frac{x^3 + 1}{\sqrt{1 - x^2}}\, dx = \int_0^{\pi/2} \frac{(\sin y)^3 + 1}{\cos y} \cos y\, dy = \int_0^{\pi/2} [(\sin y)^3 + 1]\, dy \tag{7.60b}$$

The resulting integral is nonsingular and can be integrated by any numerical method.

Another strategy that is sometimes applicable is the expansion of the integrand in a power series about the singular point and retention of a few leading terms that can be integrated by standard methods. Gauss quadrature is also particularly suitable for certain types of singularities, since formulas are sometimes available that have already accounted for the singularity in the choice of the weighting function.

A frequently used procedure, if the above strategies do not work, is to replace the integration limit where the singularity exists by a quantity close to this limit, but not equal to it. Thus, $\int_a^b f(x)\, dx$ is replaced by $\int_{a+\varepsilon}^b f(x)\, dx$, where ε is a small quantity, if $f(x)$ is singular at $x = a$. Then numerical integration is carried out over the range $a + \varepsilon \leqslant x \leqslant b$, and ε is made smaller, starting with a chosen small value, until the computed integral is not significantly affected by a further reduction in ε. This method is not very efficient, particularly if equally wide intervals are used. However, the range of integration may be broken down into a region close to the point of singularity and others further away, so that a finer mesh may be used near the singularity. For example, the following integrals may be written as

$$\int_0^1 \frac{1}{\sqrt{1 - x}}\, dx = \int_0^{0.99} \frac{1}{\sqrt{1 - x}}\, dx + \int_{0.99}^{1-\varepsilon} \frac{1}{\sqrt{1 - x}}\, dx \tag{7.61a}$$

$$\int_0^\infty \frac{1}{\sqrt{x}(1 + x)}\, dx = \int_\varepsilon^{0.01} \frac{1}{\sqrt{x}(1 + x)}\, dx + \int_{0.01}^\infty \frac{1}{\sqrt{x}(1 + x)}\, dx \tag{7.61b}$$

with ε being reduced toward zero till the integrals do not vary significantly with a further reduction in ε. The upper limit in the second integral in Eq. (7.61b) is infinity and can be treated by the methods given in Section 7.6.1. The integration region $\varepsilon \leqslant x \leqslant 0.01$ may be further subdivided as $\varepsilon \leqslant x \leqslant 0.001$ and $0.001 \leqslant x \leqslant 0.01$, if necessary. Adaptive quadrature can also be used advantageously for this problem.

Example 7.6

Many physical measurements follow the symmetrical, bell-shaped curve of the Gaussian, or normal, frequency distribution, sketched in Fig. 7.6.1. Repeated measurements of the fluid velocity in a hydraulic control system are found to closely

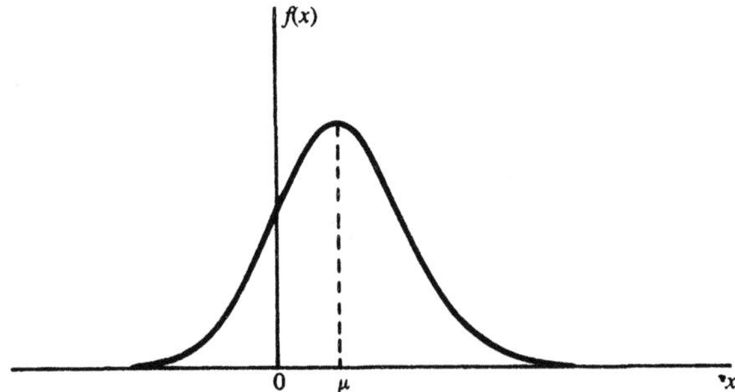

Figure 7.6.1 Sketch of the Gaussian, or normal, frequency distribution, with μ as the mean value.

approximate the Gaussian distribution

$$f(x) = \frac{1}{\sqrt{2\pi}\,\sigma} \exp\left[-\frac{1}{2}\left(\frac{x-\mu}{\sigma}\right)^2\right] \tag{7.6.1}$$

where $f(x)$ is the height of the frequency curve corresponding to a given velocity x, μ is the mean value, and σ is known as the *standard deviation*. For the given circumstance, $\mu = 10$ m/s and $\sigma = 5$ m/s. Compute the fraction of the measurements for which the velocity is larger than or equal to 0, 5.0, 10.0, and 15.0 m/s, respectively.

Solution

The normal distribution extends from $x = -\infty$ to $x = +\infty$ and is symmetrical about the mean μ. The area under the curve from, say, $x = x_1$ to $x = x_2$, gives the fraction of the total measurements for which the velocity lies between x_1 and x_2. Therefore, the integral to be computed is

$$I = \int_{x_{min}}^{\infty} f(x)\,dx \tag{7.6.2}$$

where x_{min} is the minimum value of velocity considered. For the given problem, $x_{min} = 0$, 5.0, 10.0, and 15.0 m/s, respectively. Since $\int_{-\infty}^{\infty} f(x)\,dx$ covers the entire range of measurements, this integral equals 1.0. Similarly, $\int_{\mu}^{\infty} f(x)\,dx = 0.5$, since it represents half of the measurements taken.

The given problem involves the evaluation of an improper integral, since the upper limit is infinity. Thus, following the approach given in the preceding section for such problems, we may compute the integral

$$\hat{I} = \int_{x_{min}}^{x_{max}} f(x)\,dx \tag{7.6.3}$$

by any standard method for numerical integration, such as Simpson's rule, and increase x_{max} until the value of the integral remains essentially unchanged as x_{max} is increased further. This approach is followed in the computer program for this problem; see Fig. 7.6.2. The segment width DX is chosen, and the four given values for x_{min} are successively entered. The upper limit of integration x_{max} is varied until the

```
C           SIMPSON'S ONE-THIRD RULE FOR NUMERICAL INTEGRATION
C           OF AN IMPROPER INTEGRAL WITH INFINITE UPPER LIMIT
C
C           FUNC(Y) IS THE GAUSSIAN DISTRIBUTION WHICH IS TO BE INTEGRATED
C           FROM XMIN TO INFINITY, X IS THE VELOCITY WHICH IS THE
C           INDEPENDENT VARIABLE HERE, XMAX IS THE UPPER LIMIT OF
C           INTEGRATION, DX IS THE SEGMENT WIDTH, EPS IS THE CONVERGENCE
C           CRITERION TO ENSURE THAT XMAX APPROXIMATES INFINITY, SUM IS
C           THE INTEGRAL AND N IS THE NUMBER OF SUBDIVISIONS.
C
        IMPLICIT REAL (A-H,O-Z)
C
C       DEFINE FUNCTION TO BE INTEGRATED
C
        FUNC(Y)=EXP(-((Y-10.0)**2)/50.0)/(SQRT(2.0*3.14159)*5.0)
C
C       ENTER INPUT VARIABLES
C
        PRINT *,'ENTER THE SEGMENT WIDTH DX'
        READ *,DX
        DO 4 XMIN=0.0,15.0,5.0
        WRITE(1,1)XMIN
  1     FORMAT(//5X,'XMIN=',F6.3)
        SUM1=0.0
        EPS=1.0E-5
        XMAX=XMIN+10.0
C
C       CALL SUBROUTINE FOR NUMERICAL INTEGRATION
C
  2     CALL SIMPSN(FUNC,XMIN,XMAX,DX,SUM)
        WRITE(1,3)SUM,XMAX
  3     FORMAT(/5X,'INTEGRAL =',F8.5,10X,'XMAX=',F6.2)
        SUM2=SUM
C
C       CHECK IF INTEGRAL REMAINS ESSENTIALLY UNCHANGED WITH
C       INCREASING XMAX IN ORDER TO APPROXIMATE THE UPPER LIMIT
C       OF INTEGRATION, WHICH IS INFINITY
C
        IF(ABS(SUM2-SUM1) .GE. EPS) THEN
        XMAX=XMAX+5.0
        SUM1=SUM2
        GO TO 2
        ELSE
        END IF
  4     CONTINUE
        STOP
        END
C
C
C
        SUBROUTINE SIMPSN(FUNC,XMIN,XMAX,DX,SUM)
C
C       THIS SUBROUTINE COMPUTES THE INTEGRAL BY SIMPSON'S RULE
C
        IMPLICIT REAL (A-H,O-Z)
        SUM=0.0
```

Figure 7.6.2 Computer program for integrating the Gaussian frequency distribution function $f(x)$ from a given value $x_{\min}$ of the independent variable x to $x = \infty$.

```
          X=XMIN
          N=(XMAX-XMIN)/DX
          DO 5 I=1,N-1
          X=X+DX
          S=I
          IF((S/2.0 - I/2) .GT. 0.1)THEN
          SUM=SUM+4.0*FUNC(X)
          ELSE
          SUM=SUM+2.0*FUNC(X)
          END IF
    5     CONTINUE
C
C        APPLY SIMPSON'S ONE-THIRD RULE
C
          SUM=(DX/3.0)*(FUNC(XMIN)+SUM+FUNC(XMAX))
          RETURN
          END
```

Figure 7.6.2 Continued

integral, denoted by SUM in the program, varies by less than a convergence criterion ε, taken as 10^{-5} here, with a further increase in x_{max}. Simpson's one-third rule is employed for numerical integration. We can vary the segment size DX and the convergence criterion EPS to ensure that the results obtained are not significantly dependent on the values chosen.

The numerical results obtained are shown in Fig. 7.6.3, in terms of the integral at various values of x_{max}, for each of the four given values of x_{min}. It is found that an x_{max} of 40.0 m/s is adequate for the approximation of infinity, which is the upper limit of integration in Eq. (7.6.2). The convergence criterion ε and the segment width DX were also varied. The values chosen were found to be quite satisfactory. Thus, 97.716% of the measurements yield a positive fluid velocity, that is, $x \geqslant 0$. Similarly, 84.135% of the measurements give a velocity larger than or equal to 5 m/s, and so on. Tabulated results for the integration of the normal distribution curve are available in the literature. For the four cases considered here, the values given in the literature are 0.9772, 0.8413, 0.5, and 0.1587, respectively. Clearly, these values are very close to those obtained from the given computer program. Further details on this problem and tabulated results on the area under the frequency distribution curve may be obtained from any statistics textbook.

7.6.3 Multiple Integrals

There are a few other forms of integrals that have not been discussed thus far. Among these are multiple integrals, which arise for functions that depend on more than one independent variable. Integrals over the surface area, for example, to compute the evaporation from a pond, or over the volume of a body involve multiple integrals. Similarly, the work done in the two-dimensional motion of a particle on a flat surface and the total force acting on a vertical surface, such as a building, require double

```
ENTER THE SEGMENT WIDTH DX
0.01

   XMIN= 0.000

   INTEGRAL = 0.47722            XMAX= 10.00

   INTEGRAL = 0.81858            XMAX= 15.00

   INTEGRAL = 0.95451            XMAX= 20.00

   INTEGRAL = 0.97589            XMAX= 25.00

   INTEGRAL = 0.97716            XMAX= 30.00

   INTEGRAL = 0.97716            XMAX= 35.00

   XMIN= 5.000

   INTEGRAL = 0.68269            XMAX= 15.00

   INTEGRAL = 0.81863            XMAX= 20.00

   INTEGRAL = 0.84003            XMAX= 25.00

   INTEGRAL = 0.84133            XMAX= 30.00

   INTEGRAL = 0.84135            XMAX= 35.00

   INTEGRAL = 0.84135            XMAX= 40.00

   XMIN=10.000

   INTEGRAL = 0.47728            XMAX= 20.00

   INTEGRAL = 0.49868            XMAX= 25.00

   INTEGRAL = 0.49997            XMAX= 30.00

   INTEGRAL = 0.49999            XMAX= 35.00

   INTEGRAL = 0.49999            XMAX= 40.00

   XMIN=15.000

   INTEGRAL = 0.15732            XMAX= 25.00

   INTEGRAL = 0.15863            XMAX= 30.00

   INTEGRAL = 0.15866            XMAX= 35.00

   INTEGRAL = 0.15866            XMAX= 40.00
```

Figure 7.6.3 Numerical results obtained from the integration of the Gaussian distribution from $x = x_{min}$ to $x = \infty$, for Example 7.6, at several values of x_{min}.

integrals of the form

$$I = \int_a^b \int_{g(x)}^{h(x)} f(x,y)\,dy\,dx \tag{7.62}$$

We evaluate such multiple integrals by twice applying the numerical methods discussed in this chapter, first for the inner integral over y and then for the outer integral over x. If the range of integration $a \leqslant x \leqslant b$ is divided into n segments, so that $x_i = a + i\,\Delta x$, then we may define a function $F(x)$ as

$$F(x_i) = \int_{g(x_i)}^{h(x_i)} f(x_i, y)\,dy \tag{7.63}$$

so that x is held constant at x_i. This integral may be computed as a one-dimensional integral on y. Thus, $(n + 1)$ ordinates, corresponding to $F(x_i)$, with $i = 0, 1, \ldots, n$, are generated. Since these ordinates are at evenly spaced points, we can employ Simpson's one-third rule to evaluate the integral I, provided n is even. Similarly, other numerical methods may be employed.

In conclusion, numerical integration is needed for a wide variety of engineering problems. Because of simplicity and the fourth-order accuracy, Simpson's one-third rule, used in conjunction with the three-eighths rule, is probably the most widely used method for numerical integration. Romberg integration is also a popular choice if a high level of accuracy is desired. The other methods discussed here are all used, depending on the requirements of the problem at hand.

7.7 SUMMARY

This chapter presents several available methods for the numerical integration of a given continuous function $f(x)$ over a finite range of the independent variable x. These methods, which include the rectangular, trapezoidal, Simpson's one-third, and Simpson's three-eighths rules for numerical integration, form the first four orders of the Newton-Cotes formulas. They are discussed in detail in this chapter, particularly the trapezoidal and Simpson's one-third rules, because of their wide usage. The truncation errors associated with these formulas are also derived. Simpson's one-third rule is a very popular choice in engineering problems, since it is fourth-order accurate, as compared to the trapezoidal rule which is second-order accurate. Also, when it is used in conjunction with the three-eighths rule, it imposes no constraints on the choice of the number n of the segments, or subintervals, of the integration region, except that n be 2 or larger. The trapezoidal rule is also widely used because of its simplicity. Higher-order Newton-Cotes formulas are also presented, although they are used only if a very high level of accuracy is needed. The accuracy of the numerical results can also be improved by a reduction in the step size Δx. However, at very small values of Δx, the round-off error may become significant. Higher-order formulas can then be employed more advantageously.

Various methods for the successive improvement in accuracy are also discussed, including Richardson's extrapolation, which is applicable to several other numerical procedures as well, and Romberg integration, which is a very efficient method for achieving any desired accuracy level. Romberg integration is based on the trapezoidal rule and uses a procedure similar to Richardson's extrapolation to successively eliminate the higher-order terms in the truncation error. It is presently one of the most widely used methods for the numerical integration of well-behaved functions.

Gauss quadrature uses the minimum number of function evaluations for computing the integral and is therefore particularly suitable for very complicated functions. Different formulas can be derived for a wide variety of functions and integration limits so that the results are very accurate and the number of function evaluations is minimized. Singularities can also be effectively dealt with in several cases. However, the tables of the weight factors and the zeros must be stored or computed. The programming is more involved and less versatile than that for, say, Simpson's rule. Also, this method is generally not applicable for data available at arbitrarily spaced values of the independent variable. One can use the trapezoidal and Simpson's rules in this case, using the latter if two or three adjacent segments are of the same width and the trapezoidal rule if the widths of adjacent segments are unequal. It is also possible to use curve fitting to obtain a continuous function to represent the data. Then the standard methods for numerical integration may be used. By focusing on regions where the accuracy is less than that in others and retaining fewer points in regions where the desired accuracy level has been attained, one can effectively use the method of adaptive quadrature for functions that are small in magnitude or slowly varying in certain regions.

This chapter also discusses the various methods for treating improper integrals, which exist and are finite although either the integrand blows up within the range of integration or the integration limits are infinite. Analytical procedures, such as transformation of the independent variable and integration by parts, can often be employed to eliminate the singularity. However, the most common approach is to replace the integration limit that is infinite or where the integrand is singular by a quantity that is large or close to, but not equal to, the limit. Numerical integration is then carried out and this quantity is varied until the numerical results are not significantly affected by a further variation. Although inefficient, this approach is applicable to most engineering problems involving improper integrals. Gauss quadrature can also be used advantageously for certain types of functions.

PROBLEMS

7.1. Consider the integral $\int_0^3 f(x)\,dx$ for the linear function $f(x) = 3 + 5x$. Show that the numerical results obtained by use of the trapezoidal and Simpson's one-third rules for this integral are exactly equal to the analytical value, except for the round-off error. Employ two and then four subdivisions of the integration domain.

7.2. For the parabola $f(x) = 3 + 2x + 3x^2$, show that the numerical integration $\int_1^3 f(x)\,dx$ by Simpson's rule yields the exact analytical value, except for the round-off error. What effect

7: 6, 8, 14, 22, 26

would you expect an increase in the number of subdivisions to have on the accuracy of the numerical results? Explain.

7.3. Consider the integral $\int_0^2 f(x)\,dx$, where $f(x) = 3 + 5x + 2x^2 + x^3$. Estimate the total truncation error and the maximum truncation error per step for evaluating this integral by the rectangular and trapezoidal rules. Consider the three segment sizes $\Delta x = 0.1, 0.2$, and 0.5.

7.4. Calculate the truncation error per step at $x = 0.5$ and $x = 1.0$ for the integral $\int_0^{1.5} (x^5 + 2x^4 + x^3 + 4x^2 + 2x + 6)\,dx$, taking $\Delta x = 0.1$, 0.25, and 0.5. Consider the trapezoidal rule and both Simpson's rules for numerical integration. Discuss the effect, on the error, of a reduction in segment size and also of the numerical method employed.

7.5. The truncation error in the numerical evaluation of the integral $I = \int_a^b f(x)\,dx$ by the trapezoidal rule is given by Eq. (7.20). If f''_{av} is approximated as $[f'(b) - f'(a)]/(b - a)$, obtain the resulting estimate of the error, and add it to the formula for numerical integration by the trapezoidal rule to obtain a more accurate scheme known as the *trapezoidal rule with end correction.*

7.6. Apply the procedure outlined in Prob. 7.5 to the rectangular rule, and compare the resulting formula with that for the trapezoidal rule.

7.7. Show that if Richardson's extrapolation is applied to the trapezoidal rule, the formula obtained is the same as that for Simpson's one-third rule. Also apply Richardson's extrapolation to the rectangular rule, and discuss the resulting formula for numerical integration.

7.8. The temperature T at the wall of a furnace varies periodically over the day as

$$T(\tau) = 125 + 50 \sin \frac{2\pi}{24}(\tau - 6)$$

where τ is the time in hours measured from midnight and T is in °C. The ambient temperature T_a is 25°C, and the surface area A of the wall is $10\,\text{m}^2$. If the heat transfer coefficient h is given as $20\,\text{W/m}^2\cdot{}^\circ\text{C}$, the heat transfer from the wall is given by $\int [T(\tau) - T_a]hA\,d\tau$. Using the trapezoidal rule, compute this integral as accurately as possible for the time interval $\tau = 6$ to $\tau = 12$. Also evaluate the integral analytically and compare the result with the computed value.

7.9. In chemical engineering, we frequently need to evaluate the amount of heat required to raise the temperature of a given material from a value T_1 to T_2. If $C(T)$ is the specific heat of the material, the amount of energy needed is $\int_{T_1}^{T_2} mC(T)\,dT$, where m is the mass of the material, since the specific heat is the energy required to raise the temperature of unit mass of the material by unit temperature. The average specific heat C_{av} is given by $[\int_{T_1}^{T_2} C(T)\,dT]/(T_2 - T_1)$. The specific heat of a material is given, in J/kg·K, as follows:

$$C(T) = 200 + 7.5\frac{T}{T_0} + 2.8\left(\frac{T}{T_0}\right)^2 + 0.42\left(\frac{T}{T_0}\right)^3$$

where T_0 is the reference temperature of 100 K. For 1 kg of the material, compute the total energy, in joules, needed to raise the temperature from 100 K to 1000 K. Also determine the average specific heat over this temperature range. Use the trapezoidal rule, and reduce the segment size, starting with 100 K, until the results remain essentially unchanged with further reduction.

7.10. The pressure p on a 10 m high structure due to the wind is given by the expression

$$p(x) = \frac{150x}{1 + e^x}$$

where x is measured in meters from the bottom of the structure and the pressure is in newtons/m^2. If the structure is 2 m wide, the total force due to wind is given by the integral

$$\int_0^{10} 2 \cdot \left(\frac{150x}{1 + e^x} \right) dx$$

Compute this integral as accurately as possible by the trapezoidal rule.

7.11. Consider the expression for blackbody radiation given by Eq. (3.3.1). The integral of this expression over all wavelengths, that is, $\int_0^\infty E_{b,\lambda}\, d\lambda$, gives the total energy radiated by a blackbody per unit area and time. Using Simpson's rule, compute this integral at $T = 1000$ K as accurately as possible. The analytical result is given in the literature as σT^4, where σ is known as the *Stefan-Boltzmann constant* and has a value of 5.67×10^{-8} W/m^2·K^4. Compare your numerical result with the analytical value at 1000 K.

7.12. Using Simpson's rule, repeat the problem given in Example 7.1, and compare the results obtained with those given in the example. Discuss the observed differences between the trapezoidal and Simpson's rules for this problem.

7.13. Using Simpson's three-eighths rule, compute the total momentum flow in the problem outlined in Example 7.2. The momentum flow is given by the integral $\int_0^R \rho[U(x)]^2 2\pi x\, dx$, where ρ is the fluid density, given as 1 kg/m^3 for the fluid considered.

7.14. The root mean square (RMS) value of an electric current $I(\tau)$, where I varies periodically with time τ, is given by the expression

$$I_{\text{RMS}} = \frac{1}{\tau_c} \sqrt{\int_0^{\tau_c} I^2\, d\tau}$$

where τ_c is the time period for one cycle in the variation of $I(\tau)$. If $I(\tau)$ is given as $5e^{-\tau} \sin 4\pi\tau$, with $\tau_c = 0.5$ s, compute the RMS value, using Simpson's rule.

7.15. The force $F(x)$ exerted per centimeter on a vertical plate immersed in flowing water is given by the expression

$$F(x) = 1.5x^3 e^{-x}$$

where x is measured from the top of the plate and $F(x)$ is in newtons/cm. If the plate is 10 cm high, the total force F_T, in newtons, on the plate is given by

$$F_T = \int_0^{10} F(x)\, dx$$

Employing Romberg integration, compute F_T to a convergence criterion of 10^{-4}.

7.16. The meniscus of a liquid film supported by surface tension can often be represented as

$$h(x) = Ae^{-a^2x^2}$$

where $h(x)$ is the height as a function of horizontal distance x and A and a are constants. The total volume of liquid supported by surface tension is then given by the integral $W \int_0^L h(x)\, dx$, where W is the width of the meniscus and L is its length. If W, L, h, and x are all in centimeters, compute this volume for $A = 0.8$, $a = 2.0$, $W = 1$, and $L = 1$ cm, using Romberg integration.

7.17. The velocity $v(\tau)$ of a moving particle is given as $v(\tau) = 5(1 - e^{-\tau/10})$. Using Romberg integration, compute the total distance S traveled by the particle from $\tau = 10$ s to $\tau = 20$ s. Note that the distance traveled between $\tau = \tau_1$ and $\tau = \tau_2$ is simply given by the integral $\int_{\tau_1}^{\tau_2} v(\tau)\,d\tau$.

7.18. We wish to evaluate the integral $\int_0^{6\pi} \sin^2 x\,dx$ by means of Romberg integration. Are any difficulties encountered in the application of Romberg integration to this problem? If so, suggest methods to overcome them.

7.19. Using Romberg integration, repeat the problem given in Example 7.2. Compare the results obtained with those given in the example, and comment on the numerical accuracy and the computational effort involved in the two methods used for this problem.

7.20. The velocity v of a moving body is measured at several time intervals τ and is tabulated as follows:

τ (s)	0	1	2	3	5	7	8	10
v (m/s)	10	11.5	14.8	21.1	47.5	100.3	139.6	250.0

Using this uneven distribution of data points, compute the distance traveled x as a function of time τ, and find the total distance traveled by the body in 10 s.

7.21. The pressure p in a gas being compressed by a moving piston is measured at various positions x of the piston, where x is measured from the starting position of the piston. The data are tabulated as follows:

x (m)	0	0.1	0.2	0.4	0.5	0.8	0.9	1.0
p (N/m^2)	4.0	4.23	4.53	5.34	5.88	8.03	8.96	10.0

The total work done over this distance of 1 m by the piston is given by the integral $A\int_0^1 p(x)\,dx$, where A is the cross-sectional area of the piston. Compute this integral, taking $A = 0.5$ m^2.

7.22. The fluid velocity V is measured at several radial locations r for flow in a circular pipe of radius 1 cm. The velocities in cm/s are tabulated as follows:

r (cm)	0	0.2	0.5	0.6	0.8	0.9	1.0
V (cm/s)	1.0	0.96	0.75	0.64	0.36	0.19	0.0

The volume flow in the pipe is given by the integral $\int_0^R V(r)2\pi r\,dr$, where R is the radius of the pipe. Using the data given, compute this integral.

7.23. The turbulent flow in a pipe of diameter 0.2 m is given by the velocity distribution

$$V = U\left(1 - \frac{r}{R}\right)^{1/6}$$

where U is the velocity at the axis, r is the radial distance from the axis, and R is the radius of the pipe. Using the four-point Gauss-Legendre integration formula, compute the volume flow rate in the pipe. Take $U = 2$ m/s.

7.24. Using the two-point, as well as the four-point, Gauss-Legendre integration scheme, solve Prob. 7.16. Then compare the results obtained and the computational effort involved for the two methods, Romberg integration and Gauss quadrature, considered for this problem.

7.25. Using the trapezoidal rule, compute the value of the improper integral

$$\int_0^\infty \frac{7.5\,dx}{\sqrt{x}(1+x)}$$

as accurately as possible.

7.26. Using Simpson's rule, compute the improper integral

$$\int_0^\infty \frac{2\,dx}{1+e^{-x}+x^2}$$

as accurately as possible.

7.27. Using any convenient integration scheme, determine the integral

$$\int_0^\infty \frac{dx}{e^x+e^{-x}}$$

and compare the numerical value obtained with the analytical result of $\pi/4$.

7.28. An integral commonly encountered in the estimation of the moisture lost at the surface of a wet body, such as paper or cloth, is of the form $\int_0^1 x^{-1/4}\,dx$, which is singular at $x = 0$. Compute this integral by varying the lower limit, starting with 0.1, and then reducing it to values as small as needed to make the neglected area under the curve negligible. Also compare the numerical result obtained with the analytical value which is easily determined.

7.29. Determine the integral

$$\int_0^1 \frac{5(1+x^2)}{x^{0.2}}\,dx$$

which arises in a mass transfer calculation for a chemical engineering process. Employ any suitable method.

7.30. The work done W by a force $F(x,y)$ in a two-dimensional dynamics problem is given by the integral

$$W = \int_0^1 \int_0^2 (x^2 - 2xy + 3xy^2 + y^3)\,dy\,dx$$

Using the trapezoidal rule as outlined in Eqs. (7.62) and (7.63), calculate this integral.

7.31. The area A enclosed by a curve on a plane is given by the integral

$$A = \int_0^2 \int_y^{2y} (x^2 + y^2)\,dx\,dy$$

Using the trapezoidal rule, compute this integral.

7.32. Calculate the integral $\int_0^1 e^x\,dx$ by using four important methods for numerical integration, namely, the trapezoidal rule, Simpson's one-third rule, Romberg integration, and Gauss quadrature. Compare the results obtained and the computational efforts involved in all of these methods. Comment on the conclusions that may be drawn from such a comparison.

8

Numerical Solution of Ordinary Differential Equations

8.1 INTRODUCTION

Ordinary differential equations, which involve functions of a single independent variable and their derivatives, arise in many diverse engineering problems. Several physical laws, such as those concerned with the transport of mass, momentum, and energy, are expressed in terms of differential equations. In many cases, only one independent variable, such as time or distance, exists in the problem, because of the nature of the problem or because of approximations made with respect to the other variables. In a few circumstances, the functional dependence on two or more independent variables can be expressed, by suitable transformations, in terms of a single variable. Many problems of engineering interest are, therefore, governed by ordinary differential equations. These equations arise, for instance, in heat and mass transfer, dynamics of particles, vibrations of systems, electrical circuitry, and chemical kinetics. Although analytical methods may be employed for the solution of some ordinary differential equations, numerical techniques are generally needed for most of the equations that arise in engineering applications.

A general ordinary differential equation may be written in terms of the independent variable x and the dependent function $y(x)$ as

$$f\left(x, y, \frac{dy}{dx}, \frac{d^2y}{dx^2}, \ldots, \frac{d^ny}{dx^n}\right) = 0 \tag{8.1}$$

where the highest derivative is of order n. Then the equation is known as an *nth-order ordinary differential equation*. The highest-order derivative is often separated to obtain the above equation as follows:

$$\frac{d^ny}{dx^n} = F\left(x, y, \frac{dy}{dx}, \ldots, \frac{d^{n-1}y}{dx^{n-1}}\right) \tag{8.2}$$

A function $y(x)$ that satisfies this equation is said to be a *solution* of the equation. There may be many functions $y(x)$ that satisfy a given differential equation. To obtain a unique solution, which is obviously of interest in a physical problem, n conditions on $y(x)$ and/or on its derivatives must be given at specified values of x. These conditions must be independent; that is, one condition may not be derived from a combination of the others.

If all of the n conditions are specified at the same value of x, say, $x = x_1$, then the problem represented by the ordinary differential equation (ODE) and the given conditions is termed an *initial-value problem*. If the conditions are specified at more than one value of the independent variable x, the problem is termed a *boundary-value problem*. In the former case, there is a definite starting point, and one can obtain the solution by varying x in order to march outward from this starting point. An example of an initial-value problem is given by Eq. (1.1), which is of the form $dy/dx = Ay + B$, where A and B are constants, and by the initial condition $y = y_0$ at $x = x_0$. This equation governs, for instance, the temperature of a small, heated metal sphere being cooled by a stream of cold air and the charge in the capacitor which forms part of the electrical circuit shown in Fig. 1.2. Boundary-value problems are more involved, since conditions specified at different values of x are to be satisfied, generally by iteration. Because at least two conditions are needed for specification at different x values, the ODE must be at least of second order for a boundary-value problem. Several of the methods employed for boundary-value problems are based on those for initial-value problems, often employing the root-solving procedures of Chapter 4 to satisfy the given boundary conditions.

The nth-order equation, given by Eq. (8.2), can be reduced to a system of n first-order equations by defining $(n - 1)$ new variables Y_i, where $i = 1, 2, \ldots, (n - 1)$, as follows:

$$\begin{aligned} Y_1 &= \frac{dy}{dx} \\ Y_2 &= \frac{d^2y}{dx^2} \\ &\vdots \\ Y_{n-1} &= \frac{d^{n-1}y}{dx^{n-1}} \end{aligned} \tag{8.3}$$

Therefore, the given ordinary differential equation may be written as the following n first-order equations:

$$\begin{aligned} \frac{dy}{dx} &= Y_1 \\ \frac{dY_{i-1}}{dx} &= Y_i \qquad \text{where } i = 2, 3, \ldots, (n-1) \\ \frac{dY_{n-1}}{dx} &= F(x, y, Y_1, Y_2, \ldots, Y_{n-1}) \end{aligned} \tag{8.4}$$

As an example of the above procedure, consider the following third-order, nonlinear, equation that governs the flow over a two-dimensional wedge:

$$\frac{d^3f}{dx^3} + f\frac{d^2f}{dx^2} + \beta\left[1 - \left(\frac{df}{dx}\right)^2\right] = 0 \tag{8.5}$$

where x is the dimensionless distance from the wedge surface; f is the nondimensional stream function, which is related to the velocity field; and β is a constant that gives the wedge angle $\beta\pi$, in radians. Figure 8.1 gives a graphical representation of this physical problem. Equation (8.5), which is typical of several fluid flow circumstances encountered in aeronautical, chemical, and mechanical engineering, may be written as three first-order equations by defining two variables, F_1 and F_2, as

$$F_1 = \frac{df}{dx} \qquad F_2 = \frac{d^2f}{dx^2} \tag{8.6}$$

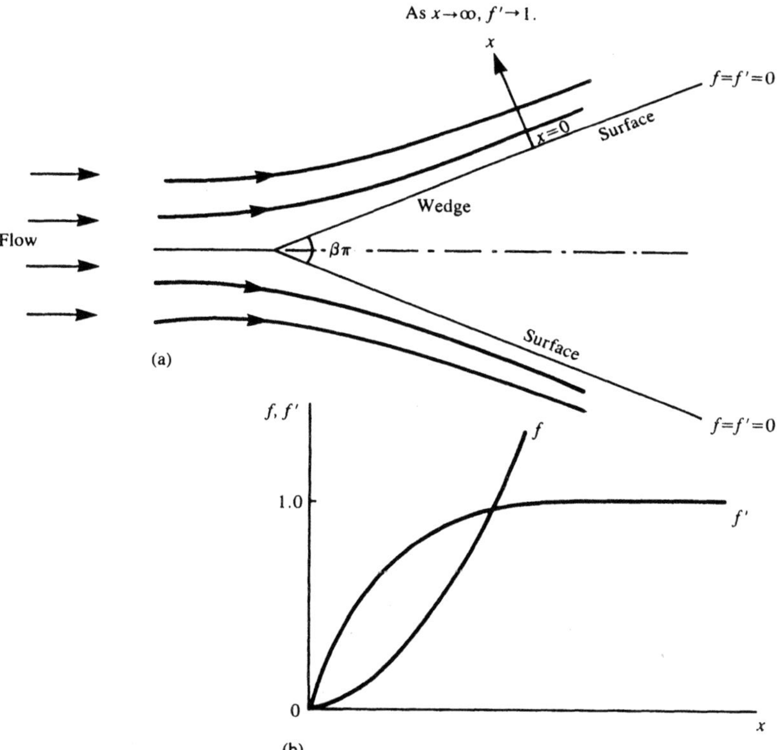

Figure 8.1 Graphical representation of the physical problem governed by Eq. (8.5). (a) Sketch of the flow over a two-dimensional wedge; (b) qualitative sketch of the variation of the functions f and f' with x.

to yield

$$\frac{df}{dx} = F_1$$

$$\frac{dF_1}{dx} = F_2 \tag{8.7}$$

$$\frac{dF_2}{dx} = -fF_2 - \beta(1 - F_1^2)$$

The velocity component parallel to the wedge surfaces is given by $F_1 = df/dx$. Figure 8.1(b) shows, qualitatively, the distributions of f and F_1. This problem is considered again, in greater detail, in Example 8.5. Therefore, a given nth-order ordinary differential equation can generally be reduced to a system of n first-order equations.

A first-order ordinary differential equation may be written in the form

$$\frac{dy}{dx} = F(x, y) \tag{8.8}$$

The methods used for solving a system of first-order equations are based on those for a single equation, and, since most higher-order equations can generally be reduced to a system of first-order equations, most of the available methods are directed at solving a single first-order equation, given by Eq. (8.8). In many problems of engineering interest, the differential equations obtained are nonlinear, with the dependent variable and its derivatives appearing as nonlinear functions in the equation. Equation (8.5) is an example of a nonlinear ordinary differential equation. However, if the equation can be written in the form given by Eq. (8.2), the solution procedure is essentially the same for linear and nonlinear equations. Still, nonlinear problems generally involve a greater computational effort and, in boundary-value problems, may lead to convergence difficulties. For linear equations, one can often use the superposition of solutions to simplify the computational scheme. Linear, homogeneous, boundary-value problems arise in some engineering applications, such as the natural vibration of systems. These situations lead to eigenvalue problems which often require special solution techniques.

In view of the above discussion, ordinary differential equations may be classified as first-order or higher-order, single equation or system of equations, initial-value or boundary-value, linear or nonlinear, and homogeneous or inhomogeneous. Although there are often large differences in the analytical solution of these different types of equations, the applicable numerical procedures are quite similar. The classification of the problem as initial-value or boundary-value is important, since different techniques for solving the equation or for satisfying the boundary conditions are generally needed. The solution of a single first-order equation is particularly important since it forms the basis for solving other types of equations.

Analytical solutions of ordinary differential equations may be obtained in a few simple cases, particularly for linear equations. As discussed in Chapter 1, analytical results, whenever available, are useful in the validation and testing of the numerical scheme, which may be first employed for the simple problem whose analytical solution is known. Once the procedure has been tested for correctness and accuracy, one may proceed to the solution of more involved problems that cannot be solved analytically. If no relevant analytical results are available, one considers the numerical results obtained in terms of the physical or chemical nature of the problem to determine whether the results follow expected trends. In several engineering problems, some experimental data may be available on the problem being solved numerically and may be employed for the verification of the method and the numerical results.

Frequently, several methods can be employed for a given ordinary differential equation. Although each method has its particular advantages over other methods, and also certain disadvantages, many numerical methods are generally applicable to a given problem, and the choice of the method frequently becomes a matter of personal preference. Generally, one solves higher-order equations by reducing them to a system of first-order equations, as outlined above. Boundary-value problems are often solved by shooting methods, which are based on the methods applicable for initial-value problems. Consider, for example, the problem shown in Fig. 8.1 and governed by Eq. (8.5). This is a boundary-value problem, with the boundary conditions given as follows:

$$\begin{aligned} &\text{At } x = 0: \quad f = 0 \quad \text{and} \quad F_1 = 0 \\ &\text{As } x \to \infty: \quad F_1 \to 1 \end{aligned} \tag{8.9}$$

Therefore, the conditions are specified at two values of the independent variable x. If F_2 at $x = 0$ is given instead of the third condition, an initial-value problem will be obtained. Therefore, F_2 at $x = 0$ may be guessed, the equation solved as an initial-value problem, and a correction scheme employed to iteratively vary the guessed value of F_2 until the third condition in Eq. (8.9) is satisfied to a desired tolerance level. Such an approach is known as a shooting method and is frequently employed, although finite difference methods for solving boundary-value problems directly are also available. Therefore, much of the discussion in this chapter is directed at the first-order initial-value problem, followed by a consideration of other problems and techniques.

There are mainly two types of methods available for solving the first order initial-value problem given by Eq. (8.8). In the first case, the desired solution at a given value of x is obtained in terms of the function $F(x, y)$, evaluated at various x values between x and $x - \Delta x$, where Δx is the chosen increment in x. The values for $x < x - \Delta x$ are not needed, and the methods are, therefore, self-starting, since only the initial condition is needed to obtain the solution at the next step, $x = \Delta x$. Euler's method and the Runge-Kutta methods fall in this category. The methods that constitute the second category require information at values of x less than $x - \Delta x$ and are not self-starting. These are known as *multistep methods* and require other methods to yield the solution for the first few steps beyond the initial condition. Included in this category are Adams multistep formulas and the predictor-corrector methods, such as those by Hamming and Milne. The multistep methods are among the most effective and efficient numerical techniques available for solving ordinary differential equations.

This chapter discusses various methods that may be employed for solving first-order initial-value problems, outlining the important advantages and limitations of each method. The solution of a system of ordinary differential equations is considered next. The solution of boundary-value problems is considered in terms of shooting methods, using the techniques for initial-value problems, and of finite difference methods. The techniques applicable for eigenvalue problems are also discussed.

8.2 EULER'S METHOD

Let us consider the solution of the first-order ordinary differential equation

$$\frac{dy}{dx} = F(x,y) \tag{8.8}$$

with the initial condition

$$y(x_0) = y_0 \tag{8.10}$$

where y_0 is the value of $y(x)$ at a given value of the independent variable, $x = x_0$. A numerical solution of this differential equation involves obtaining the numerical values of the function $y(x)$ at discrete values of x, termed *node points*, for $x > x_0$. If Δx represents a uniform step size, that is, a constant difference between successive values of x at which the numerical solution is to be obtained, the node points x_i are defined by

$$x_i = x_0 + i\,\Delta x \qquad \text{where } i = 0, 1, 2, \ldots \tag{8.11}$$

The numerical values of the solution at these points may be denoted by $y_0, y_1, \ldots, y_n, \ldots$. Therefore, the numerical scheme must provide a means of evaluating y_{i+1} from the given or computed solution at the preceding grid points. If interest lies in determining the solution for $x < x_0$, instead of $x > x_0$, x_i may be taken as $x_i = x_0 - i\,\Delta x$; or a simple transformation of the independent variable may be employed to yield a new variable that increases as x decreases. Therefore, we shall consider only the case of increasing x here.

8.2.1 Computational Formula and Interpretation of the Method

Euler's method is one of the simplest methods for solving ordinary differential equations. However, it is very seldom used since several more efficient methods are available. The main reason for studying this method is that it is simple and allows a consideration of many of the basic features of the numerical solution of ordinary differential equations without the additional complexity of other methods. The computational formula for solving Eq. (8.8) by Euler's method is

$$y_{i+1} = y_i + \Delta x\, F(x_i, y_i) \qquad \text{with } i = 0, 1, 2, \ldots \tag{8.12}$$

Therefore, the solution can be obtained for increasing x, starting with $x = x_0$. This implies that the method is self-starting. Figure 8.2 shows the geometric interpretation of Euler's method. The exact solution is denoted by $y(x)$, and a qualitative comparison between the computed results and the exact solution is shown. The tangent to the curve at $x = x_i$ has a slope of $F(x_i, y_i)$ and approximates the true curve for $x_i \leqslant x \leqslant x_{i+1}$. At the initial point, $x = x_0$, the line tangent to the graph of $y(x)$, as shown, approximates the numerical solution for $0 \leqslant x \leqslant \Delta x$. As x increases, the

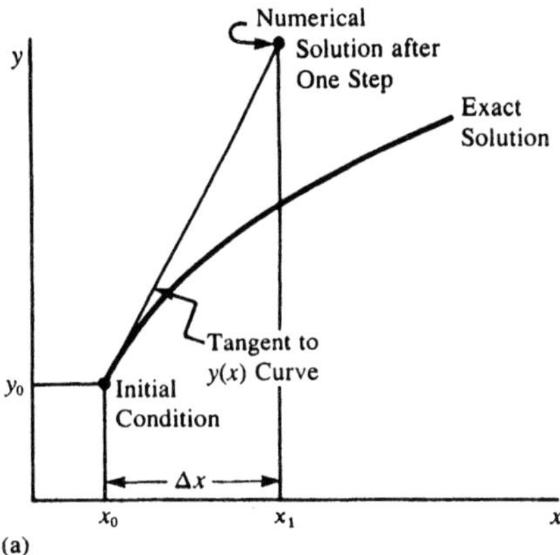

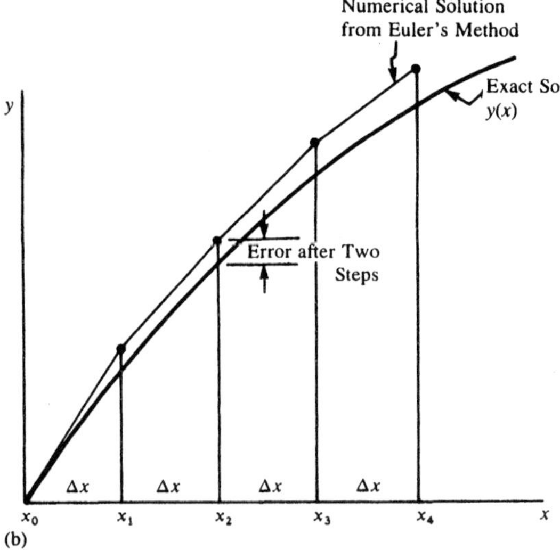

Figure 8.2 Graphical interpretation of Euler's method. (a) Numerical solution and error after the first step; (b) accumulation of error with increasing independent variable x.

numerical results increasingly deviate from the exact solution, due to accumulation of error.

There are several other ways of interpreting Euler's method. If the function $y(x)$ is assumed to be analytic near x_i, it may be expanded in a Taylor series, using Eq. (3.2), as follows:

$$y_{i+1} = y_i + \Delta x \left.\frac{dy}{dx}\right|_{x_i} + \frac{(\Delta x)^2}{2} \left.\frac{d^2y}{dx^2}\right|_{\xi_i} \qquad \text{where } x_i < \xi_i < x_{i+1} \tag{8.13}$$

Euler's method is obtained if the last term, which then is the truncation error for this computational step, is dropped. Therefore, this formulation for deriving Euler's method allows a determination of the error, as discussed in detail later in this section. We may also use numerical differentiation, with a forward difference approximation for the derivative, to represent Eq. (8.8) as

$$F(x_i, y_i) = \left.\frac{dy}{dx}\right|_{x_i} \simeq \frac{y_{i+1} - y_i}{\Delta x} \tag{8.14}$$

This equation also gives the formula for Euler's method. Similarly, numerical integration may be applied to the given differential equation to give

$$y_{i+1} = y_i + \int_{x_i}^{x_{i+1}} F(x,y)\, dx \tag{8.15}$$

Therefore, the change in y is represented by the area under the $F(x,y)$ curve. One can obtain an approximation to the integral by taking $F(x,y)$ as constant over the interval. This is the rectangular rule for numerical integration, as discussed in the preceding chapter. Thus,

$$\int_{x_i}^{x_{i+1}} F(x,y)\, dx \simeq (x_{i+1} - x_i)F(x_i, y_i) = \Delta x\, F(x_i, y_i)$$

which gives

$$y_{i+1} = y_i + \Delta x\, F(x_i, y_i) \tag{8.12}$$

Figure 8.3 shows a few steps of this numerical integration to obtain the function $y(x)$.

Both the Taylor series formulation and the numerical integration procedure can be employed to generate more accurate methods. The former leads to single-step methods such as the Runge-Kutta formulas, and the latter to multistep methods, particularly the predictor-corrector methods. Euler's method does not yield a high level of accuracy in the solution and is, therefore, rarely used. However, because the method is so simple, it is used in some engineering applications to obtain an initial estimate of the physical variables, by solving the governing differential equations with a relatively small step size Δx. The method also serves to illustrate the basic considerations that arise in the numerical solution of ordinary differential equations.

Solution of a System of Equations. Euler's method may easily be extended to yield a solution of a system of first-order equations. Consider the following system of

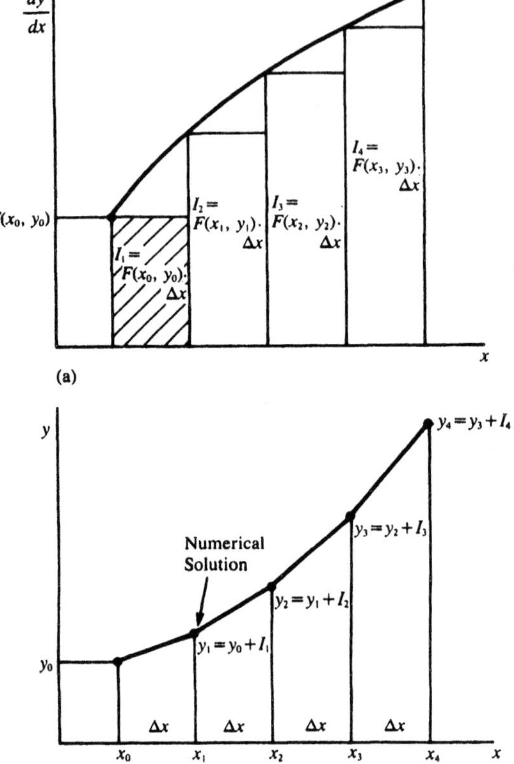

Figure 8.3 Sketch of a few steps in the numerical integration of the differential equation by Euler's method to yield the numerical solution y.

three equations:

$$
\begin{aligned}
\frac{dY_1}{dx} &= F_1(x, Y_1, Y_2, Y_3) \qquad \text{with } Y_1(x_0) = Y_{1,0} \\
\frac{dY_2}{dx} &= F_2(x, Y_1, Y_2, Y_3) \qquad \text{with } Y_2(x_0) = Y_{2,0} \qquad (8.16) \\
\frac{dY_3}{dx} &= F_3(x, Y_1, Y_2, Y_3) \qquad \text{with } Y_3(x_0) = Y_{3,0}
\end{aligned}
$$

where Y_1, Y_2, and Y_3 are three dependent variables whose values are given at $x = x_0$. We obtain the numerical solution of these equations from Euler's method by employing the computational formulas

$$
\begin{aligned}
Y_{1,i+1} &= Y_{1,i} + \Delta x\, F_1(x_i,\ Y_{1,i},\ Y_{2,i},\ Y_{3,i}) \\
Y_{2,i+1} &= Y_{1,i} + \Delta x\, F_2(x_i,\ Y_{1,i},\ Y_{2,i},\ Y_{3,i}) \qquad (8.17) \\
Y_{3,i+1} &= Y_{3,i} + \Delta x\, F_3(x_i,\ Y_{1,i},\ Y_{2,i},\ Y_{3,i})
\end{aligned}
$$

Thus, we obtain the numerical solution by proceeding in the direction of increasing x, starting with the initial conditions at $x = x_0$, and successively calculating the three independent variables Y_1, Y_2, and Y_3 at each step.

8.2.2 Errors, Convergence, and Stability

It is important to examine the errors associated with the numerical solution of a differential equation in order to determine the accuracy of the results obtained. As discussed in Chapter 2, several types of errors arise in numerical computation. Among the most important of these are the round-off and the truncation errors. The round-off error arises due to the retention of a finite number of significant figures by the computer. The round-off error is, therefore, a function of the computer and may be reduced by the use of double precision in the computation. The truncation error arises due to the finite difference approximation of a function. The infinite series that represents the function is generally truncated after a few terms to develop the scheme for the numerical solution of the differential equation. Therefore, the dropping of the remaining terms leads to the truncation error. A very important aspect in the error analysis of numerical methods is the growth or accumulation of errors as computation progresses, since this consideration is related to the stability of the scheme, as discussed below.

To find the truncation error in Euler's method, let us assume that the exact solution to the differential equation, $y(x_i)$, is known at x_i and is employed in Eq. (8.12) to compute the solution at x_{i+1}. Then

$$y_{i+1} = y(x_i) + \Delta x\ F[x_i, y(x_i)] \tag{8.18}$$

If the exact solution is analytic near x_i, we may represent it by a Taylor series as follows:

$$y(x_i + \Delta x) = y(x_i) + \Delta x \left.\frac{dy}{dx}\right|_{x_i} + \frac{(\Delta x)^2}{2}\left.\frac{d^2y}{dx^2}\right|_{x_i} + \frac{(\Delta x)^3}{3!}\left.\frac{d^3y}{dx^3}\right|_{x_i} + \cdots \tag{8.19}$$

From Eqs. (8.18) and (8.19),

$$y(x_i + \Delta x) - y_{i+1} = \frac{(\Delta x)^2}{2}\left.\frac{d^2y}{dx^2}\right|_{x_i} + O[(\Delta x)^3] \tag{8.20}$$

since $(dy/dx)_{x_i} = F[x_i, y(x_i)]$. Here, $y(x_i + \Delta x) - y_{i+1}$ may be termed the *truncation error* from x_i to x_{i+1}, starting with the exact solution at x_i. The leading term of the error is of the order of $(\Delta x)^2$ and may be denoted as $O[(\Delta x)^2]$, with the remaining terms being represented as $O[(\Delta x)^3]$.

Equation (8.20) gives the truncation error per step in Euler's formula. However, this is not the total error in the numerical solution at x_{i+1}, since the exact solution $y(x_i)$ is not known at x_i, except for the first step where the initial condition is given as exact. The value of y_i obtained by Euler's method contains the error accumulated in previous steps. The total error at a given value of x_i will be the product of the error per step and the number of steps. Since the number of steps is $x_{i+1}/\Delta x$, the total error is proportional to Δx and may, therefore, be denoted as $O(\Delta x)$ for Euler's method. For a detailed derivation of the total error, see Hornbeck (1975). Because the total error is of the first order in Δx, Euler's method is termed a *first-order method.*

In the above discussion, we have not considered the round-off error, which is inevitably present in any numerical solution. We can reduce the truncation error by making the step size Δx smaller. As illustrated above, this error decreases linearly with Δx. However, a reduction in Δx also results in an increase in the number of steps to obtain the solution over a given range in x. This results in an increase in the computing time and the round-off error, as shown qualitatively in Fig. 8.4. An optimum value of Δx at which the error is minimum is, therefore, expected. Because of the round-off error, the numerical solution will always differ from the exact solution of the differential equation. However, neglecting the round-off error, if the numerical solution approaches the exact solution, as the step size Δx approaches zero, the numerical method applied to a given differential equation is said to be *convergent.* The numerical techniques discussed in this chapter are convergent when applied to most differential equations, and, therefore, the convergence of the scheme is generally assumed. This definition of convergence is different from that employed in earlier chapters to indicate a negligible change in the solution from one step to the next during an iterative computational scheme. To distinguish between these two definitions, we shall refer to an iterative process as being *iteratively convergent.*

A very important consideration in the numerical solution of differential equations is that of stability of the numerical method. Although several definitions of stability are used in the literature, the most commonly employed definition simply considers a numerical method to be unstable if it yields an unbounded solution when the exact solution is bounded. Instability arises due to the amplification of the error, and, under certain conditions, an unbounded growth may arise. The stability of a

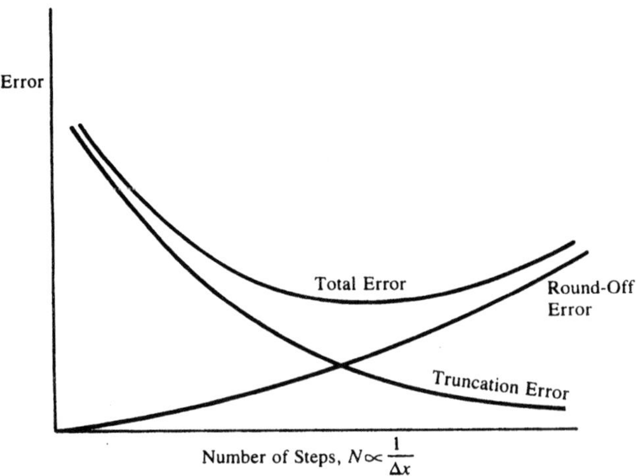

Figure 8.4 Qualitative representation of the variation of the truncation, round-off, and total errors, in the numerical solution of an ordinary differential equation, with the total number of steps N, which varies inversely as the step size Δx.

numerical scheme depends on both the method and the differential equation. For instance, as discussed in detail by Ferziger (1981), if Euler's method is applied to the differential equation $dy/dx = -\alpha y$, where α is a positive constant, the scheme is conditionally stable. If $|1 - \alpha \Delta x| > 1$, Euler's method gives rise to an increasing solution, whereas a decaying solution is given by analysis. Therefore, the scheme is stable only for a certain range of values of $\alpha \Delta x$ and not for all values. Oscillations that increase in amplitude with increasing x are observed, indicating the presence of instability (see Section 2.3.4). Numerical schemes which are stable for any value of the step size and other governing parameters are said to be *unconditionally stable*. Similarly, there are unconditionally unstable schemes that are unstable for all values. However, the computational scheme can be analyzed in only a few cases to determine its stability characteristics. A common approach employed in practice is to obtain the numerical solutions with two significantly different step sizes. If the two results are substantially different, numerical instability may be assumed to be present. If the two solutions are close to each other, then the scheme is probably stable.

In the numerical solution of a differential equation, it is important to consider the questions of accuracy, convergence, and stability, as outlined above. The exact solution is generally available only for a few simple cases. However, a comparison between the numerical solution for these cases and the exact solution will yield important information on the accuracy and correctness of the numerical results. It is also important to vary the step size Δx after the corresponding numerical solution has been obtained. By varying the step size, one can often determine whether the scheme is convergent and stable. The numerical results should be essentially independent of the step size. Several of these considerations were also discussed earlier in Chapter 2. The following example illustrates the use of Euler's method in solving a first-order initial-value problem.

Example 8.1

An electrical circuit consists of an inductance L, a resistance R, and an emf source E, as shown in Fig. 8.1.1. Initially, the switch is open and there is no current in the circuit. At time $\tau = 0$, the switch is closed and the current builds up. After 0.5 s, the switch is again opened and the current decreases with time to zero. Using Euler's method, solve this problem to obtain the variation of the current with time for (a) $E = 20$ volts,

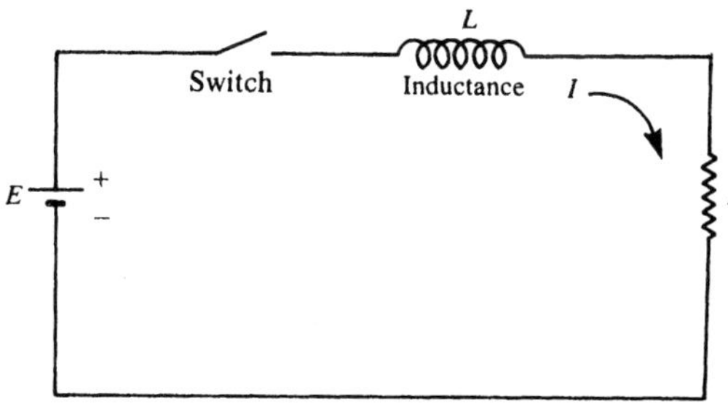

Figure 8.1.1 Electrical circuit considered in Example 8.1.

$L = 5$ henries, and $R = 10$ ohms, and (b) $E = 20$ volts, $L = 10$ henries, and $R = 5$ ohms.

Solution

We obtain the differential equation that governs the current I for the first part of the problem, when the switch is closed, by adding the voltage changes around the circuit and setting the sum equal to zero. The voltage across the inductance is $L(dI/d\tau)$, and that across the resistance is RI. [For the analysis of such electrical circuits, see, for instance, Halliday and Resnick (1986) and Ogata (1978).] Thus,

$$L\frac{dI}{d\tau} + RI - E = 0 \tag{8.1.1}$$

or

$$\frac{dI}{d\tau} = \frac{E}{L} - \frac{R}{L}I \qquad \text{with } I = 0 \text{ at } \tau = 0 \tag{8.1.2}$$

where τ is the time, in seconds, elapsed following the closing of the switch. We obtain the equation that applies for the second phase when the switch is reopened by setting $E = 0$:

$$\frac{dI}{d\tau} = -\frac{R}{L}I \qquad \text{with } I = I_1 \text{ at } \tau = 0.5 \tag{8.1.3}$$

The initial condition for this equation is the current I_1 at $\tau = 0.5$, where I_1 is obtained from the numerical solution of Eq. (8.1.2).

Therefore, the problem involves the solution of two first-order ordinary differential equations. We must first solve Eq. (8.1.2) to obtain the current I from $\tau = 0$ to $\tau = 0.5$ s. Then we solve Eq. (8.1.3) to obtain the current until it becomes essentially zero. The problem is a simple one and can be solved analytically. Here, we will consider its solution by the simple one-step, self-starting, Euler's method and compare the numerical results with the analytical solution.

For the two sets of data given for this problem, the equations are obtained as

$$\frac{dI}{d\tau} = 4 - 2I \qquad \text{for } 0 \leqslant \tau \leqslant 0.5 \tag{8.1.4a}$$

$$\frac{dI}{d\tau} = -2I \qquad \text{for } \tau > 0.5 \tag{8.1.4b}$$

and

$$\frac{dI}{d\tau} = 2 - 0.5I \qquad \text{for } 0 \leqslant \tau \leqslant 0.5 \tag{8.1.5a}$$

$$\frac{dI}{d\tau} = -0.5I \qquad \text{for } \tau > 0.5 \tag{8.1.5b}$$

The computational formula for Euler's method is given by

$$I_{i+1} = I_i + \Delta\tau\ F(\tau_i, I_i) \qquad \text{with } i = 0, 1, 2, \ldots \tag{8.1.6}$$

where the subscript i represents the computed values after the ith step, and $(i+1)$ those after the $(i+1)$th step. Here, F represents the function on the right-hand side of the equations. In the present case, F depends only on I, which in turn is a function of time τ. Also, $\tau = i\,\Delta\tau$, where $\Delta\tau$ is the time step and I_0 is the current at $\tau = 0$.

Figure 8.1.2 gives the computer program in FORTRAN 77 for this problem. The

```
C************************************************************************
C
C   THIS PROGRAM NUMERICALLY SOLVES A FIRST-ORDER ORDINARY DIFFERENTIAL
C
C   EQUATION USING EULER'S METHOD
C
C* * * * * * * * * * * * * * * * * * * * * * * * * * * * * * * * * * * *
C
C   IN THE FOLLOWING PROGRAM
C
C        T STANDS FOR TIME T            EE STANDS FOR THE E.M.F. OF THE SOURCE
C
C        EI STANDS FOR CURRENT          ER STANDS FOR RESISTANCE
C
C        EPS IS  THE CONVERGENCE CRITERION
C
C        DT IS STEP SIZE  IN T          EL STANDS FOR INDUCTANCE
C
C************************************************************
      IMPLICIT REAL (A-H,O-Z)
      OPEN(UNIT=10,FILE='ET')
      OPEN(UNIT=11,FILE='EI')
C
C    FILE 'ET' CONTAINS VALUES OF TIME T
C    FILE 'EI' CONTAINS VALUES OF EI AT CORRESPONDING T
C
C    INPUT PARAMETERS
C
      PRINT*,'INPUT PARAMETERS'
      PRINT*,'EE=   ','ER=   ','EL=   '
      READ*,EE,ER,EL
      PRINT*,'STEP SIZE DT ='
      READ*,DT
      PRINT*,'CONVERGENCE CRITERION EPS ='
      READ*,EPS
C
C    SET INITIAL CONDITIONS:
C
      T=0.0
      EI = 0.
      WRITE(10,*)T
      WRITE(11,*)EI
C
C    CALCULATE VALUES FOR THE NEXT STEP USING EULER'S METHOD
```

Figure 8.1.2 Computer program in FORTRAN 77 for solving the first-order ordinary differential equations that arise in Example 8.1.

```
C
 11      EI=EI + DT*(EE/EL - ER*EI/EL)
         T=T + DT
C
C   AT T = 0.5 SECONDS, THE SWITCH IS OPENED, THUS REMOVING THE
C   VOLTAGE SOURCE FROM THE CIRCUIT
C
         IF(T.LE.0.5) THEN
C
         WRITE(10,*)T
         WRITE(11,*)EI
         GO TO 11
         END IF
         PRINT*,'TIME T=',T,'SEC','  CURRENT I=',EI,'AMPS'
         PRINT*,'AT THIS STAGE, THE SWITCH IS OPENED'
C
C   COMPUTE VALUES FOR TIME GREATER THAN 0.5 S
C
 100     EI = EI + DT*(-ER*EI/EL)
         IF(EI.GT.EPS) THEN
         WRITE(10,*)T
         WRITE(11,*)EI
         T = T + DT
         GO TO 100
         END IF
C
C   OUTPUT RESULTS
C
 199     PRINT*,'TOTAL TIME TAKEN TO REACH STEADY STATE= ',T,'SEC'
         PRINT*,'THE VALUES OF TIME T ARE IN THE FILE ET'
         PRINT*,'THE VALUES OF CURRENT EI ARE IN THE FILE EI'
         PRINT*,'**************'
         CLOSE(UNIT=10)
         CLOSE(UNIT=11)
         STOP
         END
C  **********************************************************************
```

Figure 8.1.2 Continued

various symbols employed are defined in the program. The given parameters, E, L, and R, the time step, $\Delta\tau$, and the convergence criterion, EPS, may be entered into the program by the user interactively. The user employs the initial condition $I(0) = 0$ to start the computational scheme. Equation (8.1.4a) or Eq. (8.1.5a) is solved until $\tau = 0.5$. At this point, the computed value of the current is employed as the initial condition for Eq. (8.1.4b) or Eq. (8.1.5b), and the computation is continued until I is less than or equal to the convergence criterion EPS. Also, EPS is varied from 10^{-2} to 10^{-4}, in order to ensure that the results obtained are essentially independent of the value chosen. A value of 10^{-3} was found to be adequate, since decreasing EPS further did not affect the results significantly.

Figure 8.1.3 shows the numerical results obtained. In both cases, the current I rises sharply from zero as the switch is closed. The maximum value of the current is obtained at $\tau = 0.5$ s, beyond which the current decreases because of the reopening of

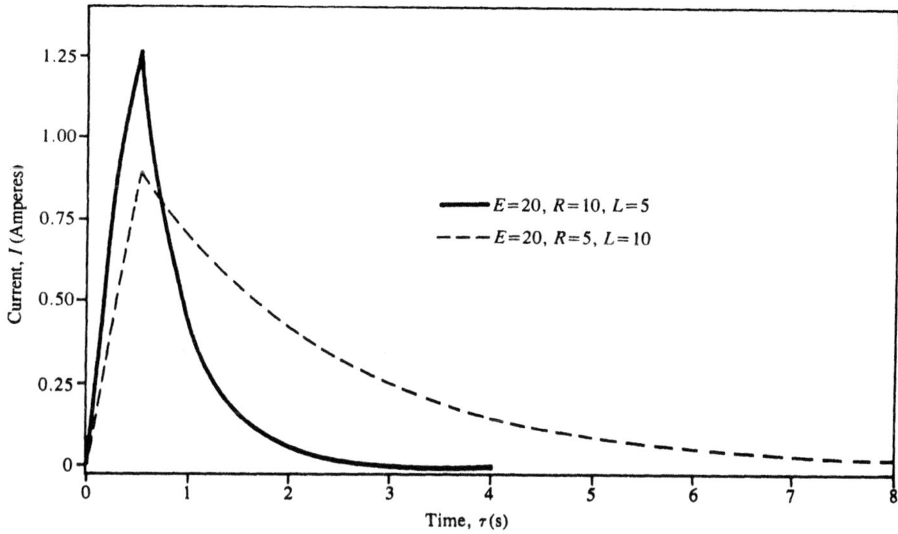

Figure 8.1.3 The computed variation of the current I with time τ for the two cases considered in Example 8.1. The time step $\Delta\tau$ is taken as 0.01 s for these results.

the switch. A larger maximum value of the current will be obtained if the switch is kept closed for a longer period of time. The second set of parameters results in a slower increase in the current, following the closing of the switch, and also a slower decrease after the switch has been reopened, as compared to the results for the first set. This behavior is expected since the derivative $dI/d\tau$, in the governing equations, is smaller in the second case.

Equation (8.1.1) can be solved analytically to give

$$I = \frac{E}{R}\left[1 - \exp\left(-\frac{R\tau}{L}\right)\right] \tag{8.1.7}$$

Similarly, Eq. (8.1.3) may be solved to give

$$I = \frac{E}{R}\left[1 - \exp\left(-\frac{R\tau_1}{L}\right)\right]\exp\left[-\frac{R(\tau - \tau_1)}{L}\right] \tag{8.1.8}$$

where τ_1 is the time at which the switch is reopened. The numerical results are found to agree quite well with the analytical solution. The current at $\tau = 0.5$ is obtained from Eq. (8.1.7) for the two cases as 1.264 and 0.885 amperes, respectively. These values are close to those obtained numerically; see Fig. 8.1.3.

Figure 8.1.4 shows the program in BASIC for solving the same problem. As illustrated in the examples given in earlier chapters, the two programs are very similar in form. The program in BASIC is relatively simpler because of the several features

```
100 REM    EULER'S METHOD FOR SOLVING A FIRST-ORDER O.D.E.
110 REM
120 REM    INPUT PARAMETERS
130     INPUT "EE=?    ",EE
140     INPUT "ER=?    ",ER
150     INPUT "EL=?    ",EL
160     INPUT "DT=?    ",DT
170     INPUT "EPS=?   ",EPS
180 REM
190 REM    SET INITIAL CONDITIONS
200 REM
210     T=0
220     EI=0
230 REM
240 REM    CALCULATE VALUES FOR THE NEXT STEP
250 REM
260     EI=EI+DT*(EE/EL - ER*EI/EL)
270     T=T+DT
280 REM
290 REM    AT T=0.5 S, THE SWITCH IS OPENED
300 REM
310     IF T > .5 THEN 330
320     GOTO 260
330     PRINT "TIME T=";T;"SEC","CURRENT I=";EI;"AMPS"
340     PRINT "AT THIS STAGE, THE SWITCH IS CLOSED"
350     EI=EI+DT*(-ER*EI/EL)
360     IF EI < = EPS THEN 390
370     T=T+DT
380     GOTO 350
390     PRINT "TOTAL TIME TO REACH STEADY STATE =";T;"SEC"
400     END
```

(a)

RESULTS

```
RUN
EE=?   20
ER=?   10
EL=?   5
DT=?   0.01
EPS=?  0.001
TIME T= .5099998 SEC        CURRENT I= 1.286227 AMPS
AT THIS STAGE, THE SWITCH IS CLOSED
TOTAL TIME TO REACH STEADY STATE = 4.049997 SEC


RUN
EE=?   20
ER=?   5
EL=?   10
DT=?   0.01
EPS=?  0.001
TIME T= .5099998 SEC        CURRENT I= .9023159 AMPS
AT THIS STAGE, THE SWITCH IS CLOSED
TOTAL TIME TO REACH STEADY STATE = 14.08023 SEC
```

(b)

Figure 8.1.4 Computer program in BASIC for solving the problem in Example 8.1.

that allow considerable ease in input/output and in control statements. BASIC is frequently used for simple problems of engineering interest, FORTRAN being more appropriate for complicated circumstances. The two programs also illustrate the ease with which one may switch from one programming language to the other.

8.3 IMPROVEMENTS IN EULER'S METHOD

8.3.1 Heun's Method

There are several numerical schemes that are based on modifications of Euler's method and can be used for solving ordinary differential equations. One of the most important is the improved Euler's method or Heun's method. In Euler's method, the function $F(x,y)$, which represents the derivative dy/dx of the dependent variable y, is taken as constant over the interval Δx at the value computed at the start of the interval. However, the derivative changes as x increases over the interval, and the accuracy of the solution can be improved if a better approximation is used for the derivative. Using Heun's method, one achieves this improvement by first calculating y_{i+1} from Euler's method, denoting this intermediate value as y^*_{i+1}, and then using this value to obtain a better approximation to the derivative. Considering the first-order initial-value problem of Eq. (8.8), the improved Euler's method is given by

$$y^*_{i+1} = y_i + \Delta x\, F(x_i, y_i) \qquad (8.21a)$$

$$y_{i+1} = y_i + \frac{\Delta x}{2}\left[F(x_i, y_i) + F(x_{i+1}, y^*_{i+1})\right] \qquad (8.21b)$$

Therefore, Euler's method is employed twice in succession, and the average of the approximations to the derivative at the two ends of the interval is used to give a more accurate value of y_{i+1}. Since y_{i+1} is unknown, the derivative at x_{i+1} is approximated by $F(x_{i+1}, y^*_{i+1})$. The method is self-starting since only the conditions at x_i are needed to obtain y_{i+1}.

The above procedure gives one of the simplest forms of the predictor-corrector methods, which are discussed in detail later in this chapter. Equation (8.21a) is a predictor equation for the first approximation to y_{i+1}, and Eq. (8.21b) is a corrector equation to yield an improved estimate of y_{i+1}. Figure 8.5 shows the graphical representation of this method. The improvement in the estimate of y_{i+1} is seen in terms of a better approximation to the slope of the graph of $y(x)$. Also shown is the integration of $F(x,y)$, as given by Eq. (8.15), using the trapezoidal rule to obtain y_{i+1}.

One can also use Eq. (8.21b) iteratively by substituting the value of y_{i+1} obtained after the first computation in place of y^*_{i+1} to obtain the next approximation to y_{i+1}. A sequence of corrected values of y_{i+1} may thus be generated. The iterative process is terminated when the change in y_{i+1} from one iteration to the next is smaller than a prescribed convergence criterion. However, greater accuracy, as compared to that obtained with Euler's method, is expected in the solution even if Eqs. (8.21a) and (8.21b) are used only once, without iteration.

From the Taylor series expansion for $y(x)$, as given by Eq. (8.19), greater accuracy is expected in the evaluation of y_{i+1}, if the terms of order $(\Delta x)^2$ are also retained. However, since the second derivative d^2y/dx^2 is not known, it may be approximated by

$$\frac{d^2y}{dx^2} \simeq \frac{y'(x_{i+1}) - y'(x_i)}{\Delta x} = \frac{F(x_{i+1},y_{i+1}) - F(x_i,y_i)}{\Delta x}$$

where the prime denotes the first derivative with respect to x. When this expression is substituted in the truncated Taylor series, we obtain

$$y_{i+1} = y_i + \Delta x\, F(x_i,y_i) + \frac{(\Delta x)^2}{2}\, \frac{F(x_{i+1},y_{i+1}^*) - F(x_i,y_i)}{\Delta x}$$

Therefore,

$$y_{i+1} = y_i + \frac{\Delta x}{2}\, [F(x_i,y_i) + F(x_{i+1},y_{i+1}^*)] \tag{8.21b}$$

where the derivative at x_{i+1} has been approximated by $F(x_{i+1},y_{i+1}^*)$ and the intermediate value y_{i+1}^* is obtained from Eq. (8.21a). Therefore, the method retains the second-order terms. The truncation error at each step is $O[(\Delta x)^3]$, and the total, or global, error is $O[(\Delta x)^2]$. The method is, therefore, a second-order method. An improved accuracy is obtained if iteration is employed with Eq. (8.21b). Also, higher-order formulas may be derived by retaining additional terms in the Taylor series expansion, as done for the Runge-Kutta methods, which are discussed in the next section. However, to achieve this increased accuracy, we need a larger computational effort for the determination of the intermediate approximations to the derivative.

The improved Euler's method may also be obtained by the application of the trapezoidal integration formula to the differential equation, Eq. (8.8), instead of the rectangular rule which yielded Euler's method, as shown in Fig. 8.5. Integrating the differential equation, we obtain

$$y_{i+1} - y_i = \frac{\Delta x}{2}\, [F(x_i,y_i) + F(x_{i+1},y_{i+1})] \tag{8.22}$$

Again, y_{i+1} in the parentheses on the right is replaced by y_{i+1}^*, which is obtained from Eq. (8.21a). Then, Eq. (8.21b) is obtained. The truncation error per step for the trapezoidal rule was given in Chapter 7 as

$$-\frac{(\Delta x)^3}{12}\, y'''(\xi)$$

where $y'''(\xi)$ is the third derivative of y with respect to x at a point ξ which lies between x_i and x_{i+1}. If y''' is assumed to be constant over this step size, it may be evaluated at x_i

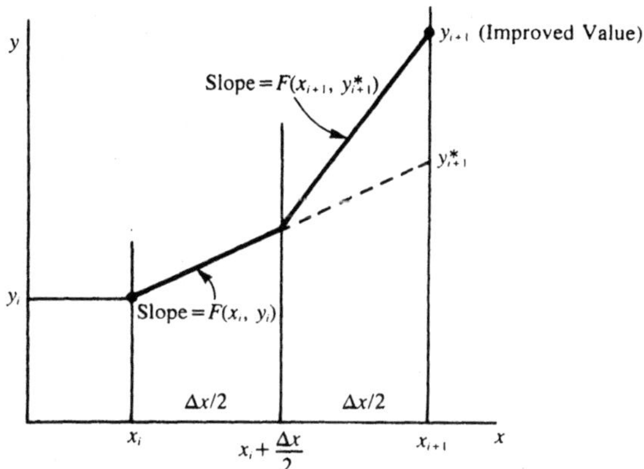

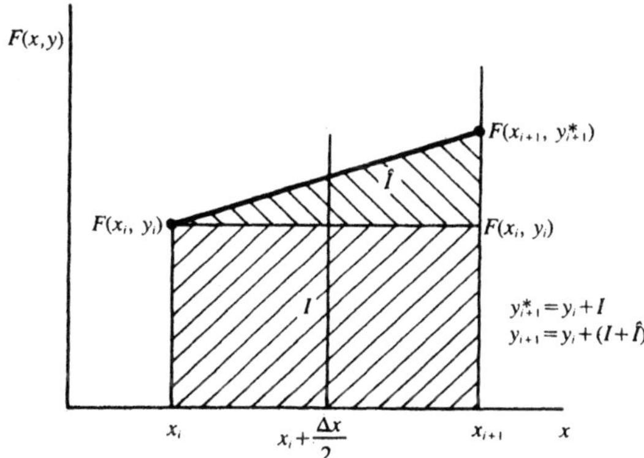

Figure 8.5 Graphical representation of the improved Euler's method, also known as *Heun's method.*

to yield an estimate of the truncation error TE per step as

$$TE = -\frac{1}{12} \cdot \frac{d^3y}{dx^3}\bigg|_{x_i} \cdot (\Delta x)^3 \tag{8.23}$$

This expression for the error applies if Eq. (8.22) is solved by iteration for y_{i+1} and not if the approximation given by Eq. (8.21b) is employed.

If the unknown y_{i+1}, appearing on the right-hand side of Eq. (8.22), is not approximated as y^*_{i+1}, we cannot solve for y_{i+1} directly, except for extremely simple cases. A nonlinear algebraic equation would generally be involved, and the iterative methods of Chapter 4 may be employed to determine y_{i+1} from this equation. Such methods, which require the solution of a nonlinear algebraic equation to obtain the new value of the function y, are known as *implicit*. Another implicit method that can be derived by application of the backward difference formula to the differential equation, Eq. (8.8), gives

$$y_{i+1} - y_i = \Delta x\, F(x_{i+1}, y_{i+1}) \tag{8.24}$$

This method, known as the *implicit Euler* or the *backward Euler method,* is first-order accurate.

Implicit methods involve more computation per step, as compared to the explicit methods discussed earlier. However, these methods generally have better numerical stability and are often unconditionally stable, as is the backward Euler method given above. In explicit methods, stability considerations may sometimes restrict the step size to a small value. In such cases, implicit methods may be preferable because of weaker restrictions on step size. However, it must be noted that the stability of a numerical scheme does not indicate the accuracy of the results. In fact, the truncation error in the implicit Euler method is equal in magnitude, but opposite in sign, to that in Euler's method.

8.3.2 Modified Euler's Method

We can obtain another modification of Euler's method by employing the midpoint, $x_i + (\Delta x/2)$, in a given step for the evaluation of the derivative $F(x,y)$. This method, often known as the *modified Euler's method* or the *improved polygon method,* is given by

$$y_{i+1/2} = y_i + \frac{\Delta x}{2} F(x_i, y_i) \tag{8.25a}$$

$$y_{i+1} = y_i + \Delta x\, F\left(x_i + \frac{\Delta x}{2}, y_{i+1/2}\right) \tag{8.25b}$$

Euler's method is employed twice, first to obtain an approximation $y_{i+1/2}$ at the midpoint and second to evaluate y_{i+1} from the derivative approximated at the midpoint. This method is self-starting, since the computation is based on the known conditions at $x = x_i$. It is second-order accurate and is also known as the *second-order Runge-Kutta method,* as discussed in the next section. The geometrical interpretation of the method is shown in Fig. 8.6, indicating the use of the midway point in determining y_{i+1}. Also, compare this figure with Fig. 8.2.

Both Heun's method and the modified Euler's method are employed for engineering applications where a very high level of accuracy is not necessary and a simple computational sdheme is desired. The former method is a simple form of the predictor-corrector methods, which are discussed in greater detail later in this chapter. The latter method is a second-order Runge-Kutta scheme and is thus a relatively less accurate version of a class of methods extensively used for practical problems. Since more accurate formulas, particularly the fourth-order methods, are much more important for engineering problems, examples of the solution of ODEs by these two methods are not included. The relevant features are demonstrated by the use of higher-order methods. However, the flow charts for both of these methods are shown

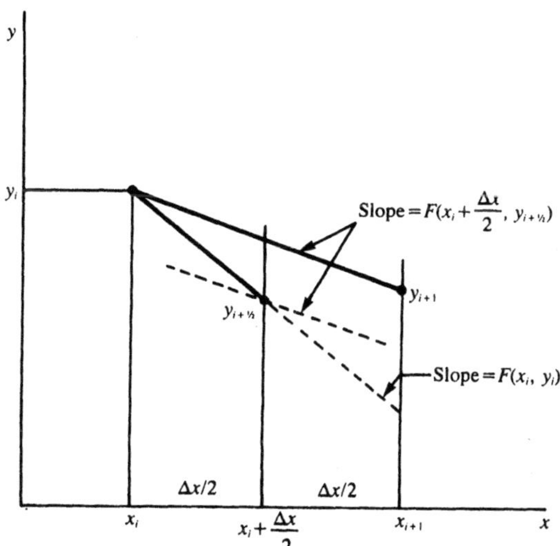

Figure 8.6 The modified Euler's method, also known as the *improved polygon method* or as the *second-order Runge-Kutta method.*

in Fig. 8.7. The corresponding computer programs can easily be written, using the program given for Example 8.1.

Several other modifications of Euler's method are available in the literature. We have considered only self-starting methods here. Some of the modifications of Euler's method are not self-starting, and one of the self-starting methods is needed to obtain the first few steps. Similarly, other implicit and explicit modifications of Euler's method may be derived. The accuracy of these methods may be improved by Richardson's deferred approach to the limit, discussed in Section 7.4.1 and also in the next section.

8.4 RUNGE-KUTTA METHODS

An important class of self-starting methods for the numerical solution of ordinary differential equations is based on retaining higher-order terms in the Taylor series expansion of the dependent variable $y(x)$, employing the computed values of the function $F(x, y)$, which represents the derivative of $y(x)$, at several values of x in the interval $x_i \leqslant x \leqslant x_{i+1}$. The resulting schemes, known as the *Runge-Kutta methods,* are widely used for the solution of various types of ordinary differential equations. In recent years, there has been an increase in the use of other methods, such as the predictor-corrector methods, which are often more efficient. However, the Runge-Kutta methods are usually employed to start these latter methods, which are not self-starting, and also for solving the complete problem in a wide variety of applications. The Runge-Kutta methods have several important advantages over other methods, besides being self-starting. They are easy to program, they have good numerical

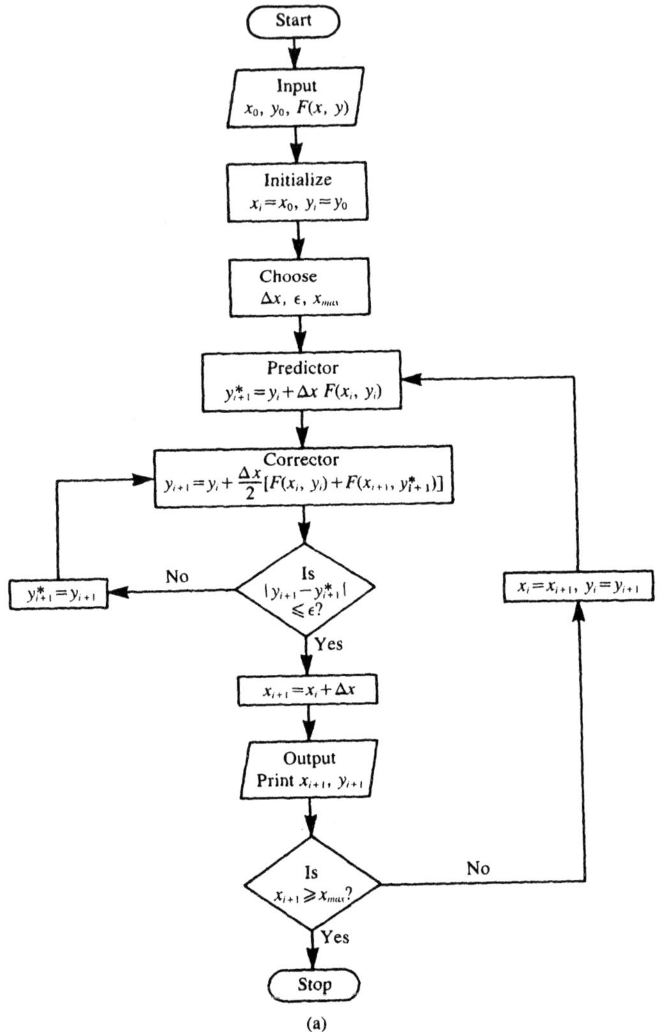

(a)

Figure 8.7 Flow charts for the solution of a first-order ordinary differential equation by (a) Heun's method and (b) modified Euler's, or improved polygon, method.

stability, and the step size can be changed easily to improve accuracy. However, for comparable accuracy, they often require more computer time than the more efficient methods. Also, the local error is not estimated easily. Nevertheless, the Runge-Kutta methods are probably the most widely used technique for solving ordinary differential equations that arise in engineering applications.

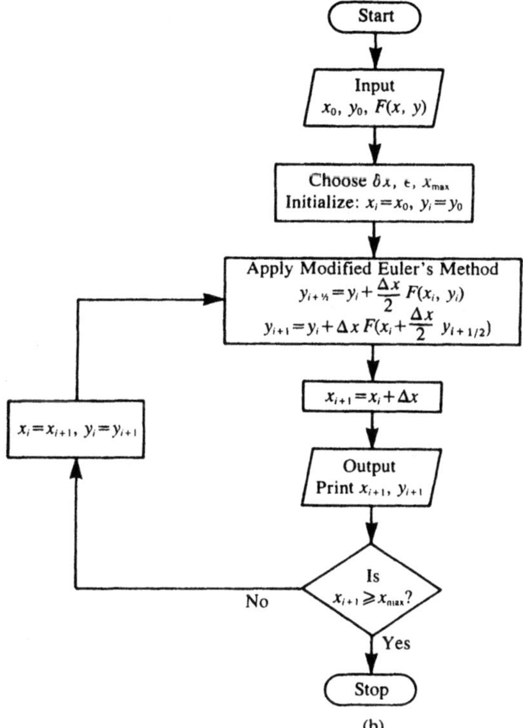

(b)

Figure 8.7 Continued

8.4.1 Computational Formulas

Runge-Kutta formulas involve a weighted average of the derivative calculated at various locations within a step size Δx. Let us consider the first-order ordinary differential equation given by

$$\frac{dy}{dx} = F(x,y) \tag{8.8}$$

with

$$y(x_0) = y_0 \tag{8.10}$$

Then $F(x,y)$ is evaluated at two values of x within the interval $x_i \leqslant x \leqslant x_{i+1}$ to yield the second-order Runge-Kutta method, which has the same accuracy as that obtained by retaining terms up to order $(\Delta x)^2$ in a Taylor series expansion for $y(x)$. Since the locations at which the function $F(x,y)$ is computed and the weighting to be used can be chosen in several ways, a family of formulas can be obtained for solving the given differential equation. The derivation of these formulas is quite involved and is based on a comparison between the terms of the Taylor series expansion for $y(x)$, about x_i, and those of the expansion for the approximations to the change in dependent variable over the given step. For instance, the second-order method may be expressed as follows:

$$y_{i+1} = y_i + C_1K_1 + C_2K_2 \tag{8.26a}$$

where

$$K_1 = \Delta x\, F(x_i, y_i) \tag{8.26b}$$

$$K_2 = \Delta x\, F(x_i + p_1\Delta x,\, y_i + p_2K_1) \tag{8.26c}$$

Thus, the K's approximate the change in y over the computational step.

The constants C_1, C_2, p_1, and p_2 are to be determined for an accurate evaluation of y_{i+1}. As discussed in detail by Ralston (1965) and Carnahan et al. (1969), one can find the relationships among these constants by comparing the expansions for $y(x)$ and K_2. One of the constants may be chosen arbitrarily. If C_2 is taken as 1/2, then it is found that $C_1 = 1/2$ and $p_1 = p_2 = 1$. This choice, therefore, leads to Heun's method, given in the preceding section. If C_2 is chosen as 1, then $C_1 = 0$ and $p_1 = p_2 = 1/2$, resulting in the modified Euler's method. Similarly, other second-order formulas may be obtained. The first-order Runge-Kutta method is simply Euler's method. One derives higher-order methods similarly by employing Taylor's expansions of $y(x)$ and retaining terms up to order $(\Delta x)^3$ for the third-order formulas and $(\Delta x)^4$ for the fourth-order. Still higher-order formulas have also been developed but are rarely used because of the large amount of computation involved at each step; see Butcher (1964) and Shanks (1966) for details.

The third-order method obtained by Kutta is given by

$$y_{i+1} = y_i + \frac{K_1 + 4K_2 + K_3}{6} \tag{8.27a}$$

where

$$K_1 = \Delta x\, F(x_i, y_i) \tag{8.27b}$$

$$K_2 = \Delta x\, F\left(x_i + \frac{\Delta x}{2},\, y_i + \frac{K_1}{2}\right) \tag{8.27c}$$

$$K_3 = \Delta x\, F(x_i + \Delta x,\, y_i + 2K_2 - K_1) \tag{8.27d}$$

The fourth-order Runge-Kutta formulas are of the general form

$$y_{i+1} = y_i + (C_1K_1 + C_2K_2 + C_3K_3 + C_4K_4) \tag{8.28}$$

where the C's are constants. Depending on the choice of the locations where the K's are determined and of the appropriate parameters that arise, several fourth-order formulas have been developed. The most widely employed formula is the classical Runge-Kutta method given by

$$y_{i+1} = y_i + \frac{K_1 + 2K_2 + 2K_3 + K_4}{6} \tag{8.29a}$$

where

$$K_1 = \Delta x\, F(x_i, y_i) \tag{8.29b}$$

$$K_2 = \Delta x\, F\left(x_i + \frac{\Delta x}{2}, y_i + \frac{K_1}{2}\right) \tag{8.29c}$$

$$K_3 = \Delta x\, F\left(x_i + \frac{\Delta x}{2}, y_i + \frac{K_2}{2}\right) \tag{8.29d}$$

$$K_4 = \Delta x\, F(x_i + \Delta x, y_i + K_3) \tag{8.29e}$$

Therefore, four evaluations of the derivative function are made within the interval $x_i \leqslant x \leqslant x_{i+1}$, in order to obtain approximations to the change in y over this step, and a suitable weighted average is employed. The truncation error per step is of order $(\Delta x)^5$, since terms up to order $(\Delta x)^4$ are retained in the expansions.

Several other fourth-order formulas are available; see Carnahan et al. (1969). A formula that was quite widely used in the past is the one developed by Gill (1951). This method minimizes the round-off error and also reduces the computer storage requirements in the solution of a system of equations. The formula was, therefore, particularly useful for small computers. It is given by

$$y_{i+1} = y_i + \frac{K_1 + (2 - \sqrt{2})K_2 + (2 + \sqrt{2})K_3 + K_4}{6} \tag{8.30a}$$

where

$$K_1 = \Delta x\, F(x_i, y_i) \tag{8.30b}$$

$$K_2 = \Delta x\, F\left(x_i + \frac{\Delta x}{2}, y_i + \frac{K_1}{2}\right) \tag{8.30c}$$

$$K_3 = \Delta x\, F\left(x_i + \frac{\Delta x}{2}, y_i + \frac{\sqrt{2} - 1}{2} K_1 + \frac{\sqrt{2} - 1}{\sqrt{2}} K_2\right) \tag{8.30d}$$

$$K_4 = \Delta x\, F\left(x_i + \Delta x, y_i - \frac{K_2}{\sqrt{2}} + \frac{\sqrt{2} + 1}{\sqrt{2}} K_3\right) \tag{8.30e}$$

The fourth-order Runge-Kutta method, given by Eqs. (8.29), continues to be a popular choice for the solution of ordinary differential equations. It is also frequently employed for boundary-value problems by incorporating a root-solving scheme, such as the secant or the Newton-Raphson method, for satisfying the boundary conditions. The main attractive features of this method are, as mentioned earlier, high accuracy level, good stability characteristics, ease in programming, applicability to a wide variety of problems, self-starting computational scheme, and ease with which the step size may be changed to improve accuracy. All of these features are common to all fourth-order Runge-Kutta formulas. The classical method is the most widely used one mainly because it has been the standard method for many years and because it is available in most computer libraries. In engineering applications, where a simple method that yields reasonably accurate results is particularly attractive, the Runge-Kutta methods are commonly employed, despite the availability of more efficient methods. However, for applications that involve a substantial computational effort, the Runge-Kutta methods are generally used only to start the scheme, and other methods, such as the predictor-corrector methods, are employed after the first few steps to obtain the solution.

8.4.2 Truncation Error and Accuracy

One of the major problems with the Runge-Kutta methods is that a quantitative estimate of the local truncation error is usually difficult to obtain. Therefore, it is often difficult to determine whether the step size is small enough to yield numerical results of desired accuracy. A common procedure is to run the computational scheme for different step sizes and to compare the results obtained. The step size is then chosen such that a further reduction in size does not significantly affect the results. However, this process is very time-consuming, and other methods for estimating the error and for improving the accuracy of the results have been developed.

The truncation error per step, TE, for a nth-order Runge-Kutta method may be written as

$$TE = A(\Delta x)^{n+1} + O[(\Delta x)^{n+2}] \tag{8.31}$$

where A is a constant that depends on the function $F(x,y)$ and its higher-order partial derivatives. If Δx is small, the error is determined largely by the first term, and the bounds for A may be determined, as discussed by Ralston (1965). One method of estimating the truncation error is to obtain the two solutions y_{i+1} and $\tilde{y}_{i+1}$ corresponding to step sizes Δx and $\Delta x/2$, respectively, by integrating between the two points x_i and x_{i+1}. If Y_{i+1} is the exact solution, an estimate of the truncation error and an improvement in the accuracy of the numerical solution may be obtained from the above expression for truncation error, considering only the dominant term. This approach, known as *Richardson's extrapolation* and discussed earlier in Chapter 7, gives

$$Y_{i+1} - y_{i+1} = A(\Delta x)^{n+1}\frac{x_{i+1} - x_i}{\Delta x}$$

$$Y_{i+1} - \tilde{y}_{i+1} = A\left(\frac{\Delta x}{2}\right)^{n+1}\frac{x_{i+1} - x_i}{\Delta x/2}$$

Therefore,

$$Y_{i+1} = \frac{2^n\tilde{y}_{i+1} - y_{i+1}}{2^n - 1} \tag{8.32}$$

which gives a more accurate approximation to the solution. Also,

$$TE = Y_{i+1} - y_{i+1} = A(x)^{n+1} = \frac{2^n(\tilde{y}_{i+1} - y_{i+1})}{2^n - 1} \tag{8.33}$$

Therefore, for the fourth-order Runge-Kutta method, TE is obtained as

$$TE = \frac{16}{15}(\tilde{y}_{i+1} - y_{i+1}) \tag{8.34}$$

One can use this estimate of the truncation error to adjust the step size and thus maintain the desired accuracy level. However, if this estimation were done at each step, the number of calculations performed would rise to approximately three times that for the Runge-Kutta method. Therefore, the error may be computed once every m steps, where m is chosen arbitrarily. Collatz (1960) gave another procedure for controlling the error, based on the calculation of the absolute value $|(K_3 - K_2)/(K_2 - K_1)|$ from Eqs. (8.29) after each step. If this quantity becomes larger than a few hundredths, then the error is too large and the step size should be reduced. Several other similar procedures have been given in the literature for limiting the error in Runge-Kutta methods.

Note from the above discussion that we can decrease the truncation error by reducing the step size Δx. However, a reduction in Δx also results in an increase in the number of steps needed for integration of the equation over a given interval. This, in turn, increases the round-off error, which depends on the number of arithmetic operations performed. The total truncation error may be estimated as a product of the error per step and the number of steps. Since the latter varies inversely with the step size, the accumulated truncation error for the nth-order Runge-Kutta method is of order $(\Delta x)^n$. Therefore, as Δx is reduced, the truncation error is decreased while the accumulated round-off error increases. An estimate for the total error due to truncation and round-off is usually very difficult to obtain. However, if analytical results are available for some simple cases, a comparison between the analytical and numerical results allows one to determine the overall accuracy of the numerical scheme.

The stability characteristics of the Runge-Kutta methods are generally good. Of particular interest in any numerical scheme is its partial instability, which may arise even when the equation being solved is not inherently unstable and which is dependent largely on the step size. If $\partial F(x,y)/\partial y$ is positive, then the error may increase without bound as x increases. On the other hand, if it is negative, then the error remains bounded for small Δx; see Fox (1962) and Ferziger (1981). An important parameter is the step factor $\Delta x\, \partial F/\partial y$, which affects the propagation of the error. The stability of various formulas has been considered, and the criteria for choosing the step factor have been given. A practical approach in most cases is to solve the problem for different step sizes, and, if close agreement is obtained between the corresponding results, the scheme may be assumed to be stable.

8.4.3 System of Equations

The Runge-Kutta methods can also be employed for solving a system of first-order ordinary differential equations and, therefore, for solving an nth-order differential equation, since it can usually be reduced to n first-order equations, as outlined earlier.

Let us consider the two simultaneous first-order equations

$$\frac{dy}{dx} = F(x,y,z) \tag{8.35}$$

$$\frac{dz}{dx} = G(x,y,z) \tag{8.36}$$

with

$$y = y_0 \quad \text{and} \quad z = z_0 \quad \text{at } x = x_0 \tag{8.37}$$

Then the classical fourth-order Runge-Kutta method for this problem is given by

$$y_{i+1} = y_i + \frac{K_1 + 2K_2 + 2K_3 + K_4}{6} \tag{8.38}$$

and

$$z_{i+1} = z_i + \frac{K_1' + 2K_2' + 2K_3' + K_4'}{6} \tag{8.39}$$

where

$$\begin{aligned}
K_1 &= \Delta x\, F(x_i,y_i,z_i) & K_1' &= \Delta x\, G(x_i,y_i,z_i)\\
K_2 &= \Delta x\, F\left(x_i + \frac{\Delta x}{2}, y_i + \frac{K_1}{2}, z_i + \frac{K_1'}{2}\right) & K_2' &= \Delta x\, G\left(x_i + \frac{\Delta x}{2}, y_i + \frac{K_1}{2}, z_i + \frac{K_1'}{2}\right)\\
K_3 &= \Delta x\, F\left(x_i + \frac{\Delta x}{2}, y_i + \frac{K_2}{2}, z_i + \frac{K_2'}{2}\right) & K_3' &= \Delta x\, G\left(x_i + \frac{\Delta x}{2}, y_i + \frac{K_2}{2}, z_i + \frac{K_2'}{2}\right)\\
K_4 &= \Delta x\, F(x_i + \Delta x,\, y_i + K_3,\, z_i + K_3') & K_4' &= \Delta x\, G(x_i + \Delta x,\, y_i + K_3,\, z_i + K_3')
\end{aligned} \tag{8.40}$$

The computation must be carried out in the sequence given above, since K_1' is needed for calculating K_2, K_2 for calculating K_3', and so on.

To illustrate the application of the above formulas to a higher-order ordinary differential equation, let us consider the following equation, which governs the vibration of a mass connected to two boundaries through a spring and a damper:

$$m\frac{d^2x}{d\tau^2} + B\frac{dx}{d\tau} + kx = P \tag{8.41}$$

Here, m is the mass of the vibrating body, B the viscous friction coefficient of the damper, k the spring constant, P an external force, x the displacement of the mass, and τ the time. The system is illustrated in Fig. 8.8. We can reduce it to two first-order equations by defining a new variable y as

$$\frac{dx}{d\tau} = y = F(x,y,\tau) \tag{8.42}$$

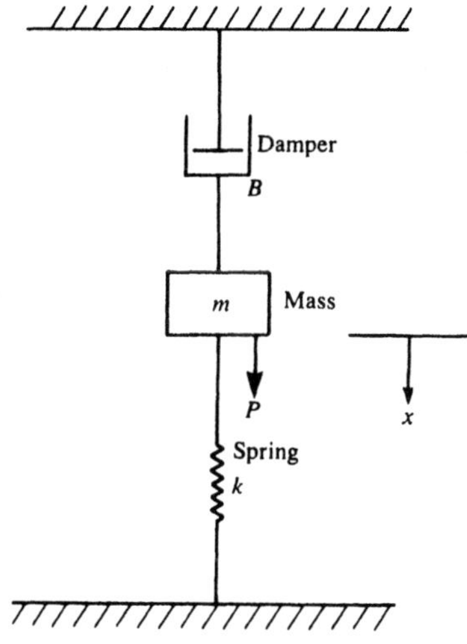

Figure 8.8 A vibrating system consisting of a vibrating mass m, a spring of stiffness k, a damper of friction coefficient B, and an external force P.

Therefore,

$$m\frac{dy}{d\tau} + By + kx = P$$

or

$$\frac{dy}{d\tau} = \frac{P - By - kx}{m} = G(x, y, \tau) \tag{8.43}$$

The two first-order equations, Eqs. (8.42) and (8.43), may now be solved by the Runge-Kutta method given above.

The recursion formulas for advancing from τ_i to τ_{i+1} are obtained as follows:

$$x_{i+1} = x_i + \frac{K_1 + 2K_2 + 2K_3 + K_4}{6} \tag{8.44}$$

$$y_{i+1} = y_i + \frac{K_1' + 2K_2' + 2K_3' + K_4'}{6} \tag{8.45}$$

where

$$\begin{aligned}
K_1 &= \Delta\tau\, y_i & K_1' &= \frac{\Delta\tau}{m}(P - By_i - kx_i) \\
K_2 &= \Delta\tau\left(y_i + \frac{K_1'}{2}\right) & K_2' &= \frac{\Delta\tau}{m}\left[P - B\left(y_i + \frac{K_1'}{2}\right) - k\left(x_i + \frac{K_1}{2}\right)\right] \\
K_3 &= \Delta\tau\left(y_i + \frac{K_2'}{2}\right) & K_3' &= \frac{\Delta\tau}{m}\left[P - B\left(y_i + \frac{K_2'}{2}\right) - k\left(x_i + \frac{K_2}{2}\right)\right] \\
K_4 &= \Delta\tau(y_i + K_3') & K_4' &= \frac{\Delta\tau}{m}[P - B(y_i + K_3') - k(x_i + K_3)]
\end{aligned} \tag{8.46}$$

Here, $\Delta\tau$ is the step size in time τ, and y represents the rate of change of the displacement x with time, that is, the velocity of the mass m. Note that the initial values of x and y are needed for starting the computation. These may be specified as follows:

$$\text{At } \tau = 0: \quad x = x_0 \quad \text{and} \quad y = \frac{dx}{d\tau} = y_0 \tag{8.47}$$

Similarly, we can use the method to solve third-, fourth-, or still higher-order equations by reducing them to a system of first-order equations. Considerations of error, accuracy, and stability are similar to those discussed for first-order equations, although these concerns become more involved for a system of equations. The following example illustrates the use of Runge-Kutta methods for solving ordinary differential equations.

Example 8.2

A projectile of mass m is shot vertically upward at a velocity of 100 m/s. The frictional force acting on the projectile due to its motion in air is given as $m(AV + BV^2)$, where A and B are constants and V is the velocity at any given time τ. Using the fourth-order Runge-Kutta method, compute the vertical position x and the velocity V of the projectile as functions of time for (a) $A = 0.01\ \text{s}^{-1}$, $B = 0.001\ \text{m}^{-1}$, and (b) $A = 0.1\ \text{s}^{-1}$, $B = 0.01\ \text{m}^{-1}$. Solve for the vertical motion until the velocity becomes zero.

Solution

The projectile is subjected to retardation due to the gravitational force, mg, where g is the magnitude of the gravitational acceleration, and the frictional force due to the motion in air. We obtain the governing differential equation for the vertical displacement x by writing the force balance, from Newton's second law, as

$$m\frac{d^2x}{d\tau^2} = -mg - m(AV + BV^2) \qquad \text{where } V = \frac{dx}{d\tau} \tag{8.2.1}$$

Here, $d^2x/d\tau^2$ is the acceleration of the projectile, taken as positive in the vertically upward direction. The gravitational force is negative since it acts downward. The frictional force is also negative since it acts in a direction opposite to that of the motion. Therefore, the equation to be solved is

$$\frac{d^2x}{d\tau^2} = -g - \left[A\frac{dx}{d\tau} + B\left(\frac{dx}{d\tau}\right)^2\right] = F(x,\tau) \tag{8.2.2}$$

This equation is solved until the velocity drops to zero and the projectile reaches its maximum height, before starting the downward motion. The initial conditions for Eq. (8.2.2) are as follows:

$$\text{At } \tau = 0: \quad x = 0 \quad \text{and} \quad \frac{dx}{d\tau} = 100 \tag{8.2.3}$$

Since SI units are being used, $g = 9.8\ \text{m/s}^2$. Also, x, τ, and V are in m, s, and m/s, respectively.

The second-order equation, given above, may be broken down into two first-order equations as

$$\frac{dx}{d\tau} = V \tag{8.2.4}$$

$$\frac{dV}{d\tau} = -g - (AV + BV^2) \tag{8.2.5}$$

Then the initial conditions for starting the Runge-Kutta scheme are as follows:

$$\text{At } \tau = 0: \quad x = 0 \quad \text{and} \quad V = 100 \tag{8.2.6}$$

Figure 8.2.1 gives the FORTRAN 77 program for this problem. The various

```
C***********************************************************************
C
C  THIS PROGRAM NUMERICALLY SOLVES A SECOND-ORDER ORDINARY DIFFERENTIAL
C
C  EQUATION USING THE 4TH ORDER RUNGE-KUTTA METHOD
C
C* * * * * * * * * * * * * * * * * * * * * * * * * * * * * * * * *   * *
C
C  IN THE FOLLOWING PROGRAM
C
C        T STANDS FOR TIME                    DT STANDS FOR STEP SIZE IN T
C
C        X STANDS FOR DISPLACEMENT            V STANDS FOR VELOCITY
C
C        A AND B ARE THE CONSTANTS APPEARING IN THE DIFFERENTIAL EQUATION
C
C        G IS THE ACCELERATION DUE TO GRAVITY = 9.8 m/(sec**2)
C
C**********************************************************************
         IMPLICIT REAL (A-H,O-Z)
         OPEN(UNIT=14,FILE='RT')
         OPEN(UNIT=15,FILE='RX')
         OPEN(UNIT=16,FILE='RV')
C
C  VALUES OF T ARE WRITTEN IN FILE RT
C  VALUES OF X ARE WRITTEN IN FILE RX
C  VALUES OF V ARE WRITTEN IN FILE RV
C
C
C  INPUT PARAMETERS
C
         PRINT*,'INPUT PARAMETERS'
         PRINT*,'A=   ','B=   '
         READ*,A,B
         PRINT*,'DT='
```

Figure 8.2.1 FORTRAN program for solving the second-order ordinary differential equation of Example 8.2 by the fourth-order Runge-Kutta method.

```
      READ*,DT
      G=9.8
C
C  SET INITIAL CONDITIONS
C
      T=0.
      X=0.
      V=100.
      WRITE(14,*)T
      WRITE(15,*)X
      WRITE(16,*)V
C
 11   Q=X
      Z=V
C
C  Q AND Z ARE VALUES OF X AND V, RESPECTIVELY, AT THE PREVIOUS TIME STEP
C
C  CALCULATE VALUES FOR THE NEXT STEP USING 4TH ORDER RUNGE-KUTTA METHOD
C
      RK1X = DT*Z
      RK1V = DT*(-G -A*Z -B*(Z**2))
      RK2X = DT*(Z + RK1V/2.)
      RK2V = DT*(-G -A*(Z + RK1V/2.) -B*(Z + RK1V/2.)**2)
      RK3X = DT*(Z + RK2V/2.)
      RK3V = DT*(-G -A*(Z + RK2V/2.) -B*(Z + RK2V/2.)**2)
      RK4X = DT*(Z + RK3V)
      RK4V = DT*(-G -A*(Z + RK3V) -B*(Z +RK3V)**2)
      X = Q +(RK1X +2.*RK2X + 2.*RK3X + RK4X)/6.
      V = Z +(RK1V + 2.*RK2V + 2.*RK3V + RK4V)/6.
      T = T + DT
C
C  STOP THE CALCULATIONS WHEN V BECOMES ZERO
C
      IF(V.GT.0.) THEN
      WRITE(14,*)T
      WRITE(15,*)X
      WRITE(16,*)V
      GO TO 11
      END IF
C
C  OUTPUT RESULTS
C
      PRINT*,'THE VELOCITY HAS BECOME ZERO OR NEGATIVE'
      PRINT*,'TOTAL TIME TAKEN TO REACH MAXIMUM HEIGHT =',T,'SEC'
      PRINT*,'TOTAL HEIGHT REACHED BY THE PROJECTILE = ',X,'METERS'
      CLOSE(UNIT=14)
      CLOSE(UNIT=15)
      CLOSE(UNIT=16)
      STOP
      END
C
C*************************************************************************
```

Figure 8.2.1 Continued

symbols employed are defined in the program. The input parameters are entered by the user. The fourth-order Runge-Kutta scheme is written, using the formulas given in Eqs. (8.38) through (8.40). Then, in terms of the nomenclature used in the text, the recursion formulas are as follows:

$$\begin{aligned}
K_1 &= \Delta\tau\, V_i & K_1' &= \Delta\tau(-g - AV_i - BV_i^2) \\
K_2 &= \Delta\tau\left(V_i + \frac{K_1'}{2}\right) & K_2' &= \Delta\tau\left[-g - A\left(V_i + \frac{K_1'}{2}\right) - B\left(V_i + \frac{K_1'}{2}\right)^2\right] \\
K_3 &= \Delta\tau\left(V_i + \frac{K_2'}{2}\right) & K_3' &= \Delta\tau\left[-g - A\left(V_i + \frac{K_2'}{2}\right) - B\left(V_i + \frac{K_2'}{2}\right)^2\right] \\
K_4 &= \Delta\tau\,(V_i + K_3') & K_4' &= \Delta\tau[-g - A(V_i + K_3') - B(V_i + K_3')^2] \\
x_{i+1} &= x_i + \frac{K_1 + 2K_2 + 2K_3 + K_4}{6} & V_{i+1} &= V_i + \frac{K_1' + 2K_2' + 2K_3' + K_4'}{6}
\end{aligned} \tag{8.2.7}$$

Therefore, the displacement x and the velocity V at the next time step, denoted by the subscript $(i + 1)$, are obtained in terms of the values at the present time step, denoted by subscript i. The computation is carried out, starting with the initial conditions, until the velocity becomes less than or equal to zero. The exact point where it becomes zero may be obtained by interpolation.

Figures 8.2.2 and 8.2.3 show the numerical results obtained for the two sets of input data given. The time step $\Delta\tau$ can be taken as larger in the first case, since the

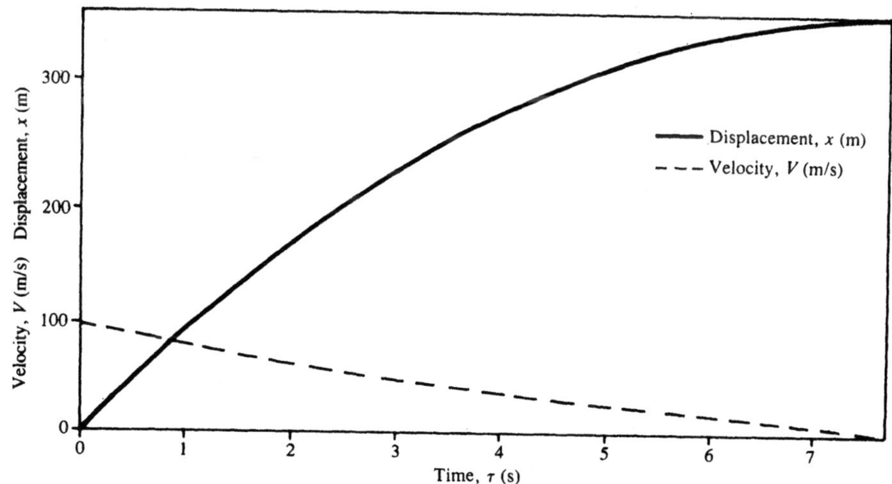

Figure 8.2.2 Computed variation of the velocity V and the displacement x with time τ for Example 8.2, for $A = 0.01\ s^{-1}$, $B = 0.001\ m^{-1}$, and $\Delta\tau = 0.05$ s.

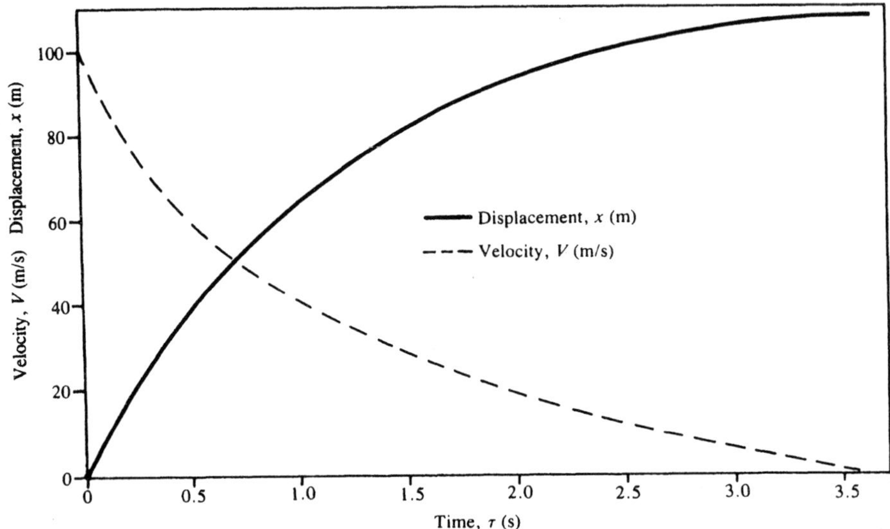

Figure 8.2.3 Computed results for $A = 0.1\ \mathrm{s}^{-1}$ and $B = 0.01\ \mathrm{m}^{-1}$ in Example 8.2, with $\Delta\tau = 0.01$ s.

variation with time is slower due to the smaller frictional force. However, $\Delta\tau$ must be varied to ensure that the numerical results are not significantly altered if $\Delta\tau$ is made smaller. For the two cases shown, $\Delta\tau$ values of 0.05 s and 0.01 s are taken, respectively, and found to be satisfactory, since a further reduction in $\Delta\tau$ resulted in a negligible change in the results. As expected, the maximum height, or displacement, is found to be larger when A and B are smaller. Also, the time taken to reach this height is larger. The problem may easily be solved analytically for $A = B = 0$ (that is, no frictional drag) to obtain the maximum height as 510.2 m, and the time taken to reach it as 10.2 s. These values compare well with the results shown.

The program is quite simple because of the self-starting feature of the method. The program, written for the present initial-value problem, may be used to obtain results for arbitrary values of the input parameters. It may be modified to solve other second-order initial-value problems. The use of this method for boundary-value problems requires a correction scheme, as discussed later in this chapter.

8.5 MULTISTEP METHODS*

In the methods considered so far for the solution of Eq. (8.8), we computed the value of y_{i+1} by using the known conditions at x_i and the approximations to the derivative at various other points in the interval $x_i \leqslant x \leqslant x_{i+1}$. However, a large number of function evaluations are needed for each step, where this number is equal to the order of the method. Therefore, interest lies in methods that give comparable accuracy with a smaller number of function evaluations per step. The multistep methods form an important class of efficient methods that use the information at mesh points preceding x_i, along with that at x_i and x_{i+1}, to yield y_{i+1}. Several multistep formulas have been derived. Formulas in which y_{i+1} is given explicitly in terms of known values of the dependent variable y and of the function $F(x, y)$ at x_i, x_{i-1}, and so on, are termed *open formulas*. Similarly, finite difference formulas that include unknown values of y and F usually require iteration to solve for y_{i+1}. Such formulas are termed *implicit* or *closed*.

If the formula requires the values of F at mesh points preceding x_i, the method is not self-starting, since at the initial condition, $x = x_0$, the only known condition is the given value y_0 of the variable y, from which the function $F(x, y)$ may be evaluated at this point. The conditions prior to x_i are not known. Therefore, a self-starting method, such as a Runge-Kutta formula with the same order of accuracy as that of the multistep formula under consideration, must be employed in the first few steps, until the information needed to proceed with the given multistep method is obtained.

8.5.1 Adams Multistep Methods

The multistep formulas may easily be derived from the Taylor series expansion of the dependent variable y, written as

$$y_{i+1} = y_i + \Delta x\, F_i + \frac{(\Delta x)^2}{2} F_i' + \frac{(\Delta x)^3}{3!} F_i'' + \cdots \tag{8.48}$$

where the primes denote differentiation with respect to x. We obtain this expansion from Eq. (8.19) by noting that

$$\left(\frac{dy}{dx}\right)_{x_i} = F_i \qquad \left(\frac{d^2y}{dx^2}\right)_{x_i} = F_i'$$

and so on. If the series in Eq. (8.48) is truncated after the second term, Euler's method is obtained. This may be considered as the first open multistep formula. We can generate a series of higher-order formulas from Eq. (8.48) by replacing the derivatives by their finite difference approximations from Chapter 3. If backward differences are used, the formulas obtained are known as *Adams-Bashforth* or *Adams open formulas*. Therefore, if F_i' is approximated as

$$F_i' = \frac{F_i - F_{i-1}}{\Delta x} + O(\Delta x) \tag{8.49}$$

the equation for y_{i+1} becomes

$$y_{i+1} = y_i + \Delta x\, F_i + \frac{(\Delta x)^2}{2}\,\frac{F_i - F_{i-1}}{\Delta x} + O[(\Delta x)^3]$$

or,

$$y_{i+1} = y_i + \Delta x\left[\frac{3}{2}F_i - \frac{1}{2}F_{i-1}\right] + O[(\Delta x)^3] \tag{8.50}$$

Therefore, a second-order formula, known as the *second open Adams formula*, is obtained. If F' is taken as constant over the interval $x_i \leqslant x \leqslant x_{i+1}$, the truncation error per step in Eq. (8.49) may be approximated as $\Delta x F_i''/2$, which leads to a truncation error of $5(\Delta x)^3 F_i''/12$ in Eq. (8.50). The total error is $O[(\Delta x)^2]$, as discussed later. The method is known as a second-order method because of the order of the total truncation error.

Similarly, we can derive the third-order formula by employing the following backward difference approximations for F_i'' and F_i':

$$F_i'' = \frac{F_i - 2F_{i-1} + F_{i-2}}{(\Delta x)^2} + O(\Delta x) \tag{8.51a}$$

$$F_i' = \frac{3F_i - 4F_{i-1} + F_{i-2}}{2\Delta x} + O[(\Delta x)^2] \tag{8.51b}$$

which leads to the equation

$$y_{i+1} = y_i + \Delta x \left[\frac{23}{12} F_i - \frac{16}{12} F_{i-1} + \frac{5}{12} F_{i-2} \right] + O(\Delta x)^4 \tag{8.52}$$

Again, if F''' is assumed to be constant over the given step Δx, the error incurred in Eq. (8.51a) may be written as $\Delta x\, F_i'''$, and that in Eq. (8.51b) as $(\Delta x)^2\, F_i'''/3$; see Chapter 3. Then the above expressions for F_i' and F_i'' may be used in Eq. (8.52) to yield a truncation error of $3(\Delta x)^4\, F_i'''/8$.

The general formula for the Adams-Bashforth method may be written as

$$y_{i+1} = y_i + \Delta x \sum_{m=1}^{n} \beta_{nm} F_{i+1-m} + O[(\Delta x)^{n+1}] \tag{8.53}$$

Thus, the truncation error per step is $O[(\Delta x)^{n+1}]$. The order of the method is n, since the number of steps needed to solve Eq. (8.8) up to a given value of x varies as $1/\Delta x$, leading to a total truncation error of order $(\Delta x)^n$. In Eq. (8.53), n may be varied to obtain Adams-Bashforth methods of different order. Table 8.1 gives the corresponding values of the β's, which are the coefficients for these explicit methods, for n up to 6. One can determine the truncation error for higher-order formulas by retaining the error terms in the finite difference approximations for the derivatives, as outlined above. Note that these methods are explicit and, thus, no iteration is needed for computing y_{i+1}.

We derive the *Adams-Moulton or Adams closed formulas* by employing a backward Taylor series expansion of $y(x)$, obtained as

$$y_i = y_{i+1} - \Delta x\, F_{i+1} + \frac{(\Delta x)^2}{2} F_{i+1}' - \frac{(\Delta x)^3}{3!} F_{i+1}'' + \cdots \tag{8.54}$$

which gives

$$y_{i+1} = y_i + \Delta x\, F_{i+1} - \frac{(\Delta x)^2}{2} F_{i+1}' + \frac{(\Delta x)^3}{3!} F_{i+1}'' + \cdots \tag{8.55}$$

TABLE 8.1 The Values of the Coefficient β_{nm} for the Adams-Bashforth Method, for n up to 6.

n	β	$m = 1$	2	3	4	5	6
1	β_{1m}	1					
2	$2\beta_{2m}$	3	−1				
3	$12\beta_{3m}$	23	−16	5			
4	$24\beta_{4m}$	55	−59	37	−9		
5	$720\beta_{5m}$	1901	−2774	2616	−1274	251	
6	$1440\beta_{6m}$	4277	−7923	9982	−7298	2877	−475

If the series is truncated after the second term, an implicit formula for y_{i+1} is obtained as follows:

$$y_{i+1} = y_i + \Delta x \, F_{i+1} + O[(\Delta x)^2] \tag{8.56}$$

The truncation error in this equation can easily be shown to be $-(\Delta x)^2 F_i'/2$. Again, if the derivatives of F in the above series are replaced by backward difference approximations, higher-order formulas are obtained. Therefore, we derive the second-order formula by employing

$$F'_{i+1} = \frac{F_{i+1} - F_i}{\Delta x} + O(\Delta x) \tag{8.57}$$

which gives

$$y_{i+1} = y_i + \Delta x \left[\frac{1}{2} F_i + \frac{1}{2} F_{i+1}\right] + O[(\Delta x)^3] \tag{8.58}$$

This formula is the same as the trapezoidal rule for integration discussed in Chapter 7, and, therefore, as shown before, the truncation error per step is $-(\Delta x)^3 F_i''/12$. Note that since F_{i+1} is involved in Eqs. (8.56) and (8.58), these formulas are implicit and iteration is needed to solve for y_{i+1}.

Both the first-order and the second-order Adams-Moulton formulas are self-starting, since F_{i-1}, F_{i-2}, and so on, are not involved. However, the third- and higher-order formulas are not self-starting and need a self-starting method, such as Runge-Kutta, to solve for the first few steps. The general expression for the Adams closed formulas is

$$y_{i+1} = y_i + \Delta x \sum_{m=0}^{n-1} \beta^*_{nm} F_{i+1-m} + O[(\Delta x)^{n+1}] \tag{8.59}$$

where n is the order of the method. Table 8.2 gives the coefficients β^*_{nm} for various values of n up to 6.

The Adams open and closed formulas are an important class of multistep

TABLE 8.2 The Values of the Coefficient β^*_{nm} for the Adams-Moulton Formulas, for n up to 6.

n	β^*	$m = 0$	1	2	3	4	5
1	β^*_{1r}	1					
2	$2\beta^*_{2m}$	1	1				
3	$12\beta^*_{3m}$	5	8	-1			
4	$24\beta^*_{4m}$	9	19	-5	1		
5	$720\beta^*_{5m}$	251	646	-264	106	-19	
6	$1440\beta^*_{6m}$	475	1427	-798	482	-173	27

formulas. Although they are rarely used separately, combinations of the two sets of formulas yield some of the most efficient predictor-corrector methods for solving ordinary differential equations, as discussed in the next section. One can also derive the Adams formulas by applying numerical integration to the differential equation; see Carnahan et al. (1969). The truncation error can be obtained quite easily for these methods. A smaller number of computations of the function F are needed in these formulas, as compared to those in the corresponding Runge-Kutta method, since the values of F at the preceding points are obtained from calculations performed for the earlier steps. The closed formulas are solved by iteration and the resulting error can be shown to be much less than that in an open formula of the same order. Also, the implicit methods generally possess better stability characteristics; see Ferziger (1981).

8.5.2 Additional Considerations

Several other multistep methods have been developed. A simple method that is often considered in studying the stability of multistep methods is the midpoint method, discussed in detail by Atkinson (1978) and given by

$$y_{i+1} = y_{i-1} + 2\,\Delta x\, F(x_i, y_i) \tag{8.60}$$

This method is second-order accurate and is not self-starting, since y_{i-1} is involved. A feature common to the various multistep methods is the existence of multiple solutions to the difference equation. For a convergent method, one of the solutions closely approximates the exact solution and is known as the *fundamental solution.* The other solutions, known as *parasitic solutions,* are not related to the exact solution of the differential equation. If these parasitic solutions grow with each computational step, instability arises in the numerical scheme. The growth of the parasitic solution is often exponential and oscillatory, leading to a rapid overpowering of the fundamental solution. In a stable scheme, the parasitic solution remains small compared to the fundamental solution. In practice, the numerical solution is obtained at two significantly different step sizes, and stability is assumed if the results are in reasonably close agreement. A reduction in step size also often leads to stability in a previously unstable solution.

The Runge-Kutta methods are often more stable than the corresponding multistep methods. Also, the starting method for computing the first few steps in multistep methods can substantially affect the final results. Therefore, it is important to vary the starting method to ensure that the results are essentially independent of the method used. Despite these disadvantages, multistep methods are frequently employed, particularly as predictor-corrector methods, and have, in many cases, replaced the Runge-Kutta scheme, which was the standard solution procedure for many years.

8.6 PREDICTOR-CORRECTOR METHODS

The predictor-corrector methods combine the advantages of accuracy and stability of the implicit formulas with the simplicity of the explicit formulas. The main problem with the implicit equations is the time-consuming iterative procedure necessary for obtaining the solution. However, if a first estimate y_{i+1}^* of the new value of the dependent variable is provided by the application of an explicit method at each step, the number of iterations needed for convergence to the solution by the use of the implicit method may be minimized. Therefore, an explicit formula is taken as the "predictor" to give a first estimate of the solution, followed by the use of an implicit formula as the "corrector" to obtain a better approximation to the solution.

8.6.1 Basic Features

The predictor-corrector methods are generally more efficient than the Runge-Kutta methods of the same order. The fourth-order Runge-Kutta method requires four function evaluations for advancing the solution by one step. The corresponding predictor-corrector method requires only one function evaluation for the predictor, since the other values are available from earlier computations, and, if fewer than three iterations are needed for the corrector, this method will require less computer time than the Runge-Kutta formulas of the same order. A detailed comparison between these two families of methods for solving ordinary differential equations is given by Hall et al. (1972). Another important advantage of the predictor-corrector methods is the ability to estimate the truncation error at each step. This ability allows one to choose a suitable step size for achieving the desired accuracy level and also to estimate the accuracy of the converged solution. Because of these advantages, the predictor-corrector methods are among the most popular ones at the present time. However, the problem of starting the scheme is usually present, since most schemes are based on multistep formulas such as those discussed in Section 8.5 and are not self-starting.

The improved Euler's method, given by Eqs. (8.21), is one of the simplest predictor-corrector methods. Euler's method is used as the predictor to give the first estimate y_{i+1}^* of the dependent variable as

$$y_{i+1}^* = y_i + \Delta x\, F(x_i, y_i) \tag{8.21a}$$

The trapezoidal rule for numerical integration is then used as the corrector to give

$$y_{i+1} = y_i + \frac{\Delta x}{2}[F(x_i,y_i) + F(x_{i+1},y_{i+1}^*)] \tag{8.21b}$$

This method is self-starting, unlike most other predictor-corrector methods, and is commonly employed in problems of engineering interest. As mentioned before, the method is also known as *Heun's predictor-corrector method.* The truncation error per step for the predictor is

$$-\frac{(\Delta x)^2}{2}\frac{d^2y}{dx^2}$$

For the corrector, the error is

$$-\frac{(\Delta x)^3}{12}\frac{d^3y}{dx^3}$$

when the corrector is taken as Eq. (8.22) and is solved by iteration. The error is larger if Eq. (8.21b) is employed, without iteration.

We apply iteration to Eq. (8.22) by substituting the computed value obtained into the right-hand side of the equation to obtain an improved value of y_{i+1}. The process is repeated until a specified convergence criterion is satisfied. The above corrector equation may also be employed with other multistep formulas as the predictor. A commonly employed non-self-starting scheme is obtained by the use of the midpoint method, given in Eq. (8.60), with Eq. (8.21b) as the corrector. Note from the above expressions for the truncation errors that the use of the corrector considerably improves the accuracy of the solution.

Several other predictor-corrector methods are employed in the solution of ordinary differential equations. The basic characteristics of these methods are the same as those outlined above. A predictor is employed to yield a first estimate of the dependent variable y_{i+1} at the next step, and this value is used to start the iterative solution of the corrector for an improved value of y_{i+1}. Among the most important and widely used predictor-corrector methods are the Adams method, Milne's method, and Hamming's method. None of these methods is self-starting, and generally a Runge-Kutta scheme, of the same order as the given predictor-corrector method, is employed for the first few steps. One can also use a Taylor series expansion of the variable $y(x)$ to obtain the values of y_1, y_2, y_3, and so on, needed to start the method. For instance, y_1 may be written as

$$y_1 = y_0 + \Delta x\, F(x_0,y_0) + \frac{(\Delta x)^2}{2} F'(x_0,y_0) + \frac{(\Delta x)^3}{3!} F''(x_0,y_0) + \cdots \tag{8.61}$$

The values of the function F and its derivatives are obtained from the given differential equation, Eq. (8.8). The number of terms retained in the series should be chosen to yield an error of the same order as that obtained by the given predictor-corrector method. However, the higher-order derivatives of the given differential equation may be quite involved. In such cases, the Runge-Kutta method is preferable for obtaining the starting values.

8.6.2 Adams Method

We can obtain a family of predictor-corrector methods by employing the Adams-Bashforth formulas as the predictor and the Adams-Moulton methods of same order as the corrector (see Tables 8.1 and 8.2). If the fourth-order formulas are chosen, the predictor equation is

$$y_{i+1}^{(0)} = y_i + \frac{\Delta x}{24}[55F_i - 59F_{i-1} + 37F_{i-2} - 9F_{i-3}] \quad (8.62a)$$

and the corresponding corrector formula is

$$y_{i+1} = y_i + \frac{\Delta x}{24}[9F_{i+1} + 19F_i - 5F_{i-1} + F_{i-2}] \quad (8.62b)$$

The superscript (0) indicates the predicted value that forms the first estimate for the corrector. The values of y and F for the first three steps of Δx, beyond the initial condition, are needed for starting this method. A fourth-order Runge-Kutta method may be employed to generate these values. Then the predictor gives an estimate for the next value of y, and, using this value as the first estimate, one iterates the corrector until convergence, in terms of a specified criterion, has been achieved. Similarly, the Adams formulas of different orders may be used to generate other predictor-corrector methods. The truncation errors associated with this method are discussed later.

8.6.3 Milne's Method

This method is based on integrating the differential equation, Eq. (8.8), to obtain

$$y_{i+1} = y_{i-3} + \int_{x_{i-3}}^{x_{i+1}} F(x,y)\,dx \quad (8.63)$$

The integral may be viewed as area under the curve from x_{i-3} to x_{i+1}, as shown in Fig. 8.9. If $F[x,y(x)]$ is approximated by the quadratic expression $ax^2 + bx + c$, where a, b, and c are constants, we can determine the constants by employing the value of the function at x_i, denoted by F_i, the value at x_{i-1}, denoted by F_{i-1}, and so on. If the above integral is then carried out, we obtain

$$y_{i+1}^{(0)} = y_{i-3} + \frac{4}{3}\Delta x[2F_i - F_{i-1} + 2F_{i-2}] \quad (8.64)$$

This is the predictor equation for Milne's method. The corrector is simply obtained from Simpson's integration rule, given by Eq. (7.25), as

$$y_{i+1} = y_{i-1} + \frac{\Delta x}{3}[F_{i-1} + 4F_i + F_{i+1}] \quad (8.65)$$

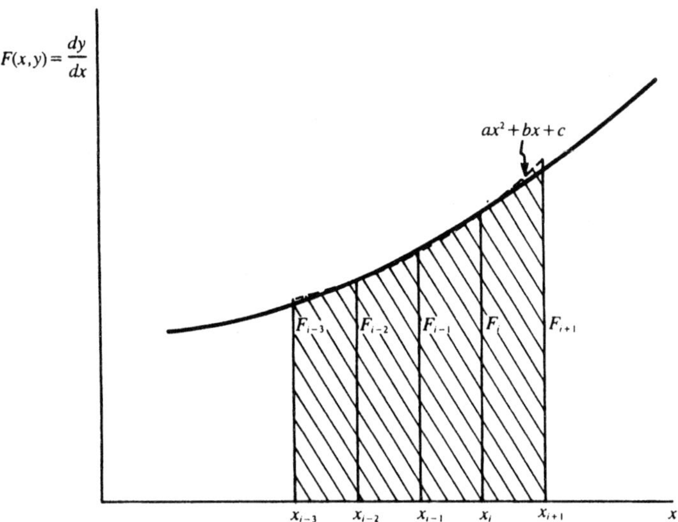

Figure 8.9 Sketch of the numerical integration used for the derivation of the predictor equation in Milne's method.

This is a fourth-order method since the total truncation error is $O[(\Delta x)^4]$, as shown later. Higher-order schemes have also been developed; see Carnahan et al. (1969) for the sixth-order Milne's method.

8.6.4 Hamming's Method

Hamming's method is based on the use of a general class of corrector equations represented by

$$y_{i+1} = a_i y_i + a_{i-1} y_{i-1} + a_{i-2} y_{i-2} + \Delta x(b_{i+1} F_{i+1} + b_i F_i + b_{i-1} F_{i-1} + b_{i-2} F_{i-2}) \tag{8.66}$$

where the a's and b's are constants. This equation includes the correctors employed for the fourth-order Adams and Milne's predictor-corrector methods. We can determine the constants by employing Taylor series expansions for all the variables and functions that appear in the above equation. Terms are retained up to $O[(\Delta x)^4]$, and the coefficients of similar terms on both sides of the equation are set equal. This results in the number of unknowns being larger than the relationships between them so that a few constants must be chosen arbitrarily. Hamming (1959) studied the stability of this corrector and chose the parameters to obtain better stability characteristics than those of the corrector used in Milne's method. The resulting equation is

$$y_{i+1} = \frac{1}{8}(9y_i - y_{i-2}) + \frac{3}{8}\Delta x(F_{i+1} + 2F_i - F_{i-1}) \tag{8.67}$$

The method employs the same predictor as that used in the fourth-order Milne's predictor-corrector method. A modifier equation is also used in order to reduce the error in the predicted value of y_{i+1}. The computational formulas for Hamming's method are, therefore, given in the order in which they are used as follows:

$$\text{Predictor:} \qquad y_{i+1}^{(0)} = y_{i-3} + \frac{4}{3}\Delta x(2F_i - F_{i-1} + 2F_{i-2}) \tag{8.68a}$$

$$\text{Modifier:} \qquad \bar{y}_{i+1}^{(0)} = y_{i+1}^{(0)} - \frac{112}{121}(y_i^{(0)} - y_i) \tag{8.68b}$$

$$\text{Corrector:} \qquad y_{i+1} = \frac{1}{8}(9y_i - y_{i-2}) + \frac{3}{8}\Delta x(F_{i+1} + 2F_i - F_{i-1}) \tag{8.68c}$$

The superscript (0) refers to the initial estimate that may be employed for starting the iteration process in the corrector equation. The estimate from the predictor is employed in the modifier, which provides the first estimate $\bar{y}_{i+1}^{(0)}$ for computing F_{i+1} on the right-hand side of the corrector for the first iteration. In practice, the step size is chosen so that only one or two iterations are needed for convergence. In fact, the method is generally used without iteration.

8.6.5 Accuracy and Stability of Predictor-Corrector Methods

We have given the general formulas employed in several important predictor-corrector methods. The basic characteristics of all the methods are quite similar, and any one of these can generally be employed for a given ordinary differential equation. The choice of a particular method is frequently made on the basis of personal preference, since the difference in the computational procedure and in the numerical results is generally small. However, the truncation errors associated with each formula are different from those that arise in other formulas. Similarly, the stability and convergence characteristics are different. These differences are sometimes important in the choice of the method for solving a given problem and are discussed here.

Truncation Errors. Let us first consider the truncation error at each step in the application of the formulas discussed above. Proceeding as outlined earlier for the second- and third-order Adams-Bashforth methods, we obtain the error that arises in the fourth-order formula as

$$(E_{i+1})_P = \frac{251}{720}(\Delta x)^5 F''''(\xi) \qquad x_{i-3} < \xi < x_{i+1} \tag{8.69}$$

where $(E_{i+1})_P$ is the estimate of the truncation error in the predictor for the $(i + 1)$th step. The corresponding error $(E_{i+1})_c$ in the corrector, which is the fourth-order Adams-Moulton formula, is

$$(E_{i+1})_c = -\frac{19}{720}(\Delta x)^5 F''''(\xi) \qquad x_{i-2} < \xi < x_{i+1} \tag{8.70}$$

Note that the error in the corrector is much smaller than that in the predictor.

If we assume the value at x_i to be exact and if the round-off error in the calculations for the $(i + 1)$th step are taken as negligible, as done before for estimating the truncation error per step, the exact solution at x_{i+1} may be written as

$$y(x_{i+1}) = y_{i+1}^{(0)} + \frac{251}{720}(\Delta x)^5 F_i'''' \tag{8.71}$$

or

$$y(x_{i+1}) = y_{i+1} - \frac{19}{720}(\Delta x)^5 F_i'''' \tag{8.72}$$

where $y_{i+1}^{(0)}$ is the predicted value from the predictor, and y_{i+1} is the converged value from the corrector. From these equations, we obtain the estimate of the trucation error per step, after the application of the corrector equation, by determining $(\Delta x)^5 F_i''''$ and then using Eq. (8.70) as follows:

$$E_{i+1} \simeq -\frac{19}{270}[y_{i+1} - y_{i+1}^{(0)}] \tag{8.73}$$

In addition to the assumptions given above, this estimate is based on the assumptions that F'''' is essentially constant over the interval $x_{i-3} \leqslant x \leqslant x_{i+1}$ and that the truncation errors per step in the predictor and the corrector are given by Eqs. (8.69) and (8.70), respectively. This estimate of the error may be used for determining whether the desired accuracy level is being maintained in the computation.

Similar estimates may be obtained for Milne's method and for Hamming's method. We can obtain the truncation error in the predictor of Milne's method by considering the approximation employed for the function F in the integral of Eq. (8.63). The corrector is based on Simpson's rule, the error for which was obtained in Chapter 7. The resulting truncation errors per step are thus obtained as

$$(E_{i+1})_P = \frac{14}{45}(\Delta x)^5 F''''(\xi) \qquad x_{i-3} < \xi < x_{i+1} \tag{8.74}$$

and

$$(E_{i+1})_c = -\frac{1}{90}(\Delta x)^5 F''''(\xi) \qquad x_{i-1} < \xi < x_{i+1} \tag{8.75}$$

Following the above procedure, the estimate for the truncation error per step, after convergence of the corrector equation, is given by

$$E_{i+1} \simeq -\frac{1}{29}[y_{i+1} - y_{i+1}^{(0)}] \tag{8.76}$$

The same approach may be applied to Hamming's method. Then the truncation errors are obtained (Carnahan et al., 1969) as follows:

$$(E_{i+1})_P = \frac{14}{45}(\Delta x)^5 F''''(\xi) \qquad x_{i-3} < \xi < x_{i+1} \tag{8.77}$$

and

$$(E_{i+1})_c = -\frac{1}{40}(\Delta x)^5 F''''(\xi) \qquad x_{i-2} < \xi < x_{i+1} \tag{8.78}$$

This results in the estimate for the truncation error per step as

$$E_{i+1} \simeq -\frac{9}{121}[y_{i+1} - y_{i+1}^{(0)}] \tag{8.79}$$

Note from the above expressions that the truncation error per step in the corrector is the smallest for Milne's method, followed by that in Hamming's method and that in the fourth-order Adams predictor-corrector method. The errors in the last two methods are close to each other and more than twice that in Milne's method. However, these methods have better stability characteristics than Milne's method, as discussed below. The truncation error may be estimated at each step to ensure that the numerical results have the desired accuracy. The step size can, therefore, be reduced if the error is too large, or it can be increased, to save computer time, if the error is too small. This ability to estimate the truncation error at each step is one of the important advantages of predictor-corrector methods over Runge-Kutta methods. However, a change in the step size is much more involved in predictor-corrector methods, since values of the derivative function F are needed at evenly spaced x values preceding x_i.

Step Size. An approach frequently adopted in changing the step size in predictor-corrector methods is simply to restart the computation scheme, with the last computed value of y as the initial condition for the new step size. Therefore, the starting method will again be needed for generating the values for the first few steps, and then the predictor-corrector method may be employed. Another approach is to use interpolation to obtain F values at the new spacing for $x \leqslant x_i$, using the computed values for the previous step size. A polynomial fit, as discussed in Chapter 6, may be employed to obtain a curve that passes through the available F values at the previous spacing. The required F values at the new spacing may then simply be obtained by interpolation, as shown qualitatively in Fig. 8.10. The estimate of the truncation error per step may be employed in the subsequent calculations to determine whether a change in the step size is again needed.

In predictor-corrector methods, the step size should be small enough to ensure convergence of the corrector equation in only one or two iterations. This is necessary in order to maintain the advantage of these methods, over one-step methods of comparable accuracy, of fewer derivative evaluations per step. However, a larger step size is desirable for reducing the computation for a given range of the independent variable x and, therefore, also for reducing the round-off error. The step size may be changed on the basis of the truncation error, as outlined above, or if more than two iterations are needed for the convergence of the corrector.

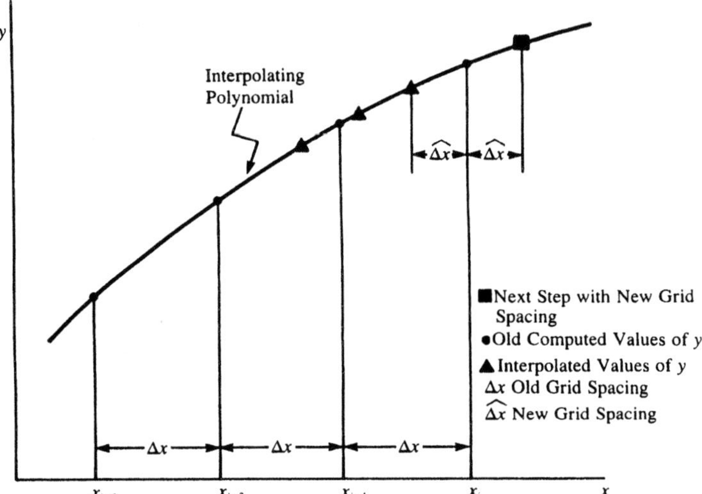

Figure 8.10 Interpolation of the preceding numerical results in order to vary the step size in predictor-corrector methods.

Stability. The stability characteristics of the various predictor-corrector methods discussed here have been studied in the literature, as considered in detail by Ralston (1962), Carnahan et al. (1969), and Atkinson (1978). If the corrector is iterated to convergence, then the stability of the predictor is not of much concern since it simply provides a first estimate. In fact, most predictors do not possess good stability characteristics. The stability of the corrector is, however, important, and the growth of error is studied to determine the overall stability of the scheme. If the corrector is not iterated but is employed only once, the stability of both the predictor and the corrector equations must be studied.

The corrector equations for the Adams method and Hamming's method have good stability characteristics. These methods are, therefore, often preferred over Milne's method, whose corrector equation is unstable for some differential equations. The stability analysis of multistep methods involves a consideration of the growth of the parasitic solutions, mentioned earlier. Milne's method is stable if the exact solution of the differential equation decays with increasing x and is generally known as a *marginally stable method.*

Most of the stability analyses consider simple linear equations, and the results obtained are often extended to more involved equations, particularly nonlinear equations. However, such an extension of the conclusions of the simple stability analyses may not be applicable in many cases. A practical approach employed in most engineering applications is to obtain the results for different step sizes. If the computed values do not differ significantly from each other, then the computational scheme is assumed to be stable. Also, analytical solutions may be available for a few simple circumstances. Then one can use a comparison of the numerical results with these solutions to study the accuracy of the results obtained and the stability characteristics of the method. Also, a reduction in step size usually results in an improvement in the stability of the numerical scheme.

8.6.6 Simultaneous Equations

It must be pointed out that, although the multistep and the predictor-corrector methods have been discussed for the first-order initial-value problem given by Eqs. (8.8) and (8.10), these methods may easily be extended to a system of simultaneous first-order equations. Since higher-order equations may be reduced to a system of first-order equations, as outlined in Section 8.1, these methods may be used for solving higher-order equations as well. A starting method, such as the Runge-Kutta method, is used to generate the required values for the first few steps, for each of the n dependent variables Y_j, where $j = 1, 2, \ldots, n$. The appropriate formulas are then applied to each equation in sequence at each step to obtain the values $(Y_j)_{i+1}$ at x_{i+1}. At each step, the stored function values from the preceding steps are employed to compute the new values, which are then stored for use in the next step. The procedure is a simple extension of the method for solving a single first-order equation.

There are several engineering problems that involve a wide range of scale, say, in length or time. In fluid flow and heat transfer, for instance, a small length scale may often be important in a given region, while a much larger length scale characterizes the remaining region. Similarly, in process control and chemical kinetics, a wide range of time or rate constants may arise. A system of equations that is associated with widely different time constants or eigenvalues is known as *stiff*, and special techniques are often needed to solve such a system. The step size must be small enough to treat the smallest scale or the fastest component of the process. An nth-order differential equation will, in general, have n scales. Employing an extremely small step size to take into account the smallest scale, although the other components can be treated with much larger step sizes, is obviously inefficient. A very small step size will result in large computer time and also large round-off errors. A major problem lies in maintaining a smooth behavior of the solution at large values of the independent variable, since large eigenvalues can lead to instability in this region. The problem of stiffness is similar to that of ill-conditioning encountered in matrices, discussed in Chapter 5. Several special techniques have been developed to solve stiff problems. The most popular among these is Gear's method. For further details, see Gear (1971), Hall and Watt (1976), and Ferziger (1981).

8.6.7 Concluding Remarks

The preceding discussion indicates that the predictor-corrector methods are among the most efficient methods available for the solution of ordinary differential equations. In addition, the truncation error at each step is determined during the computation and may be employed for maintaining the desired accuracy level by changing the step size whenever the error is excessive. However, these methods are more involved than the self-starting methods, such as Runge-Kutta formulas, which continue to be a very popular choice for the solution of the ordinary differential equations that arise in engineering problems.

The choice of a predictor-corrector method, from among those considered here, is often not an easy one, since the Adams method, Milne's method, and Hamming's method are all quite comparable in terms of efficiency and accuracy. Hamming's method avoids the instability problems of Milne's method and is, therefore, usually preferred. Also, it very seldom requires iteration and is generally used without iteration, making it a relatively more efficient method to use. However, if stability problems do not arise, Milne's method is superior because of its higher accuracy level. In general, personal preference and prior experience with the method are strong criteria for choosing it. Otherwise, Hamming's method may be chosen, despite the slight additional complexity in programming. The second-order predictor-corrector methods, although somewhat simpler to program, are rarely used because of the resulting lower accuracy in the results.

Example 8.3

A metal block of volume V and surface area A is initially at temperature T_i. At the surface, a constant energy input q, per unit area and time, is imposed at time $\tau = 0$ by thermal radiation, while the surface also loses energy by convection to air at temperature T_a surrounding the block. If the temperature in the block is assumed to be uniform at any given time and is denoted by $T(\tau)$, energy balance leads to the following governing equation for the temperature:

$$\rho C V \frac{dT}{d\tau} = qA - hA(T - T_a) \tag{8.3.1}$$

where ρ and C are the density and specific heat, respectively, of the metal. The parameter h is termed the *convective heat transfer coefficient*, and its value depends on the flow of air around the block and the temperatures involved. If the temperature difference $(T - T_a)$ is taken as the dependent variable θ, the above equation may be written as

$$\frac{d\theta}{d\tau} = \frac{qA}{\rho C V} - \frac{hA}{\rho C V}\theta = A - B\theta \tag{8.3.2}$$

where A and B are parameters defined as $A = qA/\rho CV$ and $B = hA/\rho CV$. Using the fourth-order Adams predictor-corrector method, solve this problem to obtain $\theta(\tau)$ if θ at $\tau = 0$ is given as 100°C. Consider two circumstances, given as (a) $A = 10°\text{C/s}$, $B = 0.05\ \text{s}^{-1}$, and (b) $A = 2°\text{C/s}$, $B = 0.03\ \text{s}^{-1}$.

Solution

The given problem involves solving the following one-dimensional ordinary differential equations:

$$\frac{d\theta}{d\tau} = 10 - 0.05\theta \tag{8.3.3}$$

$$\frac{d\theta}{d\tau} = 2 - 0.03\theta \tag{8.3.4}$$

with the initial condition

$$\theta = 100°\text{C} \qquad \text{for } \tau = 0 \tag{8.3.5}$$

The given equations are quite simple and may be solved analytically. However, they may be used to demonstrate the application of the Adams predictor-corrector method to ordinary differential equations. Then we can compare the numerical results with the analytical solution to evaluate the accuracy of the numerical scheme. In actual practice, however, A and B are usually not constants but vary with θ and τ, resulting in much more complicated problems for which the analytical solution may not be available.

The formulas for the fourth-order Adams predictor-corrector method are given by Eq. (8.62). A starting method is needed to generate the θ values at the three time steps, $\Delta\tau$, $2\,\Delta\tau$, and $3\,\Delta\tau$, so that Eq. (8.62a) can be employed to obtain the predicted value at $\tau = 4\,\Delta\tau$, where $\Delta\tau$ is the time step. The fourth-order Runge-Kutta scheme, given by Eq. (8.29), is employed to obtain these starting values. The value at $\tau = 0$ is, of course, the initial condition $\theta(0) = 100$. The predicted value is corrected iteratively, using Eq. (8.62b), until a specified convergence criterion has been satisfied. Thus, the value of θ at the next time step, denoted by $i + 1$, is obtained from the known values at the previous four time steps. Using this new computed value, the computation proceeds to the next time step. Thus, $\theta(\tau)$ is obtained with increasing time, starting with the initial condition. The computation is carried out until the temperature θ does not change significantly from one time step to the next, indicating the attainment of steady-state conditions.

Figure 8.3.1 shows the FORTRAN 77 program for solving a first-order ordinary differential equation by the Adams predictor-corrector method. Calculations are performed for four time steps, using the four previous values, which are initially generated by the fourth-order Runge-Kutta method. Then the newly computed values at four time intervals are employed to compute the temperatures at the next four intervals. A time step $\Delta\tau$ of 0.05 s is taken. For this value of $\Delta\tau$, only one or two iterations were needed for the convergence of the corrector. The computation is terminated when

$$\left|\frac{T_{i+1} - T_i}{T_i}\right| \leqslant \varepsilon \tag{8.3.6}$$

```
C***********************************************************************
C
C  THIS PROGRAM NUMERICALLY SOLVES A FIRST-ORDER ORDINARY DIFFERENTIAL
C
C  EQUATION USING ADAMS PREDICTOR-CORRECTOR METHOD
C
C* * * * * * * * * * * * * * * * * * * * * * * * * * * * * * * * * * * *
C
C   IN THE FOLLOWING PROGRAM
C
C      T STANDS FOR TIME                 TH STANDS FOR TEMPERATURE
C
C      DT STANDS FOR STEP SIZE IN T
C
C      EPS IS  THE CONVERGENCE CRITERION
C
C      A AND B ARE THE COEFFICIENTS IN THE DIFFERENTIAL EQUATION
C
C      THAT ARE GIVEN BY  A = Q*AR/(RHO*C*V) AND B =H*AR/(RHO*C*V)
C
C      WHERE
C
C        Q : SURFACE HEAT FLUX        RHO : DENSITY
C
C        C : SPECIFIC HEAT            V : VOLUME
C
C        H : SURFACE HEAT TRANSFER COEFFICIENT      AR : SURFACE AREA
C
C*******************************************************************
        IMPLICIT REAL*4 (A-H,O-Z)
        DIMENSION TH(8),THST(8)
        OPEN(UNIT=33,FILE='ADT')
        OPEN(UNIT=34,FILE='ADTH')
C
C  FILE 'ADT' CONTAINS VALUES  OF T
C  FILE 'ADTH' CONTAINS VALUES OF TH AT CORRESPONDING T
C
C  INPUT PARAMETERS
C
        PRINT*,'INPUT PARAMETERS'
        PRINT*,'A =     ', 'B =     '
        READ*,A,B
        PRINT*,'STEP SIZE DT ='
        READ*,DT
        PRINT*,'CONVERGENCE CRITERION EPS ='
        READ*,EPS
        PRINT*,'CONVERGENCE CRITERION FOR CORRECTOR ='
        READ*,CONVIT
C
C  SET INITIAL CONDITIONS
C
        T=0.0
        TH(1)=100.0
        WRITE(33,*)T
        WRITE(34,*)TH(1)
```

Figure 8.3.1 FORTRAN program for solving the first-order ordinary differential equation of Example 8.3 by the Adams predictor-corrector method.

```
C  THIS METHOD REQUIRES 4 INITIAL VALUES (i.e. TH(1),TH(2),TH(3),TH(4))
C  THESE ARE GENERATED IN THE FOLLOWING DO-LOOP USING 4TH ORDER
C  RUNGE-KUTTA METHOD WHICH HAS THE SAME ACCURACY AS THAT OF ADAMS
C  METHOD.
C
        DO 1 I=1,3
        RK1=DT*(A-B*TH(I))
        RK2=DT*(A-B*(TH(I)+RK1/2.0))
        RK3=DT*(A-B*(TH(I)+RK2/2.0))
        RK4=DT*(A-B*(TH(I)+RK3))
        TH(I+1)= TH(I)+ (RK1+2.0*RK2 + 2.0*RK3 +RK4)/6.0
        T=T+DT
        WRITE(33,*)T
        WRITE(34,*)TH(I+1)
 1      CONTINUE
C
C  CALCULATE VALUES FOR THE NEXT FOUR POINTS
C  USING ADAMS PREDICTOR-CORRECTOR METHOD
C
 55     DO 2 I=4,7
C
C  PREDICTOR STEP:
C
        TH(I+1)= TH(I) +DT*(55.*(A - B*TH(I)) - 59.*(A - B*TH(I-1))
     $  + 37.*(A - B*TH(I-2)) - 9.*(A- B*TH(I-3)))/24.
C
C  CORRECTOR STEP:
C
C  THE CORRECTOR STEP IS ITERATED UNTIL DESIRED CONVERGENCE IS ACHIEVED
C  THE ARRAY THST STORES THE VALUES OF TH AT PREVIOUS ITERATION
C
        THST(I+1)=TH(I+1)
 66     TH(I+1)= TH(I) + DT*(9.*(A - B*TH(I+1)) +19.*(A - B*TH(I))
     $  -5.*(A - B*TH(I-1)) + (A - B*TH(I-2)))/24.
        DTH = ABS(TH(I+1) - THST(I+1))
        IF (DTH.GT.CONVIT) THEN
        THST(I+1) = TH(I+1)
        GO TO 66
        END IF
        T=T+DT
        WRITE(33,*)T
        WRITE(34,*)TH(I+1)
        S=ABS(TH(I+1)-TH(I))/(TH(I)*DT)
C
C  HERE S IS FRACTIONAL CHANGE IN TEMPERATURE, TH(I), PER UNIT TIME STEP
C
C  CHECK FOR CONVERGENCE
C
        IF(S.LE.EPS) GO TO 99
 2      CONTINUE
C
C  THUS, KNOWING THE VALUES OF TH AT PREVIOUS 4 POINTS, THE VALUES
C  OF TH AT 4 POINTS AHEAD(IN TIME) ARE CALCULATED.
C  IN ORDER TO AVOID UNNECESSARY STORAGE, THE PREVIOUS VALUES OF TH ARE
C  REPLACED BY NEW VALUES. THIS IS DONE IN THE FOLLOWING DO-LOOP.
C
        DO 3 I=1,4
        TH(I)=TH(I+4)
 3      CONTINUE
        GO TO 55
```

Figure 8.3.1 Continued

```
99      PRINT*,'SOLUTION HAS CONVERGED'
        PRINT*,'THE STEADY STATE VALUE OF TEMPERATURE IS',TH(5)
        PRINT*,'TOTAL TIME TAKEN TO REACH STEADY STATE=',T,'  SEC.'
        CLOSE(UNIT=33)
        CLOSE(UNIT=34)
        STOP
        END
C**********************************************************************
```

Figure 8.3.1 Continued

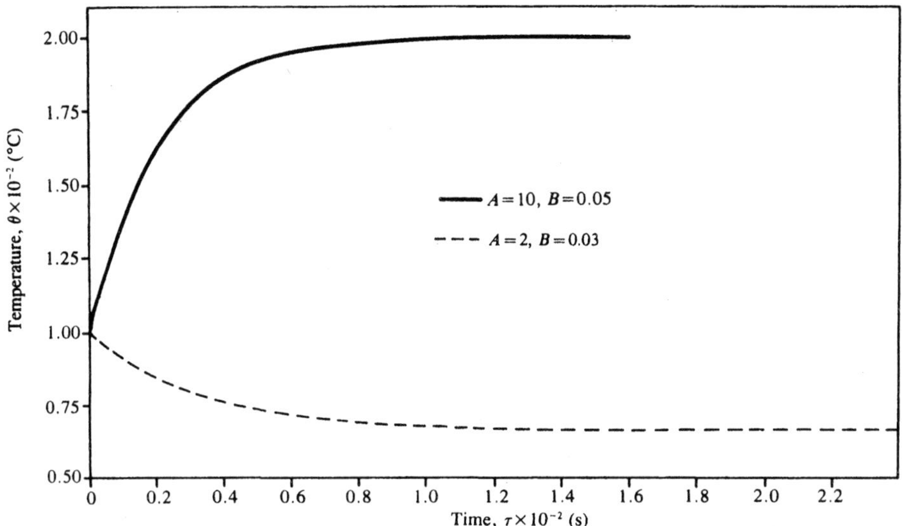

Figure 8.3.2 Computed variation of temperature θ with time τ for the two cases of Example 8.3. The results are obtained with $\varepsilon = 10^{-4}$ and $\Delta\tau = 0.05$ s.

where ε is a chosen convergence criterion for determining steady state. A value of 10^{-3} was employed for ε. This value was varied to ensure a negligible effect, on the numerical results, due to a further reduction in ε.

Figure 8.3.2 shows the computed variation of temperature θ with time τ for the two cases.The initial temperature difference $(T - T_a)$ is 100°C. In the first case, for which $A = 10$°C/s and $B = 0.05\ \text{s}^{-1}$, the energy input is larger than the convective energy loss at $\tau = 0$. Thus, $d\theta/d\tau = 5$ at $\tau = 0$ from Eq. (8.3.3). This positive initial value of the slope results in a temperature increase with time. Finally, a constant value of $\theta = 200$°C is attained at steady state. In the second case, the energy input is less than the loss, and the metal block cools down to a temperature of $\theta = 67$°C at steady state. The steady-state temperatures can easily be obtained analytically from Eqs. (8.3.3) and (8.3.4). At steady state, θ stops changing with time, and, therefore,

$d\theta/d\tau = 0$. If $d\theta/d\tau$ is set equal to zero in these equations, we obtain $\theta = 200°C$ and 66.67°C in the two cases. These values agree closely with the numerical results obtained. The computed variation of θ with τ was also compared with the analytical solution, and a close agreement between the two was obtained. For further details on the physical aspects of this problem and other similar ones, see Incropera and Dewitt (1981).

Example 8.4

A metal piece of mass m is released at zero velocity in a liquid and allowed to fall freely under gravity. The frictional force, or drag, acting on the piece is $m(AV + BV^2)$, where A and B are constants and V is the downward velocity. Using Hamming's method, compute the velocity V as a function of time τ for $B = 0.1\,\text{m}^{-1}$, at two values of A given as $A = 2\,\text{s}^{-1}$ and $4\,\text{s}^{-1}$.

Solution

The governing ordinary differential equation for this problem may be derived on the basis of the discussion in Example 8.2 as

$$\frac{dV}{d\tau} = g - (AV + BV^2) \tag{8.4.1}$$

with the following initial condition:

$$\text{At } \tau = 0: \quad V = 0 \tag{8.4.2}$$

Here, g is the acceleration due to gravity and is equal to 9.8 m/s² in SI units. Therefore, the equations to be solved are

$$\frac{dV}{d\tau} = 9.8 - (2V + 0.1V^2) = F(\tau,V) \tag{8.4.3}$$

$$\frac{dV}{d\tau} = 9.8 - (4V + 0.1V^2) = F(\tau,V) \tag{8.4.4}$$

The numerical scheme is similar to that for the Adams predictor-corrector method, outlined in Example 8.3. We use the fourth-order Runge-Kutta method to obtain the values of the velocity at the first three time steps, the value at $\tau = 0$ being given by the initial condition. Once the values at time steps 0, $\Delta\tau$, $2\,\Delta\tau$, and $3\,\Delta\tau$ have been computed, the predicted value at $\tau = 4\,\Delta\tau$ is obtained from Eq. (8.68a), written for the present problem as

$$V_{i+1}^{(0)} = V_{i-3} + \frac{4}{3}\Delta\tau(2F_i - F_{i-1} + 2F_{i-2}) \tag{8.4.5}$$

As this equation shows, to compute the predicted value of V at a given time step, we must know the values of the function F at the previous four steps. This predicted value is used in the modifier, to obtain an improved estimate $\bar{V}_{i+1}^{(0)}$ as

$$\bar{V}_{i+1}^{(0)} = V_{i+1}^{(0)} + \frac{112}{121}(V_i - V_i^{(0)}) \tag{8.4.6}$$

The step size $\Delta\tau$ was chosen as 0.05 s, and it was found that only one iteration was needed for the convergence of the corrector, given by

$$V_{i+1} = \frac{1}{8}(9V_i - V_{i-2}) + \frac{3}{8}\Delta x(F_{i+1} + 2F_i - F_{i-1}) \tag{8.4.7}$$

The two values $\bar{V}_{i+1}^{(0)}$ and V_{i+1} were found to be very close, within a chosen convergence criterion, for this value of $\Delta\tau$. If a larger time step is chosen, more than one iteration may be needed.

Figure 8.4.1 gives the computer program for the solution of this problem. The

```
C***********************************************************************
C
C   THIS PROGRAM NUMERICALLY SOLVES A FIRST-ORDER ORDINARY DIFFERENTIAL
C
C   EQUATION USING HAMMING'S PREDICTOR-CORRECTOR METHOD
C
C* * * * * * * * * * * * * * * * * * * * * * * * * * * * * * * * * * *
C
C   IN THE FOLLOWING PROGRAM
C
C     THE ARRAY VP STORES PREDICTED VALUES OF VELOCITY
C     THE ARRAY VM STORES MODIFIED VALUES OF VELOCITY
C     THE ARRAY VC STORES CORRECTED VALUES OF VELOCITY
C     THE ARRAY V STORES FINAL VALUES OF VELOCITY
C     T  STANDS FOR TIME
C     DT STANDS FOR STEP SIZE IN T
C     EPS IS THE CONVERGENCE CRITERION
C     A AND B ARE THE COEFFICIENTS APPEARING IN THE DIFFERENTIAL EQUATION
C     G IS THE ACCELERATION DUE TO GRAVITY = 9.8 (m/sec**2)
C
C* * * * * * * * * * * * * * * * * * * * * * *
        IMPLICIT REAL*4 (A-H,O-Z)
        DIMENSION V(8),VP(8),VM(8),VC(8)
        OPEN(UNIT=30,FILE='HT')
        OPEN(UNIT=31,FILE='HV')
C
C  FILE 'HT' CONTAINS VALUES  OF T
C  FILE 'HV' CONTAINS VALUES OF V AT CORRESPONDING T
C
C  INPUT PARAMETERS
        PRINT*,'INPUT PARAMETERS'
        PRINT*,'A =    ','B =    '
        READ*,A,B
        PRINT*,'STEP SIZE DT ='
        READ*,DT
        PRINT*,'CONVERGENCE CRITERION EPS ='
        READ*,EPS
        G = 9.8
```

Figure 8.4.1 Computer program for solving the first-order ODE considered in Example 8.4 by Hamming's predictor-corrector method.

```
C
C   SET INITIAL CONDITIONS
C
          T=0.0
          V(1)=0.0
          WRITE(30,*)T
          WRITE(31,*)V(1)
C
C   THIS METHOD REQUIRES 4 INITIAL VALUES (i.e. V(1),V(2),V(3),V(4))
C   THESE ARE GENERATED IN THE FOLLOWING DO-LOOP USING 4TH ORDER
C   RUNGE-KUTTA METHOD WHICH HAS THE SAME ACCURACY AS THAT OF HAMMING'S
C   METHOD.
C
          DO 1 I=1,3
          RK1=DT*(G-A*V(I)-B*V(I)**2)
          RK2=DT*(G-A*(V(I) + RK1/2.) -B*(V(I)+RK1/2.0)**2)
          RK3=DT*(G-A*(V(I) + RK2/2.) -B*(V(I)+RK2/2.0)**2)
          RK4=DT*(G-A*(V(I) + RK3) -B*(V(I)+RK3)**2)
          V(I+1)= V(I)+ (RK1+2.0*RK2 + 2.0*RK3 +RK4)/6.0
          T=T+DT
          WRITE(30,*)T
          WRITE(31,*)V(I+1)
  1       CONTINUE
C
C   CALCULATIONS FOR THE NEXT FOUR POINTS
C   USING HAMMING'S PREDICTOR CORRECTOR METHOD
C
          VP(4)=V(4)
          VM(4)=V(4)
          VC(4)=V(4)
C
 55       DO 2 I=4,7
C
C   PREDICTOR STEP:
C
          VP(I+1) = V(I-3) +4.0*DT*(2.*(G - A*V(I) - B*V(I)**2)
     $    - (G - A*V(I-1)- B*V(I-1)**2)
     $    + 2.*(G - A*V(I-2) - B*V(I-2)**2))/3.
C
C   MODIFIER STEP:
C
          VM(I+1) = VP(I+1) -(112./121.)*(VP(I) -VC(I))
C
C   CORRECTOR STEP:
C
          VC(I+1) =(9.*V(I) - V(I-2) + 3.*DT*(G - A*VM(I+1)- B*VM(I+1)**2
     $    +2.*(G - A*V(I) - B*V(I)**2) -(G - A*V(I-1) - B*V(I-1)**2)))/8.
C
          V(I+1) = VC(I+1)
C
          T=T+DT
          WRITE(30,*)T
          WRITE(31,*)V(I+1)
          S=ABS(V(I+1)-V(I))/(V(I)*DT)
C
C   WHERE S IS FRACTIONAL CHANGE IN VELOCITY, V(I), PER UNIT TIME STEP
C
C   CHECK FOR CONVERGENCE
```

Figure 8.4.1 Continued

```
C
        IF(S.LE.EPS) GO TO 99
C
  2     CONTINUE
C
C  THUS, KNOWING THE VALUES OF V AT PREVIOUS 4 POINTS, THE VALUES
C  OF V AT 4 POINTS AHEAD(IN TIME) ARE CALCULATED.
C  IN ORDER TO AVOID UNNECESSARY STORAGE, THE PREVIOUS VALUES OF V ARE
C  REPLACED BY NEW VALUES. THIS IS DONE IN THE FOLLOWING DO-LOOP.
C
        DO 3 I=1,4
        V(I)=V(I+4)
        VP(I)=VP(I+4)
        VC(I)=VC(I+4)
        VM(I)=VM(I+4)
  3     CONTINUE
        GO TO 55
  99    PRINT*,'SOLUTION HAS CONVERGED'
        PRINT*,'THE VALUE OF THE TERMINAL VELOCITY IS',V(5),'M/S'
        PRINT*,'VALUES OF T ARE IN FILE HT'
        PRINT*,'VALUES OF V ARE IN FILE HV'
        CLOSE(UNIT=30)
        CLOSE(UNIT=31)
        STOP
        END
C**********************************************************************
```

Figure 8.4.1 Continued

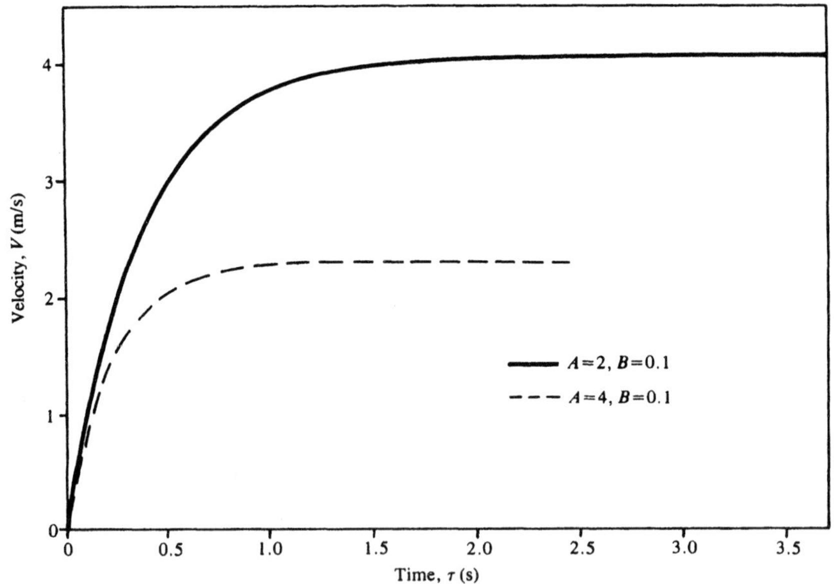

Figure 8.4.2 Computed variation of velocity V with time τ for $B = 0.1\ \text{m}^{-1}$ and $A = 2\ \text{s}^{-1}$ and $4\ \text{s}^{-1}$, as given in Example 8.5.

computation is terminated when the velocity V stops changing, indicating the attainment of the terminal velocity. This condition is determined by the convergence criterion

$$|V_{i+1} - V_i| \leqslant \varepsilon \tag{8.4.8}$$

where ε is a chosen small quantity. It was taken as 10^{-3} and reduced to smaller values to ensure that the numerical results remain essentially unchanged. Figure 8.4.2 shows the computed velocity variation with time for the two cases. The velocity starts at zero, as given by the initial condition, rises sharply, and then gradually approaches the terminal velocity. The terminal velocity is attained when the net force on the body is zero, resulting in $dV/d\tau$ becoming zero. Therefore, from Eq. (8.4.3), the terminal velocity for the first case is given by the root of the equation

$$9.8 - (2V + 0.1V^2) = 0 \tag{8.4.9}$$

which gives V as 4.07 m/s. Similarly, the value for the second case is obtained as 2.31 m/s. The numerical results agree closely with these values.

A comparison of the programs for the Adams method and Hamming's method shows that the two are fairly similar in the general approach. Both need a starting method. However, Hamming's method is generally used without iteration. The error may be monitored to ensure that the step size is not too large. Hamming's method is popular for problems that require a high level of accuracy. As mentioned in the text, it is generally more efficient than a Runge-Kutta scheme of the same order.

8.7 BOUNDARY-VALUE PROBLEMS

So far, we have considered the solution of initial-value problems, in which all the conditions to be satisfied by the solution are specified at one value of the independent variable x. Integration of the differential equation is started at this point, which is often specified as $x = 0$. Also, if the conditions are all specified at a given nonzero value of x, say, $x = a$, a change of variable to $\bar{x} = x - a$ can be employed to impose the conditions at zero value of the transformed independent variable $\bar{x}$. In engineering applications, we are also frequently concerned with problems in which the conditions are imposed at two, or more, different values of the independent variable. Such problems, known as *boundary-value problems*, arise, for instance, in mass diffusion through a porous plate, conduction heat transfer in extended surfaces, vibration of strings, fluid flow over a surface, and deflection of a beam under a given loading. Since at least two conditions, specified at two different values of the independent variable, are necessary for a boundary-value problem, we are concerned with differential equations of second or higher order.

A simple example of a two-point boundary-value problem is the second-order equation

$$\frac{d^2y}{dx^2} = F(x, y, y') \tag{8.80}$$

with the boundary conditions

$$y(a) = A \quad \text{and} \quad y(b) = B \tag{8.81}$$

Here, y is the dependent variable, and y' the first derivative. Two conditions on y are specified at two values, a and b, of the independent variable x. Therefore, the solution must satisfy the given boundary conditions at $x = a$ and $x = b$. We cannot start at an initial point and march, with increasing x, to obtain the desired solution, since the derivative y' is not known at $x = a$ or $x = b$.

There are two main approaches to the numerical solution of such boundary-value problems. The first approach reduces the problem to an initial-value problem and uses trial and error to satisfy the boundary conditions. Methods based on this approach are known as *shooting methods*, since the adjustment of initial conditions to satisfy the conditions at the other location is similar to shooting at a target. In this case, the previously discussed methods for solving initial-value problems are employed, with a root-solving method from Chapter 4, to satisfy the given boundary conditions. The second approach is based on obtaining a finite difference approximation to the differential equations and then solving the resulting algebraic equations by the methods discussed in Chapter 5. Both approaches have their advantages and disadvantages, and the choice is often made on the basis of personal preference. Let us first consider shooting methods, employing the discussion given in the preceding sections of this chapter.

8.7.1 Shooting Methods

The basic approach is to convert a boundary-value problem into an equivalent initial-value problem, by applying all the conditions at one value of the independent variable x and using guessed values for those that are unknown. For instance, the problem given in Eq. (8.80) can be reduced to the following initial-value problem:

$$\frac{d^2y}{dx^2} = F(x, y, y') \tag{8.80}$$

$$y(a) = A \quad \text{and} \quad y'(a) = P \tag{8.82}$$

where P is an unknown that must be chosen so that the condition $y(b) = B$ is satisfied in order to yield the solution to the given boundary-value problem. Once a value of P has been chosen, the initial-value problem given by Eqs. (8.80) and (8.82) may be solved by any of the methods discussed earlier in this chapter. However, the value of $y(b)$ will not, in general, agree with the value B demanded by the boundary condition at $x = b$. An iterative adjustment of the initial slope P is, therefore, needed to satisfy the boundary condition $y(b) = B$ within a specified convergence criterion. Figure 8.11

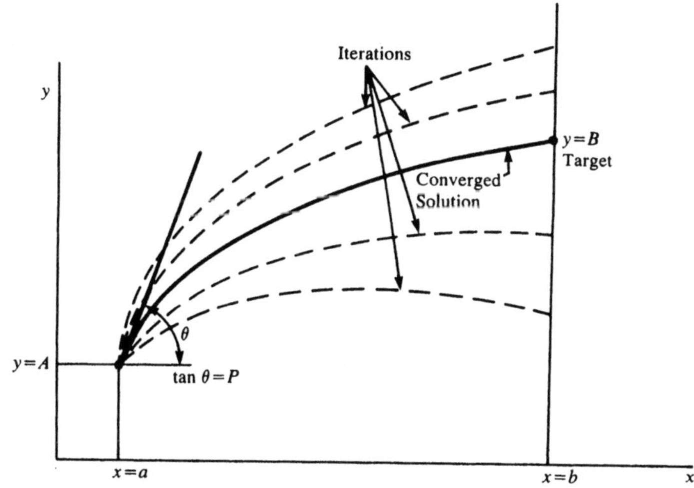

Figure 8.11 Sketch of the iterations to the converged solution, employing a shooting method for solving a boundary-value problem.

illustrates this process of correcting the initial slope until the solution of the equivalent initial-value problem satisfies the given boundary condition at $x = b$.

From the above treatment, the boundary-value problem reduces to the solution of an equivalent initial-value problem, with the initial slope P being obtained by iteratively solving the equation

$$y_b(P) = B \quad \text{or} \quad f(P) = y_b(P) - B = 0 \tag{8.83}$$

where y_b is the value of the dependent variable at $x = b$, and the parentheses indicate its dependence on P. This is a problem in root solving and a suitable method may be obtained from the various methods discussed in Chapter 4. The secant method, which was found to converge very rapidly in most cases, and the bisection method, which always converges if the interval containing the root is known, are both quite suitable for this application, provided $f(P)$ changes sign over the interval. Considering the secant method, if two solutions of the initial-value problem are obtained using P_{i-1} and P_i as two estimates of the initial slope, yielding the corresponding values of y at $x = b$ as $y_b(P_{i-1})$ and $y_b(P_i)$, then the next approximation to the root of Eq. (8.83) is obtained by recognizing that P replaces x in Eq. (4.9) as

$$P_{i+1} = \frac{P_{i-1}y_b(P_i) - P_i y_b(P_{i-1}) + (P_i - P_{i-1})B}{y_b(P_i) - y_b(P_{i-1})} \tag{8.84}$$

The initial-value problem is then solved with P_{i+1} as the initial slope, and $y_b(P_{i+1})$ is obtained. If the convergence criterion, specified as, say, $|y_b(P_{i+1}) - B| \leqslant \varepsilon$, where ε is a specified small quantity, is not satisfied, a new approximation to the root is obtained from Eq. (8.84). The iterative process is continued until the specified convergence criterion is met.

The Newton-Raphson method can also be used if one numerically determines the derivative $d[y_b(P)]/dP$ by solving the initial-value problem for two estimates, P_i and $P_i + \Delta P$, of the initial slope, where ΔP is a small change in P. This procedure was demonstrated in Example 4.4. Once the derivative has been determined, the next estimate P_{i+1} of the root is given by

$$P_{i+1} = P_i - \frac{y_b(P_i) - B}{\left[\dfrac{dy_b(P)}{dP}\right]_{P_i}} \tag{8.85}$$

This technique is frequently employed in shooting methods because of the good convergence characteristics of the Newton-Raphson method; see Example 8.5. Also, this method applies for complex solutions and also if $f(P)$ is tangent to the P-axis, resulting in multiple roots, as discussed in Section 4.5.

Shooting methods frequently employ efficient methods, such as the predictor-corrector methods, for solving the initial-value problem, since several iterations may be needed before convergence is achieved. However, Runge-Kutta methods are also often used because of their self-starting feature. Automatic step changes, on the basis of error estimates, are generally not used since the solution is needed at exactly $x = b$. Therefore, a fixed step size Δx is preferred. However, the step size may be changed from one solution of the initial-value problem to the next, ensuring that $x = b$ is exactly attained, in order to improve the accuracy of the results or to reduce the computer time. Although we have considered a simple second-order boundary-value problem here, the technique is applicable to higher-order equations and to more involved problems. Both linear and nonlinear differential equations can be solved by this approach. However, superposition can be used for linear equations, making shooting particularly simple in this case.

Linear Equations. Consider a linear second-order differential equation of the form

$$\frac{d^2y}{dx^2} = g_1(x)\frac{dy}{dx} + g_2(x)y + g_3(x) \tag{8.86}$$

with boundary conditions

$$y(a) = A \quad \text{and} \quad y(b) = B \tag{8.87}$$

Again, we can recast this problem as an initial-value problem by taking $y'(a) = P$, instead of the condition at $y = b$. We obtain two solutions to this initial-value problem by taking the initial slope as P_1 and P_2. With these solutions denoted as $Y_1(x)$ and $Y_2(x)$, respectively, a linear combination of these solutions is also a solution to the differential equation. Therefore,

$$y(x) = c_1 Y_1(x) + c_2 Y_2(x) \tag{8.88}$$

is also a solution, which satisfies the initial condition $y(a) = A$. The relationship between c_1 and c_2 may be found by substituting Eq. (8.88) into the differential equation, Eq. (8.86), to give

$$c_1\left[\frac{d^2Y_1}{dx^2} - g_1(x)\frac{dY_1}{dx} - g_2(x)Y_1\right] + c_2\left[\frac{d^2Y_2}{dx^2} - g_1(x)\frac{dY_2}{dx} - g_2(x)Y_2\right] = g_3(x) \tag{8.89}$$

Since both $Y_1(x)$ and $Y_2(x)$ independently satisfy the differential equation, the quantities within the parentheses are both $g_3(x)$. Therefore,

$$c_1 + c_2 = 1 \tag{8.90}$$

The value of $y(x)$ at $x = b$ is obtained as

$$y(b) = c_1Y_1(b) + c_2Y_2(b)$$

This equation may be set equal to B, to satisfy the given boundary condition. Then c_1 and c_2 may be obtained from Eq. (8.90) and the following equation:

$$c_1Y_1(b) + c_2Y_2(b) = B \tag{8.91}$$

We obtain the desired solution $y(x)$ by substituting the computed values of c_1 and c_2 into Eq. (8.88):

$$y(x) = \frac{1}{Y_1(b) - Y_2(b)}\{[B - Y_2(b)]Y_1(x) + [Y_1(b) - B]Y_2(x)\} \tag{8.92}$$

Therefore, this solution satisfies the given differential equation and both boundary conditions. Iteration is not needed. This technique can also be used for higher-order, linear boundary-value problems. If the number of unknown conditions at the initial point, $x = a$, were n, we would need to obtain $(n + 1)$ solutions and to take a linear combination of these numerical solutions to obtain the required solution to the ordinary differential equation.

8.7.2 Finite Difference Methods

In this approach, one reduces the solution of an ordinary differential equation to the solution of a system of algebraic equations by obtaining a finite difference approximation to the differential equation at a number of mesh points. The interval $a \leqslant x \leqslant b$, over which the numerical solution is to be obtained, is divided into n equally spaced subintervals of length Δx, as shown in Fig. 8.12. Then the values of x at the mesh, or node, points are denoted by x_i where

$$x_i = a + i\,\Delta x \qquad \text{for } i = 0, 1, 2, \ldots, n \tag{8.93}$$

Also,

$$x_n = a + n\,\Delta x = b \quad \text{and} \quad x_0 = a \tag{8.94}$$

Figure 8.12 Subdivision of a given interval into a finite number of subintervals for employing finite difference methods to solve an ordinary differential equation.

Therefore,

$$\Delta x = \frac{b-a}{n} \tag{8.95}$$

We wish to obtain the solution y_i at these node points. The given boundary conditions are employed to compute y_0 and y_n or to obtain algebraic equations from which these may be determined. The differential equation is replaced by its finite difference approximation at the interior mesh points, resulting in $(n-1)$ algebraic equations. One can solve this resulting set of simultaneous algebraic equations by employing the methods discussed in Chapter 5 to obtain the dependent variable y at the mesh points. The finite difference methods for solving ordinary differential equations are more involved than the shooting methods, discussed earlier. Consequently, shooting methods are much more frequently employed, and finite difference methods are used as an alternative technique in case problems are encountered with shooting.

As discussed in Chapter 3, there are several finite difference approximations to the first- and higher-order derivatives of the dependent variable y. If central differences are employed, the first and second derivatives of y at the ith mesh point are approximated by

$$\frac{dy}{dx} = \frac{y_{i+1} - y_{i-1}}{2\,\Delta x} + O[(\Delta x)^2] \tag{8.96}$$

and

$$\frac{d^2y}{dx^2} = \frac{y_{i+1} - 2y_i + y_{i-1}}{(\Delta x)^2} + O[(\Delta x)^2] \tag{8.97}$$

Therefore, the second-order differential equation, given by Eq. (8.80), is approximated at the ith node by

$$\frac{y_{i+1} - 2y_i + y_{i-1}}{(\Delta x)^2} = F\left[x_i, y_i, \frac{y_{i+1} - y_{i-1}}{2\,\Delta x}\right] \tag{8.98}$$

When this approximation is applied at all the interior points, $(n-1)$ equations in

$(n-1)$ unknowns are obtained. The boundary conditions yield

$$y_0 = A \quad \text{and} \quad y_n = B \tag{8.99}$$

These relationships are used in the system of equations generated by Eq. (8.98), wherever y_0 and y_n appear.

If the differential equation is linear, the algebraic equations obtained are also linear. Similarly, a nonlinear differential equation results in nonlinear algebraic equations and a homogeneous differential equation in a homogeneous system. These different types of systems were considered in Chapter 5, and the corresponding direct and iterative methods were discussed. The same may be used for solving the resulting simultaneous algebraic equations.

If the given ordinary differential equation is linear, the finite difference equations obtained from Eq. (8.98) are linear and tridiagonal, since the ith equation contains only y_{i-1}, y_i, and y_{i+1}. Such a system can easily be solved by a direct method such as Gaussian elimination, and several efficient algorithms have been developed for the purpose; see Example 5.2. Because the resulting system of equations is tridiagonal, the computation of the unknowns can be carried out efficiently and with a small round-off error. The computer program is also very simple. As a consequence of these advantages, the finite difference methods are frequently employed for solving linear boundary-value problems in ordinary differential equations; see Example 8.6. These methods are also extensively used for partial differential equations, as discussed in the next chapter. Nonlinear algebraic equations, arising from a nonlinear ordinary differential equation, are generally linearized. This is done by using the values from the previous iteration for the nonlinear terms, as shown in Section 5.8.2. The result is a linearized tridiagonal system of equations, which is solved with iteration, to yield the solution y_i at the nodal points. The Newton-Raphson method may also be used if the number of nodes is small. However, shooting methods are generally easier to use for nonlinear ordinary differential equations than finite difference methods.

In many cases, the boundary conditions are more complicated than the simple ones considered above. Frequently, a relationship between the derivative y' and the function y is given as, say,

$$a_1 y' + a_2 y = a_3 \qquad \text{at } x = a \quad \text{or} \quad x = b \tag{8.100}$$

Then, one approach is to use one-sided forward or one-sided backward finite difference approximations at the two boundaries for the derivative. At $x = a$, y' may be approximated by

$$y' \simeq \frac{y_1 - y_0}{\Delta x} \tag{8.101}$$

where y_0 is the value of y at $x = a$ and y_1 that at the adjacent nodal point. When substituted in Eq. (8.100), this equation gives the relationship between y_0 and y_1 as

$$a_1 \frac{y_1 - y_0}{\Delta x} + a_2 y_0 = a_3 \tag{8.102}$$

However, the truncation error in the approximation of Eq. (8.101) is $O(\Delta x)$, whereas the error in the finite difference equations for the interior points is $O[(\Delta x)^2]$. To improve the accuracy of the finite difference equation for the boundary point, a fictitious point x_{-1} is taken outside the computational region, as shown in Fig. 8.13. Then y' is approximated by central differences to an accuracy of $O[(\Delta x)^2]$ as follows:

$$y' \simeq \frac{y_1 - y_{-1}}{2\,\Delta x} \tag{8.103}$$

Therefore, the finite difference equation for the boundary point at $x = a$ is

$$a_1 \frac{y_1 - y_{-1}}{2\,\Delta x} + a_2 y_0 = a_3 \tag{8.104}$$

The finite difference equation, Eq. (8.98), is applied at $x = a$, and the unknown y_{-1} is eliminated between the equation thus obtained and the above equation, resulting in a relationship of accuracy $O[(\Delta x)^2]$ between y_0 and y_1.

The above finite difference formulation has a truncation error of order $(\Delta x)^2$. Higher accuracy in the numerical results can be achieved by the use of smaller step size or higher-order methods. Richardson extrapolation, Eq. (8.32), may also be used to improve the accuracy. However, higher-order methods involve more than one point on either side of the ith grid point. Therefore, a tridiagonal system is not obtained, and the treatment of the boundary conditions also becomes more involved; see Ferziger (1981) and Jaluria and Torrance (1986) for details.

The finite difference approximation to the differential equation can be obtained in many ways, leading to different sets of algebraic equations. The given equation may also be reduced to a system of first-order equations and the finite difference approximations applied to these (Keller, 1968). Equations of order higher than two can be reduced to an equivalent system of first- or second-order equations. A

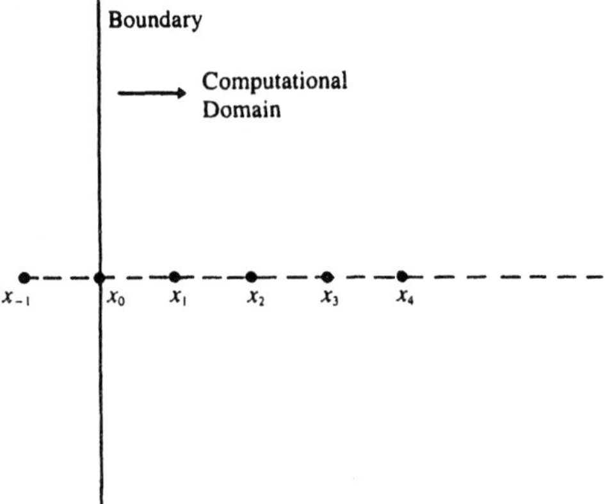

Figure 8.13 Use of a fictitious point x_{-1} outside the computational region for approximating a gradient condition at the boundary.

tridiagonal system is obtained if a system of second-order equations is employed with finite difference approximations of $O[(\Delta x)^2]$. In circumstances where the tridiagonal system is not obtained, the Gauss-Seidel and successive over-relaxation iterative methods may be used. Similarly, iterative methods are used for nonlinear equations, as discussed in Section 5.8.

Eigenvalue Problems. Homogeneous ordinary differential equations, which give rise to eigenvalue problems, are frequently encountered in elasticity theory, vibrations, stability analysis, and mechanics of materials. In this case, the solution can be obtained only for certain values of the parameters of the system. These values, examples of which are the natural frequencies of vibration of a given structure, are related to the characteristic quantities, known as *eigenvalues*, of the problem. The eigenvalue problem for homogeneous algebraic equations was discussed in Chapter 5. One generally solves the differential equation by reducing it to an equivalent system of algebraic equations, using finite difference approximations.

Consider, for example, the following equation, which arises in the natural vibration of beams:

$$\frac{d^2y}{dx^2} + a^2y = 0 \qquad \text{with } y(0) = 0 \qquad \text{and} \quad y(L) = 0 \tag{8.105}$$

where a is a constant and L is the length of the beam. If the second-order central difference approximation, Eq. (8.97), is used for the second derivative, the finite difference equation is

$$y_{i+1} + [a^2(\Delta x)^2 - 2]y_i + y_{i-1} = 0 \qquad \text{for } i = 1, 2, \ldots, n-1 \tag{8.106}$$

where n is the number of subintervals and $y_0 = y_n = 0$. This system of equations may be written in matrix form as follows:

$$(A - \lambda I)y = 0 \tag{8.107}$$

where A is a tridiagonal coefficient matrix, λ is the eigenvalue, I is an identity matrix, and y is the unknown vector of the values at the nodal points.

The above system of homogeneous equations may be solved to obtain the eigenvalues and the corresponding eigenvectors. The analytical solution gives the eigenvalues as $\lambda_n = -(n\pi/L)^2$, where $n = 1, 2, \ldots$, and the eigenvectors as $\sin n\pi x$. A good approximation to the lowest eigenvalue is generally obtained by taking only a few grid points, typically around ten. However, a much larger number of subintervals is needed to accurately determine the higher eigenvalues. The power method discussed in Chapter 5 is particularly suitable for determining the smallest eigenvalue and the corresponding eigenvector. Larger eigenvalues may also be obtained in ascending order, as outlined earlier. Other methods, such as the QL algorithm, may also be used very efficiently, since the finite difference approximation leads to symmetric tridiagonal matrices in many cases.

Example 8.5

The flow of a fluid over a two-dimensional wedge, as shown in Fig. 8.1, is governed by the third-order ordinary differential equation

$$\frac{d^3f}{dx^3} + f\frac{d^2f}{dx^2} + \beta\left[1 - \left(\frac{df}{dx}\right)^2\right] = 0 \qquad (8.5.1)$$

where x is the dimensionless distance away from the wedge in a direction normal to either edge as shown, f is known as the *dimensionless stream function,* so that df/dx is the dimensionless velocity in the direction parallel to either side of the wedge; and β is a constant related to the wedge angle. Thus, $\beta = 0$ gives the flow over a flat plate, as also considered in Prob. 8.16 with a somewhat different definition of f. Also, the stream function f lies between 0 and 1.0. The boundary conditions for this problem are as follows:

$$\text{At } x = 0: \quad f = 0 \quad \text{and} \quad \frac{df}{dx} = 0 \qquad (8.5.2)$$

$$\text{As } x \to \infty: \quad \frac{df}{dx} \to 1 \qquad (8.5.3)$$

Note that the boundary conditions are given in terms of both f and its derivative.

Despite the complexity of the physical phenomenon involved, this problem is chosen because it presents a third-order, nonlinear, boundary-value, ordinary differential equation system. Thus, it permits the illustration of several important concepts in the solution of boundary-value problems. Furthermore, such equations are frequently encountered in fluid flow phenomena of interest to several engineering disciplines. Using the fourth-order Runge-Kutta method, with the shooting technique, solve this problem for $\beta = 0$ and $\beta = 0.5$.

Solution

The given equation may be broken down into a system of three first-order equations, as outlined in Eq. (8.7). Therefore, the three equations are written as

$$\frac{df}{dx} = y$$

$$\frac{dy}{dx} = v \qquad (8.5.4)$$

$$\frac{dv}{dx} = -fv - \beta(1 - y^2)$$

where y and v are the first and second derivatives of f, respectively. The two conditions in Eq. (8.5.2) are given at $x = 0$. However, the third condition in Eq. (8.5.3) applies as $x \rightarrow \infty$, making the system a boundary-value problem. An initial-value problem is obtained if the value of $v = d^2f/dx^2$ is guessed at $x = 0$. The Runge-Kutta method may then be applied to Eq. (8.5.4) with the following boundary conditions:

$$\text{At } x = 0: \quad f = 0 \qquad y = 0 \qquad v = A \tag{8.5.5}$$

where A is a guessed value which must be adjusted by means of a correction scheme until the condition given by Eq. (8.5.3) is satisfied.

Figure 8.5.1 presents the computer program for this problem. The initial-value

```
C*********************************************************************
C
C   THIS PROGRAM SOLVES A THIRD-ORDER ORDINARY DIFFERENTIAL EQUATION
C
C   USING 4TH ORDER RUNGE-KUTTA METHOD, WITH SHOOTING TECHNIQUE.
C
C* * * * * * * * * * * * * * * * * * * * * * * * * * * * * * * * *
C
C     IN THE FOLLOWING PROGRAM:
C
C     X STANDS FOR INDEPENDENT VARIABLE
C     Y STANDS FOR DERIVATIVE OF DEPENDENT VARIABLE F
C     V STANDS FOR DERIVATIVE OF Y
C     A IS GUESSED VALUE OF V AT X  = 0
C     DA IS INCREMENT IN A
C     Z IS THE VALUE OF A IN THE PREVIOUS ITERATION
C     H IS STEP SIZE
C     EDGE IS THE MAXIMUM VALUE OF X
C     YEND IS THE KNOWN VALUE OF Y AT X = INFINITY
C     B IS THE COEFFICIENT APPEARING IN THE DIFFERENTIAL EQUATION
C     N IS THE NUMBER OF STEPS
C     EPS IS THE CONVERGENCE CRITERION
C
C* * * * * * * * * * * * * * * * * * * * * * * * * * * * * * * * *
      IMPLICIT REAL*8 (A-H,O-Z)
      PARAMETER (ND=4000)
      DIMENSION V(ND),Y(ND),F(ND)
C
C     STORING COMPUTED DATA FOR GRAPHICS
C
      OPEN(UNIT=36,FILE='X')
      OPEN(UNIT=37,FILE='YV')
      OPEN(UNIT=38,FILE='Y')
      OPEN(UNIT=39,FILE='F')
C
C     FILE X CONTAINS VALUES OF X
C     FILE YV CONTAINS VALUES OF V
C     FILE Y CONTAINS VALUES OF Y
C     FILE F CONTAINS VALUES OF F
C
C     INPUT PARAMETERS
C
      PRINT*,'INPUT PARAMETERS'
C
C     USUAL RANGE OF B IS FROM 0 TO 2
C
```

Figure 8.5.1 Computer program for solving the third-order ODE of Example 8.5, employing the fourth-order Runge-Kutta method with shooting.

```
      PRINT*,' USUAL RANGE OF B IS FROM 0 TO 2'
      PRINT*,'   '
      PRINT*,'B='
      READ*,B
      PRINT*,'STEP SIZE H ='
      READ*,H
      PRINT*,'DIMENSION OF Y, V AND F IS ',ND
      PRINT*,'
      PRINT*,'NUMBER OF STEPS N ='
C
      PRINT*,'( N SHOULD BE LESS THAN OR EQUAL TO'
      PRINT*,' DIMENSION OF Y, V AND F )'
C
      READ*,N
      PRINT*,'CONVERGENCE CRITERION EPS ='
      READ*,EPS
      EDGE = H*N
      PRINT*,'EDGE=',EDGE
C
C     SET INITIAL CONDITIONS
C
      Y(1)=0.
      F(1)=0.
C
C     SET BOUNDARY CONDITIONS
C
      YEND = 1.
C
  100 PRINT*,'INITIAL GUESS FOR V(1)=A='
C
C     'A' IS THE INITIAL GUESS FOR THE VALUE OF V AT X=0 (i.e. V(1))
C     THIS IS REQUIRED FOR THE SOLTUION TO MARCH FORWARD.
C     THIS PROGRAM MAY NOT WORK FOR ANY ARBITRARILY GUESSED VALUE OF A.
C     ONE HAS TO HAVE SOME IDEA ABOUT THE VALUE OF A.
C     THIS MAY BE FOUND OUT BY EXPERIMENTING WITH THIS PROGRAM.
C
      READ*,A
      IF(A.LE.0.0)THEN
             PRINT*,'TRY ANOTHER VALUE OF A'
             GO TO 100
          ELSE
      END IF
   40 V(1)=A
      PRINT*,'V(1)=',V(1)
C
C     CALCULATIONS ARE STARTED AT X = 0 (i.e. I = 1) AND STOPPED
C     AT X = EDGE (i.e. I = N)
C
C     IN THE DO-LOOP BELOW, TWO CALCULATIONS ARE CARRIED OUT, ONE WITH
C     V(1) = A AND ANOTHER WITH V(1) = A + DA
C
      DO 101 IT = 1,2
C
C     COMPUTE VALUES OF F, Y, V, AT NEXT I USING 4TH ORDER RUNGE-KUTTA METHOD
C
      DO 1 I=1,N-1
      RK1F=H*Y(I)
      RK1Y=H*V(I)
      RK1V=H*(-F(I)*V(I) - B*(1. - Y(I)**2))
```

Figure 8.5.1 Continued

```
C
          RK2F=H*(Y(I) + RK1Y/2.)
          RK2Y=H*(V(I) + RK1V/2.)
          RK2V=H*(-(F(I) + RK1F/2.)*(V(I) + RK1V/2.)
     $             - B*(1. - (Y(I) + RK1Y/2.)**2))
C
          RK3F=H*(Y(I) + RK2Y/2.)
          RK3Y=H*(V(I) + RK2V/2.)
          RK3V=H*(-(F(I) + RK2F/2.)*(V(I) + RK2V/2.)
     $             - B*(1. - (Y(I) + RK2Y/2.)**2))
C
          RK4F=H*(Y(I) + RK3F)
          RK4Y=H*(V(I) + RK3V)
          RK4V=H*(-(F(I) + RK3F)*(V(I) + RK3V)
     $             - B*(1. - (Y(I) + RK3Y)**2))
C
          F(I+1)= F(I) + (RK1F + 2.*RK2F + 2.*RK3F + RK4F)/6.
          Y(I+1)= Y(I) + (RK1Y + 2.*RK2Y + 2.*RK3Y + RK4Y)/6.
          V(I+1)= V(I) + (RK1V + 2.*RK2V + 2.*RK3V + RK4V)/6.
C
  1       CONTINUE
          PRINT*,'Y(N)=',Y(N)
          PRINT*,'F(N)=',F(N)
          PRINT*,'V(N)=',V(N)
C
C      SINCE THE VALUE OF A (i.e. V(1)) IS INITIALLY GUESSED,
C      IT MUST BE CORRECTED.
C      TO ARRIVE AT THE CORRECT VALUE OF 'A' FROM THE INITIALLY GUESSED
C      VALUE OF 'A', NEWTON-RAPHSON CORRECTION METHOD IS EMPLOYED.
C      CHANGE IN THE VALUE OF Y(N) FOR A SMALL CHANGE IN 'A'  IS CALCULATED.
C      THE KNOWN BOUNDARY CONDITION ON Y, AT X = INFINITY, IS APPLIED AT
C      X = EDGE (i.e. Y(N) = YEND).
C      THE VALUE OF 'A' IS ADJUSTED ACCORDINGLY.
C
          IF(IT.EQ.1)THEN
                 S1=Y(N)
                 DA=A/50000.
                 A= A+DA
                 V(1)=A
             ELSE
                 S2=Y(N)
          END IF
  101     CONTINUE
C
C      S1 IS THE COMPUTED VALUE OF Y(N) STARTING WITH V(1) = A
C      S2 IS THE COMPUTED VALUE OF Y(N) STARTING WITH V(1) = A + DA
C      S12 STANDS FOR (CHANGE IN Y(N)/CHANGE IN 'A')
C
          S12 = (S2-S1)/DA
          IF(ABS(S12).LE.EPS)GO TO 50
C
C      CORRECT VALUE OF 'A' IS ESTIMATED USING NEWTON-RAPHSON METHOD
C
          A= V(1) +(YEND - Y(N))/S12
          IF(A.LE.0.0)THEN
                 PRINT*,'TRY ANOTHER GUESS FOR A'
                 GO TO 100
             ELSE
          END IF
```

Figure 8.5.1 Continued

```
C
C     CHECK FOR CONVERGENCE
C
        IF(ABS(Z-A).GT.EPS) THEN
        Z=A
        GO TO 40
        END IF
C
C     OUTPUT RESULTS
C
 50     X=-4*H
        DO 20 I=1,N,4
        X=X+H*4
        WRITE(36,*)X
        WRITE(37,*)V(I)
        WRITE(38,*)Y(I)
        WRITE(39,*)F(I)
  20    CONTINUE
        PRINT*,'THE SOLUTION HAS CONVERGED'
        PRINT*,'THE DERIVATIVE OF Y AT X=0 IS V(1)=',V(1)
        PRINT*,'THE MAXIMUM VALUE OF X=',EDGE
        PRINT*,'MAXIMUM VALUE OF X (i.e. EDGE) SHOULD BE VARIED'
        PRINT*,'UNTIL VALUE OF  V(1) REMAINS UNCHANGED'
        CLOSE(UNIT=36)
        CLOSE(UNIT=37)
        CLOSE(UNIT=38)
        CLOSE(UNIT=39)
  55    STOP
        END
C**********************************************************************
```

Figure 8.5.1 Continued.

problem is first solved with a chosen value of A and then with an incremented value $A + \Delta A$. The value of $y = df/dx$ at EDGE, which represents $x \to \infty$, is thus determined for these two computations. This allows us to compute the derivative $dy_\infty(A)/dA$, where y_∞ represents the value of y for $x \to \infty$. The desired value of y_∞ is 1.0 and the Newton-Raphson correction scheme, given by Eq. (8.85), may be employed. This gives an improved value A_{im} of the guess as

$$A_{\text{im}} = A - \frac{y_\infty(A) - 1}{\dfrac{dy_\infty(A)}{dA}} \tag{8.5.6}$$

where

$$\frac{dy_\infty(A)}{dA} = \frac{y_\infty(A + \Delta A) - y_\infty(A)}{\Delta A} \tag{8.5.7}$$

The improved value of A is now taken and the above procedure repeated. This process is carried out until the value of A stops changing from one iteration, denoted by superscript l, to the next, denoted by superscript $(l + 1)$; that is,

$$|A^{(l+1)} - A^{(l)}| \leqslant \varepsilon \tag{8.5.8}$$

where ε is a chosen convergence criterion. The various symbols used are defined in the program. Thus, YEND = 1.0 in the present case, and $B = \beta$. The step size H and the number of steps N to obtain the maximum value of x, EDGE, may be chosen by the user.

The numerical results obtained are shown in Figs. 8.5.2 and 8.5.3, in terms of the

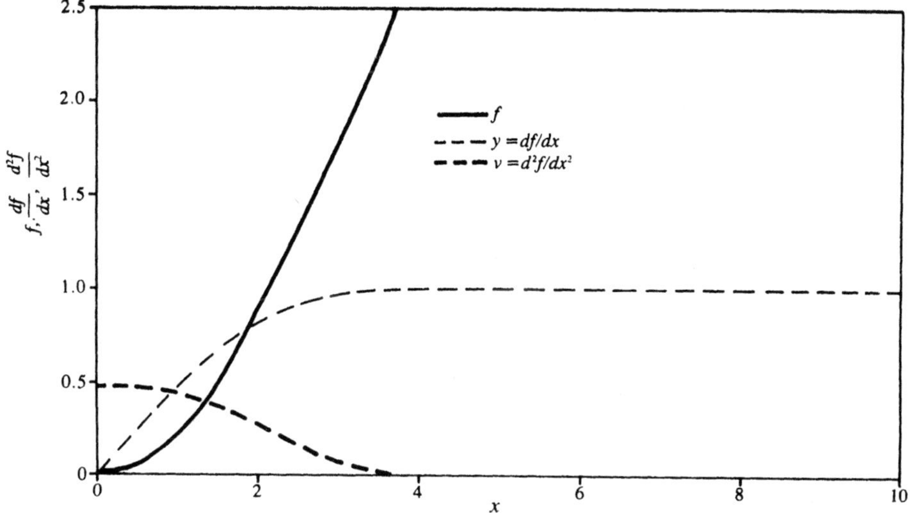

Figure 8.5.2 Computed variation of the functions f, $y = (df/dx)$, and $v = (dy/dx)$, in Example 8.5, with the independent variable x, at $\beta = 0$. Here, Δx is taken as 5×10^{-3}, and ε as 10^{-6}.

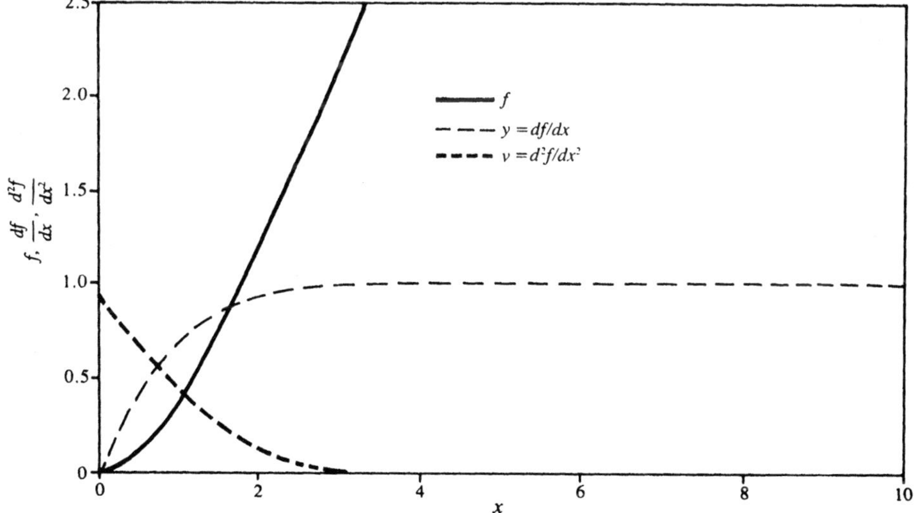

Figure 8.5.3 Computed results at $\beta = 0.5$ for Example 8.5.

variation of f, y, and v with the independent variable x. Physically, y versus x represents the velocity distribution, which goes from zero at the wedge surface to 1.0 far away from the wedge. The dimensionless velocity gradient v is related to the frictional force due to the fluid. At $\beta = 0$, from Eq. (8.5.1), $d^3f/dx^3 = 0$ at $x = 0$, since $f = 0$ at $x = 0$. This is reflected in the zero gradient, at $x = 0$, of the curve of v versus x. The results are shown for EDGE = 10, H = 0.005, and EPS = 10^{-6}. All of these values were varied to confirm that the results are not significantly affected by such a variation. The choice of the value of EDGE is an important consideration. Since the boundary condition is given for $x \to \infty$, a large value of x is needed for satisfying this condition. The initial choice of EDGE may be based on results available from earlier computations. Otherwise, an arbitrary value is taken and gradually varied until the results are not affected by a further variation, as discussed earlier in Section 2.5.

This problem is obviously a fairly complicated one and has been chosen to demonstrate several of the important features discussed above. It is also an important problem in fluid mechanics. Several such problems arise in different flow configurations and convective heat transfer circumstances. Some of these are given as problems at the end of the chapter. Gebhart (1971) may be consulted for further details on the physical background of this problem.

Example 8.6

A rod of length L has its two ends at temperatures T_1 and T_2. It loses energy by convection at the lateral surface, as shown in Fig. 8.6.1. If the temperature is assumed to be uniform over any given cross section, the temperature T in the rod is a function only of x, the distance from one end. Then the governing equation for $T(x)$ is

$$kA \frac{d^2T}{dx^2} = hp(T - T_a) \tag{8.6.1}$$

where k is a property known as *thermal conductivity* of the rod material, A is the area of cross section of the rod, p is its circumferential perimeter, h is the convective heat transfer coefficient and T_a is the ambient air temperature. The equation may be nondimensionalized by the use of dimensionless temperature θ and distance X,

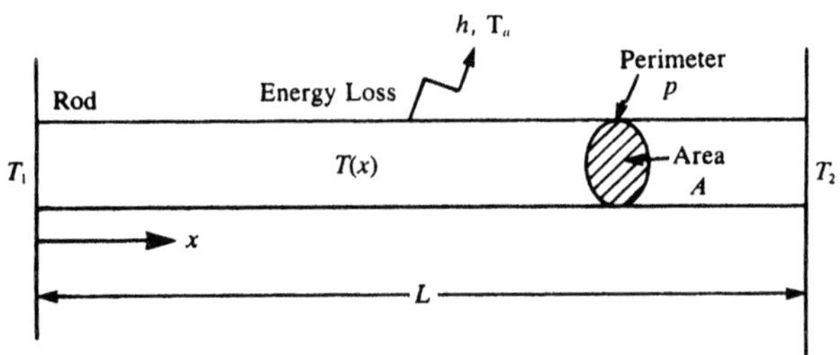

Figure 8.6.1 Conduction heat transfer in a rod, as considered in Example 8.6.

defined as follows:

$$\theta = \frac{T - T_a}{T_1 - T_a} \qquad X = \frac{x}{L} \tag{8.6.2}$$

The governing equation then becomes

$$\frac{d^2\theta}{dX^2} = \frac{hpL^2}{kA}\theta = P^2\theta \qquad \text{where } P = \sqrt{\frac{hp}{kA}}\cdot L \tag{8.6.3}$$

Here, P is a dimensionless parameter that characterizes the problem. Solve this equation by the finite difference method for $(T_2 - T_a)/(T_1 - T_a) = 0.5$ and $P = 0, 0.5, 1.0, 5.0$, and 25.0.

Solution

By nondimensionalizing the governing equation, we can apply the numerical results obtained to a wide range of physical parameters, given here as k, A, L, h, p, and T_a. The equation to be solved is

$$\frac{d^2\theta}{dX^2} = P^2\theta \tag{8.6.4}$$

with the following boundary conditions:

$$\begin{aligned} &\text{At } X = 0: \quad \theta = 1.0 \\ &\text{At } X = 1: \quad \theta = 0.5 \end{aligned} \tag{8.6.5}$$

These conditions follow from the definitions of θ and X and from the given temperature ratio $(T_2 - T_a)/(T_1 - T_a)$.

In order to use the finite difference approach for this problem, we take N nodal points over the interval, as shown in Fig. 8.6.2. Therefore, the interval $0 \leqslant X \leqslant 1$ is divided into $(N - 1)$ equal subintervals of length ΔX. Then the value of X at the node points is denoted by X_i, where $X_i = (i - 1)\Delta X$, for $i = 1, 2, \ldots, N$. Also, $\Delta X = 1/(N - 1)$, since the total dimensionless length is 1.0. The finite difference equation is

$$\frac{\theta_{i+1} - 2\theta_i + \theta_{i-1}}{(\Delta X)^2} = P^2\theta_i \quad \text{for } i = 2, 3, \ldots, N - 2 \tag{8.6.6}$$

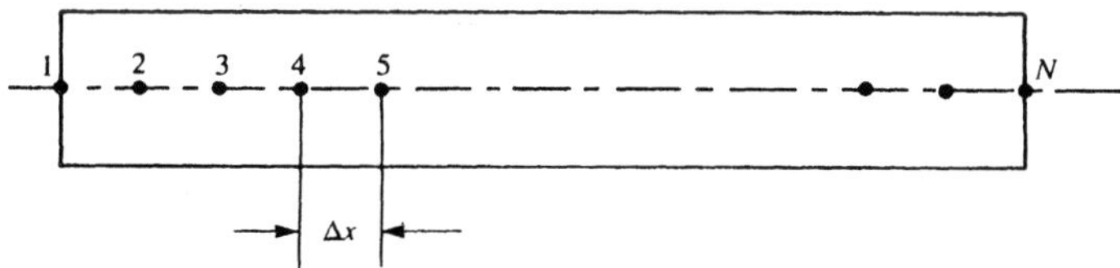

Figure 8.6.2 Subdivision of the rod in Example 8.6 for employing the finite difference approach.

or

$$\theta_{i+1} - [2 + P^2(\Delta X)^2]\theta_i + \theta_{i-1} = 0 \tag{8.6.7}$$

This equation is applied at each interior nodal point, to give rise to $(N - 2)$ equations which form a tridiagonal system. Also, $\theta = 1.0$ for $i = 1$, and $\theta = 0.5$ for $i = N - 1$.

Figure 8.6.3 shows the FORTRAN 77 computer program for solving this

```
C*****************************************************************
C   THIS PROGRAM SOLVES A SECOND-ORDER ORDINARY DIFFERENTIAL EQUATION
C
C    USING THE FINITE DIFFERENCE METHOD
C
C* * * * * * * * * * * * * * * * * * * * *  * * * * * * * * * * * * * * *
C
C      NOMENCLATURE:
C
C        X : DIMENSIONLESS  x COORDINATE
C        TH: ARRAY FOR TEMPERATURE (DIMENSIONLESS)
C        P : CONSTANT APPEARING IN THE DIFFERENTIAL EQUATION
C        A : LOWER DIAGONAL OF THE TRIDIAGONAL MATRIX
C        B : MAIN DAIGONAL  OF THE TRIDIAGONAL MATRIX
C        C : UPPER DIAGONAL OF THE TRIDIAGONAL MATRIX
C        D : ARRAY FOR RIGHT HAND SIDE COLUMN MATRIX
C        T : ARRAY CONTAINING SOLUTIONS OF THE TRIDIAGONAL SYSTEM
C        BN: DUMMY ARRAY IN THE SUBROUTINE 'TRIDIAG'
C        N : NUMBER OF GRID POINTS
C        DX: GRID SIZE
C*****************************************************************
        IMPLICIT REAL (A-H,O-Z)
        PARAMETER(N=51)
        DIMENSION A(N-2),B(N-2),C(N-2),D(N-2),T(N-2),TH(N),BN(N-2)
        OPEN(UNIT=50,FILE='IX')
        OPEN(UNIT=51,FILE='ITH')
C
C   FILE IX CONTAINS VALUES OF X
C   FILE ITH CONTAINS VALUES OF TEMPERATURE, TH
C
C   INPUT PARAMETERS
C
        PRINT*,'P='
        READ*,P
        DX=1./(N-1)
        PRINT*,'DX = ',DX
C
C   VALUE OF N CAN BE CHANGED. THE DIMENSION STAEMENT SHOULD BE
C   MODIFIED ACCORDINGLY
C
C   SET THE BOUNDARY CONDITIONS AT THE TWO EXTREME GRID POINTS
C
        TH(1)=1.0
        TH(N)=0.5
C
C   GENERATION  OF THE TRIDIAGONAL MATRIX AND
C   THE RIGHT HAND SIDE COLUMN MATRIX
C
```

Figure 8.6.3 Computer program in FORTRAN for solving the second-order ODE of Example 8.6 by the finite difference method.

```
      A(1)=0.0
      DO 1 I=2,N-2
 1    A(I)=1.0
      DO 2 I=1,N-3
 2    C(I)=1.0
      C(N-2)=0.0
      DO 3 I=1,N-2
 3    B(I)= -(2.0 +(P**2)*(DX**2))
      D(1)= -TH(1)
      D(N-2)= -TH(N)
      DO 4 I=2,N-3
 4    D(I)=0.0
C
C  THE TRIDIAGONAL MATRIX THUS GENERATED IS OF DIMENSION  N-2
C
C  THE SUBROUTINE 'TRIDIAG' SOLVES THE TRIDIAGONAL SYSTEM.
C  THE SOLUTIONS ARE PLACED IN THE ARRAY 'T'
C
      CALL TRIDIAG(A,B,C,D,T,N-2,BN)
C
C  BACK SUBSTITUTION FROM MATRIX 'T' TO 'TH'
C
      DO 5 I=2,N-1
 5    TH(I)=T(I-1)
C
C  OUTPUT RESULTS
C
      X=-DX
      DO 6 I=1,N
      X=X+DX
      WRITE(50,*)X
      WRITE(51,*)TH(I)
 6    CONTINUE
      CLOSE(UNIT=50)
      CLOSE(UNIT=51)
      PRINT *,'THE VALUES OF X ARE STORED IN FILE IX'
      PRINT *,'THE VALUES OF TEMPERATURE ARE STORED IN FILE ITH'
      STOP
      END
C* * * * * * * * * * * * * * * * * * * * * * * * * * * * * * * * * * *
C
C  THE FOLLOWING SUBROUTINE SOLVES THE TRIDIAGONAL SYSTEM USING
C  THE THOMAS ALGORITHM
C
C  A,B,C ARE THE DIAGONAL ELEMENTS AS MENTIONED IN THE MAIN PROGRAM
C  F CONTAINS THE CONSTANTS ON THE RIGHT HAND SIDE AND T THE SOLUTION
C
C* * * * * * * * * * * * * * * * * * * * * * * * * * * * * * * * * * *
      SUBROUTINE TRIDIAG(A,B,C,F,T,M,BN)
      DIMENSION A(M),B(M),C(M),F(M),T(M),BN(M)
      PRINT*,'SOLVING TRIDIAG'
C
      DO 1 I=1,M
 1    BN(I)=B(I)
      DO 2 I=2,M
      D = A(I)/BN(I-1)
      BN(I)= BN(I) -C(I-1)*D
      F(I)=F(I)-F(I-1)*D
```

Figure 8.6.3 Continued

```
2     CONTINUE
      T(M)=F(M)/BN(M)
      DO 3 I=1,M-1
      J=M-I
      T(J)=(F(J) - C(J)*T(J+1))/BN(J)
3     CONTINUE
      RETURN
      END
```

Figure 8.6.3 Continued

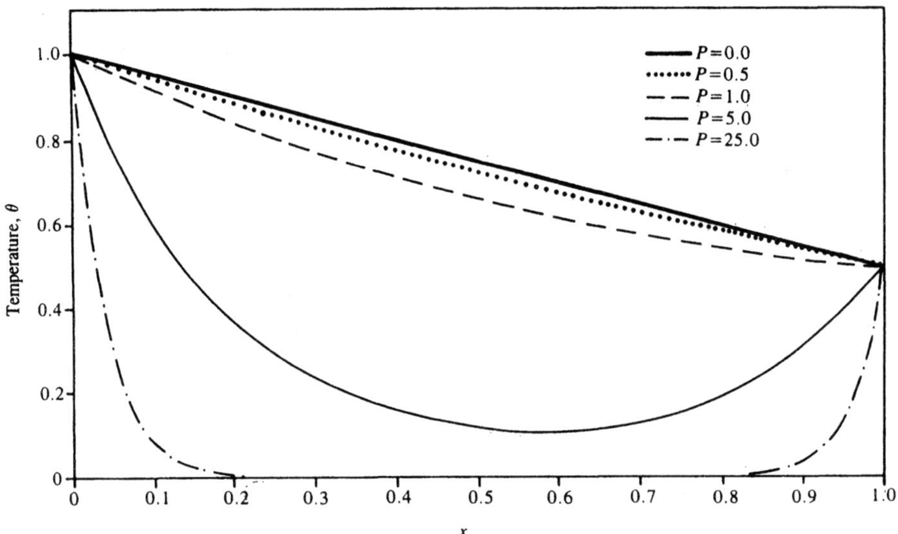

Figure 8.6.4 Computed temperature distributions for Example 8.6, with the number of grid points N taken as 51.

second-order ordinary differential equation. The nomenclature employed in the program is indicated at the beginning. The parameter P and the number of grid points N are chosen by the user. The boundary conditions are employed to set the values at the two extreme grid points. Equation (8.6.7) is used to generate the tridiagonal matrix, whose three terms in each row are denoted by A(I), B(I), and C(I), respectively. The boundary conditions give $\theta(1)$ and $\theta(N)$, which yield the constants on the right-hand side of the first and last equations of the set. Subroutine TRIDIAG is used to solve this system of equations and give the temperatures at the nodal points. See Example 5.2 for details on the solution of a tridiagonal system of algebraic equations.

Figure 8.6.4 shows the computed temperature distributions. The number of grid points N is taken as 51 for these results. At $P = 0$, the distribution is linear. This result is expected, as seen from Eq. (8.6.3), which gives $d^2\theta/dX^2 = 0$ for $P = 0$ and, thus a linear variation for this case. As P becomes larger, the distribution deviates from the linear variation. At $P = 25.0$, the temperature θ is found to be zero, or $T = T_a$, over a substantial portion of the rod. This indicates that the heat inputs at the two ends of the rod are lost in short distances near these ends, resulting in uniform temperature at T_a over much of the rod. These results were also compared with the analytical solution of Eq. (8.6.3), and a good agreement between the two was obtained. For further details on the physical aspects of this problem and other similar ones, see Incropera and Dewitt (1981). Such problems commonly arise in heat and mass transfer processes of interest in chemical and mechanical engineering systems.

8.8 SUMMARY

The solution of ordinary differential equations is discussed in this chapter. The methods for solving the first-order initial-value problem are presented in detail since higher-order equations and boundary-value problems are generally solved by reducing them to an equivalent system of first-order equations, which are then solved as initial-value problems. The methods considered include Euler's method and its modifications, Runge-Kutta methods, multistep methods, and predictor-corrector methods. Euler's method is an inaccurate, although simple, method and is rarely used. It is considered here mainly because of its simplicity which allows the presentation of the basic concepts involved in solving ordinary differential equations. The modified Euler's method and the improved Euler's method, which is also known as *Heun's method*, are second-order accurate and are often used in engineering applications if a very high level of accuracy is not needed.

The Runge-Kutta methods are among the most popular numerical schemes for solving ordinary differential equations. Although usually less efficient than the corresponding predictor-corrector methods, they have the advantages of being self-starting and simpler to program. The Runge-Kutta methods are very widely used in engineering problems, particularly if the problem is to be solved only a few times with different parametric values. If a substantial computational effort is involved, the predictor-corrector methods will be more appropriate, since these methods are more efficient and allow a simpler estimation of the truncation error, per step, which can be employed for a better control on the accuracy of the numerical results. The multistep methods, such as the Adams-Bashforth and Adams-Moulton methods, are seldom used by themselves, since a combination of the open and closed formulas leads to the predictor-corrector methods which have the advantages of both formulas. However, multistep and predictor-corrector methods are generally not self-starting, and a Taylor series expansion of the dependent variable or a Runge-Kutta formula of the same order must be employed to compute the first few steps. Among the predictor-corrector methods discussed here are the Adams method, Milne's method, and Hamming's method. All three methods are comparable in accuracy, although Milne's method has the smallest truncation error. However, it is unstable for certain equations. Hamming's method uses a modifier based on the truncation error to improve the estimate of the dependent variable obtained from the predictor and, therefore, generally requires no iteration or only one iteration for convergence of the corrector. It also has very good stability characteristics. The choice of a predictor-corrector formula for a given application is often a matter of personal preference, since all three methods are quite comparable in accuracy and efficiency.

Boundary-value problems are frequently solved by converting them into equivalent initial-value problems. Some of the conditions needed at the initial point are guessed, the differential equation is solved, and the guessed values are adjusted until the boundary conditions specified at other values of the independent variable are satisfied. Methods based on this approach are termed *shooting methods*, and the predictor-corrector methods are particularly suitable for such a solution, since several iterations may be involved and an efficient scheme is desirable. Finite difference methods may also be used for solving boundary-value problems. These methods reduce the differential equation to a system of algebraic equations, which can be solved by the standard methods discussed in Chapter 5. Generally, finite differencing results in a tridiagonal system which can easily be solved by Gaussian elimination. This approach is particularly suitable for linear differential equations which give rise to linear algebraic equations. Nonlinear differential equations lead to nonlinear systems which must be solved iteratively. Homogeneous equations arise if the differential equation is homogeneous, and the corresponding eigenvalue problem may be solved by employing the methods given in Chapter 5.

Higher-order initial-value problems are solved by reducing them to a system of first-order equations, which are then solved by the various methods discussed here. The solution of a system of simultaneous differential equations is a simple extension of the procedure for a single equation. Accuracy and stability of the numerical scheme are important considerations in the solution of a given problem. The step size may initially be chosen on the basis of an estimate of the truncation error, if available. However, it is important to vary the step size in order to ensure a negligible dependence of the results obtained on the chosen value. If the results for two significantly different step sizes are close, the numerical scheme is probably stable. In some cases, the available estimate of the error and the constraints for stability may be employed for the choice of the method and the step size.

Whenever possible, the numerical results must be compared with the analytical results available for simpler cases in order to check the accuracy and correctness of the computational procedure. Richardson's extrapolation may also be used for improving the accuracy of the results. It is usually better to go to a higher-order formula for greater accuracy than continue to reduce the step size, since the round-off error and the computer time increase as the step size is reduced. Several methods have been developed in recent years to attain a high level of accuracy while retaining the efficiency of predictor-corrector methods; see Gear (1971), Lambert (1973), Shampine and Gordon (1975), and Ferziger (1981) for details.

PROBLEMS

8.1. The differential equation $dy/dx = F(x,y)$, with $y = 0$ at $x = 0$, is to be solved by Euler's method. If the function $F(x,y)$ is zero or infinite at $x = 0$, how would you start the computation? Is this approach also applicable to the fourth-order Runge-Kutta scheme?

8.2. A stone is thrown vertically upward in air at a velocity of 50 m/s. A frictional drag AV^2, where A is a constant and V is the velocity, acts on the stone in the direction opposite to that of the motion. The stone rises until the velocity becomes zero and then accelerates downward to the ground. Its velocity is governed by the following equations, while going upward and while coming down, respectively:

$$\frac{dV}{d\tau} = -g - AV^2$$

$$\frac{dV}{d\tau} = g - AV^2$$

where τ is time and g is the gravitational acceleration, given as 9.8 m/s^2. If A is given as $10^{-3}\ \mathrm{m}^{-1}$, solve the first equation to obtain the time when velocity becomes zero. Using this as the initial condition for the second equation, find the velocity attained by the falling stone in the time taken for the upward motion. Use Euler's method and vary the time step $\Delta\tau$ to ensure that the results are not significantly affected by the step size chosen.

8.3. Problem 8.2 may also be formulated in terms of the vertical distance x by noting that $V = dx/d\tau$. The resulting equation for the upward motion is

$$\frac{d^2x}{d\tau^2} = -g - A\left(\frac{dx}{d\tau}\right)^2$$

The following initial conditions are given:

$$\text{At } \tau = 0: \quad x = 0 \qquad \frac{dx}{d\tau} = 20\ \text{m/s}$$

Using Euler's method, solve this problem to find the maximum height to which the stone rises.

8.4. Repeat Prob. 8.2 with $A = 0$, and compare the numerical results obtained with the corresponding values from the exact, analytical solution of the differential equation and with those obtained earlier for $A = 10^{-3}\ \mathrm{m}^{-1}$.

8.5. A copper sphere of diameter 5 cm is initially at temperature 200°C. It cools in air by convection and radiation. The temperature T of the sphere is governed by the equation

$$\rho C V_0 \frac{dT}{d\tau} = -[\varepsilon\sigma(T^4 - T_\infty^4) + h(T - T_\infty)]A$$

where ρ is the density of copper, C its specific heat, V_0 the volume of the sphere, τ the time, ε a property of the surface known as *emissivity*, σ a constant known as the Stefan-Boltzmann constant, T_∞ the ambient temperature, A the surface area of the sphere, and h the convective heat transfer coefficient. The initial condition is as follows:

$$\text{At } \tau = 0: \quad T = 200°\text{C}$$

Using Heun's method, without iteration, solve this differential equation to find the temperature variation with time, until the temperature drops below 50°C. Use the following values:

$$\rho = 9000 \text{ kg/m}^3 \qquad C = 400 \text{ J/kg·K} \qquad \varepsilon = 0.5$$

$$\sigma = 5.67 \times 10^{-8} \text{ W/m}^2\text{·K}^4 \qquad T_\infty = 25°\text{C} \qquad h = 15 \text{ W/m}^2\text{·K}$$

Employ time steps of 0.5 min and 1.0 min, and compare the results obtained in the two cases.

8.6. In Prob. 8.5, if the surface emissivity ε is low and the convective heat transfer coefficient h is high, radiation may be neglected to obtain the governing equation as

$$\rho C V_0 \frac{dT}{d\tau} = -hA(T - T_\infty)$$

Using the modified Euler's (second-order Runge-Kutta) method with a time step of 1 s, solve this problem for $h = 100$ W/m²·K. Also solve this equation mathematically, and compare the numerical results with the analytical solution. Comment on the error in the numerical solution.

8.7. Apply Richardson's extrapolation to Euler's method, and obtain expressions for the error and for an improved numerical solution.

8.8. Solve the nonlinear ordinary differential equation

$$\frac{dy}{dx} = y^2 + x^2 \qquad \text{with } y(0) = 1$$

by Euler's method, using step size Δx values of 0.1, 10^{-2}, 10^{-3}, and 10^{-4}. Compare the results obtained, and discuss the effect of decreasing the step size on the resulting error.

8.9. A rod of length L is attached at one end to a horizontal support and swings about this support, as shown in the figure. The motion of this pendulum is governed by

$$\frac{d^2\phi}{d\tau^2} + \frac{g}{L}\sin\phi = 0$$

where ϕ is the angle that the rod makes with the vertical at any given time τ and g is the gravitational acceleration. The initial conditions are given as follows:

$$\text{At } \tau = 0: \quad \phi = \frac{\pi}{6} \qquad \frac{d\phi}{d\tau} = 0$$

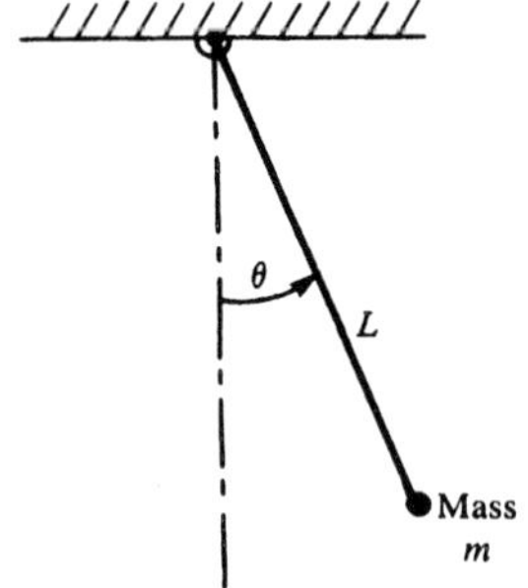

Problem 8.9

If L is given as 0.2 m and g is 9.8 m/s^2, obtain the numerical solution over twice the period of oscillation, using Heun's method, with iteration, to solve the corrector equation. The frequency f of the rod is given by

$$f = \sqrt{\frac{g}{4L\pi^2}}$$

8.10. A fourth-order predictor-corrector method is used to solve a differential equation from $x = 0$ to $x = B$, with a step size Δx. However, at $x = B$, the error per step is found to be too large, and it is decided to reduce the step size to $\Delta x/2$. Using extrapolation of the computed values, outline how this change may be carried out.

8.11. The height H of water in a tank, whose cross-sectional area is A, is a function of time τ due to an inflow q_{in} and an outflow q_{out}. The governing differential equation arises from a mass balance as

$$A\frac{dH}{d\tau} = q_{in} - q_{out}$$

where q_{in} and q_{out} are the volume flow rates, and the density of water is taken as constant. The initial height, at $\tau = 0$, is zero. Compute the time taken for the height to rise to 2 m. The area A is given as 0.03 m^2, and $q_{in} = 6 \times 10^{-4}$ m^3/s. The outflow is given by $q_{out} = 3 \times 10^{-4}\sqrt{H}$ m^3/s. Solve this problem by the fourth-order Runge-Kutta method. What is the height attained at steady state, and how long does it take to reach this value?

8.12. Solve Prob. 8.11 numerically with an initial height of 4 m and $q_{in} = 0$. Determine the time taken for the height to drop to 2 m. Also solve the equation mathematically, and compare this computed value of the time with the analytical result.

8.13. A projectile of mass 0.2 kg is initially at rest. It is accelerated by the application of a constant vertical thrust of 10 N for a period of 5 s. The frictional drag on the projectile is given as $AV^{2.5}$, where $A = 10^{-2}$ m^{-1} and V is the velocity. The gravitational acceleration is 9.8 m/s^2. Following the discussion in Prob. 8.2, obtain the differential equations for the upward motion. Solve these equations by the use of a fourth-order predictor-corrector method to obtain the velocity and height as functions of time τ, till the projectile returns to the ground.

8.14. A vibrating system consists of a body of mass m attached to a wall through a spring, of spring constant k, and a damper, of damping coefficient C, as shown. The displacement x of the mass from its static-equilibrium position is governed by the equation

$$m\ddot{x} + C\dot{x} + kx = 0$$

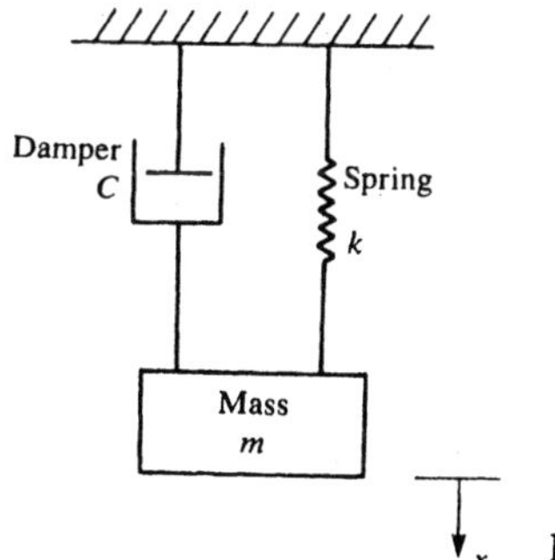

Problem 8.14

where $\ddot{x}$ is the second derivative of x with respect to time τ, and $\dot{x}$ is the first derivative. If the initial displacement $x(0)$ is given as 0.1 m and the initial velocity $\dot{x}(0)$ as zero, use Milne's method to compute the displacement x as a function of time for $0 \leqslant \tau \leqslant 2$ s. Take $m = 5$ g, $C = 20$ Ns/m, and $k = 500$ N/m.

8.15. Solve Prob. 8.14 with the damper absent, that is, $C = 0$, and compare the results with those obtained earlier, with the damper present.

8.16. The flow of a fluid over a flat plate, aligned with the flow, is governed by the equation

$$2f''' + ff'' = 0$$

where the primes represent differentiation with respect to an independent variable η, that is, $f'' = d^2f/d\eta^2$. The dimensionless velocity in the direction along the plate is given by f', which is to be computed. The dimensionless quantity f is termed the *stream function*. The boundary conditions for this problem are as follows:

$$\text{At } \eta = 0: \quad f = 0 \qquad f' = 0$$

$$\text{At } \eta = \infty: \quad f' = 1$$

Employing the fourth-order Runge-Kutta method, solve this boundary-value problem. For the application of the second boundary condition, take $\eta = 8$ as being large enough to represent infinity (see also Example 8.5).

8.17. If the flat plate in Prob. 8.16 is heated, the temperature in the flow adjacent to the plate is governed by

$$\theta'' + \frac{Pr}{2} f\theta' = 0$$

where θ is the dimensionless temperature; f is the stream function from Prob. 8.16; and Pr is a parameter, known as *Prandtl number*, which is a characteristic of the fluid. The boundary conditions are as follows:

$$\text{At } \eta = 0: \quad \theta = 1$$

$$\text{At } \eta = \infty: \quad \theta = 0$$

Using the fourth-order Runge-Kutta method, solve this problem to obtain θ as a function of η for $Pr = 1.0$.

8.18 Use the finite difference approach to solve Prob. 8.16. Would you expect this method to be more efficient, in computer time, than the shooting methods? Also, which method is expected to be more accurate? Discuss.

8.19. The conduction heat transfer in an extended surface, known as a *fin*, yields the following equation for the temperature T, if the temperature distribution is assumed to be one-dimensional in x, where x is the distance from the base of the fin, as shown in the figure:

$$\frac{d^2T}{dx^2} - \frac{hp}{kA}(T - T_\infty) = 0$$

Here, p is the perimeter of the fin, being $2\pi R$ for a cylindrical fin of radius R; A is the cross-sectional area, being πR^2 for a cylindrical fin; k is the thermal conductivity of the material; T_∞ is the ambient fluid temperature; and h is the convective heat transfer coefficient. The

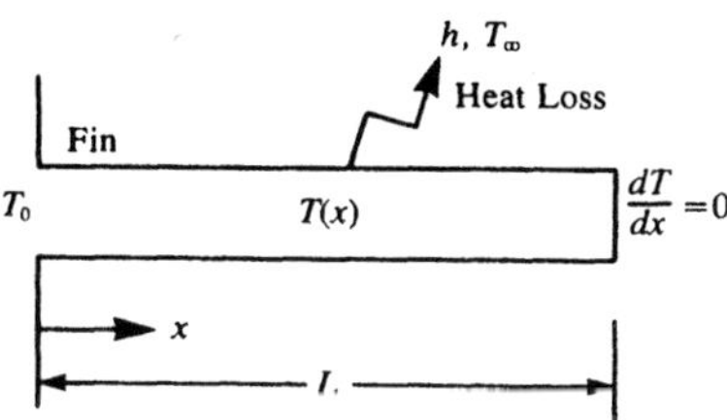

Problem 8.19

boundary conditions are as follows:

$$\text{At } x = 0: \quad T = T_0$$

$$\text{At } x = L: \quad \frac{dT}{dx} = 0$$

where L is the length of the fin. Solve this equation to obtain $T(x)$ for $R = 1$ cm, $h = 20$ W/m^2·K, $k = 15$ W/m·K, $L = 25$ cm, $T_0 = 80$°C, and $T_\infty = 20$°C. Use the Adams predictor-corrector method (see also Example 8.6).

8.20. Solve Prob. 8.19 with the following conditions, which make it an initial-value problem:

$$\text{At } x = 0: \quad T = T_0 \qquad \frac{dT}{dx} = -10^4 \text{ K/m}$$

8.21. A block of wood, resting on a horizontal surface, is attached to a wall through a spring, as shown. The displacement x of the block from its equilibrium position is governed by

$$\frac{d^2x}{d\tau^2} = -\frac{k}{m}x \pm \mu g$$

where τ is time, k the spring constant, m the mass of the block, μ the coefficient of friction, and g the gravitational acceleration. The frictional force is positive if the velocity $dx/d\tau$ is negative, and vice-versa. Taking $m = 2$ kg, $k = 200$ N/m, $g = 9.8$ m/s^2, and $\mu = 0.4$, compute x as a function of time, for time τ up to 3 s. The initial conditions are as follows:

$$\text{At } \tau = 0: \quad x = 0.1 \text{ m} \qquad \frac{dx}{d\tau} = 0$$

Employ the fourth-order Runge-Kutta method.

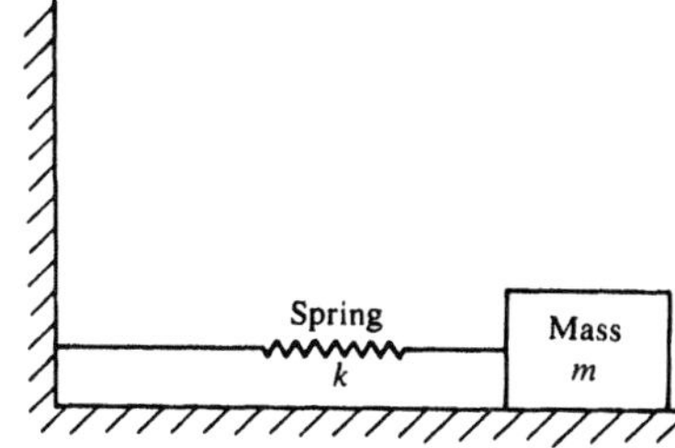

Problem 8.21

8.22. Solve the following initial-value problem by Hamming's method, and compare the predicted and corrected values for x up to 4.0:

$$\frac{d^2y}{dx^2} + 3\frac{dy}{dx} + 2y = 0$$

$$\text{At } x = 0: \quad y = 1 \qquad \frac{dy}{dx} = 0$$

Take $\Delta x = 0.1$ and study the effect of the chosen convergence criterion on the number of iterations needed for the corrector. Compare the converged results with those obtained from the corrector without iteration. Discuss the implications of these comparisons.

8.23. Solve the following linear differential equation by the method of superposition:

$$\frac{d^2y}{dx^2} + 2y = 6$$

$$\text{At } x = 0: \quad y = 1$$

$$\text{At } x = 10: \quad y = 5$$

8.24. In the electrical circuit shown, the switch is open and a current I exists, where $I = E/(R_1 + R_2)$. At time $\tau = 0$, the switch is closed. The current $I(\tau)$ is then governed by the equation

$$L\frac{dI}{d\tau} + R_1 I = E \qquad \text{with } I(0) = \frac{E}{R_1 + R_2}$$

where L is the inductance in the circuit. Solve this equation by a Runge-Kutta formula to obtain I as a function of time, until the current is close to steady state. Compare the numerical results obtained with the analytical solution:

$$I(\tau) = \frac{E}{R_1}\left[1 - \frac{R_2}{R_1 + R_2} e^{-(R_1/L)\tau}\right]$$

Take $E = 10$ volts, $R_1 = R_2 = 5$ ohms, and $L = 1$ henry.

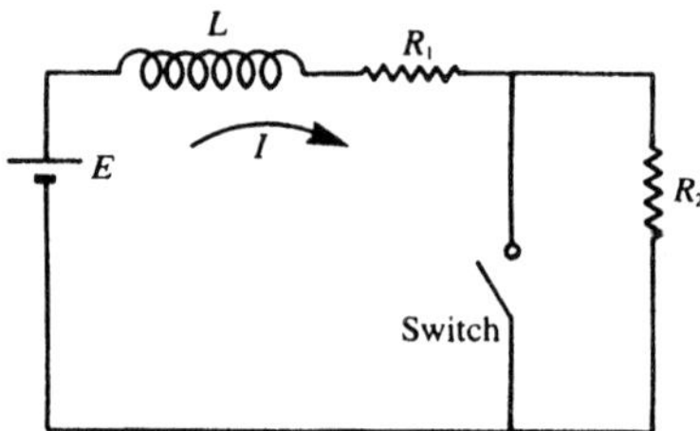

Problem 8.24

8.25. The capacitor C, in the electrical circuit shown, has an initial charge of q_0, and at time $\tau = 0$, the switch is closed. The governing equation for charge $q(\tau)$ is

$$L\frac{d^2q}{d\tau^2} + \frac{1}{C}q = 0 \qquad \text{with } q(0) = q_0 \text{ and } \frac{dq}{d\tau} = 0 \quad \text{at } \tau = 0$$

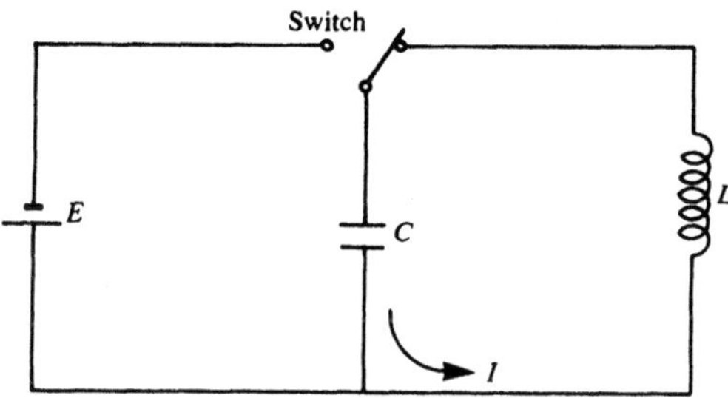

Problem 8.25

Using a Runge-Kutta formula, solve this equation for $q_0 = 0.1$ coulomb, $C = 10^{-4}$ farad, and $L = 0.01$ henry. The frequency of oscillation of this circuit is given by $\sqrt{1/LC}$. Obtain $q(\tau)$ over a few cycles.

8.26. The temperature $T(x)$ in a moving rod, shown in the figure, which loses energy to the environment, at temperature T_∞, is given byt the equation

$$\frac{d^2T}{dx^2} - \frac{1}{\alpha}U\frac{dT}{dx} - \frac{2h}{kR}(T - T_\infty) = 0$$

where x is the distance from a die out of which the material emerges at temperature T_0, U is the velocity of the material, h is the convective heat transfer coefficient, R is the radius of the material, and α and k are material properties known as *thermal diffusivity* and *thermal conductivity*, respectively. The boundary conditions are as follows:

$$\text{At } x = 0: \quad T = T_0$$

$$\text{At } x = \infty: \quad T = T_\infty$$

Employing any shooting method, compute $T(x)$. Take $U = 1$ mm/s, $h = 20$ W/m^2·K, $\alpha = 10^{-4}$ m^2/s, $k = 100$ W/m·K, $T_0 = 600$ K, $T_\infty = 300$ K, and $R = 0.02$ m. For the second boundary condition, start with a large value of x, say, 1 m, to represent ∞, and then vary this value until the results are not significantly affected by a further increase.

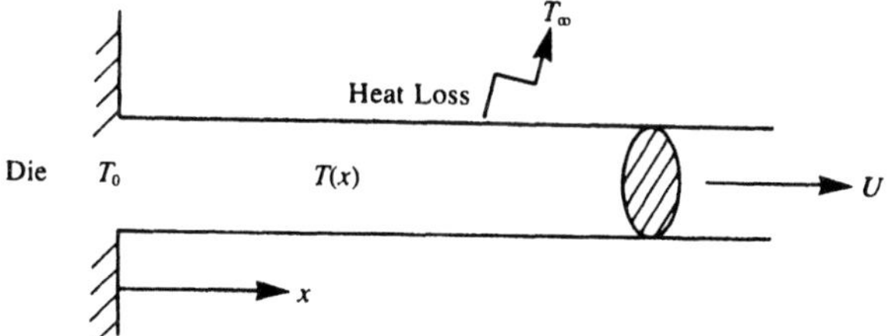

Problem 8.26

8.27. Solve the following boundary-value problem by a shooting method:

$$x^2\frac{d^2y}{dx^2} + 2x\frac{dy}{dx} + xy = 0$$

$$\text{At } x = 1: \quad y = 1$$

$$\text{At } x = 5: \quad y = 6$$

8.28. Formulate Prob. 8.27 for solution by the finite difference method, and outline the numerical scheme that may be adopted to solve the resulting algebraic equations.

8.29. The deflection y of a beam, shown in the figure and loaded axially with force P, is governed by

$$\frac{d^2y}{dx^2} + \frac{P}{EI}y = 0 \qquad \text{with } y(0) = y(L) = 0$$

where E is the modulus of elasticity of the material, I is known as its area moment of inertia, L is the length of the rod, and x is the distance from one end. We are interested in finding the smallest value of P for this eigenvalue problem, since this gives the first failure mode of the rod. Taking $P/(EI)$ as λ, solve this problem by the power method, taking five subdivisions of the rod, to obtain the smallest eigenvalue. Find the corresponding critical load P, if $EI = 1.5 \times 10^6\ \text{Nm}^2$ and $L = 1$ m.

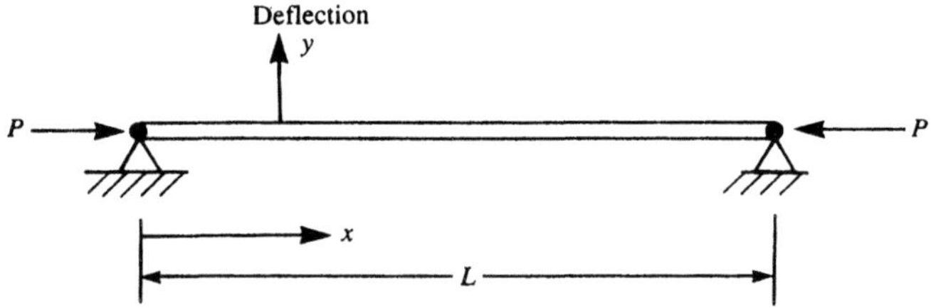

Problem 8.29

8.30. If a heated vertical plate is placed in a quiescent ambient medium, a flow is generated adjacent to the plate because the fluid becomes buoyant due to heating. The dimensionless stream function f, which gives the velocity as f', and temperature θ in this flow are governed by the coupled system of equations

$$f''' + 3ff'' - 2(f')^2 + \theta = 0$$
$$\theta'' + 3Prf\theta' = 0$$

with the boundary conditions

$$f(0) = f'(0) = 1 - \theta(0) = f(\infty) = \theta(\infty) = 0$$

where the quantity within the parentheses represents the location, in the independent variable η, where the condition is applied. The primes indicate differentiation with respect to η, and Pr is a parameter that depends on the fluid (see Prob. 8.17). Solve this system of equations by converting it into five first-order equations and employing the fourth-order Runge-Kutta method, with shooting. Obtain the velocity and temperature distributions, $f'(\eta)$ and $\theta(\eta)$, respectively, for $Pr = 1.0$. Take $\eta = 8$ as being sufficiently large to apply the conditions at infinity.

8.31. In a radiating fin (see Prob. 8.19), the temperature $T(x)$ is governed by

$$\frac{d^2T}{dx^2} - \frac{P}{kA}[h(T - T_\infty) - \varepsilon\sigma(T^4 - T_\infty^4)] = 0$$

with the following boundary conditions:

$$\text{At } x = 0: \quad T = T_0$$

$$\text{At } x = L: \quad \frac{dT}{dx} = 0$$

Here, ε is a property of the surface known as *emissivity*, and σ is a constant (see Prob. 8.5). Employing the finite difference method, solve this problem for the parametric values given in Prob. 8.19, with ε given as 0.5 and $\sigma = 5.67 \times 10^{-8}$ W/m^2·K^4.

8.32. Solve Prob. 8.29 by taking two, three, or four subdivisions of the rod for generating the finite difference equations and obtaining the polynomial from its characteristic determinant. Compare the computed value of the smallest eigenvalue λ_{min} with the analytical result π^2/L^2.

8.33. Using the finite difference method, repeat Prob. 8.26. Discuss the accuracy given by the two approaches. When would the finite difference approach be the preferred one for such problems?

8.34. Study the stability of the second-order Runge-Kutta method by considering the differential equation $dy/dx = -ay$, where a is a constant. Also determine the truncation error.

8.35. Outline a numerical scheme for solving the differential equation $dy/dx = Ax^{-1/2}$, where A is a constant and $y = 0$ at $x = 0$, without using the analytical solution.

8.36. Consider the differential equation $d^2y/dx^2 = ay$, which is to be solved by finite difference methods. Discuss the nature of the resulting algebraic equations and put them in matrix form. Outline the numerical scheme for solving this set of equations and give reasons for your choice.

8.37. Show that the modified Euler's method is second-order accurate.

8.38. Solve the problem of Example 8.2 by Milne's predictor-corrector method, and compare the results with those obtained earlier by the Runge-Kutta method. Comment on the difference between the computational effort involved in the two methods.

8.39. Consider a third-order ordinary differential equation of the form $d^3y/dx^3 = F(x, y, dy/dx)$. Develop a finite difference scheme for solving this equation if $y = A$ at $x = a$; $y = B$ at $x = b$; and $dy/dx = C$ at $x = a$. Assume that F is a linear function in the dependent variable y.

8.40. Consider a nonlinear second-order ODE of the form $d^2y/dx^2 = F(x, y, dy/dx)$, where F is nonlinear in y. Using the finite difference approach, obtain a numerical scheme for solving this equation. Assume the problem to be a boundary-value one with $y = A$ at $x = a$ and $y = B$ at $x = b$.

9

Numerical Solution of Partial Differential Equations

9.1 INTRODUCTION

In the preceding chapter, the numerical solution of ordinary differential equations, which involve a single independent variable, was discussed. However, for a wide variety of problems in science and engineering, the dependent variables are functions of two or more independent variables, such as time and the spatial coordinate distances. Consequently, the differential equations that govern such problems involve partial derivatives and are known as *partial differential equations.* These equations arise in almost all areas of engineering, for instance, in fluid mechanics, elasticity, heat transfer, electromagnetic field theory, hydraulics, neutron diffusion in nuclear reactors, and structural analysis. The numerical solution of partial differential equations is generally much more involved than that of ordinary differential equations because of the presence of several independent variables, each with its own initial and boundary conditions. Therefore, effort is often made, whenever possible, by the use of simplifying approximations and transformations, to reduce the governing partial differential equation (PDE) to an ordinary differential equation (ODE). However, this simplification is possible in only a limited number of cases. Because of the complicated nature of partial differential equations, analytical solutions are rarely obtained, and numerical methods are necessary for most problems of practical interest.

9.1.1 Classification

Many of the classifications outlined in Chapter 8 for ordinary differential equations also apply for partial differential equations. Therefore, the equations may be linear or nonlinear, homogeneous or inhomogeneous, of first or higher order, and may involve a single equation or a system of equations. The initial and boundary conditions are specified in terms of the various independent variables, making it possible for the problem to be an initial-value problem in relation to one independent variable and a boundary-value problem in relation to another variable. However, the suitable initial and boundary conditions that may be imposed for a given equation are determined by the type of the equation. Each type demands a particular set of initial and boundary conditions that must be specified to obtain a well-posed problem that is amenable to an analytical or a numerical solution.

Let us consider the general form of a second-order partial differential equation in two independent variables, given as

$$A\frac{\partial^2\phi}{\partial x^2} + B\frac{\partial^2\phi}{\partial x\,\partial y} + C\frac{\partial^2\phi}{\partial y^2} + D\left(x, y, \phi, \frac{\partial\phi}{\partial x}, \frac{\partial\phi}{\partial y}\right) = 0 \tag{9.1}$$

where A, B, and C may also be functions of the two independent variables x and y and of the dependent variable ϕ and its first-order derivatives. If ϕ appears in the first power throughout, the equation is said to be *linear*. This requires that A, B, and C be functions of only x and y and that D contain only linear functions of ϕ and its first-order derivatives. We shall concern ourselves mainly with linear partial differential equations in this chapter. Nonlinear equations are also frequently encountered, for example, in fluid flow. However, the solution of nonlinear equations is usually much more involved than linear equations, and only a brief outline of the applicable numerical techniques is given later in the chapter.

The classification of the above PDE is based on the sign of $B^2 - 4AC$. The equation is said to be *elliptic* when $B^2 - 4AC < 0$, *parabolic* when $B^2 - 4AC = 0$, and *hyperbolic* when $B^2 - 4AC > 0$. This classification is related to the nature of characteristics, which are lines along which a disturbance or information can propagate. A PDE in two dimensions, as given by Eq. (9.1), reduces to an ODE along a characteristic. A hyperbolic equation has two real and distinct characteristics, which are frequently employed to obtain the solution. In this case, a disturbance propagates at finite speed over a finite region, bounded by the two families of characteristics. For parabolic equations, the two families of characteristics merge, giving rise to an infinite propagation speed and information flow in one direction. Therefore, in a parabolic equation, the solution at a given point depends only on the results obtained along this direction up to the point and not on results in the region beyond. For time as an independent variable, this implies that the solution is affected by the occurrences in the past but not by those in the future. For elliptic equations, complex characteristics are obtained, and, therefore, no directional restrictions arise and a disturbance propagates in all directions. The solution at a given point is affected by disturbances at every other point in the region where the elliptic equation applies. There are no preferred directions, and the solution must be obtained over the entire region simultaneously. Therefore, the mathematical character of the equation is indicated by its classification, which also determines the suitable boundary conditions and the solution procedure.

9.1.2 Examples

A few examples of the three types of partial differential equations are considered here for illustration of the preceding discussion. A very common parabolic equation is the one-dimensional, unsteady, heat conduction equation, which is given as

$$\alpha \frac{\partial^2 T}{\partial x^2} = \frac{\partial T}{\partial \tau} \tag{9.2}$$

where $T(x, \tau)$ is the temperature, x is a spatial coordinate, τ is the time, and α is a constant, known as the *thermal diffusivity* of the material. This equation requires the specification of an initial condition, in time, and two boundary conditions, in x. Another important parabolic equation is the transient convective-diffusive transport equation, which may be written for a one-dimensional problem as

$$\frac{\partial \phi}{\partial \tau} + C \frac{\partial \phi}{\partial x} = D \frac{\partial^2 \phi}{\partial x^2} \tag{9.3}$$

where $\phi(x, \tau)$ is the dependent variable, such as temperature, concentration, or density, and C and D are constants. Again, an initial condition in time and two boundary conditions in x are needed. This equation applies for practical problems such as material movement in chemical reactors, extrusion of plastics, and transport of discharged effluents in a water stream.

Two elliptic equations that are frequently encountered in engineering problems are Laplace's and Poisson's equations, written, respectively, as follows:

$$\frac{\partial^2 \phi}{\partial x^2} + \frac{\partial^2 \phi}{\partial y^2} = 0 \tag{9.4}$$

$$\frac{\partial^2 \phi}{\partial x^2} + \frac{\partial^2 \phi}{\partial y^2} = \beta(x, y) \tag{9.5}$$

where $\phi(x, y)$ is the dependent variable and β may be a constant or a function of x and y. These equations arise in several areas such as fluid mechanics, electrostatics, elasticity, and conduction heat transfer. In conduction, the dependent variable becomes the temperature T and β is a distributed heat source. Boundary conditions are needed at all the edges of the solution domain. Frequently, these are specified as the value of ϕ, of its derivative, or of a linear combination of the two, at the boundaries.

A common hyperbolic equation is the wave equation, written as

$$\frac{\partial^2 \phi}{\partial \tau^2} = c^2 \frac{\partial^2 \phi}{\partial x^2} \tag{9.6}$$

where c is the propagation velocity of the wave and $\phi(x, \tau)$ is the physical dependent variable, such as the displacement of a string. Two initial conditions are needed, and the spatial domain may or may not be bounded. If bounded, boundary conditions are needed at the two boundaries of the region. This equation governs the vibration of a string as well as the behavior of waves in a given medium. Another simple hyperbolic equation is the first-order convection equation, given as

$$\frac{\partial \phi}{\partial \tau} + c\,\frac{\partial \phi}{\partial x} = 0 \tag{9.7}$$

where the physical quantity $\phi(x, \tau)$ is convected at constant velocity c. The characteristics are straight lines, given by $x - c\tau =$ constant, for this equation, which may be differentiated with respect to x or τ to obtain a second-order PDE of the form given by Eq. (9.1).

9.1.3 Basic Considerations

The above discussion illustrates that one must determine the type of the given partial differential equation before proceeding to its numerical solution. If more than two independent variables are to be considered, the equation retains the characteristics of the three types of equations discussed above, as determined by the highest derivatives in each of the independent variables. Unsteady, two-dimensional, mass diffusion, for instance, is governed by the equation

$$\frac{\partial C}{\partial \tau} = D\left(\frac{\partial^2 C}{\partial x^2} + \frac{\partial^2 C}{\partial y^2}\right) \tag{9.8}$$

where C is the concentration of a diffusing chemical species and D is known as the *mass diffusivity*. The problem retains the parabolic nature with respect to its time dependence and the elliptic behavior with respect to the spatial coordinates. Therefore, one marches in time, to obtain the concentration distribution at each time interval, using the distribution at the preceding interval. The concentration distribution at a specified time, which is generally taken as zero, is needed as the initial condition. Also, conditions involving the concentration and/or its derivatives must be specified at all the boundaries of the region. Similarly, appropriate initial and/or boundary conditions must be given for the particular type of equation being solved.

The classification of partial differential equations is discussed here in terms of second-order equations, which are the most frequently encountered equations in engineering applications. However, higher-order equations are also of interest in many problems. Fourth-order equations arise, for instance, in fluid flow and in solid mechanics. These equations can usually be solved by the methods applicable to the second-order elliptic equations, although the finite difference equations are obviously more involved. In fact, a fourth-order PDE may often be broken down into two second-order equations, which are solved simultaneously. First-order equations also occur in a few cases. Very often, these equations have real characteristics and can be solved by the methods employed for the hyperbolic equation of second order.

The main approach to the solution of a partial differential equation is based on the reduction of the equation to a system of algebraic equations, which are then solved by direct or iterative methods to yield the value of the dependent variable at a finite number of mesh points. Two techniques that are commonly employed for generating the governing system of algebraic equations are the finite difference and the finite element methods. Finite difference methods apply the approximation to the PDE at a finite number of grid points in the computational domain. Finite element methods divide the region into a finite number of subdivisions. The integral form of the PDE is then applied to each element, and the integrals are minimized or the integral statement satisfied, using interpolation functions that contain adjustable parameters, in order to satisfy the conservation principles. In both cases, a system of algebraic equations is obtained from the application of the method at the boundaries and in the interior region. The finite element method has become very important in recent years because of its versatility in treating a wide variety of boundary conditions, material property variations, and complicated geometries. However, it is generally much more involved than the finite difference method and is advantageous to use only when some of the complications mentioned above are present. We shall restrict our discussion in this chapter to relatively simple problems and largely to the finite difference methods of solving them. A brief discussion of the finite element approach is also given later in the chapter, including a comparison with the finite difference technique.

In this chapter, the three types of partial differential equations mentioned above are considered, and an introductory treatment of their solution is given. The subject is an extensive one, and various complexities arise in diverse engineering applications; see, for instance, the book by Jaluria and Torrance (1986) on the numerical solution of heat transfer problems. Here, we are interested mainly in a consideration of the basic approach to the numerical solution of these different types of equations. Therefore, a few simple equations are taken and their solution is discussed. The parabolic equations are treated first since the methods for solving them are similar to those used for ordinary differential equations and since they also form the basis for some of the methods used for elliptic equations.

9.2 PARABOLIC PARTIAL DIFFERENTIAL EQUATIONS

The solution domain in parabolic equations stretches outward indefinitely in one coordinate direction, say, z, from the given initial values, as shown in Fig. 9.1. The equation is solved for the dependent variable ϕ by advancing, or marching, in the independent variable z, starting with the initial conditions at $z = 0$. The solution must satisfy the prescribed boundary conditions, in the other independent variable x, as the solution advances in z. The solution at any z depends on the values of ϕ at smaller z but is independent of those at larger z. This implies a definite direction for disturbance propagation. A physical analogy to this circumstance is a fast-flowing river in which a disturbance at a given location travels downstream but does not affect the flow upstream. The solution procedures for parabolic equations are, therefore, based on

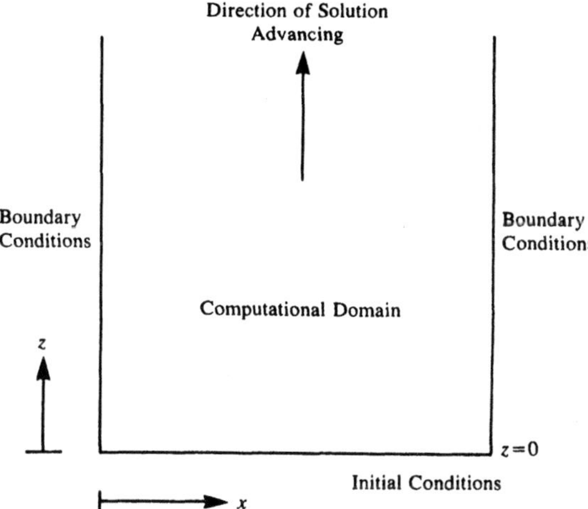

Figure 9.1 Solution domain for a parabolic partial differential equation, along with the necessary boundary and initial conditions.

marching outward from the initial conditions in one of the independent variables, while satisfying the given boundary conditions in the other.

9.2.1 Computer Solution with an Explicit Scheme

Let us consider the one-dimensional transient heat conduction problem, given by Eq. (9.2), as an example of a parabolic PDE. Therefore, the dependent variable is the temperature $T(x,\tau)$, which is governed by the equation

$$\alpha \frac{\partial^2 T}{\partial x^2} = \frac{\partial T}{\partial \tau} \tag{9.2}$$

Since the equation contains the second derivative in x and only the first derivative in time τ, two boundary conditions are needed in x and a single initial condition in τ. These may be prescribed in terms of $T(x,\tau)$ as

$$\begin{aligned} T(a,\tau) &= A \\ T(b,\tau) &= B \\ T(x,0) &= T_0 \end{aligned} \tag{9.9}$$

where $x = a$ and $x = b$ represents the boundaries of the region and T_0 is the initial condition. In general, T_0 may be a function of x, and A and B may be functions of time. However, for simplicity these are taken as constants here. The problem, as given above, is properly posed for obtaining the solution for $\tau > 0$.

The above parabolic PDE may be solved numerically by finite difference methods. A space-time grid, with Δx and $\Delta\tau$ denoting the corresponding mesh sizes, is taken, as shown in Fig. 9.2. Then, the finite difference approximations to the derivatives are applied to the given equation at each grid point. Several finite difference approximations can be obtained, depending on the representations used for the derivatives. For instance, a forward difference representation may be used for the partial derivative in time to obtain

$$\frac{\partial T}{\partial \tau} = \frac{T_{i+1,j} - T_{i,j}}{\Delta\tau} \tag{9.10}$$

where the subscript $(i + 1)$ denotes the values at time $(\tau + \Delta\tau)$, and i those at time τ. The spatial location is given by j. For the mesh shown in Fig. 9.2, $x = j\,\Delta x$ and $\tau = i\,\Delta\tau$. The truncation error is $O(\Delta\tau)$, as obtained in Chapter 3.

A central difference approximation may be employed for the second derivative to obtain the approximation at time τ and location x as

$$\left(\frac{\partial^2 T}{\partial x^2}\right)_{i,j} = \frac{T_{i,j+1} - 2T_{i,j} + T_{i,j-1}}{(\Delta x)^2} \tag{9.11}$$

with a truncation error of $O[(\Delta x)^2]$. In this expression, the second derivative in x is approximated at time τ. If the above finite difference representations are substituted into Eq. (9.2), the resulting equation may be written as

$$T_{i+1,j} = T_{i,j} + \frac{\alpha\,\Delta\tau}{(\Delta x)^2}(T_{i,j+1} - 2T_{i,j} + T_{i,j-1}) \tag{9.12}$$

where the truncation error is $O(\Delta\tau) + O[(\Delta x)^2]$. Therefore,

$$T_{i+1,j} = (1 - 2F)T_{i,j} + F(T_{i,j+1} + T_{i,j-1}) \tag{9.13}$$

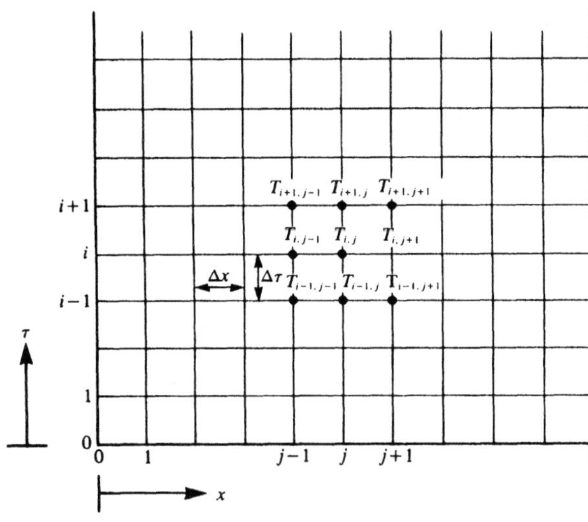

Figure 9.2 Space-time grid for the solution of a parabolic PDE by finite difference methods.

where F is a constant known as the *grid Fourier number* and is given by

$$F = \frac{\alpha \Delta\tau}{(\Delta x)^2} \tag{9.14}$$

Equation (9.13) gives the temperature at time $(\tau + \Delta\tau)$ at the grid point whose spatial coordinate is $x = j\,\Delta x$, in terms of the temperatures at time τ at the grid points with coordinates $(x - \Delta x)$, x, and $(x + \Delta x)$. The temperatures at $x = a$ and $x = b$ remain constant at A and B, respectively, because of the given boundary conditions. The initial condition gives the temperatures at the grid points at time $\tau = 0$ as T_0, where T_0 may be given as a constant or as a function of x. Using Eq. (9.13), the temperature distribution at time $\tau = \Delta\tau$ may be determined from the given temperatures at $\tau = 0$. Similarly, the distribution at the second time step $\tau = 2\,\Delta\tau$ is obtained, employing the computed values at the first time step. Therefore, the solution is obtained at increasing values of time τ. This time marching may be continued indefinitely. However, the process is generally terminated when a specified time τ_{max} has been attained or when the steady state, given by a negligible change in the solution with increasing time, is reached. In most cases, a steady-state circumstance is attained at large time, and a convergence criterion may be applied to the solution in order to terminate the computation when the solution is sufficiently close to the steady state. A convergence criterion is generally needed since the solution may approach the steady state asymptotically, thus taking infinite time, theoretically, to reach it exactly. The choice of the convergence criterion, therefore, affects the computer time taken to obtain the solution. Also, as discussed in Chapter 2, the convergence criterion must be varied to ensure a negligible dependence of the numerical results on the value chosen.

If F is chosen as 1/2, Eq. (9.13) becomes

$$T_{i+1,j} = \frac{T_{i,j+1} + T_{i,j-1}}{2} \tag{9.15}$$

which implies that the new temperature at a grid point is the average of the old temperatures at the two adjacent grid points. A graphical method, known as the *Schmidt-Binder method*, has been developed on the basis of this equation. A uniform distribution of grid points is taken, and the temperature at a point, for the next time step, is simply given by the intersection of the normal to the x-axis at this point with the line joining the graphical points representing the present temperatures at the two adjacent grid points, as shown in Fig. 9.3. Therefore, the solution to the differential equation, Eq. (9.2), may be obtained graphically over a desired time interval, starting with the initial conditions.

The computational scheme of Eq. (9.13) gives the temperatures at time $(\tau + \Delta\tau)$ in terms of known temperatures at time τ. Therefore, the temperatures at a given time step may be determined explicitly from known values at the previous time step. This method is known as the *explicit Euler method* and is also referred to as the *forward time central space* (FTCS) *method*, because of the finite difference approximations used. It provides the simplest computational procedure for computing the time-dependent temperature distribution, starting with the initial conditions.

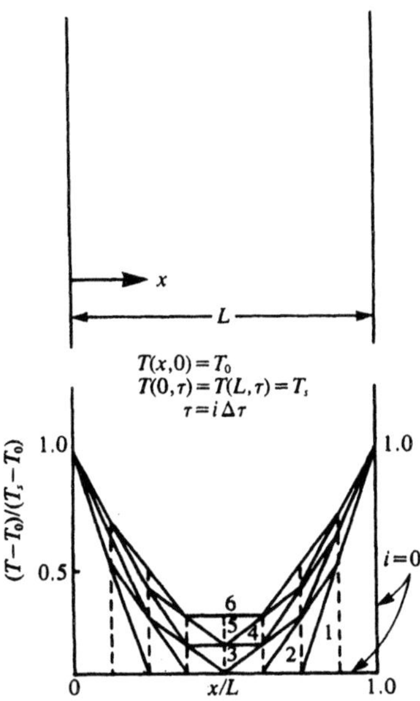

Figure 9.3 Graphical solution of transient heat conduction in a wall, with $F = 1/2$ (Schmidt-Binder method).

9.2.2 Stability of Euler's (FTCS) Method

The above method becomes unstable at large values of F, and stability is assured only if

$$F = \frac{\alpha \Delta \tau}{(\Delta x)^2} \leqslant \frac{1}{2} \tag{9.16}$$

Therefore, an amplification of the round-off and truncation errors arises if $F > 1/2$ and may lead to a meaningless solution as the computation advances in time. This condition for numerical stability is obtained by stability analysis, considering the growth of errors introduced into the solution; see, for instance, Roache (1976), Ferziger (1981), and Jaluria and Torrance (1986). A physical explanation may also be given in terms of Eq. (9.13). If $F > 1/2$, the coefficient of $T_{i,j}$ on the right-hand side of Eq. (9.13) is negative. This implies that a larger value of the temperature $T_{i,j}$ at time τ gives rise to a smaller value of the temperature $T_{i+1,j}$ at the same location at the next time step. Similarly, a smaller value of $T_{i,j}$ at time τ results in a larger value of $T_{i+1,j}$ at $(\tau + \Delta\tau)$. The result is an oscillatory and unstable solution. Figure 9.4 shows the nature of this instability in terms of the computed results for transient conduction in a plate of thickness L at different values of F (from Jaluria and Torrance, 1986). The temperature distributions are obtained by solving Eq. (9.2) with $A = B = T_s$. The results are shown in terms of dimensionless coordinate distance x/L and temperature $\theta = (T - T_s)/(T_0 - T_s)$. Only half the conduction region is shown because of symmetry. Clearly, instability arises as F increases beyond 0.5.

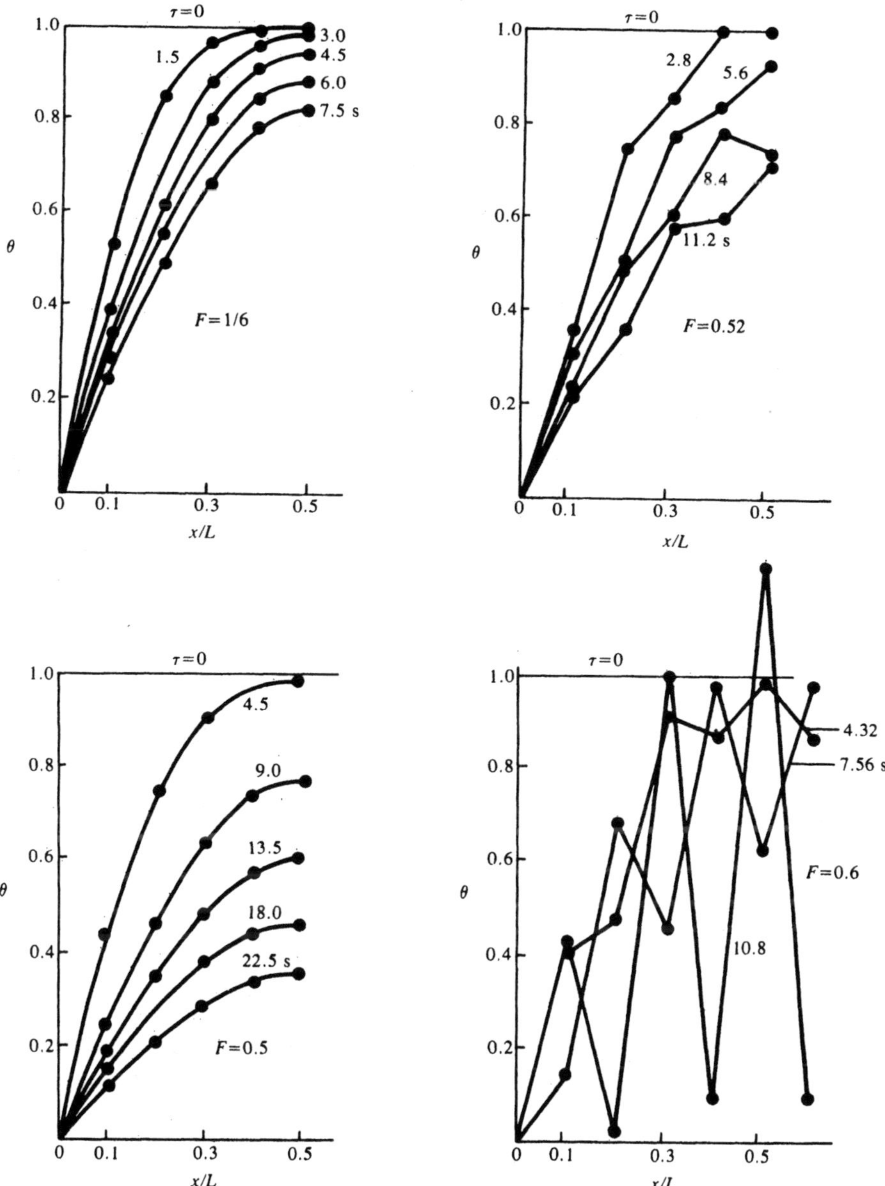

Figure 9.4 Time-dependent temperature distributions for one-dimensional conduction in a plate, governed by Eq. (9.2), at various values of the grid Fourier number F, with time τ in seconds (Jaluria and Torrance, 1986).

The major problem with the stability criterion given by Eq. (9.16) is the constraint that it imposes on the allowable time step for a given grid spacing Δx, which is generally chosen as small to keep the truncation error small. For a given value of the thermal diffusivity α, the maximum permissible time step $\Delta\tau$ is given by the stability constraint as

$$\Delta\tau \leqslant \frac{(\Delta x)^2}{2\alpha} \tag{9.17}$$

A small value of $\Delta\tau$ is desirable for keeping the truncation error, which is of order $\Delta\tau$ in this formulation, down to a desired level. However, the stability criterion generally limits the time step to a value that is much smaller than that needed for maintaining the accuracy of the solution. Therefore, the explicit method often severely constrains the time step and results in excessive computational time. Consequently, other methods have been developed which, although often more involved than the FTCS method, have better stability characteristics.

9.2.3 Implicit Methods

In the FTCS explicit method, the finite difference approximation for the second spatial derivative, $\partial^2 T/\partial x^2$, is written, in Eq. (9.11), at time τ. A family of implicit methods may be obtained by approximating this derivative at a different time, between τ and $\tau + \Delta\tau$. The resulting finite difference equation is

$$\frac{T_{i+1,j} - T_{i,j}}{\Delta\tau} = \alpha\left[\gamma\,\frac{T_{i+1,j+1} - 2T_{i+1,j} + T_{i+1,j-1}}{(\Delta x)^2} + (\gamma - 1)\,\frac{T_{i,j+1} - 2T_{i,j} + T_{i,j-1}}{(\Delta x)^2}\right] \tag{9.18}$$

where γ is a parameter that lies between 0 and 1. Therefore, the second derivative is written as a weighted average of the finite difference approximations corresponding to time levels τ and $\tau + \Delta\tau$.

If $\gamma = 0$, the FTCS explicit method, given in Section 9.2.1, is obtained. For $\gamma = 1/2$, the second derivative is evaluated midway between the two time levels, and the truncation error can be shown to become $O[(\Delta\tau)^2] + O[(\Delta x)^2]$. This method, known as the *Crank-Nicolson method*, is very popular for the solution of parabolic equations. If $\gamma = 1$, the second derivative is evaluated at time $\tau + \Delta\tau$, and the formulation is known as the *fully implicit* or the *Laasonen method*. The truncation error is the same as that for the FTCS explicit method. From Eq. (9.18), the finite difference equations for the Crank-Nicolson and the fully implicit methods are, respectively,

$$-FT_{i+1,j+1} + 2(1 + F)T_{i+1,j} - FT_{i+1,j-1} = FT_{i,j+1} + 2(1 - F)T_{i,j} + FT_{i,j-1} \tag{9.19}$$

$$-FT_{i+1,j+1} + (1 + 2F)T_{i+1,j} - FT_{i+1,j-1} = T_{i,j} \tag{9.20}$$

As seen from Eq. (9.18), a set of simultaneous linear algebraic equations must be solved for implicit methods to obtain the temperature distribution at time $\tau + \Delta\tau$. A tridiagonal system arises which is conveniently solved at each time step by Gaussian elimination, as discussed in Chapter 5, to obtain the time-dependent temperature distribution. The solution marches in time, starting with the known initial values, until steady state or a specified time τ_{max} is reached. The numerical procedure is more involved than the FTCS explicit method. However, the implicit methods generally have much better stability characteristics. The Crank-Nicolson implicit method is unconditionally stable, and much larger time steps can be taken as compared to the FTCS explicit method. The only constraint on the time step is generally because of accuracy considerations. However, oscillations that generally remain bounded do arise in the solution for certain problems at large values of F and may lead to a restriction on $\Delta\tau$, although at much larger values than that given by Eq. (9.17).

The FTCS method, on the other hand, often restricts the time step to much smaller values than those demanded by the desired accuracy of the results. The Crank-Nicolson method also has a lower truncation error in time, $O[(\Delta\tau)^2]$, as mentioned above, allowing a larger time step for given accuracy. This arises because the finite difference approximation for the time derivative is in effect obtained midway between the two time levels τ and $\tau + \Delta\tau$, making it a central difference approximation with TE of $O[(\Delta\tau)^2]$. Therefore, this method yields numerical results of greater accuracy than those from the FTCS method with a smaller computational cost. Even for nonlinear equations, which require iteration for solving the resulting set of algebraic equations, the Crank-Nicolson method is generally superior because only a few iterations are often needed for the linearized tridiagonal set. The numerical procedure may be graphically represented in terms of a computational molecule, which illustrates the grid points involved in the computation. Figure 9.5 shows the computational molecules for the FTCS, Crank-Nicolson, and fully implicit methods. These indicate the relationships between the values at the neighboring grid points. Examples 9.1 and 9.2 demonstrate the application of the FTCS and Crank-Nicolson methods, respectively.

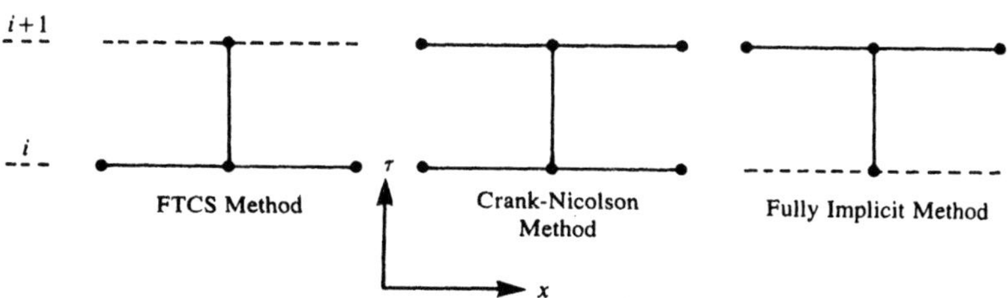

Figure 9.5 Computational molecules for the explicit Euler (FTCS), Crank-Nicolson, and fully implicit methods.

9.2.4 Other Methods and Considerations

The major advantage of implicit methods over the explicit methods is the numerical stability, which allows much larger time steps in the computation. This consideration is particularly important in regions where the solution varies slowly with time, for example, near the steady state. Because the resulting algebraic equations are in the tridiagonal form, the number of arithmetic operations needed to solve the set for each time step is only $O(n)$, n being the number of grid points where the temperatures are to be computed. This is of the same order as the number of arithmetic operations necessary for taking one time step using the explicit method. Generally, the computer time taken by the implicit methods per time step is around twice that for the FTCS method. Since, in many cases, the time step $\Delta\tau$ for the implicit methods may be taken as large as 10 to 100 times that allowed by the explicit method, due to stability considerations, a substantial reduction in computer time may be obtained by the use of implicit methods. The explicit method has the advantage of simpler programming.

Since the computation for the explicit method requires only the known values from the previous time step, the method can be used for solving nonlinear equations without much difficulty, whereas the solution of the simultaneous nonlinear equations that arise in the implicit methods requires iteration. Nonlinear equations arise in many physical problems, for instance, in the one-dimensional transient conduction problem if the material properties are not constant but functions of the temperature. Explicit methods are also advantageous to use if the boundary conditions are time-dependent.

Because of the advantage of an explicit procedure over implicit methods, for nonlinear problems, for time-dependent boundary conditions, and for other complexities in the problem, several other explicit methods with better stability characteristics than the FTCS method have been developed. One of the methods that is unconditionally stable and that has been widely used for solving parabolic partial differential equations is the Dufort-Frankel method. The corresponding finite difference equation is given by

$$\frac{T_{i+1,j} - T_{i-1,j}}{2\,\Delta\tau} = \alpha\,\frac{T_{i,j+1} - T_{i+1,j} - T_{i-1,j} + T_{i,j-1}}{(\Delta x)^2} \tag{9.21}$$

or

$$T_{i+1,j} = \frac{1 - 2F}{1 + 2F}\,T_{i-1,j} + \frac{2F}{1 + 2F}\,(T_{i,j+1} + T_{i,j-1}) \tag{9.22}$$

Therefore, a central difference approximation is used for the derivative in time, giving a truncation error of $O[(\Delta\tau)^2]$, and the temperature at the jth grid point is split into the values at two time steps τ and $\tau + \Delta\tau$. Although this method is unconditionally stable, it is not self-starting because values at $\tau - \Delta\tau$ are needed for computing those at $\tau + \Delta\tau$. Thus, it requires another method to start the computation. Also, it can behave poorly under certain conditions. However, it is still used frequently in engineering problems. Various other explicit and implicit methods are given by Carnahan et al. (1969), Smith (1978), and Ferziger (1981).

9.2.5 Multidimensional Problems

The methods discussed above for solving the one-dimensional transient problem can easily be extended to multidimensional problems. Two-dimensional, unsteady diffusion processes are governed by Eq. (9.8) for mass transfer and the following equation for heat conduction:

$$\frac{\partial T}{\partial \tau} = \alpha \left(\frac{\partial^2 T}{\partial x^2} + \frac{\partial^2 T}{\partial y^2} \right) \tag{9.23}$$

Using the FTCS explicit formulation, we obtain the finite difference equation as follows:

$$\frac{T_{i+1,j,k} - T_{i,j,k}}{\Delta \tau} = \alpha \left[\frac{T_{i,j+1,k} - 2T_{i,j,k} + T_{i,j-1,k}}{(\Delta x)^2} + \frac{T_{i,j,k+1} - 2T_{i,j,k} + T_{i,j,k-1}}{(\Delta y)^2} \right] \tag{9.24}$$

where the first subscript refers to time, the second subscript to the location in the x direction, and the third subscript to the location in the y direction. The corresponding step sizes are Δx and Δy. Therefore, the temperature distribution at the next time step, $\tau + \Delta\tau$, may be obtained in terms of the known values at time τ. Two grid Fourier numbers F_1 and F_2 arise, where

$$F_1 = \frac{\alpha \Delta \tau}{(\Delta x)^2} \quad \text{and} \quad F_2 = \frac{\alpha \Delta \tau}{(\Delta y)^2} \tag{9.25a}$$

Also, the finite difference equation is:

$$T_{i+1,j,k} = (1 - 2F_1 - 2F_2)T_{i,j,k} + F_1(T_{i,j+1,k} + T_{i,j-1,k}) + F_2(T_{i,j,k+1} + T_{i,j,k-1}) \tag{9.25b}$$

Stability considerations, similar to those for the one-dimensional problem, arise, and the implicit methods may be employed advantageously. If $\Delta x = \Delta y$, the grid Fourier number F $(=F_1 = F_2) \leqslant 1/4$ for numerical stability in the explicit FTCS method. From Eq. (9.25b), stability requires that $1 - F_1 - F_2 \geqslant 1$. As mentioned earlier, the equation retains the characteristics of a parabolic equation in time and those of an elliptic equation in the spatial coordinates. Therefore, the solution is obtained by marching in time, while satisfying the boundary conditions at the boundaries of the region. Similarly, the three-dimensional transient conduction problem may be solved. The constraint on F, for $\Delta x = \Delta y = \Delta z$, where z is the third coordinate, is obtained as $F \leqslant 1/6$ from stability considerations. For further details, see Smith (1968), Carnahan et al. (1969), and Jaluria and Torrance (1986).

The numerical solution of parabolic differential equations has been discussed here in terms of the transient heat conduction problem, since it is an important

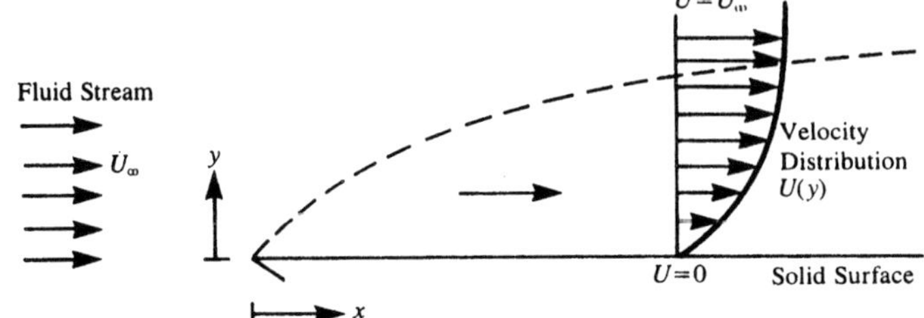

Figure 9.6 Sketch of the boundary-layer flow over a flat surface. This flow is governed by a parabolic partial differential equation.

problem and also because several other parabolic equations of engineering interest are of similar form. For instance, the equation that governs the motion of fluid due to a plate being suddenly set into motion, from rest, in an infinite medium is

$$\frac{\partial u}{\partial \tau} = \nu \frac{\partial^2 u}{\partial x^2} \tag{9.26}$$

where u is the velocity in the direction of motion x, and ν is a property of the fluid known as *kinematic viscosity*. The treatment given here may easily be applied to this equation and also extended to other forms of parabolic equations. In several cases, there is no time dependence, and the equation is parabolic in one of the spatial coordinates. An example of such a circumstance is the boundary-layer flow over a surface. In this case, the variation with time is replaced by a variation with x, the direction in which the main flow occurs; see Fig. 9.6. Information travels downstream to larger x from a given point, but is assumed not to travel upstream to smaller x, similar to the time-dependent problem. However, this problem is nonlinear, as are several problems of practical interest, and explicit methods are often advantageous to use in such cases.

Several methods that employ the useful characteristics of both implicit and explicit methods have also been developed and are among the most popular techniques for solving parabolic partial differential equations in two or more dimensions. See, for instance, the discussion on the alternating direction implicit (ADI) method in the next section. This method employs the implicit formulation in one dircction and treats the other direction explicitly, the two directions being interchanged from one step to the next. The result is a tridiagonal system at each step. The ADI method is the most important method in a class of methods known as *splitting methods*, several of which are available in the literature.

The imposition of the boundary conditions for parabolic partial differential equations has been considered here simply in terms of the value of the dependent variable, such as T, being specified at the boundaries. However, in practical problems, several other boundary conditions arise, particularly those related to the gradient of the dependent variable. Such boundary conditions and the relevant finite difference formulations are considered in the next section and also in Example 9.2.

Example 9.1

In a chemical manufacturing system, a process involves the diffusion of salt into a layer of water. The layer is of thickness L and initially has a uniform salt concentration C_0. At time $\tau = 0$, the layer is brought into contact with saline solution at one surface, and the concentration at this surface is raised to C_s, while the other surface is maintained at concentration C_0. The mass diffusivity, also known as the *diffusion coefficient*, of salt in water is denoted by D. Using the FTCS explicit method, solve this one-dimensional transient diffusion problem to obtain the time-dependent concentration distribution in the material.

Solution

The governing partial differential equation for the concentration $C(x,\tau)$ in the water layer, where x is the coordinate distance measured from the surface whose concentration is raised to C_s, is

$$D\frac{\partial^2 C}{\partial x^2} = \frac{\partial C}{\partial \tau} \tag{9.1.1}$$

Figure 9.1.1 illustrates this problem. The corresponding initial and boundary

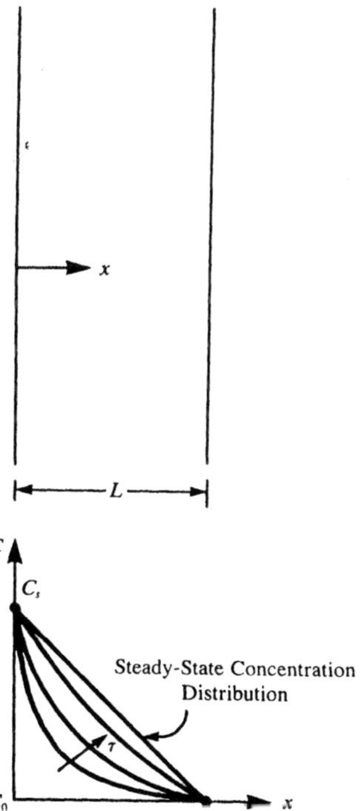

Figure 9.1.1 The physical problem, the coordinate system, and the expected transient behavior of the concentration distribution for Example 9.1.

conditions are as follows:

$$\begin{aligned} &\text{For } \tau \leqslant 0: \quad C = C_0 \quad \text{at all } x \\ &\text{For } \tau > 0: \quad C = C_s \text{ at } x = 0 \qquad \text{and} \qquad C = C_0 \quad \text{at } x = L \end{aligned} \tag{9.1.2}$$

Here, the numerical values of the physical quantities, such as concentration and the water layer thickness, are not given, so that the problem may be solved in generalized terms. Thus, the governing equation and the boundary conditions may be nondimensionalized to obtain a solution that can be used for different sets of physical quantities.

We start by defining the following dimensionless quantities:

$$X = \frac{x}{L} \qquad \bar{\tau} = \frac{D\tau}{L^2} \qquad \theta = \frac{C - C_0}{C_s - C_0} \tag{9.1.3}$$

where X, τ, and θ are the dimensionless distance, time, and concentration, respectively. We can formulate the given problem in terms of these quantities by using the above definitions to replace the physical variables by dimensionless ones. Then the governing equation is obtained as

$$\frac{\partial^2 \theta}{\partial X^2} = \frac{\partial \theta}{\partial \bar{\tau}} \tag{9.1.4}$$

The initial and boundary conditions become

$$\theta(X,0) = 0 \qquad \theta(0,\bar{\tau}) = 1 \qquad \theta(1,\bar{\tau}) = 0 \tag{9.1.5}$$

Therefore, the above dimensionless, parabolic, partial differential equation may be solved to obtain $\theta(X,\tau)$. If the concentrations, mass diffusivity, and thickness of the layer are given, the physical concentration distributions can be also determined from Eq. (9.1.3). This implies that the problem must be solved only once in dimensionless terms, instead of separately for each set of physical parameters. For this reason, the equations are often nondimensionalized and results are obtained in generalized terms, as outlined here.

For the FTCS explicit method, the forward difference approximation is used for the first derivative in time, and central difference for the second derivative in the spatial coordinate x. This yields

$$\theta_{i+1,j} = (1 - 2F)\theta_{i,j} + F(\theta_{i,j+1} + \theta_{i,j-1}) \tag{9.1.6}$$

where

$$F = \frac{\Delta \bar{\tau}}{(\Delta X)^2} \tag{9.1.7}$$

Here, the subscript i represents the time step, and j the spatial grid point. Therefore, $\bar{\tau} = i\,\Delta\tau$ and $X = j\,\Delta X$. Figure 9.1.2 gives the FORTRAN 77 computer program for solving this problem by the explicit method. An interactive program is written so that the grid size, initial concentration, and time step may be given as inputs. The total

```
C           FORWARD TIME CENTRAL SPACE (FTCS) METHOD
C
C     THIS PROGRAM SOLVES A PARABOLIC EQUATION BY THE FTCS METHOD
C
C     WHEN THE PROGRAM IS RUN, IT PROMPTS FOR THE INPUT VALUES
C     REQUIRED. TYPE IN THE INPUT VALUES AND THE OUTPUT WILL BE
C     STORED IN A FILE CALLED 'FTCS.DAT'.
C
C
C     DESCRIPTION OF THE INPUT PARAMETERS:
C
C     IL    IS THE NUMBER OF GRID POINTS.
C     DX    IS THE GRID SIZE.
C     TINIT IS THE INITIAL VALUE OF THE SOLUTION VECTOR, THETA,
C           TAKEN AS UNIFORM OVER THE WHOLE DOMAIN.
C     NLIM  IS THE MAXIMUM NUMBER OF TIME STEPS BEFORE STOPPING.
C     NSTEP IS THE NUMBER OF TIME STEPS AFTER WHICH PRINTOUT OCCURS.
C
C
C     DESCRIPTION OF OTHER VARIABLES USED:
C
C     T     IS THE SOLUTION, THETA, AT THE NTH TIME STEP.
C     TOL   IS THE SOLUTION, THETA, AT THE (N-1)TH TIME STEP.
C     DT    IS THE TIME STEP USED. THE PROGRAM USES THE MAXIMUM
C           TIME STEP ALLOWED FROM STABILITY CONSIDERATIONS.
C
C     ENTER INPUT PARAMETERS
C
      IMPLICIT REAL*8(A-H,O-Z)
      DIMENSION T(50),TOL(50)
      PRINT*,'ENTER NO. OF GRID POINTS, IL='
      READ(1,*)IL
      PRINT*,'ENTER GRID SIZE, DX='
      READ(1,*)DX
      PRINT*,'ENTER INITIAL VALUE OF CONCENTRATION TAKEN AS'
      PRINT*,'UNIFORM OVER THE WHOLE DOMAIN'
      READ(1,*)TINT
      PRINT*,'ENTER MAXIMUM NO. OF TIME STEPS BEFORE STOPPING'
      READ(1,*)NLIM
      PRINT*,'ENTER NO. OF TIME STEPS AFTER WHICH PRINTOUT OCCURS'
      READ(1,*)NSTEP
      ISTEP1=0
      ISTEP2=0
      TIME=0.
C
C     OPEN FILES FOR STORING NUMERICAL RESULTS
C
      OPEN(UNIT=10,FILE='FTCS.DAT')
C
C     SET THE INITIAL CONDITIONS
C
      DO 10 I=1,IL
      T(I)=TINT
      TOL(I)=TINT
   10 CONTINUE
C
C     SET THE BOUNDARY CONDITIONS
C
```

Figure 9.1.2 FORTRAN program for solving the parabolic PDE considered in Example 9.1 by the FTCS explicit method.

```
      T(1)=1.
      T(IL)=0.
C
C     CALCULATE THE MAXIMUM POSSIBLE TIME STEP TO AVOID INSTABILITY
C
      DT=DX**2/2.
      PRINT*,'TIME STEP = ',DT
      WRITE(10,120)DX,DT
      WRITE(10,130)IL
      WRITE(10,140)TIME
      WRITE(10,150)(T(I),I=1,IL)
C
C     INCREMENT THE ITERATION COUNTER AND CHECK IF THE CHOSEN
C     MAXIMUM NUMBER OF ITERATIONS IS EXCEEDED.
C
   15 ISTEP1=ISTEP1+1
      ISTEP2=ISTEP2+1
      TIME=TIME+DT
      IF(ISTEP1.GT.NLIM)GO TO 50
C
C     SAVE THE SOLUTION AT THE PREVIOUS TIME STEP
C
      DO 20 I=1,IL
      TOL(I)=T(I)
   20 CONTINUE
C
C     APPLY FTCS SCHEME AT INTERIOR POINTS
C
      DO 30 I=2,IL-1
      T(I)=TOL(I)+DT*(TOL(I+1)-2.*TOL(I)+TOL(I-1))/DX**2
   30 CONTINUE
C
C     APPLY BOUNDARY CONDITIONS
C
      T(1)=1.
      T(IL)=0.
C
C     OUTPUT THE RESULTS
C
      IF(ISTEP2.EQ.NSTEP)THEN
      WRITE(10,140)TIME
      WRITE(10,150)(T(I),I=1,IL)
      ISTEP2=0
      GO TO 15
      END IF
      GO TO 15
  120 FORMAT(/,4X,'DX=',F4.2,4X,'DT=',F4.2)
  130 FORMAT(//,4X,'IL=',I3)
  140 FORMAT(/,1X,'AT T = ',F7.3,1X,'CONCENTRATION FIELD IS:')
  150 FORMAT(1X,20(F8.4,2X))
   50 CLOSE(UNIT=10)
      STOP
      END
```

Figure 9.1.2 Continued

number of grid points is chosen so as to obtain a total dimensionless distance of 1.0. The output is printed at specified time intervals, given by a chosen number of time steps, and the computation is carried out until a specified time is attained. A convergence criterion may also be employed to stop the computation when steady-state conditions are reached, as indicated by a concentration distribution that does not vary with time. The OPEN statement is used for storing the computed results. This data file is subsequently employed for plotting the results. The variable names employed in the program are defined at the beginning of the program, and the various important steps in the computation are indicated by means of comment statements.

Figures 9.1.3 and 9.1.4 present the numerical results obtained with $F = 0.5$, which gives the maximum time step for a stable numerical scheme; see Fig. 9.4. The initial dimensionless concentration θ is zero throughout the plate, and the steady-state distribution is a linear variation from 1.0 at one surface to 0.0 at the other. We can easily obtain the steady-state result by setting the transient term in Eq. (9.1.4) equal to zero and solving the ordinary differential equation $d^2\theta/dX^2 = 0$, to give $\theta = 1 - X$ as the steady-state distribution. Figure 9.1.3 shows the concentration distribution at various time intervals. Note that the steady-state distribution is attained by $\bar{\tau} = 0.5$. Figure 9.1.4 shows the variation of the dimensionless concentration θ with time $\bar{\tau}$ at several locations within the plate. Note that the temperatures increase sharply from the initial value of 0.0, as time $\bar{\tau}$ increases. The final approach to the

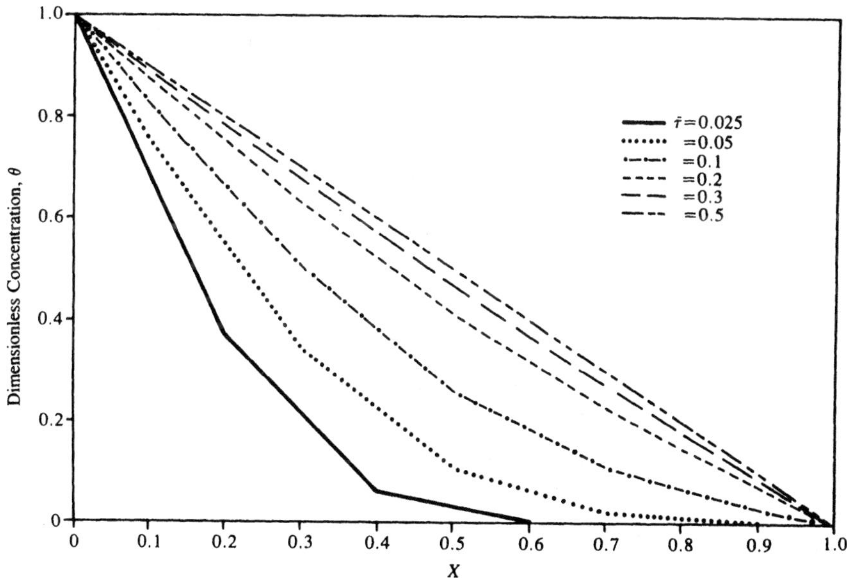

Figure 9.1.3 Computed concentration distribution at various time intervals for Example 9.1. The dimensionless time step $\Delta\bar{\tau}$ and the grid size ΔX are taken as 0.01 and 0.1, respectively.

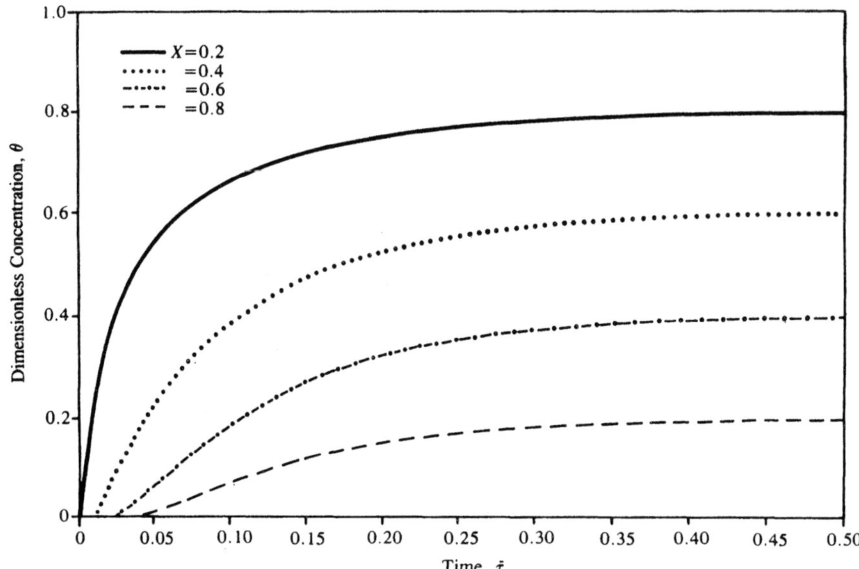

Figure 9.1.4 Computed variation of the concentration in Example 9.1, at various locations with time, indicating the approach to steady-state conditions at large time.

steady-state value is a gradual one. Also, the concentration starts changing from zero at a later time for points which are farther away from the surface $x = 0$, where the step change in concentration occurs. This indicates a finite speed for the propagation of the mass diffusion effects in the FTCS method. At each time step, only the next grid point is affected, as seen from Eq. (9.1.6) and Fig. 9.3. For details on the physical aspects of this problem and other similar ones in heat and mass transfer, standard textbooks in the area, such as Incropera and Dewitt (1981), may be consulted.

Example 9.2

A flat plate of thickness L is initially at a uniform temperature T_0. At time $\tau = 0$, the temperature at one surface is raised to T_s, while the other surface is kept perfectly insulated. The thermal diffusivity of the material is denoted by α. Solve this problem by the Crank-Nicolson method.

Solution

This problem is very similar to the one discussed in Example 9.1. The governing equation is Eq. (9.2). The given initial and boundary conditions may be written for this problem as follows:

$$\text{For } \tau \leqslant 0\text{: } T = T_0 \quad \text{at all } x$$
$$\text{For } \tau > 0\text{: } T = T_s \quad \text{at } x = 0 \qquad \text{and} \qquad \frac{\partial T}{\partial x} = 0 \quad \text{at } x = L \tag{9.2.1}$$

The last condition implies a perfectly insulated surface. The heat transfer at the surface is proportional to $\partial T/\partial x$ and is zero if the temperature gradient is zero. Dimensionless quantities similar to those defined in Eq. (9.1.3) may be employed to obtain the governing dimensionless equation. Thus, the nondimensionalization employed here is

$$X = \frac{x}{L} \qquad \bar{\tau} = \frac{\alpha\tau}{L^2} \qquad \theta = \frac{T - T_0}{T_s - T_0} \tag{9.2.2}$$

and the governing equation is

$$\frac{\partial^2\theta}{\partial X^2} = \frac{\partial\theta}{\partial\bar{\tau}} \tag{9.2.3}$$

The initial and boundary conditions become

$$\theta(X,0) = 0 \qquad \theta(0,\bar{\tau}) = 1 \qquad \frac{\partial\theta}{\partial X}(1,\bar{\tau}) = 0 \tag{9.2.4}$$

The governing equation, with the above initial and boundary conditions, is solved by the Crank-Nicolson iterative scheme. The finite difference equation for this method is obtained for Eq. (9.2.3) as

$$-F\theta_{i+1,j+1} + 2(1 + F)\theta_{i+1,j} - F\theta_{i+1,j-1} = F\theta_{i,j+1} + 2(1 - F)\theta_{i,j} + F\theta_{i,j-1} \tag{9.2.5}$$

where $F = \Delta\bar{\tau}/(\Delta X)^2$, i represents the time step, and j is the spatial grid location. This equation may be written as

$$A\theta_{j-1} + B\theta_j + C\theta_{j+1} = R \tag{9.2.6}$$

where the θ values are at the next time step, $i + 1$, and R is the expression on the right-hand side of Eq. (9.2.5). Therefore, R is a function of the θ values at the present time step i. The constants A, B, and C are the coefficients on the left-hand side of Eq. (9.2.5) and depend on the value of F chosen. No constraints arise in this problem due to stability considerations, and the time step and the grid size are chosen on the basis of desired accuracy. However, as mentioned earlier, bounded oscillations may arise in this method for some problems at large values of F. In most cases, accuracy is the main consideration in the choice of the grid size and the time step.

The FORTRAN 77 computer program for this problem is given in Fig. 9.2.1. An interactive program is written to allow the user to feed in the input parameters. A subroutine FMTDIG is used to generate the tridiagonal matrix from Eq. (9.2.5), which is divided by 2 to simplify the computation. The form of the equation is given by

```
C                CRANK-NICOLSON METHOD
C
C
C   THIS PROGRAM SOLVES 1D, UNSTEADY HEAT EQUATION BY EMPLOYING
C   IMPLICIT CRANK-NICOLSON SCHEME
C
C   THE OUTPUT WILL BE IN CN.DAT
C
C   SUBROUTINE 'FMTDIG' FORMS THE TRIDIAGONAL MATRIX.
C   'TDIG'INVERTS THE MATRIX AND SOLVES FOR TEMPERATURE.
C
C
C   DESCRIPTION OF INPUT PARAMETERS:
C
C       IL      NUMBER OF GRID POINTS.
C       DX      GRID SIZE.
C       DT      TIME STEP.
C       TINT    THE INITIAL CONDITIONS TAKEN AS UNIFORM OVER THE
C               WHOLE DOMAIN.
C       NLIM    THE MAXIMUM NUMBER OF TIME STEPS TAKEN BEFORE STOPPING

C       NSTEP   THE NUMBER OF TIME STEPS AFTER WHICH PRINTOUT OCCURS.
C
C
C   DESCRIPTION OF OTHER VARIABLES USED:
C
C       T       THE SOLUTION VARIABLE AT NTH TIME STEP.
C       TOL     THE SOLUTION VARIABLE AT (N-1)TH TIME STEP.
C
C
        DIMENSION T(50),TOL(50)
        DIMENSION A(50),B(50),C(50),R(50),SOLN(50)
        PRINT*,'ENTER NUMBER OF GRID POINTS, IL='
        READ(1,*)IL
        PRINT*,'ENTER GRID SIZE, DX='
        READ(1,*)DX
        PRINT*,'ENTER TIME STEP,DT='
        READ(1,*)DT
        PRINT*,'ENTER INITIAL CONDITIONS, TAKEN AS UNIFORM OVER THE'
        PRINT*,'WHOLE DOMAIN'
        READ(1,*)TINT
        PRINT*,'ENTER MAXIMUM NO. OF TIME STEPS BEFORE STOPPING'
        READ(1,*)NLIM
        PRINT*,'ENTER NO. OF TIME STEPS AFTER WHICH PRINTOUT OCCURS'
        READ(1,*)NSTEP
C
C   OPEN THE OUTPUT FILE
C
        OPEN(UNIT=10,FILE='CN.DAT')
        WRITE(10,100)DX,DT
  100   FORMAT(/,4X,'DX=',F4.2,2X,'DT=',F4.2)
        WRITE(10,110)IL
  110   FORMAT(/,4X,'IL=',I3,//)
        ISTEP1=0
        ISTEP2=0
        TIME=0.
C
C   SET THE INITIAL CONDITION
```

Figure 9.2.1 Computer program for solving the parabolic PDE of Example 9.2 by the Crank-Nicolson method.

```
C
      DO 10 I=1,IL
      T(I)=TINT
      TOL(I)=TINT
   10 CONTINUE
C
C  SET THE BOUNDARY CONDITIONS
C
      CALL BCOND(T,IL)
      WRITE(10,120)TIME
      WRITE(10,130)(T(I),I=1,IL)
C
C  SOLVE FOR T ON INTERIOR POINTS AT (N+1)TH TIME STEP
C
C  INCREMENT THE ITERATION COUNTERS AND CHECK FOR THE
C  MAXIMUM LIMIT OF ITERATIONS
C
   20 ISTEP1=ISTEP1+1
      ISTEP2=ISTEP2+1
      TIME=TIME+DT
      IF(ISTEP1.GT.NLIM)GO TO 40
C
C  FORM THE TRIDIAGONAL SYSTEM OF EQUATIONS
C
      CALL FMTDIG(DX,DT,IL,T,TOL,A,B,C,R)
      N=IL-2
C
C  INVERT THE TRIDIAGONAL SYSTEM OF EQUATIONS
C
      CALL TDIG(A,B,C,R,SOLN,N)
C
C  SAVE THE OLD SOLUTION
C
      DO 25 I=1,IL
      TOL(I)=T(I)
   25 CONTINUE
      DO 26 I=2,IL-1
      T(I)=SOLN(I)
   26 CONTINUE
C
C  IMPOSE THE BOUNDARY CONDITIONS
C
      CALL BCOND(T,IL)
C
C  OUTPUT THE RESULTS
C
      IF(ISTEP2.EQ.NSTEP)THEN
      WRITE(10,120)TIME
  120 FORMAT(/,1X,'AT T = ',F7.3,1X,'TEMPERATURE FIELD IS:')
      WRITE(10,130)(T(I),I=1,IL)
  130 FORMAT(1X,20(F8.4,2X))
      ISTEP2=0
      GO TO 20
      END IF
      GO TO 20
   40 CLOSE(UNIT=10)
      STOP
      END
C*****************************************************************
```

Figure 9.2.1 Continued

```
      SUBROUTINE FMTDIG(DX,DT,IL,T,TOL,A,B,C,R)
C
C   THIS SUBROUTINE FORMS THE TRIDIAGONAL MATRIX FOR THE
C   CRANK-NICOLSON METHOD. THE GENERIC FORM OF THE EQUATION IS:
C
C       A*T(I-1) + B*T(I) + C*T(I+1) = R
C
C
      DIMENSION T(50),TOL(50),A(50),B(50),C(50),R(50)
      DO 10 I=2,IL-1
      A(I-1)=-DT/(2.*DX**2)
      C(I-1)=-DT/(2.*DX**2)
   10 CONTINUE
      DO 20 I=2,IL-1
      B(I-1)=1.+DT/DX**2
      R(I-1)=TOL(I)+DT*(TOL(I+1)-2.*TOL(I)+TOL(I-1))/(2.*DX**2)
C
C   INCORPORATE THE APPROPRIATE BOUNDARY CONDITIONS:
C
C   LEFT BOUNDARY:
C
      IF(I.EQ.2)R(I-1)=R(I-1)-A(I-1)*T(I-1)
C
C   RIGHT BOUNDARY:
C
      IF(I.EQ.IL-1)THEN
      A(I-1)=A(I-1)-C(I-1)/3.
      B(I-1)=B(I-1)+4.*C(I-1)/3.
      END IF
C
C
   20 CONTINUE
      RETURN
      END
C*******************************************************************
      SUBROUTINE TDIG(A,B,C,R,SOLN,N)
C
C   THIS SUBROUTINE INVERTS A TRIDIAGONAL MATRIX BY THOMAS
C   ALGORITHM.
C
C   SOLUTION IS RETURNED IN THE ARRAY CALLED 'SOLN'.
C
      DIMENSION A(50),B(50),C(50),R(50),SOLN(50),BN(50)
      DO 10 I=1,N
      BN(I)=B(I)
   10 CONTINUE
      DO 20 I=2,N
      D=A(I)/BN(I-1)
      BN(I)=BN(I)-C(I-1)*D
      R(I)=R(I)-R(I-1)*D
   20 CONTINUE
      SOLN(N+1)=R(N)/BN(N)
      DO 30 I=1,N-1
      J=N-I
      SOLN(J+1)=(R(J)-C(J)*SOLN(J+2))/BN(J)
   30 CONTINUE
      RETURN
      END
C*******************************************************
C
```

Figure 9.2.1 Continued

```
      SUBROUTINE BCOND(T,IL)
C
C     THIS SUBROUTINE IMPLEMENTS THE APPROPRIATE BOUNDARY CONDITIONS
C
      DIMENSION T(50)
C
C     LEFT BOUNDARY:
C
      T(1)=1.
C
C     RIGHT BOUNDARY:
C
      T(IL)=4.*T(IL-1)/3.-T(IL-2)/3.
      RETURN
      END
```

Figure 9.2.1 Continued

Eq. (9.2.6). The appropriate boundary conditions, given by Eq. (9.2.4), are also incorporated to obtain the tridiagonal matrix, as discussed earlier in Example 5.2. The total number of grid points is IL, the left boundary being I = 1 and the right one I = IL in this subroutine. For the right boundary, X = 1, the second-order backward difference approximation, given by Fig. 3.6, is used so that the error is $O[(\Delta x)^2]$. Thus,

$$\left(\frac{\partial \theta}{\partial X}\right)_{i,j} = \frac{\theta_{i,j-2} - 4\theta_{i,j-1} + 3\theta_{i,j}}{2\,\Delta X} = 0 \tag{9.2.7}$$

or

$$\theta_{i,j} = \frac{4}{3}\theta_{i,j-1} - \frac{1}{3}\theta_{i,j-2} \tag{9.2.8}$$

where j is replaced by IL for the right boundary. A tridiagonal matrix, with rows from 2 to IL − 1, corresponding to the interior points in the computational domain, is obtained. This matrix is solved by subroutine TDIG to obtain the time-dependent temperature distribution, which then serves as the input for the computation of the distribution at the next time step. The temperatures at the boundaries are given by subroutine BCOND. This process is repeated until a specified time limit, or the steady-state circumstance, is attained.

Figure 9.2.2 shows the computed temperature distributions at different time intervals. The initial temperature is zero throughout the plate, and then at time $\bar{\tau} = 0$, the temperature θ at the left surface, $X = 0$, is raised to 1.0 and held at this value. The right surface, $X = 1$, is insulated. Steady-state conditions are obtained when $\theta = 1.0$ throughout the plate. This figure shows that the temperature distributions approach the steady-state distribution as time elapses. Steady state is attained when time $\bar{\tau}$ reaches a value of around 4.5. This is much larger than the time taken to reach steady state in Example 9.1; see Fig. 9.1.4. However, in the previous example, one surface was maintained at $\theta = 0$, whereas in this example, the entire plate is heated or cooled. This implies a greater transfer of energy in the present case, as compared to the mass transfer in Example 9.1. Figure 9.2.3 shows the variation of the temperature at several locations in the plate with time. The temperatures are found to rise sharply from the

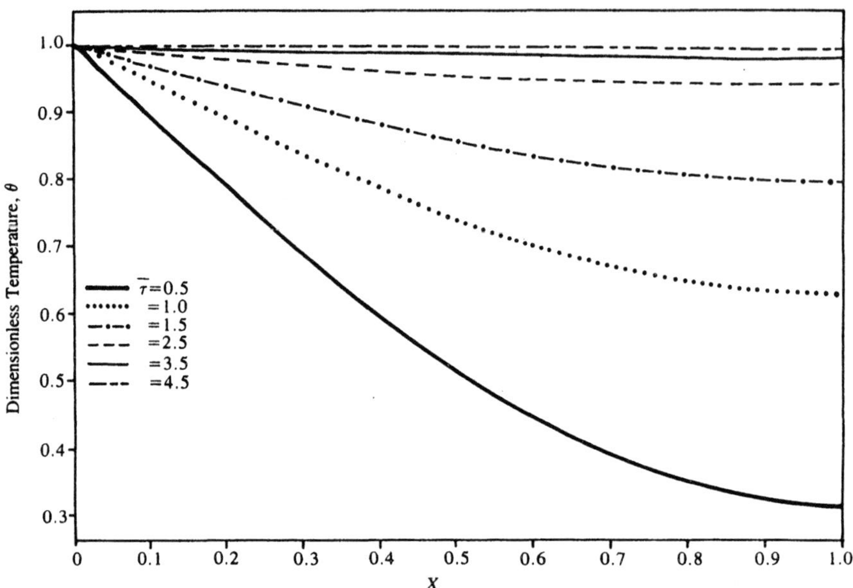

Figure 9.2.2 Computed temperature distribution at various time intervals for Example 9.2. Here, $\Delta\bar{\tau} = 0.05$ and $\Delta X = 0.1$.

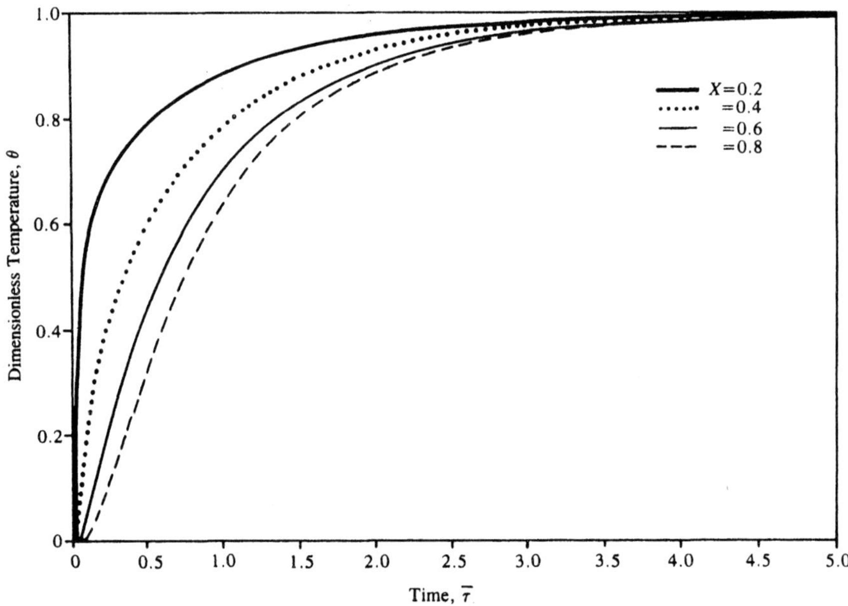

Figure 9.2.3 Variation of the temperature at several locations in the plate with dimensionless time $\bar{\tau}$ for Example 9.2. The approach to steady state is again seen at large time.

initial value of 0 and to approach the steady-state value of 1.0 gradually as time increases.

In both Examples 9.1 and 9.2, we have considered one-dimensional transient mass and heat diffusion problems in order to relate the computational procedure to the physical aspects of such problems. However, the numerical schemes discussed here can easily be extended to other physical circumstances that are governed by parabolic partial differential equations. Such problems arise, for instance, in fluid flow as given by Eq. (9.26), diffusion of moisture in porous media, neutron diffusion in nuclear reactors, and water seepage into the ground.

9.3 ELLIPTIC PARTIAL DIFFERENTIAL EQUATIONS

In an elliptic partial differential equation, a disturbance at a given point propagates in all directions, in contrast to a parabolic PDE in which there is a definite direction for the flow of information. Therefore, the solution domain in an elliptic PDE is an enclosed one, with boundary conditions specified everywhere along the edges of this domain, as shown in Fig. 9.7. The solution at each point is influenced by the solution at every other point in the region where the elliptic PDE applies. Therefore, the numerical solution at the finite number of grid points taken in the region must be obtained simultaneously. This characteristic of elliptic PDEs generally makes the numerical solution much more involved than that for parabolic PDEs, in which a marching procedure may be adopted to advance the solution in a particular direction, say, in the direction of increasing time, starting with the initial conditions. Because of the advantages of such a marching scheme, particularly in numerical stability and convergence characteristics, elliptic equations are frequently formulated as time-dependent parabolic equations, which are solved by time marching to yield the desired solution to the elliptic equations at steady state.

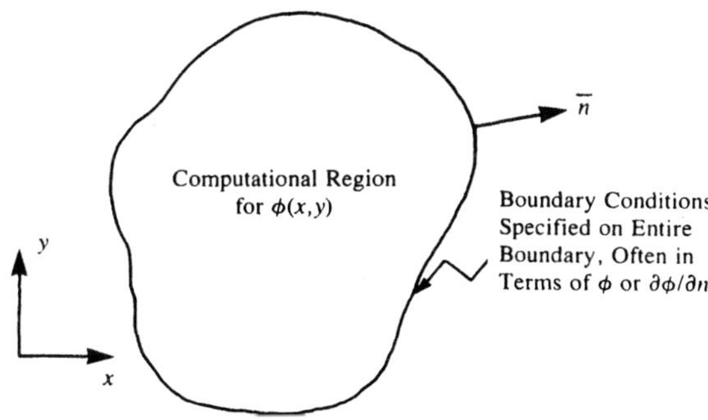

Figure 9.7 Solution domain for an elliptic PDE, along with the necessary boundary conditions.

9.3.1 Finite Difference Approach

Several important physical processes are governed by elliptic partial differential equations. These include conductive and convective heat transfer, mass transfer, the diffusion of neutrons in a nuclear reactor, deflection of a membrane or a plate, interaction of electromagnetic fields, and fluid flow. In order to discuss the numerical techniques for solving elliptic PDEs, let us consider a specific physical problem, namely, that of the two-dimensional steady-state heat conduction in the rectangular region shown in Fig. 9.8. In the absence of heat sources in the region, the temperature $T(x,y)$ is governed by Laplace's equation

$$\frac{\partial^2 T}{\partial x^2} + \frac{\partial^2 T}{\partial y^2} = 0 \tag{9.27}$$

where x and y are the coordinate axes, as indicated in Fig. 9.8. The boundary conditions are given in terms of specified values of the temperatures. Such a problem in which the value of the unknown variable, being temperature in this case, is given at the boundaries is known as a *Dirichlet problem*, and the conditions as *Dirichlet boundary conditions*. If the gradient of the variable is specified instead, the conditions are termed *Neumann boundary conditions*, considered later in this section. If a relationship between the gradient and the value of the variable is given at the boundary, the conditions are known as mixed boundary conditions. The following discussion of the numerical methods for the solution of elliptic PDEs is directed at the above Dirichlet problem. However, most of the methods considered are applicable to elliptic equations in general, as outlined later.

We wish to determine the temperature $T(x,y)$ in the interior of the region shown in Fig. 9.8 by solving the governing elliptic PDE, Eq. (9.27). The boundary conditions, also shown in Fig. 9.8, may be written as follows:

$$\begin{aligned}
&\text{At } x = 0: \ T(x,y) = T_1 \quad &&\text{for } 0 \leqslant y \leqslant H \\
&\text{At } x = L: \ T(x,y) = T_2 \quad &&\text{for } 0 \leqslant y \leqslant H \\
&\text{At } y = 0: \ T(x,y) = T_3 \quad &&\text{for } 0 < x < L \\
&\text{At } y = H: \ T(x,y) = T_4 \quad &&\text{for } 0 < x < L
\end{aligned} \tag{9.28}$$

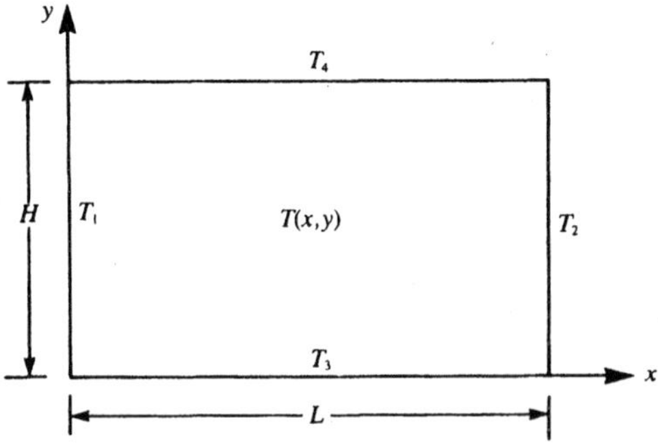

Figure 9.8 Coordinate system for steady-state heat conduction in a rectangular region.

Therefore, the value of the dependent variable $T(x,y)$ is completely specified on the boundaries of the region in which Eq. (9.27) applies. To obtain a numerical solution of the given elliptic equation by finite difference methods, we impose a grid with a mesh size of Δx by Δy on the region, as shown in Fig. 9.9. Then the numerical solution consists of determining the temperatures at the finite number of grid points in the solution domain. As done earlier for parabolic PDEs, the temperature $T(x,y)$ at a grid point (i,j) is denoted by $T_{i,j}$, where

$$x = i\,\Delta x \quad \text{and} \quad y = j\,\Delta y \tag{9.29}$$

Similarly, the temperatures at other grid points are labeled, as shown in Fig. 9.9. If the length L in the x direction is divided into m equal subdivisions, and the height H in the y direction into n equal subdivisions, then

$$\Delta x = \frac{L}{m} \quad \text{and} \quad \Delta y = \frac{H}{n} \tag{9.30}$$

Thus, i varies from 0 to m and j from 0 to n.

We may now proceed to obtain a finite difference approximation to the given elliptic PDE. The second partial derivatives at the grid point (i,j) may be approximated, in central difference form and in terms of the temperatures at the neighboring

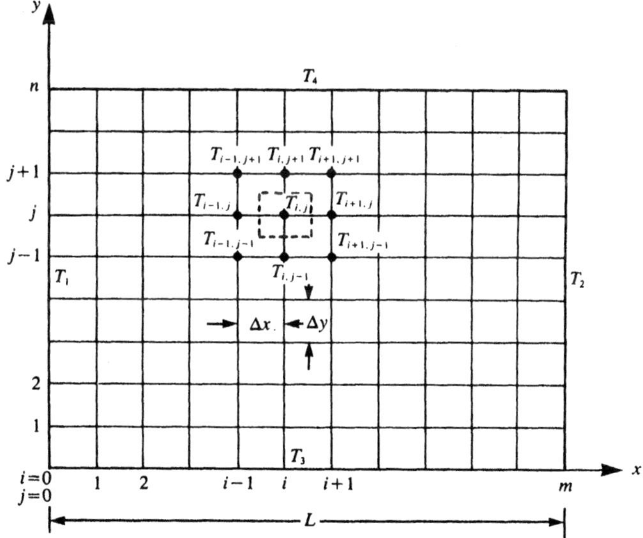

Figure 9.9 Subdivision of the computational region by means of a grid with a mesh size of Δx by Δy in the two directions x and y. The nomenclature for the labeling of the temperatures at the grid, or mesh, points is also indicated.

grid points, as follows:

$$\frac{\partial^2 T}{\partial x^2} = \frac{T_{i+1,j} - 2T_{i,j} + T_{i-1,j}}{(\Delta x)^2} \tag{9.31}$$

$$\frac{\partial^2 T}{\partial y^2} = \frac{T_{i,j+1} - 2T_{i,j} + T_{i,j-1}}{(\Delta y)^2} \tag{9.32}$$

where the truncation error in Eq. (9.31) is $O[(\Delta x)^2]$ and that in Eq. (9.32) is $O[(\Delta y)^2]$, as obtained in Chapter 3. Substituting these finite difference approximations into Eq. (9.27), we obtain

$$\frac{T_{i+1,j} - 2T_{i,j} + T_{i-1,j}}{(\Delta x)^2} + \frac{T_{i,j+1} - 2T_{i,j} + T_{i,j-1}}{(\Delta y)^2} = 0 \tag{9.33}$$

The above finite difference equation can be written at each of the interior points in the computational domain. Therefore, a system of $[(m-1)\cdot(n-1)]$ simultaneous linear equations is obtained. These equations may be solved by the methods discussed in Chapter 5 to obtain the $[(m-1)\cdot(n-1)]$ unknown temperatures at the interior grid points. Frequently, a square mesh, with $\Delta x = \Delta y$, is employed. In this case, Eq. (9.33) may be written as

$$T_{i,j} = \frac{T_{i+1,j} + T_{i-1,j} + T_{i,j+1} + T_{i,j-1}}{4} \tag{9.34}$$

which implies that the temperature at a given grid point is simply an average of the temperatures at the four adjacent grid points. The computational molecule, which indicates the effect of the values at the neighboring grid points on that at a given node, is shown in Fig. 9.10(a) for this second-order approximation of Laplace's equation. We can also obtain higher-order approximations by using a larger number of points in the neighborhood of the grid point being considered; see Fig. 3.7. Figures 9.10(b) and (c) show, for instance, the computational molecules for fourth-order approximations of Laplace's equation. The accuracy of the numerical results can be improved by the use of a higher-order difference method or by a reduction of the mesh size. However, the first approach has problems near the boundaries because of the large number of neighboring points needed for the approximation at a given nodal point. Therefore, grid refinement, with the spacing between the grid points being reduced until the numerical results are essentially independent of the mesh size, is often preferred for improving the accuracy.

As shown by Eq. (9.33), we are faced with the task of solving a large set of linear algebraic equations. If the number of points at which the numerical solution is to be obtained is M in the x direction and N in the y direction, where $M = m - 1$ and $N = n - 1$ for the problem considered above, the number of unknowns is MN. The set of equations to be solved for this problem may be written as

$$AT = B \tag{9.35}$$

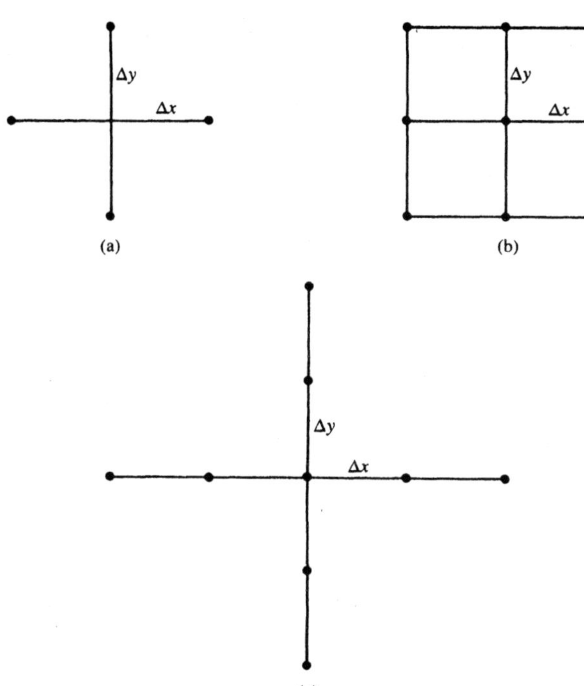

Figure 9.10 Computational molecules for various finite difference schemes for solving Laplace's equation. (a) Second-order approximation; (b) and (c) two different fourth-order approximations.

where the coefficient matrix A is of size $MN \times MN$ and B is a vector whose elements are all zero except for those that arise from the boundary conditions. The unknown temperatures constitute a vector T whose elements are $T_{1,1}, T_{1,2}, \ldots, T_{1,N}, T_{2,1}, \ldots, T_{M,N}$. Then the coefficient matrix A from Eq. (9.34) is of the following form:

$$A = \begin{bmatrix} -4 & 1 & & & 1 & & & \\ 1 & -4 & 1 & & & 1 & & \\ & 1 & -4 & 1 & & & \ddots & \\ 1 & & & & & & & 1 \\ & 1 & & & & & & \\ & & \ddots & & & & & \\ & & & 1 & & 1 & -4 & 1 \\ & & & & 1 & & 1 & -4 \end{bmatrix} \tag{9.36}$$

where only the elements at the diagonal, on either side of it and in the two distant bands shown are nonzero.

Therefore, the coefficient matrix is not tridiagonal but has two additional bands, which are one element wide and are far removed from the main diagonal. In fact, the last nonzero element in the first row and the lowest nonzero element in the first column are both at the $(N + 1)$th position. Since the coefficient matrix is very sparse, although not tridiagonal, iterative methods can be employed advantageously as compared to direct methods for solving this system of equations. The number of equations is generally large, since even for a coarse grid with $M = N = 20$, we have 400 equations. We shall first consider iterative methods for solving the system of linear equations obtained from the finite difference formulation, followed by a discussion of some direct methods that have been developed in recent years.

9.3.2 Numerical Solution by Iterative and Direct Methods

Several iterative methods for solving simultaneous linear equations were discussed in Chapter 5. These included the Jacobi, the Gauss-Seidel, and the successive over-relaxation or under-relaxation methods. It was indicated that diagonal dominance is needed for the convergence of these methods. The finite difference equation, Eq. (9.33), can be written for each grid point. Then the coefficient of $T_{i,j}$ is the largest one in magnitude and its absolute value is equal to the sum of the coefficients of the other terms. The system of equations can be arranged so that the dominant terms appear along the diagonal. As discussed in Section 5.6, the absolute value of the diagonal coefficient must be larger than the sum of the absolute values of the remaining coefficients in each row of the matrix for a diagonally dominant system that is guaranteed to converge. However, the present system of equations has adequate diagonal dominance to converge in most cases. Therefore, for the application of iterative methods, Eq. (9.33) is solved for $T_{i,j}$, which constitutes the diagonally dominant term, to give

$$T_{i,j} = \frac{T_{i+1,j} + T_{i-1,j} + (\Delta x/\Delta y)^2(T_{i,j+1} + T_{i,j-1})}{2[1 + (\Delta x/\Delta y)^2]} \tag{9.37}$$

This equation yields Eq. (9.34) if $\Delta x = \Delta y$.

In the Jacobi iteration method, we start with initial, assumed values of the dependent variable at all the grid points in the computational domain. Using this assumed initial distribution, we obtain the next approximation to the solution from Eq. (9.37) and compare it with the starting solution. If a specified convergence criterion is not satisfied, the computed results are used in Eq. (9.37) to obtain the next iteration. This process is repeated until the given convergence criterion is satisfied. Generally, the convergence criterion demands that the change in the value of the dependent variable from one iteration to the next be less than a prescribed small quantity ε, at each grid point. The computed results from two successive iterations are, therefore, stored, and the values are updated only after the completion of the computation for a given iteration. This numerical scheme is given by the recursive formula

$$T_{i,j}^{(l+1)} = \frac{T_{i+1,j}^{(l)} + T_{i-1,j}^{(l)} + (\Delta x/\Delta y)^2(T_{i,j+1}^{(l)} + T_{i,j-1}^{(l)})}{2[1 + (\Delta x/\Delta y)^2]} \quad \text{for } \begin{array}{l} 1 \leqslant i \leqslant m-1 \\ 1 \leqslant j \leqslant n-1 \end{array} \tag{9.38}$$

where the superscript refers to the number of the iteration. The starting values are denoted by the superscript (0). As discussed in Chapter 5, this method is very inefficient for conventional computers, since the old values of the unknown are replaced by the new ones only after all the values for a given iteration have been computed and since both of the iterative solution vectors must be stored.

A considerable improvement in the computational procedure is obtained by the Gauss-Seidel method, which employs the most recent values of the unknowns in the computation. Generally, a systematic traverse is used, for instance, by increasing i at a given value of j, which is itself increased by 1 after each traverse in the x direction. Then $T_{i-1,j}$ and $T_{i,j-1}$ are calculated before $T_{i,j}$ for a given iteration. Therefore, the iterative scheme for the Gauss-Seidel method is given by

$$T_{i,j}^{(l+1)} = \frac{T_{i+1,j}^{(l)} + T_{i-1,j}^{(l+1)} + (\Delta x/\Delta y)^2(T_{i,j+1}^{(l)} + T_{i,j-1}^{(l+1)})}{2[1 + (\Delta x/\Delta y)^2]} \qquad \text{for } 1 \leqslant i \leqslant m-1, \quad 1 \leqslant j \leqslant n-1 \tag{9.39}$$

The old value of an unknown is replaced by the new value as soon as it is obtained, and, therefore, only one value of each unknown needs to be stored. The programming is also simplified, since we must deal with only one iterative value of the temperature at a given grid point. The method converges if the system is diagonally dominant, which is adequately achieved in the problem being considered. Convergence is generally obtained with even weaker diagonal dominance. The systems of linear equations obtained from the finite difference approximation of the differential equations that arise in engineering problems generally have sufficient diagonal dominance, and iterative methods can be employed satisfactorily.

The convergence of the iterative scheme is given in terms of the change in the computed values from one iteration to the next. If the magnitude of this change, at each grid point, is less than a specified small number ε, which is known as the *convergence parameter*, the scheme is assumed to have converged. This convergence criterion may be given in terms of the absolute or the normalized value of the change in the temperatures. Therefore, the iterative process is terminated if

$$|T_{i,j}^{(l+1)} - T_{i,j}^{(l)}| \leqslant \varepsilon \qquad \text{for } 1 \leqslant i \leqslant m-1, \quad 1 \leqslant j \leqslant n-1$$

or

$$\left|\frac{T_{i,j}^{(l+1)} - T_{i,j}^{(l)}}{T_{i,j}^{(l)}}\right| \leqslant \varepsilon \qquad \text{for } 1 \leqslant i \leqslant m-1, \quad 1 \leqslant j \leqslant n-1 \tag{9.40}$$

The value of ε is taken as small, say, 10^{-4}, and is varied over a few orders of magnitude to ensure that the computed results are independent of its value.

Point Relaxation. The Gauss-Seidel method converges about twice as fast as the Jacobi method, on conventional computing machines with a central processing unit, for a given convergence criterion. The rate of convergence can be improved considerably by the use of the successive over-relaxation (SOR) method, discussed in

Chapter 5. This method is given by the formula

$$T_{i,j}^{(l+1)} = \omega[T_{i,j}^{(l+1)}]_{GS} + (1-\omega)T_{i,j}^{(l)} \tag{9.41}$$

where ω is a constant, known as the *relaxation factor*, and $[T_{i,j}^{(l+1)}]_{GS}$ is the value obtained from the Gauss-Seidel iteration formula, such as Eq. (9.39). For successive over-relaxation, ω lies between 1 and 2. The method diverges for $\omega > 2$; the Gauss-Seidel scheme is obtained for $\omega = 1$; and successive under-relaxation is obtained if $0 < \omega < 1$. Substituting Eq. (9.39) into Eq. (9.41), we find that the SOR method for the problem under consideration is given by

$$T_{i,j}^{(l+1)} = \omega\left[\frac{T_{i+1,j}^{(l)} + T_{i-1,j}^{(l+1)} + (\Delta x/\Delta y)^2(T_{i,j+1}^{(l)} + T_{i,j-1}^{(l+1)})}{2(1+(\Delta x/\Delta y)^2)}\right] + (1-\omega)T_{i,j}^{(l)} \quad \text{for } 1 \leqslant i \leqslant m-1,\ 1 \leqslant j \leqslant n-1 \tag{9.42}$$

There is an optimum value of the relaxation factor, ω_{opt}, at which convergence is the fastest. For a square region, with $n = m$, the Gauss-Seidel method converges about twice as fast as the Jacobi method and the SOR method, at the optimum value of the relaxation factor, 6 and 19 times faster, respectively, than the Gauss-Seidel method for $n = 10$ and $n = 30$ (Jaluria and Torrance, 1986). Therefore, if ω_{opt} is known, the SOR method is extremely efficient. However, ω_{opt} varies with the PDE, the boundary conditions, the grid spacing, the geometry of the computational domain, and so on. It is not known in most cases, and the analytical determination of its value is very involved. Therefore, one generally determines it by solving the problem at different values of ω to obtain the optimum or by employing the information available on other similar problems. If several problems of a particular type are to be solved, it would be worthwhile to spend the effort and time to determine ω_{opt}.

The rate of convergence is quite sensitive to the value of ω, and, for a value far from the optimum value, the convergence rate is close to that for the Gauss-Seidel method. For some simple cases, ω_{opt} may be obtained analytically. For Laplace's equation in a rectangular region with Dirichlet conditions (see Fig. 9.8), the optimum value is given by

$$\omega_{opt} = \frac{2}{1 + (1-a^2)^{1/2}} \tag{9.43a}$$

where

$$a = \frac{1}{1+(\Delta x/\Delta y)^2}\left[\cos\frac{\pi}{m} + \left(\frac{\Delta x}{\Delta y}\right)^2 \cos\frac{\pi}{n}\right] \tag{9.43b}$$

Therefore, for $m = n = 20$, $\omega_{opt} = 1.7295$; and for $m = n = 30$, it is 1.8107, with $\Delta x = \Delta y$ in these cases. Figure 9.11 shows the dependence of the number of iterations, for convergence, on ω for a rectangular region. The need to employ a value close to the optimum is clear. It may also be mentioned here that successive under-relaxation is generally used to improve the convergence characteristics of the iterative process, particularly for nonlinear equations which may diverge when Gauss-Seidel iteration is applied.

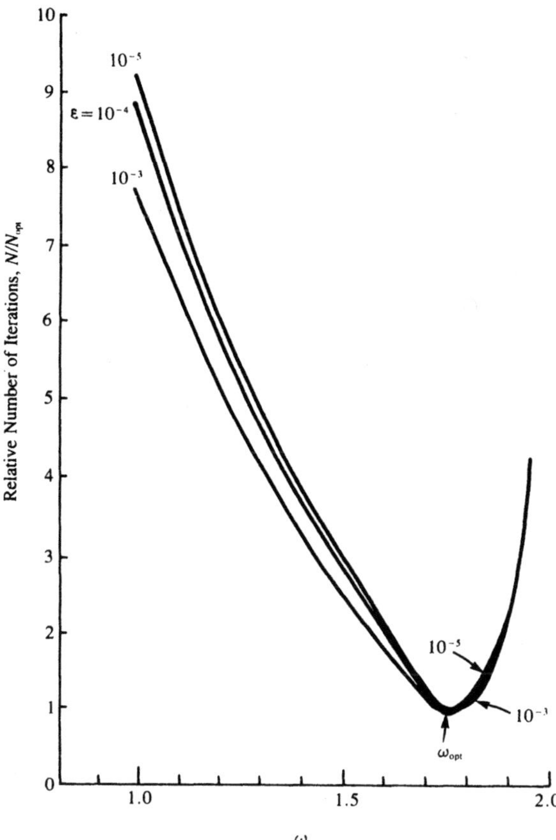

Figure 9.11 Variation of the number of iterations, for convergence of the second-order finite difference scheme for Laplace's equation in a square region, with the relaxation factor ω, for the successive over-relaxation method. Note the strong dependence on ω and the considerable reduction in number of iterations as ω varies from 1.0 (Gauss-Seidel) to the optimum value ω_{opt}. Here, the number of mesh lengths in either direction is 20.

Direct Methods. Several direct methods, based on elimination, were discussed in Chapter 5. Among the most important of these are the Gaussian elimination and the matrix decomposition methods. Many other methods, such as Gauss-Jordan and matrix inversion methods, are based on Gaussian elimination. For a tridiagonal matrix system, Gaussian elimination may be used very effectively, as demonstrated in Example 5.2. In this case, the number of arithmetic operations required are of order n, instead of n^3 for a general system of n equations. In the triangular decomposition method, such as Crout's method, the coefficient matrix A is factored into lower and upper triangular matrices, each of which may be solved by forward and backward substitution. However, except for tridiagonal systems, these direct methods are often not as efficient as the iterative methods, such as the optimized SOR method, and also give rise to much larger round-off errors. Therefore, iterative methods are frequently used for solving the large systems of algebraic equations obtained from the finite difference approximation of elliptic partial differential equations. Nonlinear algebraic equations are obtained if the elliptic PDE is nonlinear. In such cases, iteration is generally necessary for the solution of the equations, and iterative methods, such as the Gauss-Seidel and relaxation methods, are particularly appropriate.

Recently, specialized direct methods for solving finite difference approximations of the Poisson and Laplace equations in simple geometries have been developed. These methods include the cyclic reduction and the fast Fourier transform methods, which are emerging as the most efficient means for solving these equations in simple, two-dimensional regions for Dirichlet or Neumann boundary conditions. A discussion of these methods is beyond the scope of this book. Further details and references may be obtained from Ferziger (1981) and Jaluria and Torrance (1986). Generally, available computer software is used for the application of these methods, since the algorithms tend to be very involved.

9.3.3 Other Methods

A very efficient iterative method for solving elliptic partial differential equations is the alternating direction implicit (ADI) method, which gives rise to a tridiagonal set in each iterative step. The method employs the unknown values of the dependent variable from the current iteration along one direction and known values from the previous iteration along the other direction. In the next step, these directions are reversed. An acceleration parameter $\tilde{\omega}$, similar to that in the SOR method, is used to improve the rate of convergence. The recursion formulas for two iterative steps are

$$\frac{T_{i+1,j}^{(l+1)} - (2+\tilde{\omega})T_{i,j}^{(l+1)} + T_{i-1,j}^{(l+1)}}{(\Delta x)^2} + \frac{T_{i,j+1}^{(l)} - (2-\tilde{\omega})T_{i,j}^{(l)} + T_{i,j-1}^{(l)}}{(\Delta y)^2} = 0 \tag{9.44a}$$

$$\frac{T_{i+1,j}^{(l+1)} - (2-\tilde{\omega})T_{i,j}^{(l+1)} + T_{i-1,j}^{(l+1)}}{(\Delta x)^2} + \frac{T_{i,j+1}^{(l+2)} - (2+\tilde{\omega})T_{i,j}^{(l+2)} + T_{i,j-1}^{(l+2)}}{(\Delta y)^2} = 0 \tag{9.44b}$$

These two steps are generally considered to constitute one complete iteration. The tridiagonal sets obtained are solved by Gaussian elimination, and the iteration is repeated until convergence is attained. This method, developed by Peaceman and Rachford (1955), is used extensively for two-dimensional steady-state diffusion problems, governed by elliptic equations, and transient problems, as outlined below.

In several cases, particularly in nonlinear problems, the elliptic PDE is solved by considering an equivalent time-dependent problem, which is parabolic in time. Laplace's equation may, for instance, be solved by obtaining the transient solution of the equation

$$\frac{\partial T}{\partial \tau} = \frac{\partial^2 T}{\partial x^2} + \frac{\partial^2 T}{\partial y^2} \tag{9.45}$$

where the steady-state solution at large time is the required solution of the elliptic PDE. One could use time marching to solve this problem, using the various techniques outlined in the preceding section. The ADI method may be employed without iteration for this problem and is one of the most efficient methods for such two-dimensional transient problems. The corresponding finite difference equations, with the superscripts denoting the time step, are as follows:

$$\frac{T_{i,j}^{(l+1)} - T_{i,j}^{(l)}}{\Delta\tau} = \frac{T_{i+1,j}^{(l+1)} - 2T_{i,j}^{(l+1)} + T_{i-1,j}^{(l+1)}}{(\Delta x)^2} + \frac{T_{i,j+1}^{(l)} - 2T_{i,j}^{(l)} + T_{i,j-1}^{(l)}}{(\Delta y)^2} \tag{9.46a}$$

$$\frac{T_{i,j}^{(l+2)} - T_{i,j}^{(l+1)}}{\Delta\tau} = \frac{T_{i+1,j}^{(l+1)} - 2T_{i,j}^{(l+1)} + T_{i-1,j}^{(l+1)}}{(\Delta x)^2} + \frac{T_{i,j+1}^{(l+2)} - 2T_{i,j}^{(l+2)} + T_{i,j-1}^{(l+2)}}{(\Delta y)^2} \tag{9.46b}$$

This approach for solving an elliptic PDE is frequently employed for nonlinear equations, such as those encountered in fluid mechanics and in heat transfer. The main advantage is that time marching generally yields better stability and convergence characteristics.

9.3.4 Other Geometries and Boundary Conditions

We have considered simple rectangular regions and Dirichlet boundary conditions in the above discussion. However, there are many complexities that arise due to irregularly shaped regions and more involved boundary conditions. Since these considerations are particularly important in real physical problems, a brief discussion of these aspects is included here.

Consider a boundary at $x = 0$, as shown in Fig. 9.12 with a distribution of grid points. If the boundary condition is of Neumann type, that is $\partial T/\partial x = B$, where B is a constant, the value $T_{0,j}$ at the boundary is not known and must be obtained from a finite difference approximation of the derivative in terms of the neighboring grid points. The simplest formulation is

$$\frac{\partial T}{\partial x} \simeq \frac{T_{1,j} - T_{0,j}}{\Delta x} = B \tag{9.47}$$

which employs a forward difference in x and is accurate only to order Δx. We can derive a more accurate approximation by employing the Taylor series expansions for

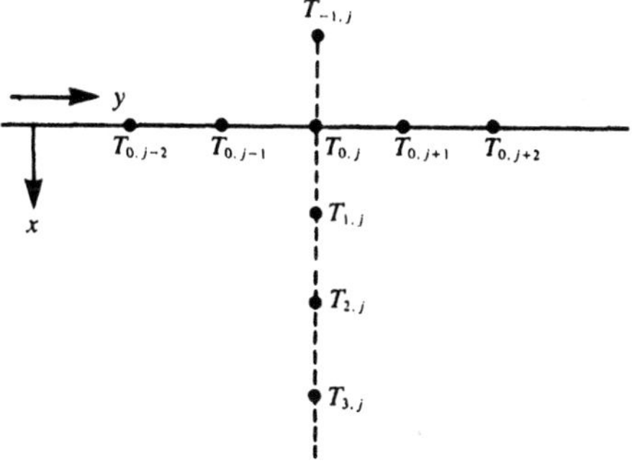

Figure 9.12 Distribution of grid points at a boundary, showing the fictitious point $T_{-1,j}$ outside the region.

the three points adjacent to the boundary, as discussed in Chapter 3. This gives

$$\frac{\partial T}{\partial x} \simeq \frac{-T_{2,j} + 4T_{1,j} - 3T_{0,j}}{2(\Delta x)} = B \tag{9.48}$$

This formulation has a truncation error of order $(\Delta x)^2$ and may be written for all the points on the boundary. Therefore, an equation for $T_{0,j}$, in terms of the values at the neighboring grid points, is obtained. Similar equations may be written at other boundaries.

Another approach is to employ a fictitious point $T_{-1,j}$ outside the boundary, as shown in Fig. 9.12, and write $\partial T/\partial x$ in the central difference approximation; that is,

$$\frac{\partial T}{\partial x} \simeq \frac{T_{i,j} - T_{-i,j}}{2\,\Delta x} \tag{9.49}$$

Then $T_{-1,j}$ is eliminated between this equation and the finite difference equation, of the given PDE, written for the grid point $(0, j)$ at the boundary. This also gives an error of $O[(\Delta x)^2]$. However, a row of unknowns at points outside the boundary is introduced, increasing the computational effort. Similarly, other, more involved, boundary conditions may be treated. The resulting equations for the surface grid points are used along with the equations for the interior region to obtain the solution.

An approach frequently employed in fluid flow and in heat and mass transfer is based on the mass, momentum, and energy balance equations for the finite regions represented by the surface grid point. For instance, a rectangular region of dimensions $\Delta y \times (\Delta x/2)$ may be placed symmetrically surrounding the grid point with temperature $T_{0,j}$. Then finite difference equations are written to balance the mass, momentum, and energy transported across the boundaries of this finite region against those stored in the region. This approach gives an accurate and physically representative equation for the dependent variable at the surface node. The equation will also be consistent with basic physical or chemical laws governing the transport processes under consideration. This approach is the preferred one for most transport phenomena of interest in engineering applications. For further details, see Jaluria and Torrance (1986).

Frequently, we are faced with an irregular region, and it becomes necessary to obtain an equation applicable to an interior grid point that lies near such a boundary. Consider a point C in a square mesh, as shown in Fig. 9.13, with points A and B at the boundary. Since these points A and B are at a distance $\beta_1\,\Delta x$ and $\beta_2\,\Delta y$, respectively, away from C, where β_1 and β_2 are constants that are both less than 1.0, the finite difference equation, such as Eq. (9.34), derived for the interior region does not apply at C. One method of determining the value at C is to use interpolation between the points A and 3, or B and 4. An average of these interpolations may also be employed to represent the value at C. Another method is to employ Taylor series expansions to derive the finite difference approximations for the derivatives at the point C in terms of the values at the points A, B, C, 3, and 4. These are then substituted in the given PDE to yield the equation that applies at C. For instance, the finite difference approxi-

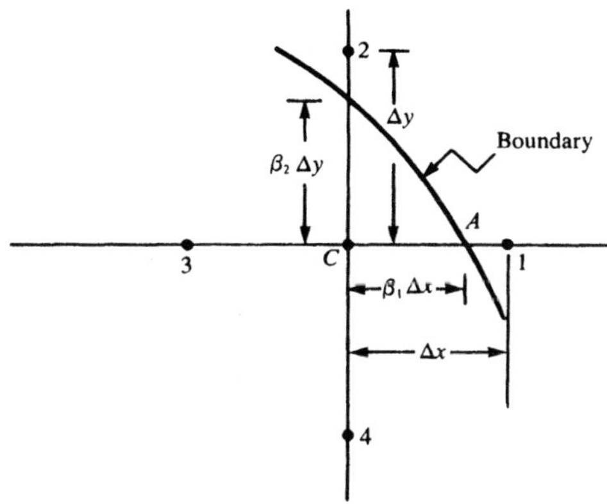

Figure 9.13 Grid points of a rectangular mesh near an irregular boundary.

mation for Laplace's equation at point C is obtained by this method as follows:

$$\frac{2T_A}{\beta_1(1+\beta_1)} + \frac{2T_B}{\beta_2(1+\beta_2)} + \frac{2T_3}{1+\beta_1} + \frac{2T_4}{1+\beta_2} - \left(\frac{2}{\beta_1} + \frac{2}{\beta_2}\right) T_C = 0 \qquad (9.50)$$

Therefore, T_C may be obtained in terms of the values at the boundary and the interior points. This applies for Dirichlet conditions. For further details and for other boundary conditions, see Forsythe and Wasow (1960) and Smith (1968).

We conclude this discussion on the finite difference solution of elliptic partial differential equations by repeating that a finite difference approximation is obtained for the given PDE, employing a chosen grid in the computational region, to yield a system of algebraic equations. For Dirichlet conditions, the values at the boundary grid points is known. For other boundary conditions, algebraic equations are obtained that relate the values at the boundaries with those at the interior grid points. The resulting system of equations may be solved by direct or iterative methods to yield the desired solution. Iterative methods are more frequently used because of the large number of equations that arise and the simplicity in programming. Iteration is usually necessary for nonlinear equations. Time marching may be employed in some cases, and specialized direct methods are also available for some simple problems. Examples 9.3 and 9.4 discuss the Gauss-Seidel and the SOR methods, respectively, for solving elliptic partial differential equations.

9.3.5 Finite Element and Other Solution Methods*

In the preceding sections, we considered the solution of parabolic and elliptic partial differential equations by means of finite difference approximations, which are applied to the governing differential equations. However, over the last decade, other methods, particularly the finite element method, have gained in popularity for practical problems in engineering. Finite difference methods are simpler to comprehend, and it is easier to develop computer programs for them. They are still widely used for engineering problems because of this ease in programming, and, therefore, we discussed them in detail here. However, practical circumstances often involve complexities, such as complicated geometries, boundary conditions, and material property variations. In such cases, the finite element approach provides a very versatile method that can be employed for a wide range of engineering problems. Frequently, available software is used, since the development of the computer program is generally involved and time-consuming.

The finite element method is based on the integral formulation of the conservation principles. The computational region is divided into a number of finite elements, several forms and types of which are available for different geometries and governing equations. Triangular elements for two-dimensional problems and tetrahedral elements for three-dimensional problems are commonly employed, as shown in Fig. 9.14. The variation of the dependent variable is generally taken in terms of a polynomial and frequently as linear within the elements. Integral equations that apply for each element are derived, and the conservation postulates are satisfied by minimization of the integrals or by reducing their weighted residuals to zero. The latter gives rise to a commonly used method known as *Galerkin's method.* Thus, the distribution of the dependent variable within the elements, and then in the entire region, is obtained. As mentioned above, the method is particularly suitable for irregular boundaries and complicated boundary conditions. Consequently, it is finding a growing interest and application with respect to practical problems in engineering. For details on finite element methods, see the books by Mitchell and Wait (1977) and Huebner and Thornton (1983).

Two other approaches have gained in importance in recent years. These are the boundary element method and the control volume approach. The former is similar to the finite element method, except that the integral formulation for the computational domain is transformed to one that applies for the bounding surface.

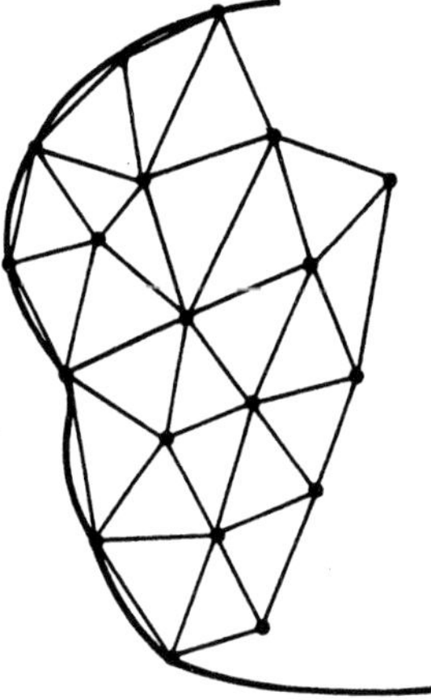

Figure 9.14 Finite element discretization, employing triangular elements.

Although somewhat limited in its applicability, this method is finding much interest for many problems of practical interest, particularly for those where the phenomena at the surface are of main concern. The method has the advantage, over finite element methods, of a smaller number of elements and unknowns. See the books by Brebbia (1977) and Banerjee and Butterfield (1981) on the background and application of this method.

The control volume approach is also based on the integral formulation. The physical region is divided into a set of nonoverlapping control volumes, such as those obtained by drawing lines parallel to the coordinate axes midway between the nodes; see the dashed lines in Fig. 9.9. The integral conservation statement is applied to each control volume, using interpolation between the node points to approximate the integrands. Thus, the volume and surface integrals are approximated, using values at the nodes. The resulting algebraic equations are similar to those obtained from the finite difference approach, which is based on the differential equations. However, the finite volume method satisfies the conservation principles more accurately and is particularly valuable for the numerical formulation of the boundary conditions. See Jaluria and Torrance (1986) for details on this method.

Example 9.3

The transverse deflection ϕ of a flexible membrane, which cannot resist any bending, is governed by the Poisson equation

$$\frac{\partial^2\phi}{\partial x^2} + \frac{\partial^2\phi}{\partial y^2} = -\frac{p}{T} \tag{9.3.1}$$

where p is the pressure on the membrane and T is the tension per unit length at the edges. For small deflections, T may be assumed to be constant. A square membrane, of 1.0 m side, is fixed at its boundaries and is subjected to a pressure of 4×10^7 N/m^2; see Fig. 9.3.1. The tension T is 10^8 N/m. Employing the Gauss-Seidel method, compute the variation of the deflection ϕ across the membrane. Take $\Delta x = \Delta y = 0.1$ m.

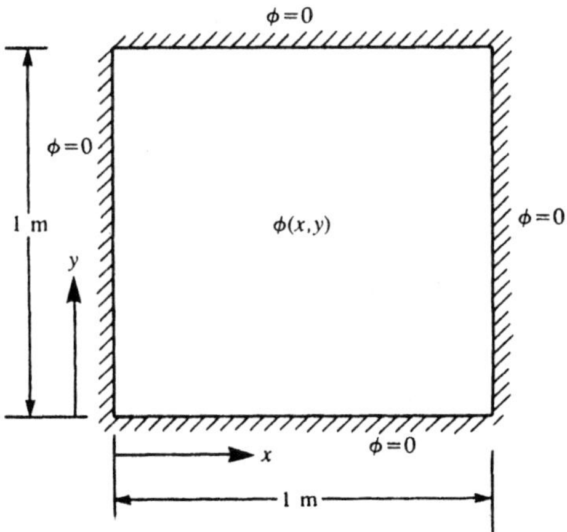

Figure 9.3.1 Coordinate system for computing the deflection $\phi(x,y)$ of a square membrane, as considered in Example 9.3.

Solution

The elliptic partial differential equation to be solved numerically is

$$\frac{\partial^2 \phi}{\partial x^2} + \frac{\partial^2 \phi}{\partial y^2} = -0.4 \tag{9.3.2}$$

with the boundary conditions

$$\begin{aligned} &\text{at } x = 0 \text{ and } x = 1.0 \text{ m: } \phi = 0 \qquad \text{for } 0 \leqslant y \leqslant 1.0 \text{ m} \\ &\text{At } y = 0 \text{ and } x = 1.0 \text{ m: } \phi = 0 \qquad \text{for } 0 < x < 1.0 \text{ m} \end{aligned} \tag{9.3.3}$$

where ϕ is also in meters. The origin is taken at the lower left corner of the membrane and x and y are along two sides of the square, as shown in Fig. 9.3.1. Taking the grid spacing to be equal in both directions, that is, $\Delta x = \Delta y$, we obtain the finite difference approximation of the governing equation as follows:

$$\phi_{i,j} = \frac{\phi_{i+1,j} + \phi_{i-1,j} + \phi_{i,j+1} + \phi_{i,j-1}}{4} + 0.4(\Delta x)^2/4 \tag{9.3.4}$$

where $x = i\,\Delta x$ and $y = i\,\Delta y$, as shown in Fig. 9.9. Also, for $\Delta x = 0.1$, the last term becomes 0.001.

The iterative scheme for the Gauss-Seidel method is obtained from Eq. (9.39) for this problem as follows:

$$\phi_{i,j}^{(l+1)} = \frac{[\phi_{i+1,j}^{(l)} + \phi_{i,j+1}^{(l)} + \phi_{i-1,j}^{(l+1)} + \phi_{i,j-1}^{(l+1)}]}{4} + 0.001 \qquad \text{for } \begin{matrix} 1 \leqslant i \leqslant n-1 \\ 1 \leqslant j \leqslant n-1 \end{matrix} \tag{9.3.5}$$

where n is the number of subdivisions in each of the two directions. Figure 9.3.2 gives the FORTRAN program for this problem. An initially uniform ϕ distribution is assumed in the computational domain, and Eq. (9.3.5) is employed to compute the values for the next iteration. Only the most recent value of ϕ at any given grid point is stored, so that the values are updated as soon as they are computed. The program allows an interactive input of parameters such as grid size; dimensions of the computational region, that is, number of grid points in each direction; initial uniform value of ϕ in the domain; and convergence criterion. The grid size and the number of grid points in the x and y directions may be taken as different, if the computational domain is not a square or if the boundary conditions are nonsymmetric. Note that the constant 0.001 in Eq. (9.3.5) must be replaced, in the program, by $0.4/[2/(\Delta x)^2 + 2/(\Delta y)^2]$ for arbitrary Δx and Δy.

In the given program, the boundary conditions are implemented in the numerical scheme by means of the subroutine BCOND. The computation is terminated if the number of iterations exceeds a specified limit or if the following convergence criterion is satisfied:

$$|\phi_{i,j}^{(l+1)} - \phi_{i,j}^{(l)}| \leqslant \varepsilon \qquad \text{for } 1 \leqslant i \leqslant n-1 \quad \text{and} \quad 1 \leqslant j \leqslant n-1 \tag{9.3.6}$$

where ε is the convergence parameter, denoted by EPSI in the program. Figure 9.3.3 shows the variation of ϕ with x at different values of y. Because of the symmetry of the given problem, a similar plot of ϕ versus y is obtained at the corresponding values of

```
C             GAUSS-SEIDEL METHOD FOR AN ELLIPTIC EQUATION
C
C       THIS PROGRAM SOLVES THE POISSON EQUATION BY EMPLOYING
C       THE GAUSS-SEIDEL ITERATIVE METHOD.
C
C       WHEN THE PROGRAM IS RUN, IT PROMPTS FOR THE INPUT VALUES
C       REQUIRED. ENTER THE INPUT VALUES AND THE OUTPUT WILL BE
C       STORED IN A FILE CALLED 'GS.DAT'
C
C
C        DESCRIPTION OF INPUT PARAMETERS:
C
C        IL      IS THE NUMBER OF GRID POINTS IN THE X DIRECTION.
C        JL      IS THE NUMBER OF GRID POINTS IN THE Y DIRECTION.
C        DX      IS THE GRID SIZE IN X DIRECTION.
C        DY      IS THE GRID SIZE IN Y DIRECTION.
C        PHIINT  IS THE INITIAL GUESS FOR PHI, TAKEN AS UNIFORM OVER
C                THE WHOLE DOMAIN.
C        MAXITERATION IS THE MAXIMUM NUMBER OF ITERATIONS BEFORE
C                TERMINATING THE COMPUTATION.
C        EPSI    IS THE CONVERGENCE CRITERION.
C
C
C        DESCRIPTION OF OTHER VARIABLES:
C
C        PHI     IS THE SOLUTION AT THE NTH TIME STEP.
C        PHIOL   IS THE SOLUTION AT THE (N-1)TH TIME STEP.
C
C
C        ENTER INPUT PARAMETERS
C
         CHARACTER*2 XFILE(5)
         CHARACTER*2 YFILE(5)
         DIMENSION PHI(11,11),PHIOL(11,11)
         PRINT*,'ENTER INITIAL GUESS FOR PHI, TAKEN AS UNIFORM OVER'
         PRINT*,'THE WHOLE DOMAIN'
         READ(1,*)PHIINT
         PRINT*,'ENTER GRID SIZE DX=, DY='
         READ(1,*)DX,DY
         PRINT *,'ENTER NO. OF GRID POINTS IL=, JL='
         PRINT*,' MAXIMUM NO. POSSIBLE IS 11 FOR BOTH IL AND JL'
         PRINT*,'UNLESS DIMENSION STATEMENT IS CHANGED'
         READ(1,*)IL,JL
         PRINT*,'ENTER MAXIMUM  NO. OF ITERATIONS TO BE CARRIED'
         PRINT*,'OUT BEFORE STOPPING'
         READ(1,*)MAXITERATION
         PRINT *,'ENTER CONVERGENCE CRITERION'
         READ(1,*)EPSI
         PRINT*,'THE INPUT VALUES ARE:'
         PRINT*,'INITIAL GUESS FOR PHI=',PHIINT
         PRINT*,'DX=',DX,'DY=',DY
         PRINT*,'IL=',IL,'JL=',JL
         PRINT*,'MAX. NO. OF ITERATIONS=',MAXITERATION
         PRINT*,'CONVERGENCE CRITERION=',EPSI
         ITERATION=0
C
C        OPEN THE DATA FILES FOR GRAPHICS
C
```

Figure 9.3.2 Computer program for solving the Poisson elliptic PDE considered in Example 9.3 by the Gauss-Seidel iterative method.

```
      XFILE(1)='X1'
      XFILE(2)='X2'
      XFILE(3)='X3'
      XFILE(4)='X4'
      XFILE(5)='X5'
      YFILE(1)='Y1'
      YFILE(2)='Y2'
      YFILE(3)='Y3'
      YFILE(4)='Y4'
      YFILE(5)='Y5'
C
C     SET INITIAL DISTRIBUTION OF PHI
C
      DO 5 I=1,IL
      DO 5 J=1,JL
      PHI(I,J)=PHIINT
    5 CONTINUE
C
C     START SOLVING FOR PHI
C
   15 ITERATION=ITERATION+1
      IF(ITERATION.GE.MAXITERATION)GO TO 40
C
C     SAVE THE FIELD AT PREVIOUS TIME STEP
C
      DO 10 I=1,IL
      DO 10 J=1,JL
      PHIOL(I,J)=PHI(I,J)
   10 CONTINUE
C
C     EMPLOY GAUSS-SIEDEL ITERATION FOR PHI AT INTERIOR POINTS
C
      DO 20 J=2,JL-1
      DO 20 I=2,IL-1
      PHIGS=(PHI(I+1,J)+PHI(I-1,J))/DX**2+(PHI(I,J+1)+PHI(I,J-1))
     $ /DY**2
      PHIGS=PHIGS/(2./DX**2+2./DY**2)
      PHI(I,J)=PHIGS+0.001
   20 CONTINUE
C
C     IMPOSE THE BOUNDARY CONDITIONS
C
      CALL BCOND(PHI,IL,JL)
C
C     CHECK FOR CONVERGENCE
C
      DO 35 I=1,IL
      DO 35 J=1,JL
      IF(ABS(PHI(I,J)-PHIOL(I,J)).GE.EPSI)GO TO 15
   35 CONTINUE
      GO TO 50
   40 PRINT*,'SOLUTION DOES NOT CONVERGE'
   50 OPEN(UNIT=10,FILE='GS.DAT')
      WRITE(10,110)EPSI
  110 FORMAT(1X,'CONVERGENCE CRITERION ='1X,E9.1)
      WRITE(10,120)ITERATION
  120 FORMAT(//,1X,'NO. OF ITERATIONS TO CONVERGE=',1X,I4,//)
      WRITE(10,130)
  130 FORMAT(1X,'PHI DISTRIBUTION IS:',//)
      WRITE(10,140)(I,I=1,IL)
```

Figure 9.3.2 Continued

```
  140     FORMAT(1X,'I=',8X,11(I2,8X))
          DO 60 J=1,JL
          WRITE(10,100)J,(PHI(I,J),I=1,IL)
   60     CONTINUE
  100     FORMAT(1X,'J=',I2,3X,11(F8.5,2X))
C
C         OUTPUT FOR GRAPHICS
C
          II=1
          DO 70 I=1,5
          II=II+1
          OPEN (UNIT=12,FILE=XFILE(I))
          DO 66 J=1,JL
          WRITE(12,*)PHI(II,J)
   66     CONTINUE
          CLOSE(UNIT=12)
   70     CONTINUE
          JJ=1
          DO 71 J=1,5
          JJ=JJ+1
          OPEN(UNIT=12,FILE=YFILE(J))
          DO 72 I=1,IL
          WRITE(12,*)PHI(I,JJ)
   72     CONTINUE
          CLOSE(UNIT=12)
   71     CONTINUE
          OPEN(UNIT=12,FILE='XX')
          DO 73 I=1,IL
          XX=FLOAT(I-1)*DX
          WRITE(12,*)XX
   73     CONTINUE
          CLOSE(UNIT=12)
          OPEN(UNIT=12,FILE='YY')
          DO 74 J=1,JL
          YY=FLOAT(J-1)*DY
          WRITE(12,*)YY
   74     CONTINUE
          CLOSE(UNIT=12)
          STOP
          END
C********************************************************
          SUBROUTINE BCOND(PHI,IL,JL)
C
C         THIS SUBROUTINE IMPLEMENTS THE APPROPRIATE BOUNDARY
C         CONDITIONS.
C
          DIMENSION PHI(11,11)
C
C         SET THE CONDITIONS AT I=1 AND I=IL SURFACES
C
          DO 25 J=1,JL
          PHI(1,J)=0.
          PHI(IL,J)=0.
   25     CONTINUE
C
C         SET THE CONDITIONS AT J=1 AND J=JL SURFACES
C
          DO 30 I=1,IL
          PHI(I,1)=0.
          PHI(I,JL)=0.
   30     CONTINUE
          RETURN
          END
```

Figure 9.3.2 Continued

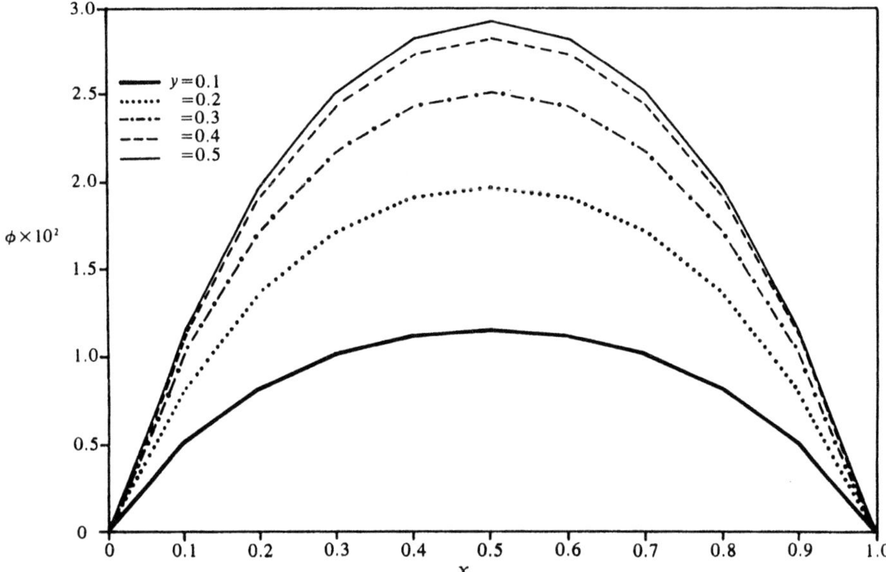

Figure 9.3.3 Computed distributions of the deflection $\phi(x,y)$ at various values of the coordinate distance y, for Example 9.3. The grid spacing $\Delta x = \Delta y = 0.1$ m.

x. Note in this figure that, as imposed by the boundary conditions, ϕ is zero at the boundaries and is maximum midway between the boundaries. Thus, the maximum deflection is at the center of the square region and is around 2.9 cm for the given values of the physical variables. The deflection ϕ increases as y increases from 0 at the boundary to 0.5 m at the midway point and then decreases toward the far boundary at $y = 1.0$ m. Because of symmetry, one could also consider only one-fourth of the membrane, employing the zero slope conditions of $\partial\phi/\partial x = 0$ at $x = 0.5$ m and $\partial\phi/\partial y = 0$ at $y = 0.5$ m.

The given program can also be used for the solution of other engineering problems that are governed by Poisson's equation. Examples of these are steady-state, two-dimensional conduction in a long bar with a heat source, mass diffusion in a porous medium with a mass source due to chemical reaction, and neutron diffusion in a material with a neutron source or sink.

Example 9.4

(a) The temperature in a long bar of square cross section is governed by the Laplace equation. The temperature at one surface is T_1, while the other three surfaces are maintained at temperature T_2; see Fig. 9.4.1. Using the successive over-relaxation method, compute the temperature distribution in the bar. Also determine the optimum value of the relaxation factor ω.

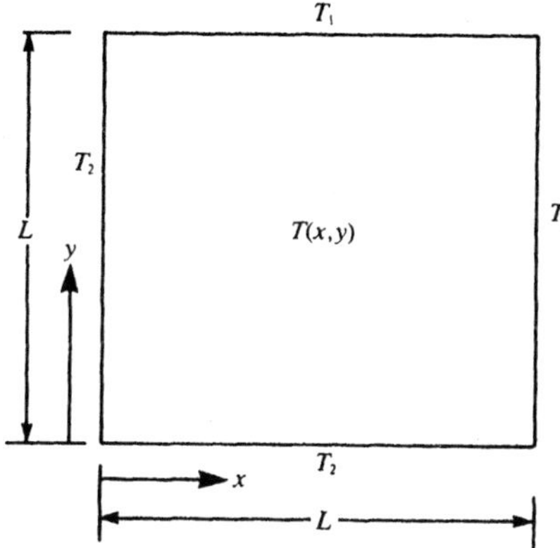

Figure 9.4.1 Physical problem considered in Example 9.4(a), along with the coordinate system.

(b) Using the program developed in Part (a), solve the equation $\nabla^2\psi = 0$, with the boundary conditions shown in Fig. 9.4.2. This equation governs the flow of a fluid in the absence of viscous, or frictional, and rotational effects. Here, ψ is known as the *dimensionless stream function.* It is related to the flow rate and, hence, to the flow field. Compute the ψ distribution in the flow region due to the inflow and outflow as shown in Fig. 9.4.2, and obtain streamlines, or contours of constant ψ, corresponding to $\psi = 0, 0.05, 0.1, 0.25, 0.5, 0.75$, and 1.0.

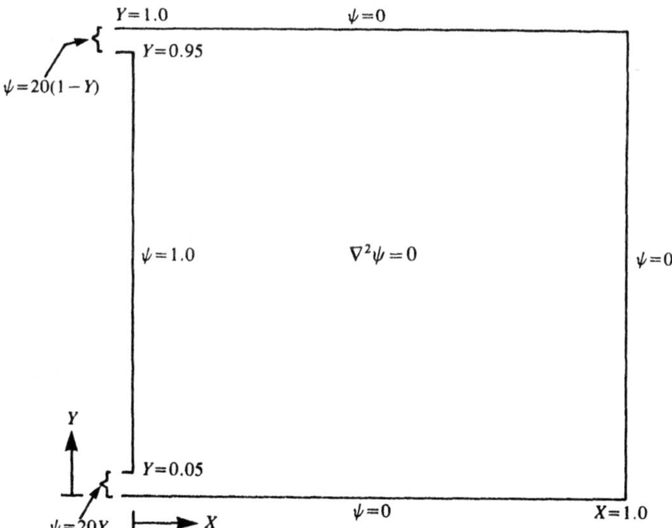

Figure 9.4.2 Flow problem governed by Laplace's equation, as considered in Example 9.4(b).

Solution

(a) We can formulate the given problem in dimensionless terms by defining the nondimensional temperature ϕ and coordinate distances X and Y as

$$\phi = \frac{T - T_2}{T_1 - T_2} \qquad X = \frac{x}{L} \qquad Y = \frac{y}{L} \tag{9.4.1}$$

where $T(x,y)$ is the temperature at an arbitrary location, given by the coordinates x and y in the computational domain, and L is the length or width of the region, as shown in Fig. 9.4.1. The governing equation is obtained as

$$\frac{\partial^2 \phi}{\partial X^2} + \frac{\partial^2 \phi}{\partial Y^2} = 0 \tag{9.4.2}$$

with the following boundary conditions:

$$\begin{aligned} &\text{At } X = 0 \text{ and } X = 1.0\text{:} && \phi = 0 && \text{for } 0 \leqslant Y \leqslant 1.0 \\ &\text{At } Y = 0\text{:} && \phi = 0 && \text{for } 0 \leqslant X < 1.0 \\ &\text{At } Y = 1.0\text{:} && \phi = 1.0 && \text{for } 0 < X < 1.0 \end{aligned} \tag{9.4.3}$$

If a rectangular region of length L and width W is considered instead, the above nondimensionalization may again be used, so that the governing equation remains unchanged. However, L/W will appear as a parameter in the boundary conditions in that case.

For the SOR method, the iterative scheme is obtained from Eq. (9.42) as follows:

$$\phi_{i,j}^{(l+1)} = \omega \left[\frac{\phi_{i+1,j}^{(l)} + \phi_{i-1,j}^{(l+1)} + \phi_{i,j+1}^{(l)} + \phi_{i,j-1}^{(l+1)}}{4} \right] + (1 - \omega)\phi_{i,j}^{(l)} \qquad \begin{aligned} &\text{for } 1 \leqslant i \leqslant n - 1 \\ &1 \leqslant j \leqslant n - 1 \end{aligned} \tag{9.4.4}$$

where the grid spacings Δx and Δy are taken as equal and n is the number of subdivisions in each direction, as shown in Fig. 9.9 for a rectangular region.

For successive over-relaxation, ω lies between 1 and 2. We solve the problem for several values of ω to determine the optimum value. The Gauss-Seidel method is obtained for $\omega = 1$. Figure 9.4.3 gives the computer program for the SOR method. The program allows one to choose the grid sizes in the two directions, as well as the corresponding dimensions by choosing the total number of grid points. The initial, guessed distribution of ϕ is taken as uniform throughout the computational domain. The value of this initial guess may be specified. One uses a convergence criterion on $\phi_{i,j}$ to check for convergence; see Eq. (9.3.6). One also specifies a maximum number of iterations in order to terminate the computation if convergence is not attained. The boundary conditions are implemented by subroutine BCOND. The convergence parameter EPSI and the grid size Δx were varied to ensure that the results obtained were not significantly dependent on the values chosen.

Figure 9.4.4 shows the variation of the number of iterations for convergence with the relaxation factor ω. The optimum value is found to be around 1.55. Note that

```
C      SUCCESSIVE OVER-RELAXATION METHOD FOR AN ELLIPTIC EQUATION
C
C       THIS PROGRAM SOLVES THE LAPLACE EQUATION BY EMPLOYING
C       THE SUCCESSIVE OVER-RELAXATION METHOD.
C
C       WHEN THE PROGRAM IS RUN, IT PROMPTS FOR THE INPUT VALUES
C       REQUIRED. ENTER THE INPUT VALUES AND THE OUTPUT WILL BE
C       STORED IN A FILE CALLED 'SOR.DAT'
C
C
C        DESCRIPTION OF INPUT PARAMETERS:
C
C        IL       IS THE NUMBER OF GRID POINTS IN THE X DIRECTION.
C        JL       IS THE NUMBER OF GRID POINTS IN THE Y DIRECTION.
C        DX       IS THE GRID SIZE IN X DIRECTION.
C        DY       IS THE GRID SIZE IN Y DIRECTION.
C        RP       IS THE RELAXATION PARAMETER.
C        PHIINT   IS THE INITIAL GUESS FOR PHI, TAKEN AS UNIFORM OVER
C                 THE WHOLE DOMAIN.
C        MAXITERATION IS THE MAXIMUM NUMBER OF ITERATIONS BEFORE
C                 TERMINATING THE COMPUTATION.
C        EPSI     IS THE CONVERGENCE CRITERION.
C
C
C        DESCRIPTION OF OTHER VARIABLES:
C
C        PHI      IS THE SOLUTION AT THE NTH TIME STEP.
C        PHIOL    IS THE SOLUTION AT THE (N-1)TH TIME STEP.
C
C
C        ENTER INPUT PARAMETERS
C
         CHARACTER*2 XFILE(5)
         CHARACTER*2 YFILE(5)
         DIMENSION PHI(11,11),PHIOL(11,11)
         PRINT*,'ENTER INITIAL GUESS FOR PHI, TAKEN AS UNIFORM OVER'
         PRINT*,'THE WHOLE DOMAIN'
         READ(1,*)PHIINT
         PRINT*,'ENTER GRID SIZE DX=, DY='
         READ(1,*)DX,DY
         PRINT *,'ENTER NO. OF GRID POINTS IL=, JL='
         PRINT*,' MAXIMUM NO. POSSIBLE IS 11 FOR BOTH IL AND JL'
         PRINT*,'UNLESS DIMENSION STATEMENT IS CHANGED'
         READ(1,*)IL,JL
         PRINT *,'ENTER THE RELAXATION PARAMETER'
         READ(1,*)RP
         PRINT*,'ENTER MAXIMUM  NO. OF ITERATIONS TO BE CARRIED'
         PRINT*,'OUT BEFORE STOPPING'
         READ(1,*)MAXITERATION
         PRINT *,'ENTER CONVERGENCE CRITERION'
         READ(1,*)EPSI
         PRINT*,'THE INPUT VALUES ARE:'
         PRINT*,'INITIAL GUESS FOR PHI=',PHIINT
         PRINT*,'DX=',DX,'DY=',DY
         PRINT*,'IL=',IL,'JL=',JL
         PRINT*,'MAX. NO. OF ITERATIONS=',MAXITERATION
         PRINT*,'CONVERGENCE CRITERION=',EPSI
         ITERATION=0
```

Figure 9.4.3 FORTRAN program for solving the elliptic PDE of Example 9.4(a) by the successive over-relaxation method.

```
C
C       OPEN THE DATA FILES FOR GRAPHICS
C
        XFILE(1)='X1'
        XFILE(2)='X2'
        XFILE(3)='X3'
        XFILE(4)='X4'
        XFILE(5)='X5'
        YFILE(1)='Y1'
        YFILE(2)='Y2'
        YFILE(3)='Y3'
        YFILE(4)='Y4'
        YFILE(5)='Y5'
C
C       SET INITIAL DISTRIBUTION OF PHI
C
        DO 5 I=1,IL
        DO 5 J=1,JL
        PHI(I,J)=PHIINT
    5   CONTINUE
C
C       START SOLVING FOR PHI
C
   15   ITERATION=ITERATION+1
        IF(ITERATION .GE. MAXITERATION)GO TO 40
C
C       SAVE THE FIELD AT PREVIOUS TIME STEP
C
        DO 10 I=1,IL
        DO 10 J=1,JL
        PHIOL(I,J)=PHI(I,J)
   10   CONTINUE
C
C       EMPLOY SOR ITERATIVE METHOD FOR PHI AT INTERIOR POINTS
C
        DO 20 J=2,JL-1
        DO 20 I=2,IL-1
        PHIGS=(PHI(I+1,J)+PHI(I-1,J))/DX**2+(PHI(I,J+1)+PHI(I,J-1))
     $  /DY**2
        PHIGS=PHIGS/(2./DX**2+2./DY**2)
        PHI(I,J)=RP*PHIGS+(1.0-RP)*PHIOL(I,J)
   20   CONTINUE
C
C       IMPOSE THE BOUNDARY CONDITIONS
C
        CALL BCOND(PHI,IL,JL)
C
C       CHECK FOR CONVERGENCE
C
        DO 35 I=1,IL
        DO 35 J=1,JL
        IF(ABS(PHI(I,J)-PHIOL(I,J)).GE.EPSI)GO TO 15
   35   CONTINUE
        GO TO 50
   40   PRINT*,'SOLUTION DOES NOT CONVERGE'
   50   OPEN(UNIT=10,FILE='SOR.DAT')
        WRITE(10,110)EPSI
  110   FORMAT(1X,'CONVERGENCE CRITERION ='1X,E9.1)
        WRITE(10,115)RP
  115   FORMAT(//,1X,'RELAXATION PARAMETER RP=',F5.2)
```

Figure 9.4.3 Continued

```
      WRITE(10,120)ITERATION
 120  FORMAT(//,1X,'NO. OF ITERATIONS TO CONVERGE=',1X,I4,//)
      WRITE(10,130)
 130  FORMAT(1X,'PHI DISTRIBUTION IS:',//)
      WRITE(10,140)(I,I=1,IL)
 140  FORMAT(1X,'I=',8X,11(I2,8X))
      DO 60 J=1,JL
      WRITE(10,100)J,(PHI(I,J),I=1,IL)
  60  CONTINUE
 100  FORMAT(1X,'J=',I2,3X,11(F8.5,2X))
C
C     OUTPUT FOR GRAPHICS
C
      II=1
      DO 70 I=1,5
      II=II+1
      OPEN (UNIT=12,FILE=XFILE(I))
      DO 66 J=1,JL
      WRITE(12,*)PHI(II,J)
  66  CONTINUE
      CLOSE(UNIT=12)
  70  CONTINUE
      JJ=1
      DO 71 J=1,5
      JJ=JJ+1
      OPEN(UNIT=12,FILE=YFILE(J))
      DO 72 I=1,IL
      WRITE(12,*)PHI(I,JJ)
  72  CONTINUE
      CLOSE(UNIT=12)
  71  CONTINUE
      OPEN(UNIT=12,FILE='XX')
      DO 73 I=1,IL
      XX=FLOAT(I-1)*DX
      WRITE(12,*)XX
  73  CONTINUE
      CLOSE(UNIT=12)
      OPEN(UNIT=12,FILE='YY')
      DO 74 J=1,JL
      YY=FLOAT(J-1)*DY
      WRITE(12,*)YY
  74  CONTINUE
      CLOSE(UNIT=12)
      STOP
      END
C*****************************************************************
      SUBROUTINE BCOND(PHI,IL,JL)
C
C     THIS SUBROUTINE IMPLEMENTS THE APPROPRIATE BOUNDARY
C     CONDITIONS.
C
      DIMENSION PHI(11,11)
C
C     SET THE CONDITIONS AT I=1 AND I=IL SURFACES
C
      DO 25 J=1,JL
      PHI(1,J)=0.
      PHI(IL,J)=0.
  25  CONTINUE
C
```

Figure 9.4.3 Continued

```
C       SET THE CONDITIONS AT J=1 AND J=JL SURFACES
C
        DO 30 I=1,IL
        PHI(I,1)=0.
        PHI(I,JL)=1.0
   30   CONTINUE
        RETURN
        END
```

Figure 9.4.3 Continued

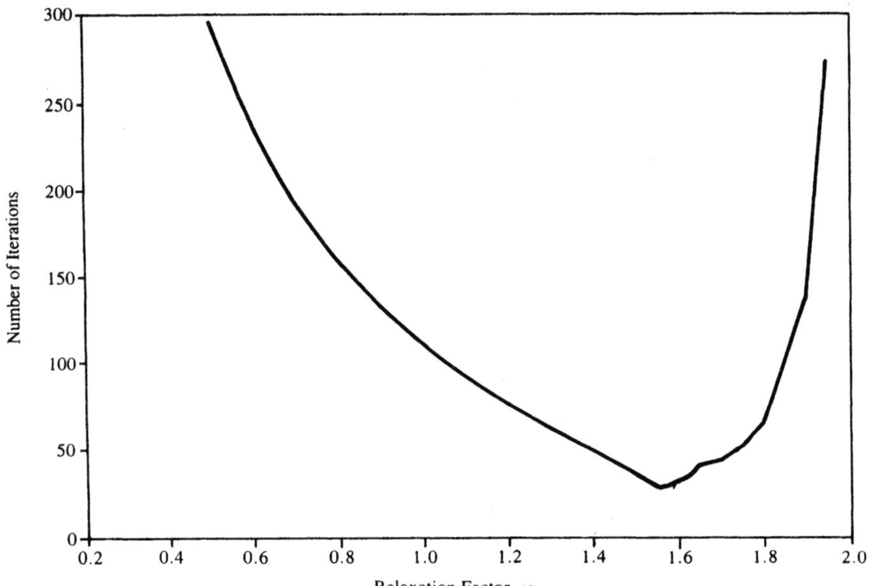

Figure 9.4.4 Variation of the number of iterations, for convergence in Example 9.4(a), with the relaxation factor ω, employing $\Delta X = \Delta Y = 0.1$.

the number of iterations increases sharply as ω is varied away from the optimum value. Therefore, the SOR method is advantageous to use if the value of ω is close to the optimum. Also, see Fig. 5.5.2, which illustrates the application of the SOR method to a system of linear equations. Figures 9.4.5 and 9.4.6 show the temperature distributions in the region. Three surfaces are at temperature $\phi = 0$, and the fourth one is at $\phi = 1.0$. The temperatures decrease as one moves away from the hot surface. The distributions are symmetric about $X = 0.5$, as expected.

(b) The $\psi(X,Y)$ distribution in the flow region is governed by Eq. (9.4.2) with ϕ replaced by ψ. The problem, as presented in Fig. 9.4.2, is given in nondimensional terms, so that ψ varies from 0 to 1, as do the coordinate distances X and Y. The

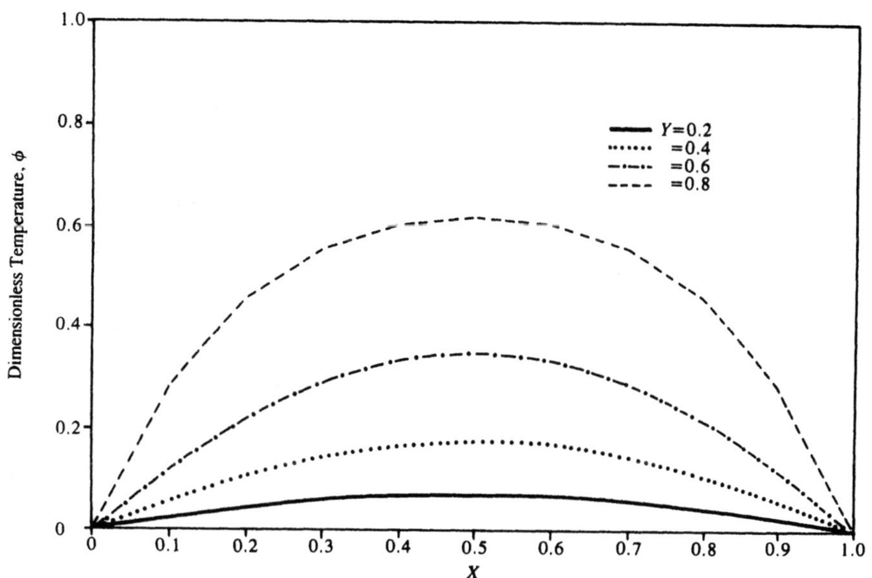

Figure 9.4.5 Computed temperature distributions in Example 9.4(a), with $\omega = 1.6$.

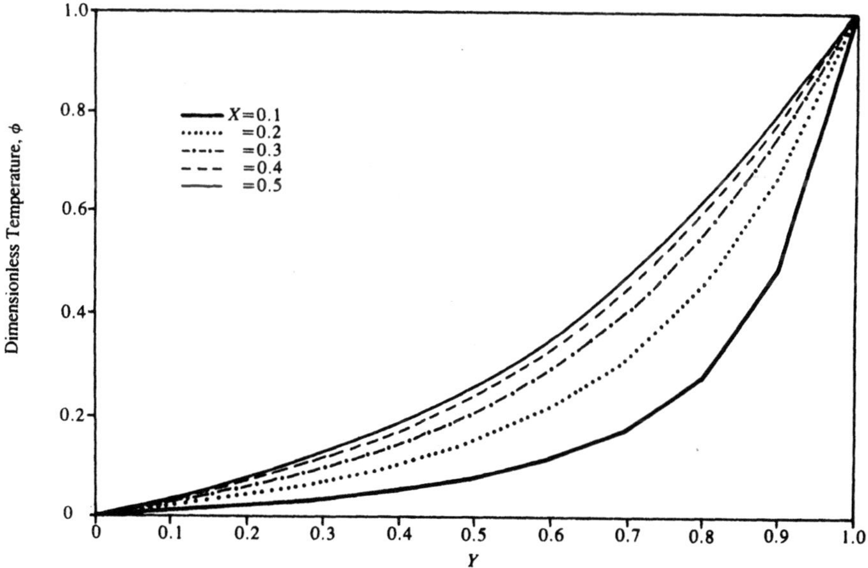

Figure 9.4.6 Computed temperature variation with Y at various values of X in Example 9.4(a), with $\omega = 1.6$.

boundary conditions are written as follows:

At $Y = 0$ and

$$
\begin{array}{llll}
\text{At } Y = 0 \text{ and } Y = 1.0: & \psi = 0 & \text{for } 0 \leqslant X \leqslant 1.0 & \\
\text{At } X = 1.0: & \psi = 0 & \text{for } 0 < Y < 1.0 & \\
\text{At } X = 0: & \psi = 1.0 & \text{for } 0.05 \leqslant Y \leqslant 0.95 & (9.4.5) \\
\text{At } X = 0: & \psi = 20Y & \text{for } 0 < Y < 0.05 & \\
\text{At } X = 0: & \psi = 20(1 - Y) & \text{for } 0.95 < Y < 1.0 &
\end{array}
$$

These equations imply that $\psi = 0$ on three sides of the enclosure and that it varies linearly to 1.0 at the inflow/outflow channels. Although more involved than Eq. (9.4.3), these conditions can easily be incorporated into the program by a suitable modification of the subroutine BCOND. One can then employ the main program to obtain the results for chosen values of the grid spacing ΔX and ΔY.

Figure 9.4.7 shows the contours of constant ψ, or streamlines, at ψ values of 0, 0.05, 0.1, 0.25, 0.5, 0.75, and 1.0. The computed ψ values at the nodal points are used with a simple linear interpolation scheme to determine the X, Y locations where these ψ values are attained. Using graphical procedures, similar to those for Part (a), we draw the contours by joining the various locations on the X-Y plane where a given value of ψ is obtained. The grid spacing $\Delta X = \Delta Y$ was taken as 0.025 at the start and reduced to 0.01 to confirm that the results were not significantly dependent on the value chosen. A convergence parameter ε of 10^{-5} was employed in Eq. (9.3.6), with ϕ

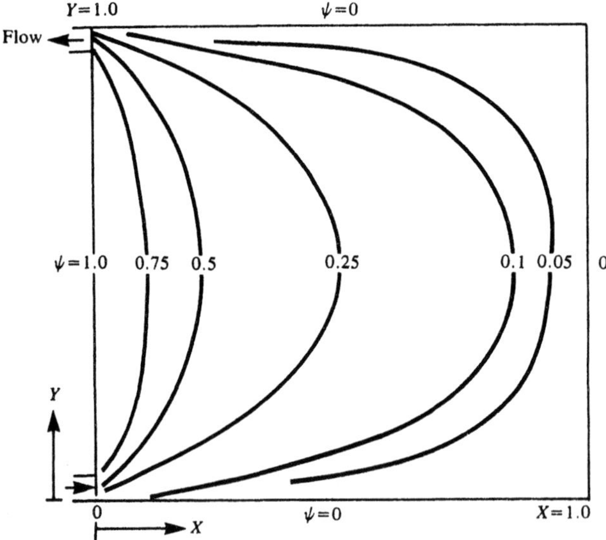

Figure 9.4.7 Computed contours of constant stream function ψ for Example 9.4(b).

replaced by ψ. Again, ε was varied to ensure a negligible effect of the value chosen on the numerical results.

This problem represents an important circumstance encountered in several engineering applications, particularly in mechanical and civil engineering. The problem concerns the flow field generated in an enclosed region due to the inflow and outflow of a fluid, assuming the viscous and rotational effects to be absent. If rotational effects are present, with viscous effects still negligible, the Poisson equation is obtained and may be solved in a similar way.

9.4 HYPERBOLIC PARTIAL DIFFERENTIAL EQUATIONS

9.4.1 Basic Aspects

Hyperbolic partial differential equations arise in several problems of engineering interest, such as vibration of rods and strings, transmission of sound in air, and supersonic flow. As discussed earlier, hyperbolic PDEs have two real and distinct characteristics. Information travels at finite speed in regions defined by these characteristics, as shown in Fig. 9.15. An observer at point 0 in Fig. 9.15(a) is affected by disturbances only in the region of dependence of the point 0, and a disturbance at 0 can be felt only in the region of influence, where both of these regions are marked by the two families of characteristics. The flow of an object in a stationary fluid or of the fluid past a stationary object at speeds greater than the speed of sound in that fluid is known as *supersonic flow* and is also governed by a hyperbolic PDE. Figure 9.15(b) shows the supersonic flow of air over an airplane, and the region of influence is given by the two characteristic lines shown. A disturbance at 0 is, therefore, felt only in the region of influence, and an observer outside this region is not affected by the presence of the airplane. For this reason, the sound of a supersonic airplane is heard only after it has passed overhead. In this figure, the angle θ is given by $\sin\theta = a/V$, where a is the speed of sound in air and V is the speed of the air flow, with $V > a$ for supersonic flow. If $V/a = 2$, for instance, the angle θ, which is known as the *angle of the Mach cone*, is 30°. This flow circumstance is analogous to that of an airplane moving at speed V in quiescent air.

9.4.2 Method of Characteristics

An important numerical technique for solving hyperbolic partial differential equations is the method of characteristics. In this method, the computational domain is divided into finite regions by the two families of characteristics, taking the grid points at the intersections of these lines, as shown in Fig. 9.15(a). The properties of characteristics are used to reduce the problem to a system of ordinary differential equations, which are solved by methods similar to those applicable to ODEs. The

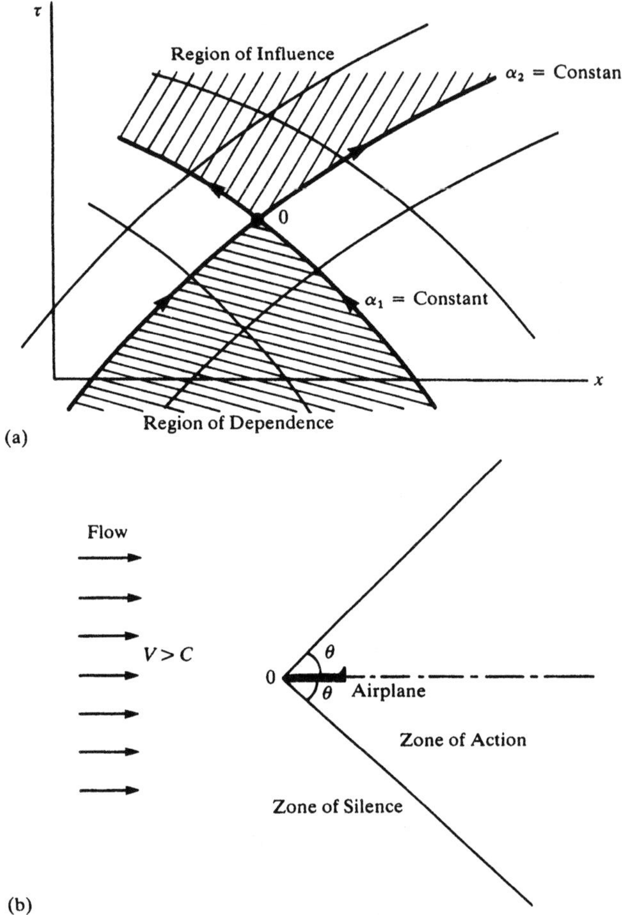

Figure 9.15 Characteristics associated with a hyperbolic PDE. (a) Information travel as limited by characteristics in the computational region; (b) supersonic flow of air over an airplane, indicating the zones of action and silence.

main advantage of this method is that the important properties of the exact solution are preserved in the numerical solution. Discontinuities, such as shock waves in supersonic flow, can easily be treated, since discontinuities can occur only along characteristics. However, this method is difficult to use in complicated geometries, because of the problem of keeping track of the characteristics, and in problems where an elliptic or parabolic PDE may apply in one portion of the computational domain and a hyperbolic PDE in the other. For further details on the methods based on characteristics for solving hyperbolic PDEs, see Smith (1968) and Ferziger (1981).

9.4.3 Finite Difference Methods

Much of the recent work on hyperbolic PDEs has been based on finite difference methods, which are quite similar to those discussed earlier for parabolic and elliptic equations. Some of these methods are outlined here. Two important hyperbolic equations that we consider are the first-order convection equation

$$\frac{\partial \phi}{\partial \tau} + c\frac{\partial \phi}{\partial x} = 0 \tag{9.7}$$

and the second-order wave equation

$$\frac{\partial^2 \phi}{\partial \tau^2} = c^2\frac{\partial^2 \phi}{\partial x^2} \tag{9.6}$$

where c is the convection velocity in the former case and the propagation velocity of the wave in the latter. Also, ϕ is a dependent variable such as temperature in the first case and deflection of a string in the second.

The initial and boundary conditions for the wave equation may be written as follows:

$$\begin{aligned} &\text{At } \tau = 0: \quad \phi = \alpha_1(x) \qquad \frac{\partial \phi}{\partial \tau} = \alpha_2(x) \\ &\text{At } x = 0: \quad \phi = \beta_1 \qquad \text{for } \tau > 0 \\ &\text{At } x = L: \quad \phi = \beta_2 \qquad \text{for } \tau > 0 \end{aligned} \tag{9.51}$$

where α_1 and α_2 are given constants, or functions of x, and β_1 and β_2 are constants. It may be pointed out that Eq. (9.7) yields Eq. (9.6), on differentiation, as follows:

$$\frac{\partial^2 \phi}{\partial \tau^2} = \frac{\partial}{\partial \tau}\left(-c\frac{\partial \phi}{\partial x}\right) = -c\frac{\partial}{\partial x}\left(\frac{\partial \phi}{\partial \tau}\right) = -c\frac{\partial}{\partial x}\left(-c\frac{\partial \phi}{\partial x}\right) = c^2\frac{\partial^2 \phi}{\partial x^2} \tag{9.52}$$

Therefore, the methods for solving the first-order equation may also be used for solving the second-order equation.

The solution domain for the wave equation is shown in Fig. 9.16. The boundary conditions are specified at two values of the spatial coordinate x, and the initial conditions are given at $\tau = 0$. The dependent variable ϕ is to be computed at all of the grid points over a given time interval. Starting with the known values at $\tau = 0$, the numerical solution progresses in the direction of increasing time, or increasing i, with the boundary conditions being satisfied at each time step. If central difference approximations for the second derivatives are substituted into the wave equation, we obtain the finite difference equation

$$\frac{\phi_{i+1,j} - 2\phi_{i,j} + \phi_{i-1,j}}{(\Delta \tau)^2} = c^2\frac{\phi_{i,j+1} - 2\phi_{i,j} + \phi_{i,j-1}}{(\Delta x)^2} \tag{9.53}$$

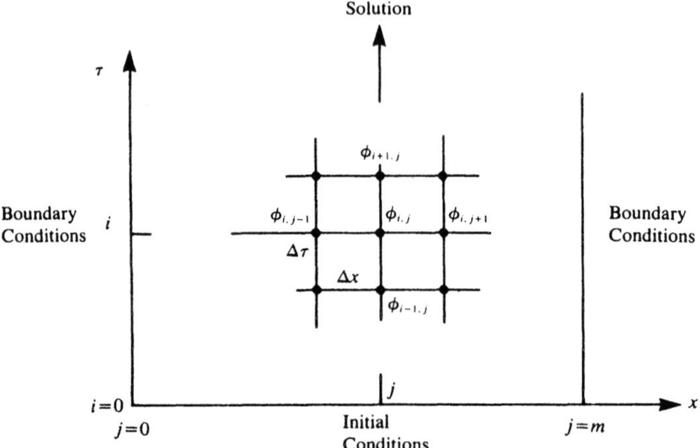

Figure 9.16 Solution domain for the wave equation, indicating the initial and boundary conditions needed and the mesh employed.

which gives

$$\phi_{i+1,j} = -\phi_{i-1,j} + c^2 \frac{(\Delta\tau)^2}{(\Delta x)^2} (\phi_{i,j+1} + \phi_{i,j-1}) + 2\left[1 - c^2 \frac{(\Delta\tau)^2}{(\Delta x)^2}\right]\phi_{i,j} \quad (9.54)$$

This equation is an explicit representation since the values to be computed at a given time step are obtained directly from known values at earlier time steps. The initial conditions are given in terms of the function ϕ and its derivative $\partial\phi/\partial\tau$. The condition on the derivative provides a relation between $\phi_{1,j}$ and $\phi_{0,j}$, and, therefore, the values of $\phi_{1,j}$ for starting the computational scheme may be determined, as shown in Example 9.5. The stability of the above representation may be considered in terms of our earlier discussion on the stability of the explicit schemes for parabolic equations. Then we would expect the numerical scheme to become unstable when the coefficient of $\phi_{i,j}$ becomes negative. Therefore, the method would be expected to be stable if

$$\frac{c\,\Delta\tau}{\Delta x} \leqslant 1 \quad (9.55)$$

This stability condition is frequently known as the *Courant condition*, and the dimensionless parameter $c\,\Delta\tau/\Delta x$ as the *Courant number*. A more detailed stability analysis of the method also yields the above stability constraint. Example 9.5 demonstrates the solution of the wave equation by this explicit method.

Similarly, the finite difference approximation for Eq. (9.7) may be written as

$$\frac{\phi_{i+1,j} - \phi_{i,j}}{\Delta\tau} = -c\,\frac{\phi_{i,j} - \phi_{i,j-1}}{\Delta x} \quad (9.56)$$

Backward differences are used so that the solution depends only on the domain of dependence, as shown in Fig. 9.17. This method is known as the *backward* or *upwind difference method.* The formulation is, therefore, of accuracy $[O(\Delta\tau), O(\Delta x)]$. Again, the necessary condition for stability is Eq. (9.55). This condition basically ensures that the solution at a given location at a given time is affected only by the values at the grid points in its domain of dependence. Example 9.6 discusses the numerical solution of this equation, along with the stability considerations.

A more accurate explicit method is the Lax-Wendroff method, which is second-order accurate in both time and space. It is also stable for Courant numbers less than unity and is frequently employed for linear hyperbolic equations. When applied to the first-order convection equation, the finite difference form of this method is obtained as follows:

$$\phi_{i+1,j} = \phi_{i,j} - \frac{c\,\Delta\tau}{2\,\Delta x}(\phi_{i,j+1} - \phi_{i,j-1}) + \frac{c^2(\Delta\tau)^2}{2(\Delta x)^2}(\phi_{i,j+1} - 2\phi_{i,j} + \phi_{i,j-1}) \quad (9.57)$$

We can also apply the method to the second-order wave equation by breaking the equation down into two coupled first-order equations as

$$\frac{\partial \phi}{\partial \tau} = c\,\frac{\partial u}{\partial x} \quad (9.58a)$$

$$\frac{\partial u}{\partial \tau} = c\,\frac{\partial \phi}{\partial x} \quad (9.58b)$$

The stability constraint is given by Eq. (9.55). It can be shown that the results are theoretically exact when the Courant number is 1.0 and that the accuracy of the solution decreases as the Courant number decreases below 1.0; see Example 9.6. Therefore, the value of this parameter is generally chosen to be around unity.

Several other explicit and implicit methods, similar to those for parabolic PDEs, have been developed for hyperbolic partial differential equations. These methods include the unconditionally stable Crank-Nicolson method, which yields a tridiagonal

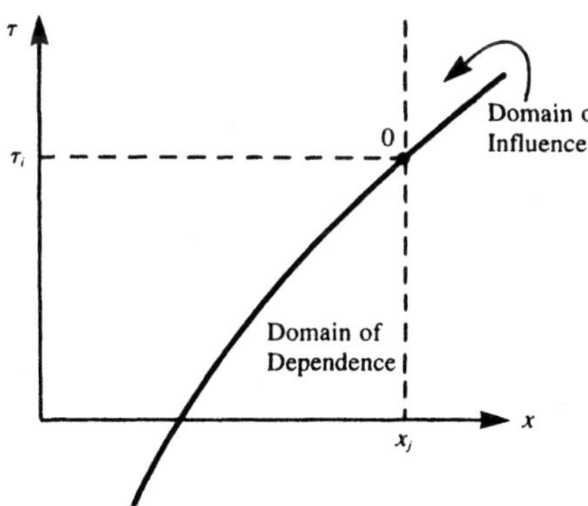

Figure 9.17 Domains of dependence and influence for the first-order convection equation, which is a hyperbolic PDE.

system of equations for the one-dimensional problem. For two or three space dimensions, splitting methods similar to the alternating direction implicit (ADI) method, outlined earlier, are employed. These methods are also unconditionally stable and give rise to tridiagonal systems, which are easily solved by Gaussian elimination. Many such methods have been developed and applied to aerodynamic applications. For further details, see Ferziger (1981).

Example 9.5

Consider the vibration of a string, 1 m in length, stretched between two supports with an initial tension of 40 N. The mass of the string is 0.04 kg/m. The string is displaced from its equilibrium position, as shown in Fig. 9.5.1, held at rest in this configuration, and then released. Compute the variation of the displacement at various points along the string with time. For approximately one period of vibration, following the release of the string from rest, determine the configuration of the string at various intermediate time intervals. Also consider the case when the string is plucked in the middle, instead of at the one-fourth point, and obtain the string configuration as a function of time.

Solution

The governing equation for this problem is the wave equation, written as

$$\frac{\partial^2 u}{\partial \tau^2} = c^2 \frac{\partial^2 u}{\partial x^2} \tag{9.5.1}$$

where u is the vertical displacement at a point on the string, indicated by coordinate distance x, τ is the time following the release of the string from rest, and c^2 is a constant. For a vibrating string, it can be shown from the derivation of the governing equation that $c^2 = T/m$, where T is the tension and m is the mass per unit length of the string. Therefore, in the given problem,

$$c^2 = \frac{T}{m} = \frac{40 \text{ N}}{0.04 \text{ kg/m}} = 1000 \text{ m}^2/\text{s}^2 \tag{9.5.2}$$

We now select the value of Δx as 0.05 m, giving 21 grid points along the string. We must consider the Courant number $C = c\,\Delta\tau/\Delta x$ to select a suitable time step. As discussed in Section 9.3, $C \leqslant 1$ for numerical stability, and a greater accuracy is obtained if C is close to 1.0, that is,

$$\frac{T}{m}\frac{(\Delta\tau)^2}{(\Delta x)^2} = 1 \tag{9.5.3}$$

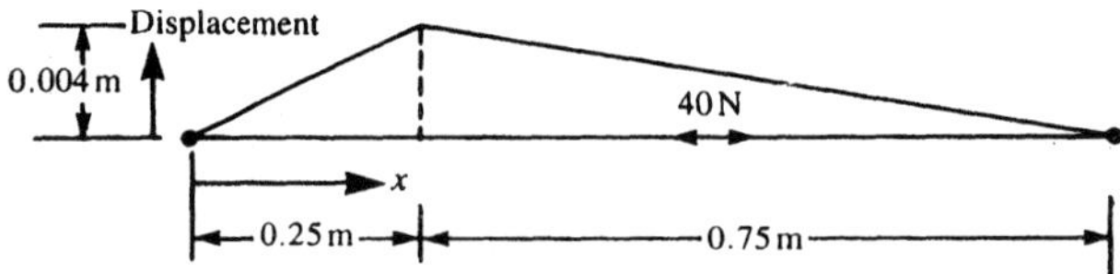

Figure 9.5.1 Physical circumstance of a vibrating string, considered in Example 9.5.

This gives the value of $\Delta\tau$ as 1.58×10^{-3} s. Therefore, $\Delta\tau$ is chosen as 0.0015 s for convenience.

The initial conditions, on $u(x,\tau)$ in meters, for the string plucked at the one-fourth point, are as follows:

$$u(x, 0) = \begin{cases} 0.016x & \text{for } 0 \leqslant x \leqslant 0.25 \\ \dfrac{0.016}{3}(1 - x) & \text{for } 0.25 < x \leqslant 1.0 \end{cases} \tag{9.5.4}$$

$$\frac{\partial u}{\partial \tau}(x, 0) = 0 \tag{9.5.5}$$

The boundary conditions are

$$u(0, \tau) = u(1, \tau) = 0 \tag{9.5.6}$$

Similarly, the initial conditions for plucking the string in the middle may be written.

The given problem may be solved by finite difference methods. If central differences are used, the finite difference equation is Eq. (9.54), which may be written for the present case as

$$u_{i+1,j} = -u_{i-1,j} + C^2(u_{i,j+1} + u_{i,j-1}) + 2(1 - C^2)u_{i,j} \tag{9.5.7}$$

where $C^2 = (T/m)(\Delta\tau)^2/(\Delta x)^2$. The initial condition given by Eq. (9.5.5) is written, using central differencing, as

$$\frac{u_{2,j} - u_{0,j}}{2\,\Delta\tau} = 0 \tag{9.5.8}$$

where $u_{0,j}$ is the value at a fictitious point one time step before the initial condition, $i = 1$. Thus,

$$u_{2,j} = u_{0,j} \tag{9.5.9}$$

Substituting this relationship into Eq. (9.5.7) for $i = 1$, we obtain

$$u_{2,j} = \frac{C^2}{2}(u_{1,j+1} + u_{1,j-1}) + (1 - C^2)u_{1,j} \tag{9.5.10}$$

Equation (9.5.10) is used for advancing from the initial configuration to the first time interval, $\tau = \Delta\tau$. Beyond that, Eq. (9.5.7) is employed. A first-order approximation may also be used for Eq. (9.5.5), giving $u_{2,j} = u_{1,j}$. The boundary conditions give the displacement at grid points 1 and 21 as zero. For the remaining points, we compute the displacement for the next time step, using the values at the adjacent points corresponding to the previous time steps. Since the values of the displacements for only the last two time steps are needed to advance the solution, only three arrays corresponding to time intervals τ, $\tau + \Delta\tau$, and $\tau + 2\,\Delta\tau$ need be considered, where the values at $\tau + 2\,\Delta\tau$ are obtained in terms of the other two arrays.

Figure 9.5.2 shows the computer program for solving this problem. The various symbols employed are defined in the program. Arrays U, U_1, and U_2 contain the

```
C              SOLUTION OF THE WAVE EQUATION
C
C        THIS PROGRAM SOLVES A SECOND-ORDER HYPERBOLIC PARTIAL
C        DIFFERENTIAL EQUATION BY THE FINITE DIFFERENCE METHOD.
C
C        SUBROUTINE INPUT PROVIDES THE INPUT DATA NECESSARY TO
C        RUN THE PROGRAM
C
C        DESCRIPTION OF VARIABLES:
C
C        DX IS THE GRID SIZE.
C        IL IS THE NUMBER OF GRID POINTS.
C        DT IS THE TIME STEP.
C        C IS THE COURANT NUMBER. CHOOSE DX AND DT SUCH THAT C IS
C          APPROXIMATELY 1.0.
C        U, U1, U2 CONTAIN THE U FIELD AT THE THREE TIME STEPS T,
C          T+DT AND T+2DT, WHERE U IS THE DEPENDENT VARIABLE.
C        NLIM = MAXIMUM NUMBER OF TIME STEPS BEFORE TERMINATION
C          OF THE CALCULATION.
C        NSTEP= NUMBER OF TIME STEPS AFTER WHICH PRINTOUT OCCURS.
C
C
         DIMENSION U(25),U1(25),U2(25)
C
C        ENTER THE INPUT VALUES
C
         CALL INPUT(DX,DT,IL,U1,NLIM,NSTEP,ASQR)
C
         OPEN(UNIT=10,FILE='HPB.DAT')
         C=ASQR*DT**2/DX**2
C
C        SET THE BOUNDARY CONDITIONS
C
         CALL BCOND(U2,IL)
C
C        INITIALIZE THE VARIABLES
C
         DO 10 I=2,IL-1
         U2(I)=U1(I)+C*(U1(I+1)-2.*U1(I)+U1(I-1))/2.
   10    CONTINUE
C
         ISTEP1=1
         ISTEP2=1
         T=DT
C
C        SOLVE FOR U AT SUCCESSIVE TIME STEPS
C        AND SAVE THE PREVIOUS VALUES
C
   30    DO 20 I=1,IL
         U(I)=U1(I)
         U1(I)=U2(I)
   20    CONTINUE
C
C        INCREMENT THE TIME
C
         T=T+DT
         ISTEP1=ISTEP1+1
         IF(ISTEP1.GT.NLIM)GO TO 50
```

Figure 9.5.2 Computer program for solving the wave equation, a hyperbolic PDE, by finite difference methods.

```
      ISTEP2=ISTEP2+1
C
C     CALCULATE NEW 'U2'
C
      DO 40 I=2,IL-1
      U2(I)=2.*U1(I)-U(I)+C*(U1(I+1)-2.*U1(I)+U1(I-1))
 40   CONTINUE
C
C     OUTPUT THE RESULTS
C
      IF(ISTEP2.EQ.NSTEP)THEN
      WRITE(10,100)T
100   FORMAT(//,1X,'AT TIME=',F8.4,1X,'U FIELD IS')
      WRITE(10,110)(U2(I),I=1,IL)
110   FORMAT(/,1X,20(F8.4,2X))
      ISTEP2=0
      GO TO 30
      END IF
      GO TO 30
 50   STOP
      END
C*****************************************************************
      SUBROUTINE INPUT(DX,DT,IL,U1,NLIM,NSTEP,ASQR)
C
C     THIS SUBROUTINE PROVIDES THE INPUT VALUES TO THE MAIN PROGRAM
C
C     DESCRIPTION OF THE VARIABLES:
C
C     DX = GRID SIZE.
C     DT = TIME STEP.
C     IL = NUMBER OF GRID POINTS.
C     NLIM = MAXIMUM NUMBER OF TIME STEPS TO BE COMPUTED.
C     NSTEP = NUMBER OF TIME STEPS AFTER WHICH PRINTOUT OCCURS.
C     ASQR = CONSTANT IN THE DIFFERENTIAL EQUATION.
C     U1 CONTAINS THE INITIAL DISTRIBUTION OF THE DEPENDENT
C        VARIABLE.
C
      DIMENSION U1(25)
      DX=0.05
      DT=0.0015
      ASQR=1000.
      IL=21
      PRINT*,'ENTER MAXIMUM NUMBER OF TIME STEPS ALLOWED'
      READ(1,*)NLIM
      PRINT*,'ENTER NO. OF TIME STEPS AFTER WHICH OUTPUT OCCURS'
      READ(1,*)NSTEP
C
C     INITIAL DISTRIBUTION OF U1
C
      DO 10 I=1,IL
      IF(I.LE.6)U1(I)=FLOAT(I-1)*DX*0.0016
      IF(I.GT.6)U1(I)=-0.04*(FLOAT(I-1)*DX-1.)/75.
      IF(I.LE.11)U1(I)=0.004*FLOAT(I-1)*DX/0.5
      IF(I.GE.11)U1(I)=0.004-2.*0.004*(FLOAT(I-1)*DX-0.5)
      PRINT*,U1(I)
 10   CONTINUE
      RETURN
      END
C*****************************************************************
```

Figure 9.5.2 Continued

```
      SUBROUTINE BCOND(U2,IL)
C
C     THIS SUBROUTINE IMPOSES THE BOUNDARY CONDITIONS
C
      DIMENSION U2(25)
      U2(1)=0.
      U2(IL)=0.
      RETURN
      END
```

Figure 9.5.2 Continued

displacements at time τ, $\tau + \Delta\tau$, and $\tau + 2\,\Delta\tau$. Subroutine INPUT provides the input values to the main program. The maximum number of time steps and the time intervals after which the output is obtained may be specified. As indicated above, $\Delta x = 0.05$ m, $\Delta\tau = 0.0015$ s, and IL = 21, where IL is the total number of grid points. The initial distribution is entered into the main program by this subroutine. Another subroutine BCOND is used to incorporate the boundary conditions.

The displacements at various locations along the string are computed as time elapses. Figure 9.5.3 shows the variation of the displacement at four locations with time. A periodic behavior is clearly seen, as expected. Also, the time period is found to be about 0.2 s. Figure 9.5.4 shows the configuration of the string at various time intervals. The string starts at its initial distribution, given in Fig. 9.5.1, and, as time elapses, the displacement at each point undergoes a periodic process. At different time

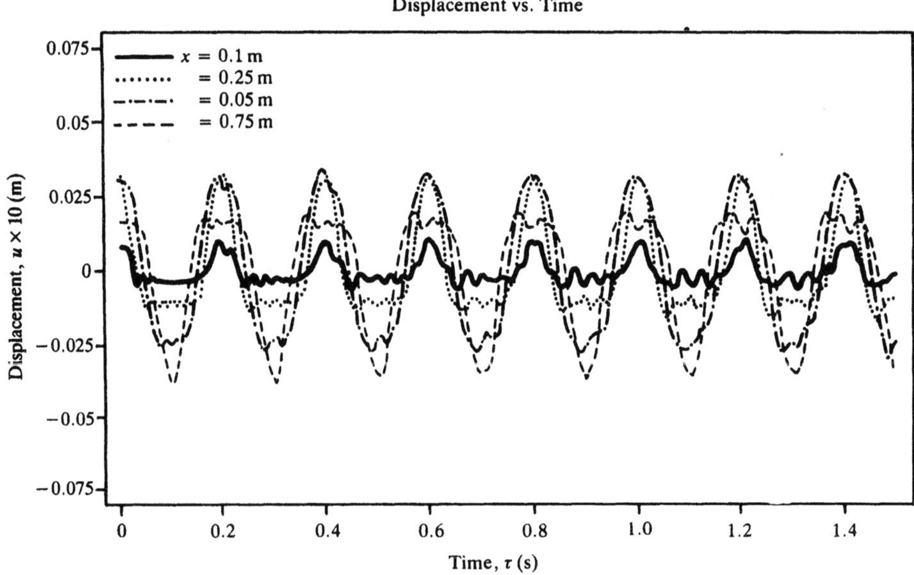

Figure 9.5.3 Computed displacements at four locations along the string as functions of time in Example 9.5.

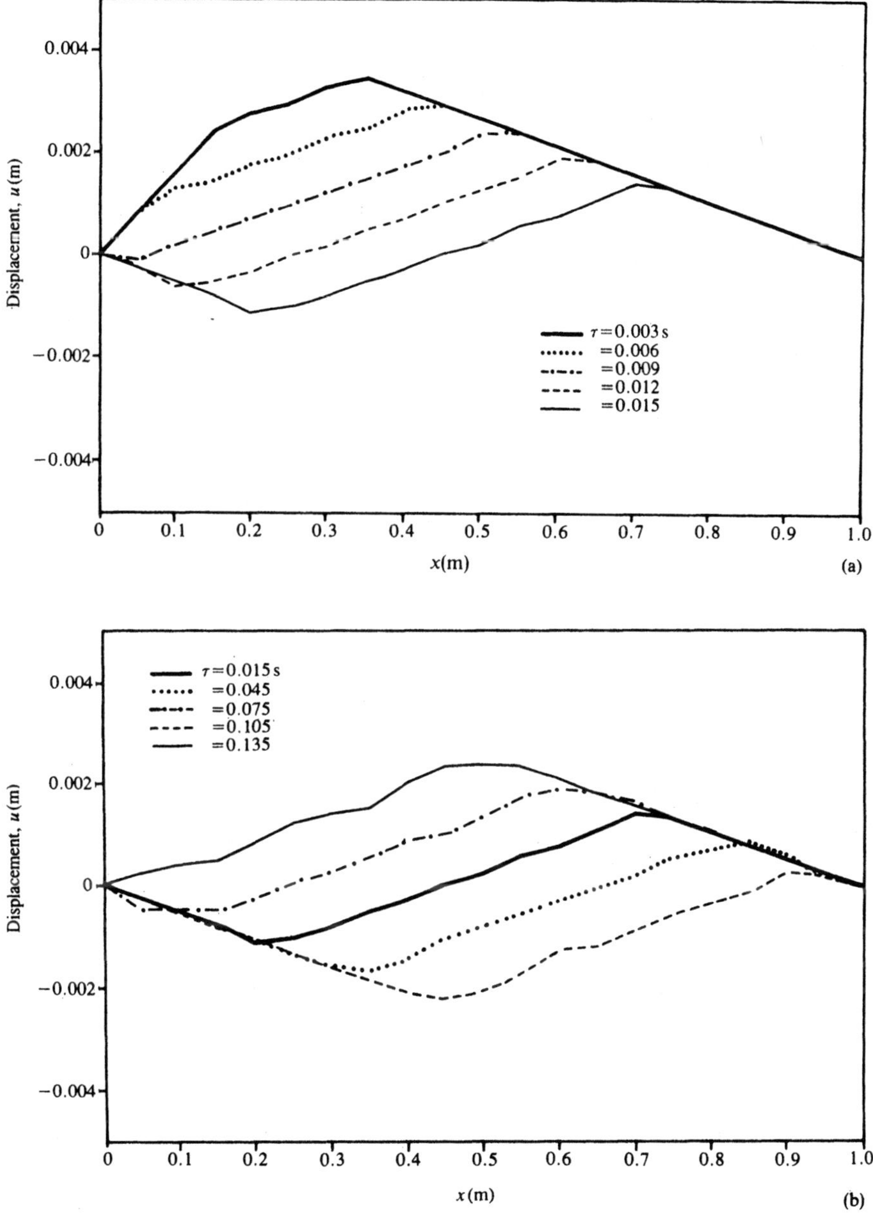

Figure 9.5.4 Calculated configuration of the string in Example 9.5 at various time intervals, when it is plucked at the one-fourth point, as shown in Fig. 9.5.1.

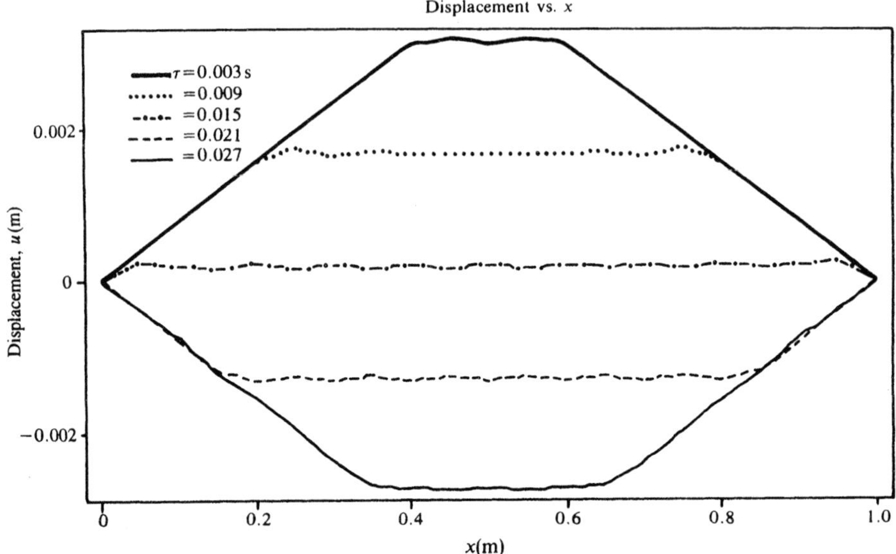

Figure 9.5.5 Calculated configuration of the string at various time intervals, when it is plucked in the middle.

intervals, different displacements exist at the various grid points, giving rise to different configurations of the string. Initially, the displacements are all positive, that is, on one side of the equilibrium position. Then, with time, the displacements become negative over portions of the string. Figure 9.5.5 shows the corresponding results for the case when the string is plucked in the middle. An expected symmetry arises in the displacement.

Example 9.6

Consider the first-order convection equation

$$\frac{\partial P}{\partial \tau} + c\,\frac{\partial P}{\partial x} = 0 \tag{9.6.1}$$

which governs the transport of a physical or chemical quantity $P(x,\tau)$ by convection. Here, x is the spatial coordinate distance, τ is time, and P represents a convected quantity such as concentration or temperature. Using Euler's method, the backward or upwind differencing method, and the Lax-Wendroff method, solve this hyperbolic equation. The initial and boundary conditions are given as follows;

$$\begin{aligned} &\text{At } \tau \leqslant 0: \quad P = 0 \qquad \text{for } x \geqslant 0 \\ &\text{At } \tau > 0: \quad P = 1 \qquad \text{for } x = 0 \end{aligned} \tag{9.6.2}$$

Take the convection velocity c as 2.5 m/s. Solve for x up to 5.0 m, taking the grid size Δx as 0.5 m. Compute the results up to time $\tau = 2.0$ s, taking the step size $\Delta\tau$ as 0.05, 0.1, and 0.2 s.

Solution

Several important transport problems in fluid flow and heat transfer are governed by first-order hyperbolic equations such as the one given here. For instance, wave propagation in a shallow water body is governed by a nonlinear first-order hyperbolic equation. The solution to the problem given here is simply the movement of the step change, at $x = 0$ and $\tau = 0$, downstream, with no change in amplitude and at the convection velocity c. Thus, we can use this analytical result to evaluate the solution of the given equation by the three methods considered.

The finite difference equation for the solution of Eq. (9.6.1) by Euler's method is

$$\frac{P_{i+1,j} - P_{i,j}}{\Delta\tau} = -c\frac{P_{i,j+1} - P_{i,j-1}}{2\,\Delta x} \tag{9.6.3}$$

where the first subscript refers to the time step and the second to the spatial location. Thus, this method uses forward difference for the derivative with respect to time and second-order central difference for the spatial derivative. The finite difference equations for the backward difference method and the Lax-Wendroff method are obtained from Eqs. (9.56) and (9.57) as

$$\frac{P_{i+1,j} - P_{i,j}}{\Delta\tau} = -c\,\frac{P_{i,j} - P_{i,j-1}}{\Delta x} \tag{9.6.4}$$

$$P_{i+1,j} = P_{i,j} - \frac{c\,\Delta\tau}{2\,\Delta x}(P_{i,j+1} - P_{i,j-1}) + \frac{c^2(\Delta\tau)^2}{2(\Delta x)^2}(P_{i,j+1} - 2P_{i,j} + P_{i,j-1}) \tag{9.6.5}$$

Thus, all three methods are explicit, and the only parameter that arises is the Courant number C, given by

$$C = \frac{c\,\Delta\tau}{\Delta x} \tag{9.6.6}$$

For the values given in the problem, $C = 0.25$, 0.5, and 1.0. As pointed out in the text, the last two methods are unstable for $C > 1.0$.

A computer program in BASIC is written for solving the given problem, as shown in Fig. 9.6.1. The input quantities and the initial and boundary conditions are entered. The numerical method for solving the problem is chosen interactively by the user. Computations for the three values of the time step $\Delta\tau$ are then carried out up to time $\tau = 2.0$ s at each step size. The computed results are printed at five locations in x, corresponding to $x = 1, 2, 3, 4$, and 5 m. The various symbols employed are defined in the program. Here, P and PN are used to denote values at the previous and present time steps, respectively. I denotes the spatial location. Calculations are needed for the ten grid points corresponding to $I = 2$ to $I = 11$. However, both Euler's method and

```
100 REM     SOLUTION OF FIRST-ORDER CONVECTION (HYPERBOLIC) EQUATION
110 REM
120 REM     X IS THE COORDINATE DISTANCE, T IS THE TIME, P(I) REPRESENTS
130 REM     THE VALUES OF THE DEPENDENT VARIABLE AT PREVIOUS TIME AND PN(I)
140 REM     THOSE AT THE PRESENT TIME STEP, N IS THE NUMBER OF SPATIAL GRID
150 REM     POINTS, C IS THE CONVECTION VELOCITY, CO IS THE COURANT NUMBER,
160 REM     DX IS THE STEP SIZE IN X, DT IS THE TIME STEP, TMAX IS THE MAXIMUM
170 REM     TIME FOR WHICH COMPUTATION IS CARRIED OUT AND T IS THE TIME.
180 REM
190 REM
200      DIM P(11),PN(11)
210 REM
220 REM     ENTER INPUT QUANTITIES
230 REM
240      DX=.5
250      C=2.5
260      N=11
270      DT=.05
275      TMAX=2
280 REM
290 REM     CHOOSE SOLUTION METHOD
300 REM
310      PRINT "ENTER 1, 2 OR 3 TO CHOOSE FROM EULER'S, BACKWARD OR UPWIND"
320      PRINT "DIFFERENCING AND LAX-WENDROFF METHODS"
330      INPUT "M=";M
340 REM
350 REM     INPUT INITIAL AND BOUNDARY CONDITIONS
360 REM
370      FOR IK=1 TO 3
375      PRINT
378      PRINT "TIME STEP=";DT
380      PRINT
390      FOR I=1 TO N
400      P(I)=0
410      PN(I)=0
420      NEXT I
430      P(1)=1
440      CO=C*DT/DX
450      T=0
460      T=T+DT
470      FOR I=2 TO N
480      IF I=N OR M=2 THEN 580
490      IF M=3 THEN 630
500 REM
510 REM     EULER'S METHOD
520 REM
530      PN(I)=P(I)-CO*(P(I+1)-P(I-1))/2
540      GOTO 640
550 REM
560 REM     BACKWARD OR UPWIND DIFFERENCE METHOD
570 REM
580      PN(I)=(1-CO)*P(I)+CO*P(I-1)
590      GOTO 640
600 REM
610 REM     LAX-WENDROFF METHOD
620 REM
630      PN(I)=P(I)-CO*(P(I+1)-P(I-1))/2 + (CO^2)*(P(I+1)-2*P(I)+P(I-1))/2
640      NEXT I
650      FOR J=2 TO 11
660      P(J)=PN(J)
670      NEXT J
```

Figure 9.6.1 Computer program in BASIC for solving the first-order hyperbolic equation given in Example 9.6.

```
680      PRINT "TIME T=";T
690      PRINT "X=1.0: P=";P(3),"X=2.0: P=";P(5),"X=3.0: P=";P(7)
695      PRINT "X=4.0: P=";P(9),"X=5.0: P=";P(11)
700      IF T <= TMAX THEN 460
710      DT=2*DT
720      NEXT IK
730      END
```

Figure 9.6.1 Continued

the Lax-Wendroff method need the value at $(I + 1)$ to calculate the value at I. One approach to avoiding the problem that arises at $I = 11$ is to compute the value at $I = 11$ using the backward difference method which does not require the value at $I = 12$. This is done in the program, and the computed results are obtained for all three methods.

The results are shown in Fig. 9.6.2 in terms of the computed variation of P at a few locations as a function of time. Interestingly, the numerical solution is exact for

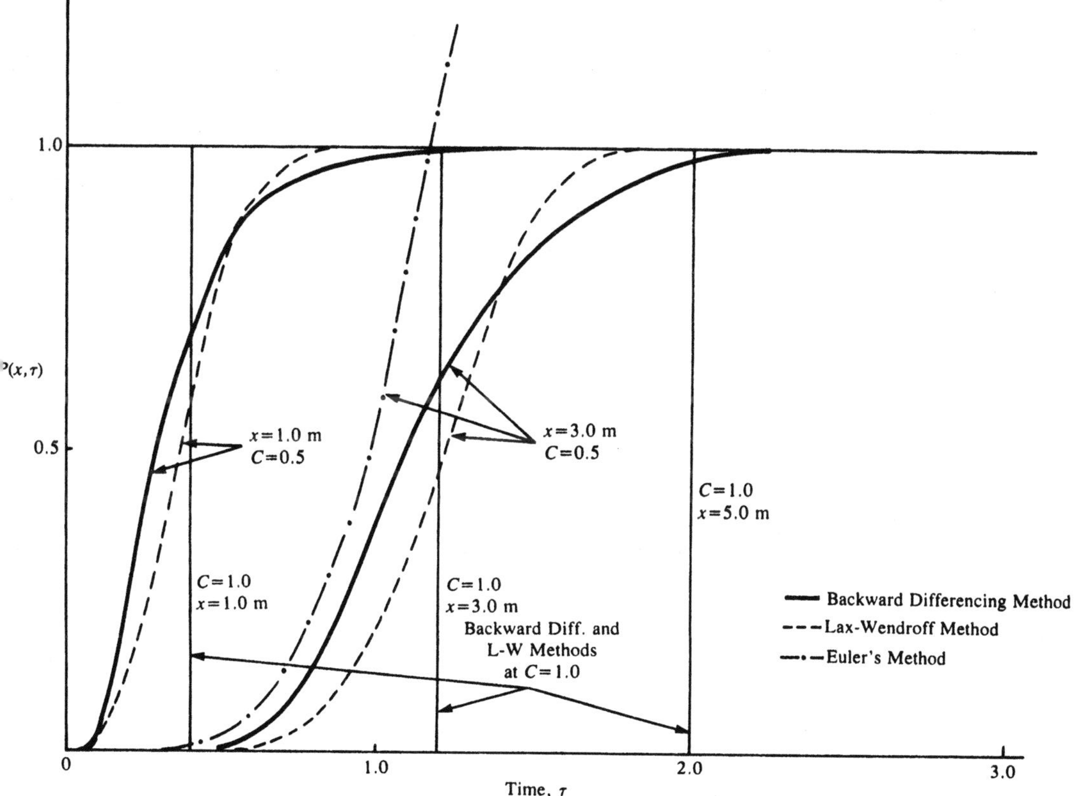

Figure 9.6.2 Numerical results in terms of the computed variation of the dependent variable P with time τ for Example 9.6.

both the upwind differencing and the Lax-Wendroff methods at Courant number $C = 1.0$. Generally, accuracy is expected to increase as $\Delta\tau$ is decreased, at a given value of Δx, because of smaller truncation error in time. However, for this particular problem, $C = 1.0$ yields the exact solution. In fact, as mentioned earlier, C is generally taken as close to 1.0 for greater accuracy. The results from both of these methods at $C = 0.5$ and $C = 0.25$ are unable to capture the expected step variation in P. However, that is not surprising since any numerical method will introduce computational errors.

The Lax-Wendroff method is expected to be more accurate due to second-order accuracy in space. Euler's method is found to be unstable. The solution oscillates at low values of C, and these oscillations grow without bound as time increases. This instability was found to be worse at large C, as expected. As shown by Ferziger (1981), this method is unconditionally unstable for this problem, and, therefore, instability arises in all cases as time elapses. For $C > 1.0$, the other two methods were also found to indicate numerical instability.

This problem is a fairly simple example of a hyperbolic equation. However, it has been chosen to demonstrate the use of three numerical techniques for solving hyperbolic equations and the constraints imposed by numerical instability.

9.5 SUMMARY

This chapter gives a brief discussion of the numerical solution of partial differential equations. The solution procedure is dependent on the type of the PDE: parabolic, elliptic, or hyperbolic. Employing simple examples of these three types of PDEs, various important numerical methods for solving them are outlined. Only linear equations are considered to illustrate the methods, since nonlinear PDEs, although important in many engineering applications, are beyond the scope of this book. However, in several cases, a nonlinear partial differential equation may be linearized and then solved by the methods discussed in this chapter. Still, the nonlinear problem is generally much more involved than the linear one. Frequently, the stability and the convergence characteristics of the numerical scheme are not known for nonlinear equations, and numerical experimentation is needed to ensure the accuracy and correctness of the solution.

The main approach to the solution of partial differential equations, considered in this chapter, is by means of finite difference methods, which give rise to a system of algebraic equations. A solution of this system of equations yields the value of the dependent variable at a finite number of grid points in the computational domain. Parabolic equations are solved by marching in one coordinate direction. Explicit methods allow the computation of values at a given step from the known values at earlier steps. However, the step size is generally constrained in explicit methods due to considerations of numerical stability. Implicit methods have better stability characteristics, but they require the solution of a system of simultaneous algebraic equations. Direct methods are usually employed if the system is tridiagonal. Otherwise, iterative methods, such as the Gauss-Seidel and the successive over-relaxation methods, are used. A tridiagonal system is obtained in one-dimensional problems, and Gaussian elimination is used for these. For multidimensional problems, splitting methods, which treat one direction as implicit and alternate between the various directions, are frequently employed, since these methods also give rise to tridiagonal systems.

Elliptic partial differential equations are often solved by iterative methods. Direct methods are applicable in a few special cases. Splitting methods, such as the alternating direction implicit (ADI) method, can also be used, with an acceleration parameter to obtain a faster convergence. In some cases, a pseudotransient term is added to the elliptic PDE. Then the resulting equation is parabolic in time and may be solved by time-marching techniques, giving the required solution at large time.

Hyperbolic partial differential equations are solved by methods similar to those for parabolic PDEs. A specialized method, known as the *method of characteristics*, is also an important method for hyperbolic equations, since it allows the treatment of discontinuities which frequently arise in these equations. In recent years, finite difference methods have become very popular for the solution of hyperbolic PDEs, and several very efficient schemes have been developed.

Another approach to the solution of partial differential equations is the finite element method. In this method, the solution domain is subdivided into finite regions, and the PDE is integrated over each region, employing weight functions with the equation. The solution and the weight functions are taken as polynomials, and the integrals or the weighted residuals are minimized, or reduced to zero, to yield a system of algebraic equations that is solved by the usual methods. Although more complicated in implementation, the finite element methods have become very popular in recent years because of their advantages in the treatment of irregular boundaries and complex boundary conditions. The finite difference approach is, in fact, a simple subset of the general approach represented by the finite element method. A brief outline of the finite element, the boundary element, and the control volume methods for solving partial differential equations is included in this chapter.

PROBLEMS

9.1. Consider the governing partial differential equation for the dependent variable $\phi(x, y)$, given as

$$\frac{\partial \phi}{\partial x} + A\frac{\partial \phi}{\partial y} = B\frac{\partial^2 \phi}{\partial y^2}$$

where A and B are constants. Determine the nature of this equation, and give a set of boundary conditions that may be applied to it.

9.2. The temperature $T(x,y)$ in a steady, two-dimensional flow with heat transfer is governed by the equation

$$u\frac{\partial T}{\partial x} + v\frac{\partial T}{\partial y} = \alpha\left(\frac{\partial^2 T}{\partial x^2} + \frac{\partial^2 T}{\partial y^2}\right) + Q$$

where $u(x,y)$ and $v(x,y)$ are the two velocity components, α is a constant known as *thermal diffusivity*, and $Q(x,y)$ is a function that gives the energy generation per unit volume in the fluid. Determine the nature of this equation, and specify suitable boundary conditions. Also, for each of the three special circumstances of (a) $u = v = 0$, (b) $\alpha = 0$, and (c) $\partial^2 T/\partial x^2 = 0$, classify the resulting reduced equations, and give the relevant boundary conditions.

9.3. In a one-dimensional diffusion problem, the governing equation is $\partial\phi/\partial\tau = C\,\partial^2\phi/\partial x^2$, where $\phi(x,\tau)$ is the dependent variable, C is a constant, x is the spatial coordinate, and τ is time. The boundary condition at $x = 0$ is given as $\partial\phi/\partial x = B$, where B is a constant. The condition at the other boundary, at $x = L$, is given as $\phi = 0$. Obtain the finite difference equation for solving this problem by the Crank-Nicolson method. Write the gradient boundary condition in terms of forward differences, using both the first-order and the second-order approximations. Is the resulting system of equations tridiagonal? If not, can it be obtained in tridiagonal form by simple elimination?

9.4. For the numerical solution of a one-dimensional transient diffusion problem, governed by

$$\frac{\partial\phi}{\partial\tau} = A\,\frac{\partial^2\phi}{\partial x^2}$$

the explicit Euler method is to be used. If the grid size Δx is taken as 0.1 m, find the maximum time step that may be employed for a stable numerical scheme, if $A = 10^{-6}\,\text{m}^2/\text{s}$. Also find the limitation on the time step if the grid size is reduced to 0.01 m.

9.5. A long bar of rectangular cross section is initially at a uniform temperature T_0. At time $\tau = 0$, the temperature at the outer surface is raised to T_s and held at this value. Write down the governing PDE and obtain the finite difference equations for solving this problem by the explicit FTCS and the Crank-Nicolson methods. Also give the equations for the relevant boundary and initial conditions. Indicate the constraints, if any, on the time step, due to stability considerations, for chosen values of Δx and Δy. What method would you employ for solving the system of algebraic equations obtained in the Crank-Nicolson method? Justify your choice.

9.6. Consider the one-dimensional conduction heat transfer in a plate of thickness 3 cm. The plate is initially at 1000°C. At time $\tau = 0$, the temperature at two surfaces is dropped to 0°C and maintained at this value. The thermal diffusivity is given as $5 \times 10^{-6}\,\text{m}^2/\text{s}$. Employing $\Delta x = 0.3$ cm, solve this problem numerically to obtain the time-dependent temperature distributions for $F = 1/6$, 0.5, 0.52, and 0.6. Does numerical instability arise for $F > 0.5$? Discuss.

9.7. Solve Problem 9.6 graphically, by the Schmidt method, for $F = 0.5$.

9.8. If a plate in a stationary fluid is suddenly set into motion, at a velocity U, the governing equation is $\partial u/\partial\tau = \nu\,\partial^2 u/\partial x^2$, where u is the local velocity, ν is a constant known as *kinematic viscosity of the fluid*, τ is time, and x is the distance out from the plate, which is at $x = 0$, as shown in the figure. The boundary conditions are, therefore, as follows:

$$\text{At } x = 0{:}\quad u = U \qquad \text{for } \tau > 0$$

$$\text{as } x \to \infty{:}\ u \to 0$$

The initial condition is the following:

$$\text{For } \tau \leqslant 0{:}\ u = 0 \qquad \text{for } x \geqslant 0$$

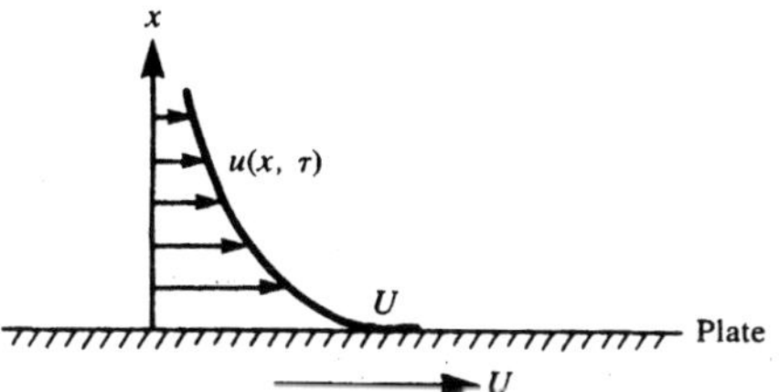

Problem 9.8

This problem is to be solved by the explicit FTCS method. The values of u are computed, at each time step, outward from the plate until u is zero. Taking $U = 1$ m/s, $\nu = 10^{-5}$ m^2/s, and $\Delta x = 0.01$ m, find the maximum time step that may be employed if numerical instability is to be avoided. Using this maximum time step, solve this problem.

9.9. For specifying the boundary condition $\partial\phi/\partial x = B$ in Prob. 9.3, a fictitious grid point is taken outside the computational domain, as shown. The boundary condition is then written in central difference form, using this point. The finite difference equation for the PDE is also written for a point at the boundary, again employing this fictitious point. The unknown value of ϕ at this grid point outside the region is eliminated by using the two equations thus obtained. The resulting equation gives the finite difference equation for the boundary condition. Compare this result with that given in Eq. (9.48).

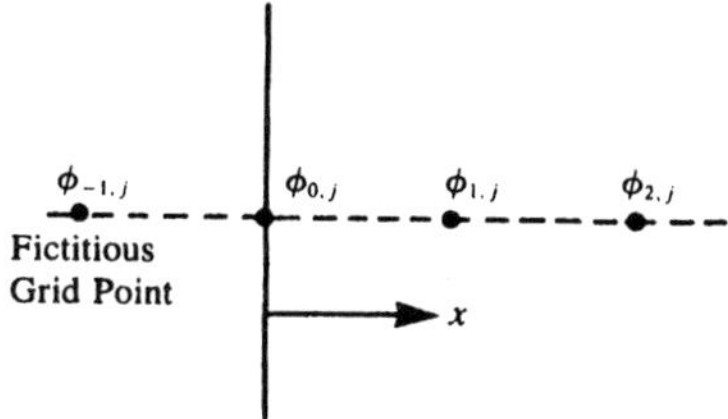

Problem 9.9

9.10. We wish to solve the following equation for $\phi(x,y)$

$$\frac{\partial^2\phi}{\partial y^2} = f(\phi)\frac{\partial\phi}{\partial x}$$

This equation is nonlinear because of the presence of the function $f(\phi)$. Formulate a simple numerical scheme, based on the discussion in the text, for solving this problem.

9.11. One-dimensional conduction in a rod of length L is governed by the equation

$$\frac{1}{\alpha}\frac{\partial T}{\partial \tau} = \frac{\partial^2 T}{\partial x^2} - H(T - T_a)$$

where x is the distance from one end, as shown, τ is time, T_a is the ambient temperature, and H is a heat loss parameter. For time $\tau < 0$, the temperature throughout the rod is T_a. At $\tau = 0$, the temperatures at the two ends, at $x = 0$ and $x = L$, are raised to 100°C and held at this value for $\tau > 0$. Using any suitable numerical method, solve this problem. Take $T_a = 15$°C, $\alpha = 10^{-6}$ m^2/s, $L = 0.4$ m, $\Delta x = 0.04$ m, and $H = 100$ m^{-2}.

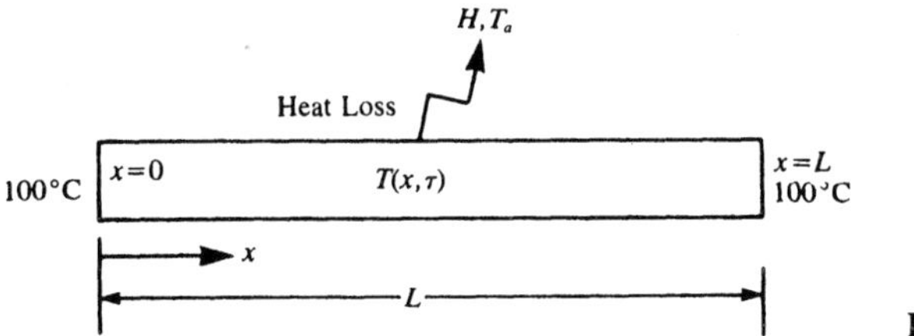

Problem 9.11

9.12. The one-dimensional diffusion of water in a porous medium is governed by the equation $D\,\partial^2 C/\partial x^2 = \partial C/\partial \tau$, where C is the concentration of water in kg/m^3 and D is the mass diffusivity in m^2/s. A long, hollow cylinder of outer diameter 0.2 m and inner diameter 0.1 m is initially dry; that is, water concentration is zero. Then, at time $\tau = 0$, the outer surface is brought in contact with water, raising the concentration there to 1000 kg/m^3, while the concentration at the inner surface is held at zero. If $D = 10^{-4}$ m^2/s, obtain the time-dependent concentration profiles in the cylinder, taking ten grid points across the annular region. Neglect the effect of curvature in the problem and treat the region as a slab.

9.13. Consider the problem discussed in Example 9.1. Study the effect of varying the grid size on the numerical results, by solving the problem with Δx half and also twice the value taken in the example. Take $\Delta x = \Delta y$. Compare the results obtained with the earlier results presented in Example 9.1, and discuss the dependence on grid size.

9.14. Steady-state mass diffusion in an enclosed region is governed by Laplace's equation for the concentration C. Consider the diffusion in a rectangular region of length 0.3 m and width 0.1 m. The third dimension is given as large. For this two-dimensional mass diffusion problem, the concentration, in nondimensional terms, is given as 1.0 at one surface and as 0.0 at the remaining three surfaces. Compute the concentration distribution in the region by the successive over-relaxation (SOR) method, and determine the optimum value of the relaxation factor. Compare this value with that obtained from the analytical expression, Eq. (9.43), given in the text.

9.15. A rectangular trampoline may be considered as a rubber membrane, of length L and width W, fastened securely at the boundary. As given in Example 9.3, the vertical deflection f is governed by the equation

$$\frac{\partial^2 f}{\partial x^2} + \frac{\partial^2 f}{\partial y^2} = -p/T$$

where p is the pressure and T the tension. Solve this problem by the Gauss-Seidel method, taking p, $T = 1.0$ m^{-1}, $L = 2.0$ m, and $W = 1.0$ m.

9.16. If fluid friction, or viscosity, is taken as negligible in a flow, the flow is termed *inviscid*. In the absence of rotational effects, the flow is then governed by the equation $\nabla^2\psi = 0$, where ψ is the stream function. A line of constant ψ is known as a *streamline*, and the velocity field may be obtained from a given ψ distribution. Consider the flow in a channel whose cross-sectional area varies as shown. The boundary conditions on ψ are also given, as linear distributions at the inflow and outflow. Compute the ψ distribution in the channel, and obtain the streamlines, or contours of constant ψ, corresponding to $\psi = 0$, 0.2, 0.4, 0.6, 0.8, and 1.0. Use the Gauss-Seidel iterative scheme. See also Example 9.4(b).

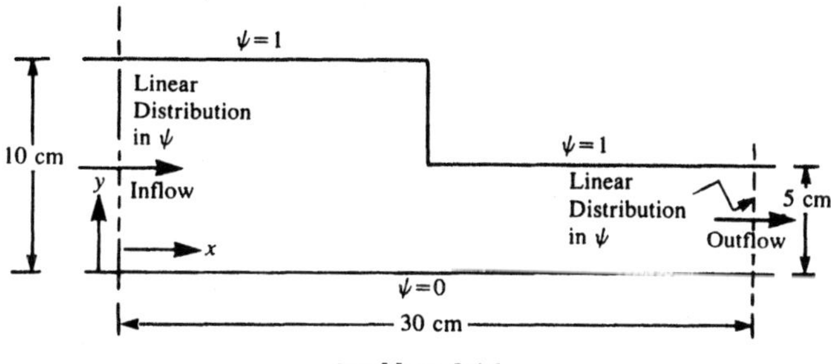

Problem 9.16

9.17. A solid cylinder of diameter D and length L has its two ends at temperature T_r, and the curved, lateral surface at temperature T_s. The temperature distribution may be assumed to be independent of the angular position. Thus, the problem becomes two-dimensional, or axisymmetric. Write the governing PDE and obtain the relevant finite difference equation for solving this problem by the SOR method. Use polar coordinates.

9.18. The flow of a very viscous fluid in a circular tube, as shown, is governed by the equation

$$\nabla^2 u = \frac{1}{\mu}\frac{dp}{dz}$$

where the vector operator ∇^2 may be written in polar coordinates, $u(x, y)$ is the velocity in the axial direction z, μ is the coefficient of viscosity, and dp/dz is the constant pressure drop along the flow. The velocity is zero at the boundary. Formulate this problem for a numerical solution by the Gauss-Seidel method. Give the governing equation and the relevant boundary conditions in finite difference form, and outline the numerical procedure.

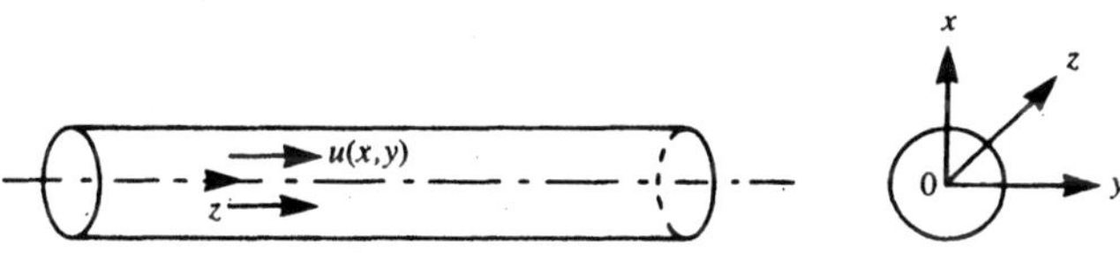

Problem 9.18

9.19. We are interested in the steady-state temperature distribution in a hollow cylinder of length L and inner and outer diameters D_i and D_o, respectively. The inner and outer surfaces are at temperature T_s, and the ends at T_0. Formulate this problem as a two-dimensional transient heat conduction problem whose solution yields the steady-state results at large time. Also give the finite difference equations for solution by the ADI method. What is the order of the truncation error in this formulation?

9.20. A rectangular finite difference mesh is used for solving Laplace's equation in a circular region. Consider the grid points near the circular boundary, and derive the applicable finite difference equation, as done in the text, taking a uniform grid distribution along both directions.

9.21. A long rod of rectangular, 10 cm × 5 cm, cross section has all the surfaces maintained at 100°C. Due to nuclear reaction, energy is generated within the material at a uniform rate $\dot{Q}$ of 5×10^7 W/m^3. The thermal conductivity k of the material is 50 W/m·K. The temperature distribution in steady-state conduction with energy generation is governed by the Poisson equation

$$\frac{\partial^2 T}{\partial x^2} + \frac{\partial^2 T}{\partial y^2} + \frac{\dot{Q}}{k} = 0$$

Using the Gauss-Seidel iterative method, solve this problem to obtain the temperature distribution. Plot the temperature variation along the two axes of the rectangular region.

9.22. Consider the one-dimensional convection equation

$$\frac{\partial \phi}{\partial \tau} + c\,\frac{\partial \phi}{\partial x} = 0$$

Determine the nature of this equation and give a set of relevant boundary conditions. Also obtain the finite difference equation for solving it by the Crank-Nicolson method.

9.23. In Example 9.5, if the initial deflection of the string results from being plucked at the one-third point instead of at the one-fourth point, compute the time-dependent displacements at various locations on the string after the string has been released.

9.24. The longitudinal vibration of a beam is governed by the equation

$$\frac{\partial^2 u}{\partial x^2} = \frac{\rho}{E}\,\frac{\partial^2 u}{\partial \tau^2}$$

where x is the coordinate along the axis, τ is time, u is the longitudinal displacement, ρ is the material density, and E is a constant known as the *elastic modulus* for the material. The two ends of the beam, at $x = 0$ and $x = L$, are fixed. A deflection u_0 is given at the midpoint of the beam and then released from rest. Using the explicit method, formulate this problem for a numerical solution. Give the relevant finite difference equations and outline the numerical procedure.

9.25. A string is fixed at its two ends. The initial deflection u is given as

$$u = 2(x - 0.2) \qquad \text{for } 0.2 < x \leqslant 0.6$$

$$u = 2(0.8 - x) \qquad \text{for } 0.6 < x \leqslant 0.8$$

$$u = 0 \qquad \text{at all other values of } x \text{ in the range 0 and 1.0}$$

Also, the time derivative of u is zero, that is,

$$\frac{\partial u}{\partial \tau}(x,0) = 0$$

Using the explicit method, solve this equation with the constant c in the governing wave equation given as 1.0. Take $\Delta x = 0.1$ and Courant number = 0.5 and 1.0. Compute the results up to $\tau = 2.0$. Discuss the observed trends in terms of the nature of hyperbolic equations.

9.26. Determine the nature of the following equation which governs the propagation of waves in a nonuniform medium:

$$\frac{\partial^2 u}{\partial \tau^2} = \frac{\partial}{\partial x}\left[c^2(x)\frac{\partial u}{\partial x}\right]$$

where u represents the displacement and c^2 varies with location. Outline a numerical method for solving this problem.

9.27. If the problem discussed in Example 9.5 is to be solved by the Crank-Nicolson method, give the resulting finite difference equations. Also outline the numerical method that may be adopted for solving these equations.

9.28. If in Example 9.5, the initial rate of change of displacement $\partial u/\partial \tau(x,0)$ is given as 0.1 m/s for $0 < x < 1$, compute the resulting displacement as a function of time at four points on the string.

9.29. If in Example 9.6, $P = 0$ for $\tau \leqslant 0$ and $e^{-\tau}$ at $x = 0$ for $\tau > 0$, solve the given hyperbolic PDE.

9.30. In a rectangular region of length L and width W, with $L/W = 2$, the Laplace equation governs the electric field ϕ. The value of ϕ is zero on three sides and is given as $\sin(\pi x/L)$ on the fourth side $y = W$. Compute the ϕ distribution in the region.

9.31. The dimensionless concentration C of a diffusing species in a square region is governed by

$$\frac{\partial^2 C}{\partial X^2} + \frac{\partial^2 C}{\partial Y^2} = \frac{\partial C}{\partial \bar{\tau}}$$

Starting with an initial value of C as zero in the entire region, calculate the transient and steady-state distributions, using the FTCS method. C is given as zero on two opposite sides of the region and as 1.0 on the other two. Both X and Y vary from 0 to 1.0. Determine if the initial conditions affect the steady-state distribution.

References

ABRAMOWITZ, M., and STEGUN, I. A., eds. *Handbook of Mathematical Functions with Formulas, Graphs and Mathematical Tables.* National Bureau of Standards, Appl. Math. Ser., vol. 55, 1964.

AHLBERG, H. J.; NILSON, E. N.; and WALSH, J. L. *The Theory of Splines and Their Applications.* New York: Academic Press, 1967.

AMAZIGO, J. C., and RUHENFELD, L. A. *Advanced Calculus and its Applications to the Engineering and Physical Sciences.* New York: Wiley, 1980.

ANTON, H. *Elementary Linear Algebra.* 4th ed. New York: Wiley, 1984.

ATKINSON, K. *An Introduction to Numerical Analysis.* New York: Wiley, 1978.

BANERJEE, P. K., and BUTTERFIELD, R. *Boundary Element Method in Engineering Science.* London: McGraw-Hill, 1981.

BREBBIA, C. A., *The Boundary Element Method for Engineers.* 3rd ed. London: McGraw-Hill, 1977.

BRENT, R. "Some Efficient Algorithms for Solving Systems of Nonlinear Equations." *SIAM J. Num. Anal.* 10: 327–44, 1973.

BRONSON, R. *Matrix Methods: An Introduction.* New York: Academic Press, 1970.

BRULÉ, J. F. *Artificial Intelligence: Theory, Logic and Application.* Blue Ridge Summit, Penn.: Tab Books, Inc., 1986.

BUTCHER, J. C. "On Runge-Kutta Processes of High Order." *J. Austr. Math. Soc.* 4: 179–94, 1964.

CARNAHAN, B. H.; LUTHER H. A.; and WILKES, J. O. *Applied Numerical Methods.* New York: Wiley, 1969.

CLARK, K. L., and MCCABE, F. G. *Micro-PROLOG: Programming in Logic.* Englewood Cliffs, N.J.: Prentice-Hall, 1984.

CLOCKSIN, W. F., and MELLISH, C. S. *Programming in PROLOG.* 2nd ed. New York: Springer-Verlag, 1984.

COAN, J. S. *Basic BASIC.* 2nd ed. Rochelle Park, N.J.: Hayden, 1978.

COLLATZ, L. *The Numerical Treatment of Differential Equations.* 3rd ed. Berlin: Springer-Verlag, 1960.

COSTALES, B. *C: From A to Z.* Englewood Cliffs, N.J.: Prentice-Hall, 1985.

DAVIS, P. J., and RABINOWITZ, P. *Numerical Integration.* Waltham, Mass: Ginn-Blaisdell, 1967.

DRAPER, N. R., and SMITH, H. *Applied Regression Analysis.* 2nd ed. New York: Wiley, 1981.

FERZIGER, J. *Numerical Methods for Engineering Application.* New York: Wiley-Interscience, 1981.

FORSYTHE, G. E., MALCOLM, M. A.; and MOLER, C. B. *Computer Methods for Mathematical Computations.* Englewood Cliffs, N.J.: Prentice-Hall, 1977.

FORSYTHE, G. and WASOW, W. *Finite Difference Methods for Partial Differential Equations.* New York: Wiley, 1960.

FOX, L. *Numerical Solution of Ordinary and Partial Differential Equations.* Oxford, England: Pergamon Press, 1962.

FRANCIS, J. G. F. "The QR Transformation." *The Computer Journal* 4: 265–71, 1961; and 332–45, 1962.

FRIEDMAN, F. L., and KOFFMAN, E. B. *Problem Solving and Structured Programming in FORTRAN.* 2nd ed. Reading, Mass.: Addison-Wesley, 1981.

GEAR, C. W. *Numerical Initial Value Problems in Ordinary Differntial Equations.* Englewood Cliffs, N.J.: Prentice-Hall, 1971.

GEBHART, B. *Heat Transfer.* 2nd ed. New York: McGraw-Hill, 1971.

GERALD, C. F., and WHEATLEY, P. O. *Applied Numerical Analysis.* 3rd ed. Reading, Mass.: Addison-Wesley, 1984.

GILL, S. "A Process for the Step-by-Step Integration of Differential Equations in an Automatic Computing Machine." *Proc. Camb. Phil. Soc.* 47: 96–108, 1951.

HALL, T. E.; ENRIGHT, W. N.; FELLEN, B. M.; and SEDGEWICK, A. E. "Comparing Numerical Methods for Ordinary Differential Equations." *SIAM J. Num. Anal.* 9: 603–37, 1972.

HALL, G., and WATT, J. M. *Modern Numerical Methods for Ordinary Differential Equations.* Oxford, England: Clarendon Press, 1976.

HALLIDAY, D., and RESNICK, R. *Fundamentals of Physics.* 2nd ed. New York: Wiley, 1986.

HAMMING, R. W. "Stable Predictor-Corrector Methods for Ordinary Differential Equations." *J. Assoc. Comput. Mach.* 6: 37–47, 1959.

HOCKNEY. R. W., and JESSHOPE, C. R. *Parallel Computers.* Bristol, England: Adam Hilger Ltd., 1981.

HORNBECK, R. W. *Numerical Methods.* New York: Quantum, 1975.

HOUSEHOLDER, A. S. *The Numerical Treatment of a Single Nonlinear Equation.* New York: McGraw-Hill, 1970.

HOUSEHOLDER, A. S. *The Theory of Matrices in Numerical Analysis.* Waltham, Mass.: Blaisdell, 1964.

HUEBNER, K. H., and THORNTON, E. A., *The Finite Element Method for Engineers.* 2nd ed. New York: Wiley, 1983.

INCROPERA, F. P., and DEWITT, D. P. *Fundamentals of Heat Transfer.* New York: Wiley, 1981.

JALURIA, Y., and TORRANCE, K. E. *Computational Heat Transfer.* New York: Hemisphere (Harper & Row), 1986.

JAMES, M. L.; SMITH, G. M.; and WOLFORD, J. C. *Applied Numerical Methods for Digital Computation.* 3rd ed. New York: Harper & Row, 1985.

KAPLAN, W. *Advanced Calculus.* 3rd ed. Reading, Mass.: Addison-Wesley, 1984.

KEISLER, H. J. *Elementary Calculus: An Infinitesimal Approach.* 2nd ed. Boston, Mass.: Prindle, Weber & Schmidt, 1986.

KELLER, H. B. *Numerical Methods for Two-Point Boundary-Value Problems.* Waltham, Mass.: Ginn-Blaisdell, 1968.

KERNIGHAN, B. W., and PLAUGER, P. J. *Software Tools in PASCAL.* Reading, Mass.: Addison-Wesley, 1981.

KERNIGHAN, B. W., and RITCHIE, D. M. *The C Programming Language.* Englewood Cliffs, N.J.: Prentice-Hall, 1978. Also, revised printing, 1984.

LAMBERT, J. D. *Computational Methods in Ordinary Differential Equations.* New York: Wiley, 1973.

LANCASTER, P. *Theory of Matrices.* New York: Academic Press, 1969.

MERCHANT, M. J. *FORTRAN 77 Language and Style.* Belmont, Calif.: Wadsworth, 1981.

MILLER, A. R. *BASIC Programs for Scientists and Engineers.* Berkeley, Calif: Sybex, 1981.

MITCHELL, A. R., and WAIT, R. *The Finite Element Method in Partial Differential Equations.* New York: Wiley, 1977.

OGATA, K. *System Dynamics,* Englewood Cliffs, N.J.: Prentice-Hall, 1978.

OSTROWSKI, A. M. *Solution of Equations and Systems of Equations.* New York: Academic Press, 1966.

PURDUM, J. C. *C Programming Guide.* 2nd ed. Que Corp., 1985.

RALSTON, A. "Runge-Kutta Methods with Minimum Error Bounds." *Math. Comp.* 16: 431–37, 1962.

RALSTON, A. *A First Course in Numerical Analysis.* New York: McGraw-Hill, 1965.

RALSTON, A., and RABINOWITZ, P. *A First Course in Numerical Analysis.* 2nd ed. New York: McGraw-Hill, 1978.

REINER, I. *Introduction to Matrix Theory and Linear Algebra.* New York: Holt, Rinehart and Winston, Inc., 1971.

REYNOLDS, W. C., and PERKINS, H. C. *Engineering Thermodynamics.* 2nd ed. New York: McGraw-Hill, 1977.

RICE, J. R. *Numerical Methods, Software and Analysis.* New York: McGraw-Hill, 1983.

ROACHE, P. J. *Computational Fluid Dynamics.* Rev. Printing. Albuquerque, N.M.: Hermosa, 1976.

RUTISHAUSER, H. "Solution of Eigenvalue Problems with the LR Transformation," National Bureau of Standards, *Appl. Math. Ser.*, vol. 49, 47–81, 1958.

SALVADORI, M. G., and BARON, M. L. *Numerical Methods in Engineering.* 2nd ed. Englewood Cliffs, N.J.: Prentice-Hall, 1961.

SEARS, F. W.; ZEMANSKY, M. W.; and YOUNG, H. D. *University Physics.* 6th ed. Reading, Mass.: Addison-Wesley, 1981.

SEITER, C., and WEISS, R. *PASCAL for BASIC Programmers.* Reading, Mass.: Addison-Wesley, 1982.

SHAMPINE, L. P., and GORDON, M. K. *Computer Solution of Ordinary Differential Equations.* San Francisco: Freeman, 1975.

SHANKS, E. B. "Solutions of Differential Equations by Evaluations of Functions." *Maths. of Computation* 20: 21–38, 1966.

SHOUP, T. E. *A Practical Guide to Computer Methods for Engineers.* Englewood Cliffs, N.J.: Prentice-Hall, 1979.

SMITH, G. D. *Numerical Solution of Partial Differential Equations.* Oxford, England: Oxford University Press, 1965.

SOKOLINIKOFF, I. S., and REDHEFFER, R. M. *Mathematics of Physics and Modern Engineering.* New York: McGraw-Hill, 1966.

STOECKER, W. F. *Design of Thermal Systems.* 2nd ed. New York: McGraw-Hill, 1980.

STROUD, A. H., and SECREST, D. *Gaussian Quadrature Formulas.* Englewood Cliffs, N.J.: Prentice-Hall, 1966.

THOMAS, G. B., and FINNEY, R. L. *Calculus and Analytic Geometry.* 6th ed. Reading, Mass.: Addison-Wesley, 1984.

TRAUB, J. F. *Iterative Methods for the Solution of Equations.* Englewood Cliffs, N.J.: Prentice-Hall, 1964.

WAITE, M.; PRATA, S.; and MARTIN, D. *C Primer Plus.* Indianapolis, Ind.: H. W. Sams & Co., 1986.

WEISS, R., and SEITER, C. *PASCAL for FORTRAN Programmers.* Reading, Mass.: Addison-Wesley, 1984.

WILKINSON, J. H. *The Algebraic Eigenvalue Problem.* Oxford, England: Oxford University Press, 1965.

WILLIAMS, G. *Elementary Linear Algebra with Applications.* Newton, Mass.: Allyn & Bacon, 1984.

WINSTON, P. H., and HORN, B. K. P. *LISP.* 2nd ed. Reading, Mass.: Addison-Wesley, 1984.

Index